MW00910923

HARMONIC ANALYSIS, GROUP REPRESENTATIONS, AUTOMORPHIC FORMS AND INVARIANT THEORY

In Honor of Roger E. Howe

LECTURE NOTES SERIES
Institute for Mathematical Sciences, National University of Singapore

Series Editors: Louis H. Y. Chen and Ka Hin Leung
Institute for Mathematical Sciences
National University of Singapore

ISSN: 1793-0758

Published

Lecture Notes Series, Institute for Mathematical Sciences,
National University of Singapore

Vol.
12

HARMONIC ANALYSIS, GROUP REPRESENTATIONS, AUTOMORPHIC FORMS AND INVARIANT THEORY

In Honor of Roger E. Howe

Editors

Jian-Shu Li
Hong Kong University of Science & Technology, Hong Kong

Eng-Chye Tan
National University of Singapore, Singapore

Nolan Wallach
University of California, San Diego, USA

Chen-Bo Zhu
National University of Singapore, Singapore

 World Scientific

NEW JERSEY · LONDON · SINGAPORE · BEIJING · SHANGHAI · HONG KONG · TAIPEI · CHENNAI

Published by

World Scientific Publishing Co. Pte. Ltd.

5 Toh Tuck Link, Singapore 596224

USA office: 27 Warren Street, Suite 401-402, Hackensack, NJ 07601

UK office: 57 Shelton Street, Covent Garden, London WC2H 9HE

British Library Cataloguing-in-Publication Data
A catalogue record for this book is available from the British Library.

**Lecture Notes Series, Institute for Mathematical Sciences,
National University of Singapore — Vol. 12**
HARMONIC ANALYSIS, GROUP REPRESENTATIONS, AUTOMORPHIC
FORMS AND INVARIANT THEORY
In Honor of Roger E. Howe

ISBN-13 978-981-277-078-3
ISBN-10 981-277-078-X

CONTENTS

FOREWORD

The Institute for Mathematical Sciences at the National University of Singapore was established on 1 July 2000. Its mission is to foster mathematical research, particularly multidisciplinary research that links mathematics to other disciplines, to nurture the growth of mathematical expertise among research scientists, to train talent for research in the mathematical sciences, and to provide a platform for interaction and collaboration between local and foreign mathematical scientists, in support of national development.

The Institute organizes thematic programs which last from one month to six months. The theme or themes of a program will generally be of a multidisciplinary nature, chosen from areas at the forefront of current research in the mathematical sciences and their applications, and in accordance with the scientific interests and technological needs in Singapore.

Generally, for each program there will be tutorial lectures on background material followed by workshops at the research level. Notes on these lectures are usually made available to the participants for their immediate benefit during the program. The main objective of the Institutes Lecture Notes Series is to bring these lectures to a wider audience. Occasionally, the Series may also include the proceedings of workshops and expository lectures organized by the Institute.

The World Scientific Publishing Company has kindly agreed to publish the Lecture Notes Series. This Volume, "Harmonic Analysis, Group Representations, Automorphic Forms and Invariant Theory: In Honor of Roger E. Howe", is the twelfth of this Series. We hope that through the regular publication of these lecture notes the Institute will achieve, in part, its objective of promoting research in the mathematical sciences and their applications.

May 2007

Louis H. Y. Chen
Ka Hin Leung
Series Editors

PREFACE

Roger E. Howe, member of the National Academy of Sciences, USA and fellow of the American Academy of Arts and Sciences, and Professor of Mathematics at Yale University, is a scholar of distinction.

Howe's major research interest is in applications of symmetry, particularly harmonic analysis, group representations, automorphic forms and invariant theory. To honor Howe's exceptional achievements both as a scholar and a teacher, an International Conference on Harmonic Analysis, Group Representations, Automorphic Forms and Invariant Theory was held at the National University of Singapore (NUS) from January 9 - 11, 2006, on the occasion of his 60th birthday.

The fact that the conference was held in Singapore and had many attendees from the Asia-Pacific region is clearly fitting due to Howe's extensive contact with and significant influence within the region. He has on many occasions visited universities in Australia, Israel, Japan, Singapore, Hong Kong and China. In particular, he was a fellow of the Institute for Advanced Studies at Hebrew University in Jerusalem in 1988, a fellow of the Japan Society for the Advancement of Science in 1993 and is currently chair of the Scientific Advisory Board of the Institute for Mathematical Sciences (IMS) in NUS.

The speakers of the conference are Michael Cowling (University of New South Wales, Australia), Stephen Gelbart (Weizmann Institute, Israel), Michael Harris (Université Paris VII, France), Masaki Kashiwara and Toshiyuki Kobayashi (both of RIMS, Kyoto University, Japan), Hanspeter Kraft (Universität Basel, Switzerland), Colette Moeglin (Université Paris VII, France), Allen Moy (Hong Kong University of Science and Technology, Hong Kong), Toshio Oshima (University of Tokyo, Japan), Wilfried Schmid (Harvard University, USA), David Vogan (MIT, USA), Gregg Zuckerman (Yale University, USA).

The eleven articles of this volume are for the most part expanded from the invited lectures of the Conference and are all reviewed. In the following, we give a brief indication of the ranges of topics represented in this volume.

The first article is by Jeffrey Adams, on the theta correspondence over the reals (this has circulated informally for many years). One can hardly fail to notice Howe's influence. A favorite subject of Howe, the Hisenberg group (in relation to rigidity) appears in the next article by Michael Cowling *et al.* The article by Evans and Wallach delights with another classical subject (the Pfaffian) and its applications in integer games. Stephen Gelbart's survey article discusses three kinds of methods for proving that an L-function is non-zero in a part of its critical strip, which will be welcomed by the L-function community. Next Michael Harris presents his fourth in a series of articles devoted to the study of special values of L-functions of automorphic forms contributing to the cohomology of Shimura varieties attached to unitary groups. Here as Harris mentions in the article, the influence of Howe's approach to theta functions as a means for relating automorphic forms on different groups is evident. The article by Kobayashi and Mano examines holomorphic extension of the minimal representation of the conformal group, which is a strong analog of Howe's oscillator semigroup and it represents an excellent contribution of explicit representation theory to classical analysis, and vice versa.

As is well-known, another area where Howe made path-breaking contributions is in representation theory of p-adic groups. The articles by Moeglin and by Moy and Tadić, address two of the central questions on p-adic groups, namely the classification of discrete series, and how to attach invariants (algebra of distributions and the Bernstein center) to p-adic group representations. The final three articles are on real Lie groups and Lie algebras, and are contributed by some of the "real" masters of Lie theory. Specifically Oshima constructs a generator system of the annihilator of certain generalized Verma modules and the quantized minimal polynomials; Vogan addresses the fundamental question of branching of a standard representation to a maximal compact subgroup; Willenbring and Zuckerman introduces a novel notion of a small algebra of a semisimple Lie algebra.

Contributed by some of the leading members of the Lie theory community and with the range and diversity of the topics, we hope that the current volume pays a fitting tribute to the originality, depth and influence of Howe's mathematical work.

We would like to take the opportunity to thank the IMS and the Faculty of Science, of NUS for sponsoring this conference. The expertise and dedication of all IMS staff, especially its director Louis Chen, contributed essentially to the success of this memorable event, which all four of us enjoyed much to organize.

May 2007

Jian-Shu Li
Hong Kong University of Science and Technology,
Hong Kong

Eng-Chye Tan
National University of Singapore,
Singapore

Nolan Wallach
University of California,
San Diego, USA

Chen-Bo Zhu
National University of Singapore,
Singapore

Photo 1: A group photo.

Photo 2: Howe with his students.

Photo 3: Howe with his co-authors.

Photo 4: Howe with senior colleagues.

Photo 5: Howe with Yalies.

THE THETA CORRESPONDENCE OVER $\mathbb{R}$

JEFFREY ADAMS

Department of Mathematics
University of Maryland
College Park, MD20742, USA
E-mail: jda@math.umd.edu

1. Introduction

These notes were written for the workshop on the Theta correspondence held at the University of Maryland, May 1994. They have circulated informally for many years. Except for minor changes and some improvements to the references they have not been changed. There have been too many developments in the field since 1994 to try to bring them up to date.

This is dedicated to Roger Howe, who introduced me to representation theory in graduate school in 1977.

2. Fock Model: Complex Lie Algebra

The Fock model is the algebraic oscillator representation: it is the $(\mathfrak{g}, \widetilde{K})$ module of the oscillator representation of $\widetilde{Sp}(2n, \mathbb{R})$. The basic references are [6], [7], and [47]. See also [11], [17] and [18]. The presentation here is due to Steve Kudla (unpublished); a similar treatment is found in [49], as well as [51]. In this section we define the oscillator representation of the complex Heisenberg and symplectic Lie algebras.

Let ψ be a non–trivial linear form on $\mathbb{C}$, i.e. fix $\lambda \in \mathbb{C}^\times$ and let $\psi(z) = \lambda z$ ($z \in \mathbb{C}$).

Definition 2.1 *Let W be a vector space with a symplectic form $\langle, \rangle$, possibly degenerate. Let*

$$\Omega(W) = T(W)/I$$

Author's work partially supported by the National Science Foundation grant numbers DMS-0554278 and DMS-0532393.

1

where T is the tensor algebra of W and I is the two–sided ideal generated by elements of the form

$$v \otimes w - w \otimes v - \psi\langle v, w \rangle \quad (v, w \in W).$$

This is an associative algebra, sometimes referred to as the *quantum algebra*.

For example if $\langle v, w \rangle = 0$ for all v, w, then $\Omega(W)$ is isomorphic to the polynomial algebra on W.

Let $\mathfrak{h}$ be the complex Heisenberg Lie algebra of dimension $2n + 1$, so $\mathfrak{h} = W \oplus L$ where $W \simeq \mathbb{C}^{2n}$ is a vector space with a non–degenerate symplectic form $\langle, \rangle$, and $L \simeq \mathbb{C}$ is the center of $\mathfrak{h}$. Write $L = \mathbb{C}\mathbb{1}$; the Lie algebra structure of $\mathfrak{h}$ is given by $[v, w] = \langle v, w \rangle \mathbb{1}$ for $v, w \in W$. We consider ψ as an element of $Hom(L, \mathbb{C})$ by $\psi(\lambda \mathbb{1}) = \psi(\lambda)$ ($\lambda \in \mathbb{C}$). Let $\mathcal{U}(\mathfrak{h})$ be the universal enveloping algebra of $\mathfrak{h}$.

Lemma 2.2

$$\Omega(W) \simeq \mathcal{U}(\mathfrak{h})/\langle \mathbb{1} - \psi(\mathbb{1}) \rangle$$

where $\langle \mathbb{1} - \psi(\mathbb{1}) \rangle$ is the two–sided ideal generated by the element $\mathbb{1} - \psi(\mathbb{1})$.

Proof. Define a map $\mathfrak{h} \to \Omega(W)$ by $w \to \overline{w} \in T(W)/I$ and $\mathbb{1} \to \overline{\psi(\mathbb{1})}$. This extends to a map $\mathcal{U}(\mathfrak{h}) \to T(W)/I$ by the universal property of $\mathcal{U}(\mathfrak{h})$, and factors to the quotient; it is easily seen to be bijective.   □

Note: The definition of $\Omega(W)$ depends on ψ. Over $\mathbb{C}$ the algebras defined by any ψ and ψ' are isomorphic. The isomorphism takes $v \in W$ to λv where $\lambda^2 = \psi(1)/\psi'(1)$. Over an arbitrary field F there are $|F^\times/F^{\times 2}|$ isomorphism classes of such algebras.

Since $T(W) = T^0(W) \oplus T^1(W) \oplus \ldots$ is graded, $\Omega(W)$ is filtered, i.e. $\Omega^0(W) \subset \Omega^1(W) \subset \ldots$ where $\Omega^k(W)$ consists of those elements which may be represented by some $X \in T^0(W) \oplus T^1(W) \oplus \cdots \oplus T^k(W)$. Then if $[A, B] = A \otimes B - B \otimes A$ is the commutator bracket in $\Omega(W)$, it follows easily that

$$[\Omega^k(W), \Omega^\ell(W)] \subset \Omega^{k+\ell-2}(W).$$

It follows that $\Omega^2(W)$ with the bracket $[,]$ is a Lie algebra, and $\Omega^1(W)$ is an ideal in $\Omega^2(W)$. Similarly $[\Omega^1(W), \Omega^1(W)] \subset \Omega^0(W)$, and $\Omega^1(W)$ is a two–step nilpotent Lie algebra.

We therefore have the exact sequences of Lie algebras

$$0 \to \Omega^0(W) \to \Omega^1(W) \to \Omega^1(W)/\Omega^0(W) \to 0 \qquad (2.3)(a)$$

$$0 \to \Omega^1(W) \to \Omega^2(W) \to \Omega^2(W)/\Omega^1(W) \to 0 \qquad (2.3)(b)$$

If $\langle , \rangle$ non–degenerate it is easy to see $\Omega^1(W) \simeq \mathfrak{h}$ and $\Omega^1(W)/\Omega^0(W) \simeq W$ (as an abelian Lie algebra). The exact sequence (2.3)(a) is isomorphic to $\mathbb{C} \to \mathfrak{h} \to W$, which splits as vector spaces (but not as Lie algebras). Something stronger holds for (2.3)(b):

Lemma 2.4 *Let* $\mathfrak{g} \simeq \mathfrak{sp}(2n, \mathbb{C})$ *be the Lie algebra of* $Sp(W) \simeq Sp(2n, \mathbb{C})$.
(a) $\Omega^2(W)/\Omega^1(W)$ *is isomorphic to* $\mathfrak{g}$.
(b) The exact sequence (1.3)(b) splits, to give an isomorphism

$$\Omega^2(W) \simeq \mathfrak{h} \oplus \mathfrak{g}$$

where the right–hand side is the semi–direct product of Lie algebras, in which $\mathfrak{g}$ *acts on* $\mathfrak{h}$ *by*

$$X \cdot (w, t) = (Xw, t)$$

for $X \in \mathfrak{g}, w \in W, t \in L$.

Define an action of $\Omega^2(W)$ on $\Omega^1(W)/\Omega^0(W)$ by $X \cdot Y\Omega^0(W) = [X, Y]\Omega^0(W)$. This is defined on $\Omega^1(W)$ since $[\Omega^2(W), \Omega^1(W)] \subset \Omega^1(W)$ and passes to the quotient since $[\Omega^2(W), \Omega^0(W)] \subset \Omega^0(W)$ (in fact this bracket is 0). By the isomorphism $\Omega^1(W)/\Omega^0(W) \simeq W$ this defines a map $\Omega^2(W) \to \operatorname{End}(W)$. If $v, w \in W$ write vw for the coset of $v \otimes w \in T(W)$, modulo(I). Explicitly for $v_1, v_2, w \in W$ we have

$$
\begin{aligned}
v_1 v_2 \cdot w &= v_1 v_2 w - w v_1 v_2 \\
&= v_1 v_2 w - v_1 w v_2 - \psi \langle w, v_1 \rangle v_2 \\
&= v_1 v_2 w - v_1 v_2 w - \psi \langle w, v_2 \rangle v_1 - \psi \langle w, v_1 \rangle v_2 \\
&= \psi \langle v_1, w \rangle v_2 + \psi \langle v_2, w \rangle v_1.
\end{aligned}
$$

It follows readily that the image of $\Omega^2(W)$ is in $\mathfrak{sp}(W)$. Furthermore this map is injective, and by dimension counting it is an isomorphism, proving (a).

Choose a standard basis $e_1, \ldots, e_n, f_1, \ldots, f_n$ of W, i.e. $\langle e_i, f_i \rangle = 1$ and all other brackets are 0. Consider the subspace of $\Omega^2(W)$ spanned by the following elements. Note that in $\Omega(W)$ all the $e_j's$ commute among themselves, similarly the $f_j's$, and $e_i f_j = f_j e_i + \psi(\delta_{i,j})$. Let

$$
\begin{aligned}
X_{i,j}^+ &= f_i f_j \quad 1 \le i \le j \le n \\
X_{i,j}^- &= e_i e_j \quad 1 \le i \le j \le n \\
Z_{i,j} &= \frac{1}{2}(e_i f_j + f_j e_i) \quad 1 \le i, j \le n.
\end{aligned}
\qquad (2.5)(a)
$$

Note that

$$Z_{i,j} = \frac{1}{2}(f_j e_i + e_i f_j) = \frac{1}{2}(f_j e_i + e_i f_j - f_j e_i + f_j e_i)$$

$$= f_j e_i + \frac{1}{2}[e_i, f_j] \qquad (2.5)(b)$$

$$= f_j e_i + \frac{1}{2}\psi(\delta_{i,j}).$$

We claim the elements (2.5)(a) span a Lie subalgebra of Ω^2 isomorphic to $\mathfrak{sp}(2n, \mathbb{C}) = \mathfrak{g}$. To see this, recall (cf. [13], for example) $\mathfrak{g}$ has the decomposition $\mathfrak{g} = \mathfrak{k} \oplus \mathfrak{p} = \mathfrak{k} \oplus \mathfrak{p}^+ \oplus \mathfrak{p}^-$ where $\mathfrak{k} \simeq \mathfrak{gl}(n, \mathbb{C})$ is the complexified Lie algebra of a maximal compact subgroup K of $Sp(2n, \mathbb{R})$. The $Z_{i,j}$ span an algebra isomorphic to $\mathfrak{k}$, with a Cartan subalgebra $\mathfrak{t}$ the span of the $Z_{i,i}$. Let ϵ_i denote the element of $\mathfrak{t}^*$ satisfying $\epsilon_i(Z_{j,j}) = \psi(\delta_{i,j})$. Then $Z_{j,k}$ ($j \neq k$) is a root vector of weight $\epsilon_j - \epsilon_k$:

$$= [f_i e_i + \frac{1}{2}\psi(1), f_j e_k] \quad \text{(by (2.5)(b))}$$

$$= [f_i e_i, f_j e_k]$$

$$= f_i e_i f_j e_k - f_j e_k f_i e_i \qquad (2.5)(c)$$

$$= f_i f_j e_i e_k + f_i e_k \psi(\delta_{i,j}) - f_j f_i e_i e_k - f_j e_i \psi(\delta_{i,k})$$

$$= \psi(\delta_{i,j} - \delta_{i,k}) f_j e_k$$

$$= (\epsilon_j - \epsilon_k)(Z_{i,i}) Z_{j,k}.$$

Similarly $X_{i,j}^{\pm}$ is a root vector of weight $\pm(\epsilon_i + \epsilon_j)$, so the $X_{i,j}^{\pm}$ span $\mathfrak{p}^{\pm}$. It follows from similar calculations that $[\mathfrak{k}, \mathfrak{p}^{\pm}] = \mathfrak{p}^{\pm}$, $[\mathfrak{p}^+, \mathfrak{p}^-] = [\mathfrak{k}, \mathfrak{k}] = \mathfrak{k}$, so this is a Lie algebra. From the description of the roots, the root system is of type C_n, proving Lemma 2.4.

For later use we note that the $Z_{i,j}$ with $i < j$ together with the $X_{i,j}^+$ are positive root vectors of $\mathfrak{t}$ in $\mathfrak{g}$.

Now let $W = X \oplus Y$ be a complete polarization of W (this is implicit in our choice of standard basis as in the proof of Lemma 2.4). Thus X and Y are Lagrangian planes, i.e. maximal isotropic subspaces of W. Since $\langle,\rangle$ restricted to Y is trivial, $\Omega(Y) \simeq \mathbb{C}[Y]$, the polynomial algebra on Y.

By the Poincare–Birkhoff–Witt theorem [24] $\Omega(W) = \Omega(Y) \oplus \Omega(W)X$. Let $e_1, \ldots, e_n$ be a basis of X, and $f_1, \ldots f_n$ be a basis of Y. This says that any element of $\Omega(W)$ may be written as a sum of monomials, where each monomial is in terms of $f_i's$ only, or has some e_j on the right. It follows immediately that

$$\Omega(W)/\Omega(W)X \simeq \Omega(Y) \simeq \mathbb{C}[Y]. \qquad (2.6)$$

This carries a representation of $\Omega(W)$ by multiplication on the left. By Lemma 1.4 this defines by restriction a representation of $\mathfrak{h}$ and $\mathfrak{g}$, or equivalently $\mathcal{U}(\mathfrak{h})$ and $\mathcal{U}(\mathfrak{g})$, on $\mathbb{C}[Y]$. The center of $\mathfrak{h}$ acts by the character ψ.

Definition 2.7 *The Fock model of the oscillator representation of* $\mathfrak{h} \oplus \mathfrak{g}$ *is the vector space* $\mathcal{F} = \mathbb{C}[Y]$, *where* $\Omega(W)$ *acts by multiplication on the left and the isomorphism (1.6), and* $\mathfrak{h} \oplus \mathfrak{g}$ *acts via the embedding in* $\Omega(W)$ *(Lemma 1.4(b)).*

This definition depends on the choice of ψ and complete polarization $W = X \oplus Y$, and we will write $\mathcal{F}_\psi$ on occasion. The isomorphism class is independent of these choices. Over an arbitrary field F the number of isomorphism classes is $|F^\times/F^{\times 2}|$ (see the note following Lemma 1.2).

To be explicit, with the basis e_i, f_j above, let X (resp. Y) be the span of the $e_i's$ (resp. $f_j's$). Then $\mathcal{F} = \mathbb{C}[f_1, \ldots, f_n]$ is the space of polynomials in the f_i. The action of $f_i \in \mathfrak{h}$ is given by $f_i \cdot f_j = f_i f_j$. To compute the action of e_j, in $\Omega(W) = T(W)/I$ write

$$\begin{aligned} e_j f_i &= e_j f_i - f_i e_j + f_i e_j \\ &= \psi(\langle e_j, f_i \rangle) + f_i e_j \qquad (2.8)(a) \\ &= \psi(\delta_{i,j}) + f_i e_j \end{aligned}$$

Passing to $\mathcal{F}$ by the isomorphism (1.6) the term $f_i e_j$ is zero, and thus

$$e_j \cdot f_i = \psi(\delta_{i,j}). \qquad (2.8)(b)$$

Thus to compute the action of the $e_i's$ multiply on the left, then move all $e_i's$ terms past all $f_j's$ to the right using the defining relation of Ω, and finally all terms with $e_i's$ on the right are 0.

Under the isomorphism with a subspace of Ω^2 the decomposition $\mathfrak{g} = \mathfrak{p}^+ \oplus \mathfrak{k} \oplus \mathfrak{p}^-$ becomes $S^2(Y) \oplus (X \otimes Y) \oplus S^2(X)$. Here $S^2(Y)$ acts by operators which raise degree by 2, $S^2(X)$ lowers degree by 2, and $X \otimes Y$ acts by degree preserving operators.

This is a version of the canonical commutation relations. To see this we identify $\mathbb{C}[Y]$ with the polynomial functions $\mathbb{C}[z_1, \ldots, z_n]$ on Y via the dual variables $z_i(f_j) = \delta_{i,j}$. Write $\mathcal{F} = \sum \mathcal{F}^n$ where $\mathcal{F}^n$ is the polynomials of degree n. Then $f_i \in W \subset \mathfrak{h}$ acts by multiplication by z_i, e_j acts by $\psi(1)\frac{d}{dz_j}$, and $L \subset \mathfrak{h}$ acts by the character ψ. The image of $\mathcal{U}(\mathfrak{h})$ is the Weyl algebra of polynomial coefficient differential operators. More precisely, identify X and Y with abelian subalgebras of $\mathfrak{h}$. Then

$$X \to \{\frac{d}{dz_i}\}$$
$$Y \to \{z_i\}.$$

The action of $\mathfrak{g}$ is given by:

$$\mathfrak{p}^+ \simeq S^2(Y) \to \{z_i z_j\}$$
$$\mathfrak{p}^- \simeq S^2(X) \to \{\frac{d^2}{dz_i dz_j}\}$$
$$\mathfrak{k} \simeq X \otimes Y \to \{\frac{1}{2}(z_i\frac{d}{dz_j} + \frac{d}{dz_j}z_i) = z_i\frac{d}{dz_j} + \frac{1}{2}\delta_{i,j}\}.$$

Lemma 2.9 *(1) $\mathcal{F}$ is an irreducible representation of $\mathfrak{h}$, with central character ψ. The space of vectors annihilated by $X \subset \mathfrak{h}$ is the one–dimensional space of constants.*

(2) As a representation of $\mathfrak{g}$, $\mathbb{C}[Y]$ is the direct sum of two irreducible representations $\mathcal{F}_\pm$ of even and odd functions. The space of vectors in $\mathcal{F}_+$ (resp. $\mathcal{F}_-$) annihilated by $S^2(X) \subset \mathfrak{g}$ is $\mathcal{F}^0$ (resp. $\mathcal{F}^1$), of dimension 1 (resp. n).

The proof is immediate.

Remark: Suppose V is a vector space with a non–degenerate *orthogonal* form $(,)$, and let $\Omega(V) = T(V)/\langle v \otimes w + w \otimes v - \psi(v,w)\rangle$ (cf. Definition 1.1). Then Ω is isomorphic to a quotient of the Cayley algebra [5] (compare Lemma 2.2) and this construction produces the *spin* module of $\mathfrak{so}(V)$. There exists a theory of dual pairs in $Spin(V)$ and the duality correspondence for the spin representation. See [20], [22], [48], [1].

Remark: The action of $\mathfrak{k} \simeq \mathfrak{gl}(n,\mathbb{C})$ on $\mathcal{F} \simeq \mathbb{C}[\mathbb{C}^n]$ is by the "standard" action on $\mathbb{C}^n$, twisted by a character. Up to a character this is the derived representation of the action of $GL(n,\mathbb{C})$ acting on polynomials by $g \cdot f(v) = f(g^{-1}v)$. This is the connection between the oscillator representation and classical invariant theory [55], [20].

For use in Section 4 we record a generalization of this construction. Formally the operators (1.8) can be made to act on spaces of functions other than polynomials. For example, it is immediate that for any $\gamma \in \mathbb{C}$ the space

$$\{Pe^{\gamma \sum z_i^2} \,|\, P \in \mathbb{C}[z_1,\dots,z_n]\}$$

is invariant under this action, and so defines a representation of $\mathfrak{h}$ and $\mathfrak{g}$.

3. Schrödinger Model

See the lectures of Brooks Roberts [43], and the references there; also see [19], [17].

We now let W_0 be a real vector space of dimension $2n$ with a non–degenerate symplectic form $\langle,\rangle$. Let $Sp(W_0) \simeq Sp(2n, \mathbb{R})$ be the isometry group of this form, and $\widetilde{Sp}(2n, \mathbb{R})$ its connected two–fold cover. Let $K \simeq U(n)$ be a maximal compact subgroup of $Sp(2n, \mathbb{R})$, and $\widetilde{K}$ its inverse image in $\widetilde{Sp}(2n, \mathbb{R})$. Then $\widetilde{K}$ is connected, and is isomorphic to the $det^{\frac{1}{2}}$ cover of K. That is, it is isomorphic to the subgroup of $U(n) \times \mathbb{C}^\times$ consisting of pairs (g, z) such that $det(g) = z^2$. This has a two–to–one map $p : (g, z) \to g$ onto $U(n)$, and a genuine character $det^{\frac{1}{2}} : (g, z) \to z$ whose square is det.

Fix a non–trivial unitary additive character ψ of $\mathbb{R}$, and let $\omega = \omega_\psi$ be the corresponding oscillator representation of $\widetilde{Sp}(2n, \mathbb{R})$ on a Hilbert space. If $W_0 = X_0 \oplus Y_0$ is a complete polarization of W_0 we may realize ω on $L^2(Y_0)$. This is the direct sum of two irreducible representations.

There is a vector Γ in ω_ψ, unique up to scalars, which is an eigenvector for the action of $\widetilde{K}$. In the usual coordinates this is the Gaussian $e^{-\frac{1}{2}\sum x_i^2}$.

For $a \in \mathbb{R}$ write $a \cdot \psi(x) = \psi(ax)$. Then the oscillator representations defined by ψ and $a^2 \cdot \psi$ are isomorphic. Up to isomorphism there are two oscillator representations of $\widetilde{Sp}(2n, \mathbb{R})$.

The $\widetilde{K}$–finite vectors $\omega_{\widetilde{K}}$ in ω are a dense subspace. The action of $\widetilde{Sp}(2n, \mathbb{R})$ lifts to an action of the real Lie algebra $\mathfrak{g}_0 = \mathfrak{sp}(2n, \mathbb{R})$ on $\omega_{\widetilde{K}}$, and this extends by complex linearity to $\mathfrak{g} = \mathfrak{sp}(2n, \mathbb{C})$. Furthermore $\widetilde{K}$ acts on this space in a way compatible with the action of $\mathfrak{g}$. The action of $\widetilde{K}$ is *locally finite*: the subspace generated by any vector under the action of $\widetilde{K}$ is finite dimensional. Finally the the multiplicity of any $\widetilde{K}$–type is finite. (See Section 9, Example 2 for a further discussion of the $\widetilde{K}$–structure of ω_ψ). This says that the $\widetilde{K}$–finite vectors in ω are a $(\mathfrak{g}, \widetilde{K})$–module [52],[27].

4. Fock Model: Real Lie Algebra

The discussion in section 2 made no reference to the real Lie algebra $\mathfrak{sp}(2n, \mathbb{R})$, or (equivalently) the group K. In this section we describe the $\mathfrak{g} = \mathfrak{sp}(2n, \mathbb{C})$ module arising as the $(\mathfrak{g}, \widetilde{K})$–module of the oscillator representation ω of $\widetilde{Sp}(2n, \mathbb{R})$.

Let W_0 and $Sp(W_0) \simeq Sp(2n, \mathbb{R})$ be as in Section 2. Let $\mathfrak{g}_0 = \mathfrak{sp}(W_0) \simeq \mathfrak{sp}(2n, \mathbb{R})$ be the real Lie algebra of G. Let $W = W_0 \otimes_\mathbb{R} \mathbb{C}$, and extend $\langle,\rangle$

to W by linearity. We then obtain $Sp(W) \simeq Sp(2n, \mathbb{C})$ and $\mathfrak{g} = \mathfrak{sp}(W) \simeq \mathfrak{sp}(2n, \mathbb{C})$ as in Section 2. Given a complete polarization $W = X \oplus Y$ of W and $\psi : \mathbb{C} \to \mathbb{C}$ we obtain the oscillator representation of $\mathfrak{g}$ on $\mathcal{F} = \mathbb{C}[Y]$. The isomorphism class of $\mathcal{F}$ as a representation of $\mathfrak{g}$ is independent of the choice of complete polarization. However as a representation of $\mathfrak{g}_0$ this depends on the position of this polarization with respect to the real structure.

Totally complex polarization: We assume first that X is a totally complex subspace of W, i.e. $X \cap W_0 = 0$. See [7], also [36]. The choice of such X is equivalent to the choice of a compatible complex structure J on W_0: i.e. an element $J \in Sp(W_0)$ satisfying $J^2 = -1$. Given such J it extends by complex linearity to W, and we let X be the i-eigenspace of J. Conversely given X there exists unique $J \in Sp(2n, \mathbb{R})$ such that $J|_X = iId$. Note that $Y = \overline{X}$ is the $-i$-eigenspace of J, and $W = X \oplus Y$ is a complete polarization of W.

Now J makes W_0 a complex vector space of dimension n (with multiplication by i given by J). We denote this space W_J: this is a complex vector space of dimension n equal as a set to W_0. Furthermore $\{v, w\} = \langle v, Jw \rangle - i\langle v, w \rangle$ is a non-degenerate Hermitian form on W_J. We assume this form is positive definite, i.e. of signature $(n, 0)$.

To be explicit, if $e_1, \dots, f_n$ is a standard basis of W_0, we can take $X = \langle e_1 + if_1, \dots, e_n + if_n \rangle$, and then $J = \begin{pmatrix} 0 & I_n \\ -I_n & 0 \end{pmatrix}$. That is $Je_i = -f_i, Jf_i = e_i$. More generally we could take X spanned by some $e_i + if_i$ and some $e_j - if_j$. The signature of $\{,\}$ is then (p, q) where p (resp. q) is the number of plus (resp. minus) signs.

Let $K = \{g \in Sp(W_0) | gJ = Jg\}$. Equivalently $\{gv, gw\} = \{v, w\}$ for all $v, w \in W_J$, so $K \simeq U(n, 0)$, and K is a maximal compact subgroup of $Sp(W_0)$. Furthermore this is equivalent to the condition $gX = X, gY = Y$, i.e. K is the stabilizer of the complete polarization. The corresponding complexified Lie algebra $\mathfrak{k}$ is isomorphic to $\mathfrak{gl}(n, \mathbb{C})$, and its intersection $\mathfrak{k}_0$ with $\mathfrak{g}_0$ is isomorphic to $\mathfrak{u}(n)$.

As in Section 2 consider the Fock model of the oscillator representation on $\mathbb{C}[Y]$. Then $\mathfrak{k} \simeq X \otimes Y$ acts by operators which preserve degree (cf. Section 2), and therefore $\mathfrak{k}$ acts locally finitely. The action of $\mathfrak{k}_0$ exponentiates to a certain two-fold cover $\widetilde{K}$ of K (see Section 2). It follows that $\mathbb{C}[Y]$ is a $(\mathfrak{g}, \widetilde{K})$-module. We have $\mathfrak{p}^+ \simeq S^2(Y)$ acts by operators of degree 2, and $\mathfrak{p}^- \simeq S^2(X)$ by operators of degree -2.

Real polarization: At the opposite extreme we could take a complete polarization of W to be the complexification of one of W_0. To distinguish this case from the previous one we write $W_0 = \mathcal{X}_0 \oplus \mathcal{Y}_0$ and $W = \mathcal{X} \oplus \mathcal{Y}$, so $\mathcal{X} = \mathcal{X}_0 \otimes_{\mathbb{R}} \mathbb{C}$, $\mathcal{X}_0 = \mathcal{X} \cap W_0$, and similarly for $\mathcal{Y}$. We can then define the representation of $\mathfrak{g}$ on $\mathcal{F} = \mathbb{C}[\mathcal{Y}]$. The subalgebra of $\mathfrak{g}$, i.e. $\mathcal{X} \otimes \mathcal{Y}$ which acts locally finitely is again isomorphic to $\mathfrak{gl}(n, \mathbb{C})$, but the intersection with $\mathfrak{g}_0$ is then $\mathfrak{gl}(n, \mathbb{R})$. Assume $exp(\psi) : \mathbb{R} \to \mathbb{C}^{\times}$ is unitary, i.e. $\psi(z) = \lambda z$ for $\lambda \in i\mathbb{R}$.

This is essentially the Schrödinger model. Recall (§3) in the Schrödinger model the Gaussian $\Gamma = e^{-\frac{1}{2} \sum x_i^2}$ is an eigenvector for the action of $\widetilde{K}$, or equivalently $\mathfrak{k}$.

In the Fock model on $\mathbb{C}[Y]$, the constants are the unique eigenspace for $\mathfrak{k} \simeq X \otimes Y$. The constants are also characterized as the kernel of the action of $X \subset \mathfrak{h}$. Therefore we look for a vector in the Schrödinger model on $\mathbb{C}[\mathcal{Y}]$ annihilated by X. Choose a basis of $\mathcal{Y}$ with dual coordinates $z_1, \ldots, z_n$ so that $\mathbb{C}[Y] \simeq \mathbb{C}[z_1, \ldots, z_n]$. Recall (1.10) we can extend the action on $\mathbb{C}[z_1, \ldots, z_n]$ to polynomials times a Gaussian.

Lemma 4.1

1. *There is a unique value of γ for which the Gaussian*

$$\Gamma = e^{-\gamma \frac{1}{2} \sum z_i^2}$$

   is annihilated by $X \subset \mathfrak{h}$.

2. *The space*

$$\mathbb{C}[\mathcal{Y}]\Gamma = \{ Pe^{-\frac{1}{2}\gamma \sum z_i^2} \mid P \in \mathbb{C}[z_1, \ldots, z_n] \}$$

   is a $(\mathfrak{g}, \widetilde{K})$–module,

3. $\mathbb{C}[Y]$ *and* $\mathbb{C}[\mathcal{Y}]\Gamma$ *are isomorphic as $\mathfrak{h}$ and as $(\mathfrak{g}, \widetilde{K})$–modules.*

4. *Given $\mathcal{Y}$ we can choose Y (totally complex) so that*

$$\Gamma = e^{-\frac{1}{2} \sum z_i^2}.$$

   Then $\mathbb{C}[Y]$ is isomorphic to the space of $\widetilde{K}$–finite vectors $L^2(\mathcal{Y}_0)_{\widetilde{K}}$ in the Schrödinger model $L^2(\mathcal{Y}_0)$ of the oscillator representation.

Proof. This essentially reduces to $SL(2)$. Let e, f be a standard basis of W_0, and let $\mathcal{X} = \mathbb{C}\langle e \rangle, \mathcal{Y} = \mathbb{C}\langle f \rangle$. Let $X = \mathbb{C}\langle e + i\nu f \rangle, Y = \mathbb{C}\langle e - i\nu f \rangle$ for some $0 \neq \nu \in \mathbb{C} - i\mathbb{R}$. Let $\lambda = \psi(1)$. With dual coordinate z on $\mathcal{Y}$ satisfying $z(gf) = 1$ we have $e \in \mathcal{X}$ acts on $\mathbb{C}[\mathcal{Y}]$ by $\lambda \frac{d}{dz}$, and $f \in \mathcal{Y}$ acts by multiplication by z. Therefore $e + \nu if$ acts by $\tau = i\nu z + \lambda \frac{d}{dz}$, and $\tau \cdot e^{\gamma z^2} = (2\gamma\lambda + i\nu)z e^{\gamma z^2}$. Thus taking

$$\gamma = \frac{-i\nu}{2\lambda}$$

gives $\tau \cdot e^{\gamma z^2} = 0$. It is immediate that γ is determined uniquely. This proves (1), and (2) follows readily from this.

It is an easy argument by induction on degree that $\mathbb{C}[\mathcal{Y}]\Gamma$ is irreducible as a representation of $\mathfrak{h}$. Since the same holds for $\mathbb{C}[Y]$, and the central characters agree, by the Stone–von Neumann theorem these are isomorphic as representations of $\mathfrak{h}$, and (3) follows immediately from this.

For (4) choose $\nu = -i\lambda$, so $\gamma = -\frac{1}{2}$ and Γ has the stated form. Finally restricting the functions in $\mathbb{C}[\mathcal{Y}]\Gamma$ to $\mathcal{Y}_0$ gives functions in $L^2(\mathcal{Y}_0)$, and this gives the isomorphism of (4).   □

Remark: Suppose $\mathcal{X} = \langle e_1, \ldots, e_n \rangle, \mathcal{Y} = \langle f_1, \ldots, f_n \rangle$ as usual, and $X = \langle e_1 + if_1, \ldots, e_n + if_n \rangle, Y = \langle e_1 - if_1, \ldots, e_n - if_n \rangle$. Also assume $\psi(1) = i$ and $\nu = 1$. Then K is the usual maximal compact subgroup of $Sp(2n, \mathbb{R})$, and this is how the Fock/Schrödinger models are usually stated.

Remark: As far as the actions of the Lie algebras are concerned, we see that both the Fock and Schrödinger models are examples of the construction of Section 2, extended to polynomials times the Gaussian in the latter case. However the full Hilbert spaces in the two cases look somewhat different. In the Fock model this consists of holomorphic functions on Y which are square integrable with respect to a certain measure [7], whereas in the Schrödinger model this is $L^2(\mathcal{Y}_0)$. The existence of these two isomorphic models accounts for the simultaneously analytic and algebraic nature of the oscillator representations.

This duality is nicely illustrated by looking at the operators in the two models in the case of $SL(2)$. We use the notation of the proof of Lemma 4.1, with $\lambda = i$ and $\nu = 1$. We write $\mathfrak{sl}(2, \mathbb{C}) = \mathfrak{k} \oplus \mathfrak{p}^+ \oplus \mathfrak{p}^-$, or alternatively as $= \mathfrak{m} \oplus \mathfrak{n} \oplus \bar{\mathfrak{n}}$.

Fock Model:

Here the space of $\widetilde{K}$–finite vectors in the oscillator representation is
$\mathcal{F} = \mathbb{C}[Y] = \mathbb{C}[z]$.

$\mathfrak{k} \oplus \mathfrak{p}$ acts by:

$$\begin{pmatrix} 0 & i \\ -i & 0 \end{pmatrix} \in \mathfrak{k} \to -z\frac{d}{dz} - \frac{1}{2}$$

$$\begin{pmatrix} 1 & i \\ i & -1 \end{pmatrix} \in \mathfrak{p}^+ \to z^2$$

$$\begin{pmatrix} 1 & -i \\ -i & -1 \end{pmatrix} \in \mathfrak{p}^- \to -\frac{d^2}{dz^2}$$

And the operators from $\mathfrak{m} \oplus \mathfrak{n} \oplus \bar{\mathfrak{n}}$:

$$\begin{pmatrix} 1 & 0 \\ 0 & -1 \end{pmatrix} \in \mathfrak{m} \to \frac{1}{2}(z^2 - \frac{d^2}{dz^2})$$

$$\begin{pmatrix} 0 & 4i \\ 0 & 0 \end{pmatrix} \in \mathfrak{n} \to -2z\frac{d}{dz} - 1 + z^2 + \frac{d^2}{dz^2}$$

$$\begin{pmatrix} 0 & 0 \\ -4i & 0 \end{pmatrix} \in \bar{\mathfrak{n}} \to -2z\frac{d}{dz} - 1 - z^2 - \frac{d^2}{dz^2}$$

Schrödinger Model:

Now the $\widetilde{K}$–finite vectors are polynomials in x times the Gaussian $e^{-\frac{1}{2}x^2}$.
Here the formulas for $\mathfrak{m} \oplus \mathfrak{n} \oplus \bar{\mathfrak{n}}$ are easy:

$$\begin{pmatrix} 1 & 0 \\ 0 & -1 \end{pmatrix} \in \mathfrak{m} \to x\frac{d}{dx} + \frac{1}{2}$$

$$\begin{pmatrix} 0 & 1 \\ 0 & 0 \end{pmatrix} \in \mathfrak{n} \to \frac{i}{2}x^2$$

$$\begin{pmatrix} 0 & 0 \\ 1 & 0 \end{pmatrix} \in \bar{\mathfrak{n}} \to \frac{i}{2}\frac{d}{dx^2}$$

whereas $\mathfrak{k} \oplus \mathfrak{p}$ acts by:

$$\begin{pmatrix} 0 & 1 \\ -1 & 0 \end{pmatrix} \in \mathfrak{k} \to \frac{i}{2}(x^2 - \frac{d^2}{dx^2})$$

$$\begin{pmatrix} 1 & i \\ i & -1 \end{pmatrix} \in \mathfrak{p}^+ \to (x\frac{d}{dx} + \frac{1}{2}) - \frac{1}{2}x^2 - \frac{1}{2}\frac{d^2}{dx^2}$$

$$\begin{pmatrix} 1 & -i \\ -i & -1 \end{pmatrix} \in \mathfrak{p}^- \to (x\frac{d}{dx} + \frac{1}{2}) + \frac{1}{2}x^2 + \frac{1}{2}\frac{d^2}{dx^2}$$

We look at the isomorphism between the Fock and Schrödinger models more closely. It takes the constants in the Fock model to the multiples of the Gaussian in the Schrödinger model. It intertwines the actions of Y, for which these vectors are cyclic, and the isomorphism is determined by these conditions.

Consider the example of $SL(2)$ in the proof of Lemma 4.1, and the action of $e - \nu i f \in Y$ on $\mathbb{C}[Y]$ and $\mathbb{C}[\mathcal{Y}]\Gamma$; the isomorphism intertwines these actions. With $\nu = -i\lambda$ let z dual to $e - \lambda f$ be the coordinate on Y and x dual to f on $\mathcal{Y}$. We conclude multiplication by z goes to $\lambda(\frac{d}{dx} - x)$ on $\mathbb{C}[\mathcal{Y}]\Gamma$. Thus the isomorphism takes

$$z^n \to \lambda^n (\frac{d}{dx} - x)^n \cdot e^{-\frac{1}{2}x^2}.$$

Note that $(\frac{d}{dx} - x)(x^n\Gamma) = (-2x^{n+1} + \text{lower order terms})\Gamma$. It follows that this isomorphism takes

$$z^n \to p_n(x)e^{-\frac{1}{2}x^2}$$

where $p_n(z) = ((-2\lambda z)^n + \text{lower order terms})e^{-\frac{1}{2}z^2}$. The following Lemma follows readily from this discussion.

Lemma 4.2 *Given ψ and $W_0 = \mathcal{X}_0 \oplus \mathcal{Y}_0$, let $L^2(\mathcal{Y}_0)$ be the corresponding oscillator representation, with $\widetilde{K}$-finite vectors $L^2(\mathcal{Y}_0)_{\widetilde{K}}$. Let $W = X \oplus Y$ be as in Lemma 3.1(4) so that there is an isomorphism of $\mathfrak{h}$ and $(\mathfrak{g}, \widetilde{K})$-modules*

$$\phi : \mathbb{C}[Y] \to L^2(\mathcal{Y}_0)_{\widetilde{K}}.$$

Then we can choose coordinates $z_1, \ldots, z_n$ on Y and $x_1, \ldots, x_n$ on $\mathcal{Y}_0$ such that if $P(z_1, \ldots, z_n)$ is a polynomial of degree d then

$$\phi(P) = (c_d P(x_1, \ldots, x_n) + Q)\Gamma$$

where Q is a polynomial of degree strictly less than d, and c_d is a constant depending only on d.

In fact $\phi(z_1^{a_1} \ldots z_n^{a_n}) = (cx_1^{a_1} \ldots x_n^{a_n} + Q)\Gamma$ and up to a constant $cx_1^{a_1} \ldots x_n^{a_n} + Q$ is a Hermite polynomial [23].

5. Duality

The results in this section are due to R. Howe [21].

We begin by listing the irreducible reductive dual pairs over $\mathbb{R}$. This follows the general scheme of [15], and is spelled out in this case in [21]. See the lectures of Brooks Roberts [43] for more details.

The irreducible pairs of type I, together with the corresponding division algebra and involution are:

1. $(O(p,q), Sp(2n,\mathbb{R})) \subset Sp(2n(p+q),\mathbb{R}),$ $(\mathbb{R},1)$

2. $(U(p,q), U(r,s)) \subset Sp(2(p+q)(r+s),\mathbb{R})$ $(\mathbb{C},^-)$

3. $(Sp(p,q), O^*(2n)) \subset Sp(4n(p+q),\mathbb{R})$ $(\mathbb{H},^-)$

4. $(O(m,\mathbb{C}), Sp(2n,\mathbb{C})) \subset Sp(4mn,\mathbb{R})$ $(\mathbb{C},1)$

The type II dual pairs are

1. $(GL(m,\mathbb{R}), GL(n,\mathbb{R})) \subset Sp(2mn,\mathbb{R})$

2. $(GL(m,\mathbb{C}), GL(n,\mathbb{C})) \subset Sp(4mn,\mathbb{R})$

3. $(GL(m,\mathbb{H}), GL(n,\mathbb{H})) \subset Sp(8mn,\mathbb{R})$

Notation is as in [13]. The group $GL(m,\mathbb{H})$ is sometimes called $U^*(2m)$ and is a real form of $GL(2m,\mathbb{C})$.

Now let (G,G') be a reductive dual pair in $Sp(2n,\mathbb{R})$. Fix a non–trivial unitary additive character ψ of $\mathbb{R}$, and let ω_ψ be the corresponding oscillator representation. Fix a maximal compact subgroup $\mathbb{K} \simeq U(n)$ of $Sp(2n,\mathbb{R})$ and let $\mathcal{F}_\psi \simeq \mathbb{C}[z_1,\ldots,z_n]$ be the corresponding Fock model.

We may assume $K = G \cap \mathbb{K}$ and $K' = G' \cap \mathbb{K}$ are maximal compact subgroups of G and G' respectively. Then $\mathcal{F}$ is a $(\mathfrak{g} \oplus \mathfrak{g}', \widetilde{K} \cdot \widetilde{K}')$ module.

Remark 5.1 For H a subgroup of $Sp(W)$, $\widetilde{H}$ will always denote the inverse image of H in $\widetilde{Sp}(W)$, unless otherwise stated. For (G,G') a dual pair $\widetilde{G}$ commutes with $\widetilde{G}'$, and so $\widetilde{G \cdot G'} = \widetilde{G} \cdot \widetilde{G}'$.

We denote an *outer* tensor product of representations π_i of G_i as $\pi_1 \check{\otimes} \pi_2$. This is a representation of $G_1 \times G_2$. In the case $G = G_1 = G_2$ the restriction to the diagonal subgroup G is the ordinary tensor product.

We will often abuse notation and write $\pi \check{\otimes} \pi'$ for a representation of $\widetilde{G} \cdot \widetilde{G}'$. Here π (resp. π') is a genuine representation of $\widetilde{G}$ (resp. $\widetilde{G}'$) and $\pi \check{\otimes} \pi'$ is a representation of $\widetilde{G} \times \widetilde{G}'$. There is a surjective map $\widetilde{G} \times \widetilde{G}' \to \widetilde{G} \cdot \widetilde{G}'$, and $\pi \check{\otimes} \pi'$ is trivial on the kernel, so it factors to a genuine representation of $\widetilde{G} \cdot \widetilde{G}'$. If $G \cap G' = 1$ then this map is two–to–one.

Definition 5.2 $\mathcal{R}(\mathfrak{g}, \widetilde{K}, \psi)$ *is the set of isomorphism classes of irreducible* $(\mathfrak{g}, \widetilde{K})$ *modules* π *such that there exists a non–zero* $(\mathfrak{g}, \widetilde{K})$*–module map (necessarily a surjection)* $\mathcal{F} \to \pi$.

Theorem 5.3 ([21], Theorem 2.1) *If* $\pi \in \mathcal{R}(\mathfrak{g}, \widetilde{K}, \psi)$, *there exists a unique* $\pi' \in \mathcal{R}(\mathfrak{g}', \widetilde{K}', \psi)$ *such that*

$$\mathrm{Hom}_{\mathfrak{g} \oplus \mathfrak{g}', \widetilde{K} \cdot \widetilde{K}'}(\mathcal{F}_\psi, \pi \check{\otimes} \pi') \neq 0.$$

In fact this Hom space is one–dimensional.

Definition 5.4 *We define a correspondence*

$$\mathcal{R}(\mathfrak{g}, \widetilde{K}, \psi) \leftrightarrow \mathcal{R}(\mathfrak{g}', \widetilde{K}', \psi)$$

by the condition $\pi \leftrightarrow \pi'$ *if*

$$\mathrm{Hom}_{\mathfrak{g} \oplus \mathfrak{g}', \widetilde{K} \cdot \widetilde{K}'}(\mathcal{F}_\psi, \pi \check{\otimes} \pi') \neq 0.$$

By Theorem 5.3 this is a bijection. We refer to this as the duality correspondence or (local) theta–correspondence, and we write $\pi' = \theta(\psi)(\pi)$ *and* $\pi = \theta(\psi)(\pi')$.

The duality correspondence is a bijection between certain $(\mathfrak{g}, \widetilde{K})$ and $(\mathfrak{g}', \widetilde{K}')$ modules. By the standard theory of $(\mathfrak{g}, K)$–modules a module π in $\mathcal{R}(\mathfrak{g}, \widetilde{K}, \psi)$ corresponds to a genuine representation of $\widetilde{G}$. Let $\widehat{\widetilde{G}}_{genuine}$ denote the irreducible admissible genuine representations of $\widetilde{G}$. Then the duality correspondence defines a bijection between subsets of $\widehat{\widetilde{G}}_{genuine}$ and $\widehat{\widetilde{G}'}_{genuine}$.

Given $\pi \in \mathcal{R}(\mathfrak{g}, \widetilde{K}, \psi)$ let $\mathcal{N}(\pi)$ be the intersection of the kernels of all $\phi \in \mathrm{Hom}_{\mathfrak{g}, \widetilde{K}}(\mathcal{F}_\psi, \pi)$. Then $\mathcal{N}(\pi)$ is stable under the action of $(\mathfrak{g} \oplus \mathfrak{g}', \widetilde{K} \cdot \widetilde{K}')$, and $\mathcal{F}_\psi / \mathcal{N}(\pi)$ is isomorphic to $\pi \check{\otimes} \pi_1'$ for some $(\mathfrak{g}', \widetilde{K}')$ module π_1'. The preceding Theorem follows immediately from

Theorem 5.5 ([21] Theorem 2.1) *The module* π_1' *is of finite length, and has a unique irreducible quotient.*

Then π' is the unique irreducible quotient of π_1'.

Theorems 5.3 and 5.5 have smooth and unitary analogues, i.e. with $\mathcal{F}$ replaced by the unitary representation of $\widetilde{Sp}(2n, \mathbb{R})$ on a Hilbert space $\mathcal{H}$, or on the smooth vectors $\mathcal{H}^\infty$ of $\mathcal{H}$. See [21].

Remark 5.6 We say a two–fold covering $p : \widetilde{G} \to G$ *splits* if the exact sequence

$$1 \to \mathbb{Z}/2\mathbb{Z} \to \widetilde{G} \to G \to 1$$

splits. All of our extensions will be central, i.e. the kernel of the covering map is contained in the center of $\widetilde{G}$, so having a splitting is equivalent to $\widetilde{G} \simeq G \times \mathbb{Z}/2\mathbb{Z}$. The choice of splitting is equivalent to the choice of this isomorphism. Given such a choice there is a bijection between the irreducible genuine representations of $\widetilde{G}$ and the irreducible representations of G. This bijection is realized by tensoring with the distinguished character $\zeta = 1\check{\otimes}sgn$ of $G \times \mathbb{Z}/2\mathbb{Z}$.

More generally even if (4.6) is not split, there may exist a genuine character ζ of $\widetilde{G}$. For example this is the case if $\widetilde{G}$ is the $det^{\frac{1}{2}}$ cover of $GL(n)$ or a subgroup such as $O(n)$ or $U(n)$. In fact this is the case for all the groups under consideration except the metaplectic group $\widetilde{Sp}(2n, \mathbb{R})$. Tensoring with ζ then determines a bijection as in the split case.

Thus, given some choices, in the duality correspondence between $\widetilde{G}_{genuine}^{\,\widehat{}}$ and $\widetilde{G}'^{\,\widehat{}}_{genuine}$ in some cases we can replace $\widetilde{G}_{genuine}^{\,\widehat{}}$ by $G^{\widehat{}}$, and similarly $\widetilde{G}'$.

Open problem: *Explicitly* compute the duality correspondence of Definition 4.4.

Here "explicitly" has several possible meanings. In some sense the best solution is to compute the bijection in terms of some parameterization of representations, such as the Langlands classification or variations of it. However other descriptions are possible.

Question: How does the duality correspondence behave with respect to unitarity?

In the stable range $\theta(\psi)$ preserves unitarity going from the smaller to the larger group: if π is unitary then so is $\theta(\psi)(\pi)$ [30]. Very little is known in general. It takes Hermitian representations to Hermitian representations [37].

6. Compact Dual Pairs

In this section we consider the duality correspondence when one member G or G' of the dual pair (G, G') is compact. We refer to such a dual pair as a *compact* dual pair. For the general dual pair (G, G') the maximal compact subgroup K is a member of a compact dual pair (K, M'), and the duality correspondence for (K, M') plays a fundamental role in determining that for (G, G').

The duality correspondence for compact dual pairs is substantially simpler than the general case. The representations of the non–compact member of the dual pair are all (unitary) highest weight modules, and hence determined by their highest weight. (In fact all unitary highest weight modules of the classical groups arise this way [9], with a small number of exceptions in the case of $O^*(2n)$ [8]). An explicit computation of most cases is found in [25], and a closely related discussion is in [20]. In fact a number of such cases, treated in an *ad hoc* manner, provided some of the initial evidence for Howe's formalism and duality conjecture [10], [12], [44].

Let (K, G') be a dual pair in $Sp(2n, \mathbb{R})$ with K compact. Let $\mathbb{K}, \psi$ and $\mathcal{F}$ be as in Section 4. We assume $K, K' \subset \mathbb{K}$ where K' is a maximal compact subgroup of G. Recall (§2) the complexified Lie algebra of $Sp(2n, \mathbb{R})$ has the decomposition $\mathfrak{p}^- \oplus \mathfrak{k} \oplus \mathfrak{p}^+$.

The complexified Lie algebra $\mathfrak{g}'$ of G' has the decomposition $\mathfrak{g}' = \mathfrak{p}'^+ \oplus \mathfrak{k}' \oplus \mathfrak{p}'^-$ given by intersection with the corresponding decomposition of $\mathfrak{sp}(2n, \mathbb{C})$.

Theorem 6.1 *Let*

$$\mathcal{H}(K) = \{P \in \mathcal{F} \mid X \cdot P = 0 \quad \forall X \in \mathfrak{p}'^-\},$$

the space of K–harmonics.

Since the action of $\mathfrak{p}'^-$ commutes with the action of $\widetilde{K}$, $\mathcal{H}(K)$ is clearly $\widetilde{K}$–invariant.

Recall that $\widetilde{K}$ preserves the degree of polynomials in $\mathcal{F}$, so the polynomials $\mathcal{F}^n$ of degree n are a $\widetilde{K}$–stable subspace.

Definition 6.2 *Let σ be an irreducible representation of $\widetilde{K}$. The degree $d(\sigma)$ of σ is defined to be the minimal degree d such that the σ isotypic component $\mathcal{F}_\sigma^d$ of $\mathcal{F}^d$ is non–zero, or ∞ if $\mathcal{F}_\sigma$ is empty.*

If $d(\sigma) \neq \infty$, consider $\mathcal{F}_\sigma$, which is invariant by $\widetilde{K}$ and $\mathfrak{g}'$. Part of the proof of Theorem 5.5 is contained in the next result. See [20] and [21, Section 3].

Theorem 6.3

1. $\mathcal{F}_\sigma$ *is irreducible as a representation of* $\widetilde{K} \times \mathfrak{g}'$. *In particular* $\mathcal{F}_\sigma \simeq \sigma \otimes \pi'$ *for some irreducible* $(\mathfrak{g}', \widetilde{K}')$–*module* π'.

2. *The correspondence* $\sigma \leftrightarrow \pi'$ *is a bijection.*

3. *The polynomials of lowest degree* $(=d(\sigma))$ *in* $\mathcal{F}_\sigma$ *are precisely* $\mathcal{H}(K)_\sigma$.

4. $\mathcal{F}_\sigma = \mathcal{U}(\mathfrak{p}'^+) \cdot \mathcal{H}(K)_\sigma$

5. $\mathcal{H}(K)_\sigma$ *is irreducible as a representation of* $\widetilde{K} \times \mathfrak{k}'$, *so* $\mathcal{H}(K)_\sigma \simeq \sigma \otimes \tau'$ *for some irreducible representation* τ' *of* $\mathfrak{k}'$.

6. *The correspondence* $\sigma \leftrightarrow \tau'$ *is a bijection.*

In other words

$$\mathcal{F} \simeq \sum \sigma \otimes \pi'$$

as a representation of $\widetilde{K} \times \mathfrak{g}'$, and $\sigma \leftrightarrow \pi'$ is a bijection. This is a special case of Theorem 5.3. Furthermore each π' contains a unique $\widetilde{K}'$–type τ' of lowest degree, which generates π' under the action of $\mathcal{U}(\mathfrak{p}'^+)$, and finally the correspondence $\sigma \leftrightarrow \tau'$ is also a bijection.

Thus π' is a highest weight module, in that it has a cyclic vector under the action of the Borel subalgebra $\mathfrak{b} = \mathfrak{b}_{\mathfrak{k}'} \oplus \mathfrak{p}'^+$, where $\mathfrak{b}_{\mathfrak{k}'}$ is a Borel subalgebra of $\mathfrak{k}'$ [24]. Such a module is completely determined by its highest weight, which is the highest weight of τ' as a representation of $\mathfrak{k}'$.

Note that $\widetilde{K}$ may be a non–trivial covering group of K, and $\widetilde{K}$ as well as K may be disconnected. However except for these technicalities, the correspondence $\sigma \leftrightarrow \pi'$ is explicitly described by the correspondence of highest weights of σ for $\widetilde{K}$ and τ' for $\mathfrak{k}'$. We proceed to do this for all irreducible compact dual pairs.

The irreducible compact dual pairs are the following.

1. $(O(p), Sp(2n, \mathbb{R})) \subset Sp(2pn, \mathbb{R})$,

2. $(U(p), U(m, n)) \subset Sp(2p(m+n), \mathbb{R})$,

3. $(Sp(p), O^*(2n)) \subset Sp(4pn, \mathbb{R})$

(An exceptional case is $(Sp(p,q), O^*(2))$ with $pq \neq 0$. Here $O^*(2) \simeq U(1)$ is compact, but Theorem 5.3 does not apply, because the decomposition preceding Definition 5.1 does not hold.)

I. $(O(p), Sp(2n, \mathbb{R}))$.

The irreducible representations of $SO(p)$ are parameterized by highest weights $(a_1, \ldots, a_{[\frac{p}{2}]})$ with $a_i \in \mathbb{Z}$ and $a_1 \geq a_2 \geq \cdots \geq a_{[\frac{p}{2}]-1} \geq |a_{[\frac{p}{2}]}|$. If p is odd we may assume $a_{[\frac{p}{2}]} \geq 0$.

Following Weyl [55] we parameterize representations of $O(p)$ by restriction from $U(p)$. Consider the irreducible representations of $U(p)$ given by Young's diagrams with p rows of length $a_1, \ldots, a_p$, or equivalently with highest weight $(a_1, \ldots, a_p)$ with $a_1 \geq a_2 \geq \cdots \geq a_p \geq 0$. We embed $O(p)$ in $U(p)$ in the usual way, i.e $O(p) = U(p) \cap GL(p, \mathbb{R})$.

Lemma 6.4 ([55]) *The irreducible representations of $O(p)$ are parametrized by Young diagrams Y with rows of length $a_1 \geq a_2 \geq \cdots \geq a_p \geq 0$ such that the sum of the lengths of the first two columns is less than or equal to p. The representation of $O(p)$ defined by Y is defined to be the irreducible summand of the representation of $U(p)$ defined by Y which contains the highest weight vector.*

Equivalently the highest weights of the representations of $U(p)$ are of the form
$$(a_1, \ldots, a_k, 0, \ldots, 0) \quad a_1 \geq \cdots \geq a_k > 0, k \leq [\frac{p}{2}]$$
or
$$(a_1, \ldots, a_k, \overbrace{1, \ldots, 1}^{\ell}, 0, \ldots, 0), \quad a_1 \geq \cdots \geq a_k \geq 1, 2k + \ell = p.$$

In (a) the length of the first column is less than or equal to $[\frac{p}{2}]$, and we say this representation has "highest weight" $(a_1, \ldots, a_k, 0, \ldots, 0; 1)$. In (b) the length of the first column is greater than $[\frac{p}{2}]$ and we denote the "highest weight" of this representation $(a_1, \ldots, a_k, 0, \ldots, 0; -1)$.

For example the *sgn* representation of $O(p)$ is the restriction of the determinant representation of $U(p)$, which has highest weight $(1, \ldots, 1)$; the corresponding "highest weight" of $O(p)$ is $(0, \ldots, 0; -1)$.

The representations given by $(a_1, \ldots, a_k, 0, \ldots, \epsilon)$ with $\epsilon = \pm 1$ differ by tensoring with *sgn*. These two representations are not isomorphic to one another unless p is even and $a_{\frac{p}{2}} \neq 0$.

The genuine irreducible representations of the $det^{\frac{1}{2}}$ cover of $U(n)$ are parametrized by highest weights $(a_1, \ldots, a_n)$ with $a_i \in \mathbb{Z} + \frac{1}{2}$ and $a_1 \geq a_2 \geq \cdots \geq a_n$.

Now the inverse image $\widetilde{K}$ of $K = O(p)$ in the metaplectic group is isomorphic to the cover defined by $det^{\frac{n}{2}}$, and is therefore trivial over the connected component $SO(p)$, and also over $O(p)$ if and only if n is even. In any event $\widetilde{O}(p)$ has a genuine character ζ, and we identify genuine representations of $\widetilde{O}(p)$ with representations of $O(p)$ by tensoring with ζ. There are two choices of ζ if n is odd. The covering $\widetilde{Sp}(2n, \mathbb{R})$ of $Sp(2n, \mathbb{R})$ is split if and only if p is even, otherwise it is the metaplectic group. The maximal compact subgroup K' of $Sp(2n, \mathbb{R})$ is isomorphic to $U(n)$. There is a choice of this isomorphism, which will be specified below. Then $\widetilde{K}'$ is the split cover of $U(n)$ if p is even, or is isomorphic to the $det^{\frac{1}{2}}$ cover $\widetilde{U}(n)$ if n is odd. Since this is connected it is equivalent to work on the Lie algebra $\mathfrak{k}'$.

The duality correspondence in this case is a correspondence between genuine irreducible representations of $\widetilde{K}$ and $\widetilde{K}'$, or equivalently $\mathfrak{k}'$. We identify genuine representations of $\widetilde{K}$ with representations of $O(p)$, and those of $\widetilde{K}'$ with $\widetilde{U}(n)$ or $\mathfrak{gl}(n, \mathbb{C})$. The choices involved are determined by the following convention.

Normalization: We choose an isomorphism of K' with $U(n)$ so that the constants $\mathcal{F}^0$ of $\mathcal{F}$ are considered the irreducible representation of $\widetilde{U}(n)$ or $\mathfrak{gl}(n, \mathbb{C})$ with highest weight $(\frac{p}{2}, \ldots, \frac{p}{2})$.

Let ζ be the character by which $\widetilde{K}$ acts on $\mathcal{F}^0$. The correspondence $\sigma \to \sigma \otimes \zeta$ defines a bijection between the irreducible representations of $O(p)$ and the genuine irreducible representations of $\widetilde{K}$.

We thus describe the duality correspondence $\sigma \leftrightarrow \tau'$ in terms of representations of $O(p)$ and $\widetilde{U}(n)$ or $\mathfrak{gl}(n, \mathbb{C})$.

This normalization depends on ψ. For example it says that the highest weights of the oscillator representation defined by ψ are $(\frac{1}{2}, \ldots, \frac{1}{2})$ and $(\frac{3}{2}, \frac{1}{2}, \ldots, \frac{1}{2})$.

Proposition 6.5 *The duality correspondence for the dual pair* $(O(p), Sp(2n, \mathbb{R}))$ *is given by the correspondence of highest weights* $\sigma \leftrightarrow \tau'$:

$$O(p): \quad \sigma = (a_1, \ldots, a_k, 0, \ldots, 0; \epsilon)$$

$$Sp(2n, \mathbb{R}): \quad \tau' = (a_1 + \frac{p}{2}, \ldots, a_k + \frac{p}{2}, \overbrace{\frac{p}{2} + 1, \ldots, \frac{p}{2} + 1}^{\frac{1-\epsilon}{2}(p-2k)}, \frac{p}{2}, \ldots, \frac{p}{2}).$$

All such highest weights occur, subject to the constraints $k \leq [\frac{p}{2}]$ *and* $k + \frac{1-\epsilon}{2}(p - 2k) \leq n$.

This means that the weight σ for $O(p)$ is the highest weight of the irreducible representation σ, and the weight for $Sp(2n, \mathbb{R})$ is the highest weight of the $\widetilde{K}'$–type of τ' of lowest degree in π'.

II. $(U(p), U(m, n))$

The inverse image $\widetilde{K}$ of $U(p)$ in $\widetilde{Sp}(2p(m+n), \mathbb{R})$ is isomorphic to the cover defined by the character $det^{\frac{m+n}{2}}$. The cover of $U(m, n)$ is the $det^{\frac{p}{2}}$ cover. Therefore the correspondence $\sigma \leftrightarrow \tau'$ is described in terms of highest weights for unitary groups or their two–fold covers, given by integers or half–integers respectively.

We assume the isomorphisms of the covers of $U(p)$ and $U(m, n)$ with the det^* covers have been chosen so the constants $\mathcal{F}^0$ have weight $(\frac{m-n}{2}, \ldots, \frac{m-n}{2})$ for $\widetilde{K}$, and weight $(\frac{p}{2}, \ldots, \frac{p}{2}) \otimes (-\frac{p}{2}, \ldots, -\frac{p}{2})$ for $\widetilde{K}'$.

Proposition 6.6 *The duality correspondence for $(U(p), U(m, n))$ is given by*

$$U(p): \quad \sigma = (a_1, \ldots, a_k, 0, \ldots, 0, b_1, \ldots, b_\ell)$$

$$+ (\frac{m-n}{2}, \ldots, \frac{m-n}{2}) \to$$

$$U(m, n): \quad \tau' = (a_1, \ldots, a_k, 0, \ldots, 0) \otimes (0, \ldots, 0, b_1, \ldots, b_\ell)$$

$$+ (\frac{p}{2}, \ldots, \frac{p}{2}) \otimes (-\frac{p}{2}, \ldots, -\frac{p}{2}).$$

All such weights occur, subject to the obvious constraints $k + \ell \leq p, k \leq m, \ell \leq n$.

III. [32] $(Sp(p), O^*(2n))$.

Both $Sp(p)$ and $O^*(2n)$ are connected, and the coverings split, so we consider only the linear groups. The maximal compact subgroup of $O^*(2n)$ is isomorphic to $U(n)$.

$$Sp(p): \quad \sigma = (a_1, \ldots, a_k, 0, \ldots, 0) \to$$
$$O^*(2n)): \quad \tau' = (a_1 + p, \ldots, a_k + p, p, \ldots, p).$$

All such weights occur, subject to the obvious constraints $k \leq p, n$.

7. Joint Harmonics

The proof of the results in Section 5 depends heavily on information about K–types. This information is also important for computing the correspondence. The results in this section are all from [21].

We continue with the setup of Section 5. The maximal compact subgroup K of G is itself a member of a dual pair (K, M') with $K \subset G$ and $M' \supset G'$ (this is an example of a see–saw dual pair [28]) and we may apply the machinery of Section 5. The same applies *mutatis mutandis* to (M, K').

Definition 7.1 *The joint harmonics* $\mathcal{H}$ *is the space of polynomials which are both K–harmonic and K'–harmonic. That is*

$$\mathcal{H} = \mathcal{H}(K) \cap \mathcal{H}(K').$$

This is a $\widetilde{K} \cdot \widetilde{K}'$ *invariant subspace of* $\mathcal{F}$.

Let $\mathcal{R}(K, \mathcal{H})$ be the isomorphism classes of irreducible $\widetilde{K}$–modules σ such that $\mathcal{H}_\sigma \neq 0$, and define $\mathcal{R}(K', \mathcal{H})$ similarly. As in the definition of the duality correspondence, we say $\sigma \in \mathcal{R}(K, \mathcal{H})$ corresponds to $\sigma' \in \mathcal{R}(K', \mathcal{H})$ if $\sigma \widetilde{\otimes} \sigma'$ is a direct summand of $\mathcal{H}$.

Definition 7.2 *Suppose* $\pi \in \mathcal{R}(\mathfrak{g}, K, \psi)$ *and σ is a $\widetilde{K}$–type occurring in π, i.e. $\pi_\sigma \neq 0$. Then we say σ is of minimal degree in π if $d(\sigma)$ is minimal among the degrees of all $\widetilde{K}$–types of π.*

Since $d(\sigma) \geq 0$, it is immediate that the set of $\widetilde{K}$–types of π of minimal degree is non–empty. The notion of $\widetilde{K}$–type of π of minimal degree depends on the dual pair, and so is not an intrinsic notion to $\mathfrak{g}$. The relationship between the $\widetilde{K}$–types of minimal degree and those which are minimal in the sense of Vogan [52] is a subtle and important one.

The following structural result is a key to the proof of Theorem 5.3, as well as to an explicit understanding of the duality correspondence.

Theorem 7.3 ([21], Section 3)

1. *The correspondence $\sigma \to \sigma'$ defined on $\mathcal{H}$ is a bijection*

$$\mathcal{R}(K, \mathcal{H}) \leftrightarrow \mathcal{R}(K', \mathcal{H}).$$

   We write $\sigma' = \theta(\psi, \mathcal{H})(\sigma)$.

2. *$\mathcal{H}$ generates $\mathcal{F}$ as a representation of $\mathfrak{g} \oplus \mathfrak{g}'$, i.e. $\mathcal{U}(\mathfrak{g} \oplus \mathfrak{g}')\mathcal{H} = \mathcal{F}$.*

3. *Suppose $\pi \in \mathcal{R}(\mathfrak{g}, \widetilde{K}, \psi)$ and σ is a $\widetilde{K}$–type of π of minimal degree (Definition 6.2). Then $\sigma \in \mathcal{R}(K, \mathcal{H})$.*

4. *In the setting of (3), let $\pi' = \theta(\psi)(\pi)$, and let $\sigma' = \theta(\psi, \mathcal{H})(\sigma)$. Then σ' is a $\widetilde{K}'$–type of π' of lowest degree.*

Thus the duality correspondence θ comes equipped with a correspondence $\theta(\psi, \mathcal{H})$ of $\widetilde{K}, \widetilde{K}'$–types of lowest degree. It is therefore important to understand $\theta(\psi, \mathcal{H})$. From the results of Section 5 this may be computed explicitly, as we now describe.

We continue to work with our dual pair (G, G') together with two auxiliary pairs (K, M') and (M, K'). We also consider the maximal compact subgroups K_M and $K_{M'}$ of M and M'; $(K_M, K_{M'})$ is a dual pair with both members compact. We have the diagram:

$$
\begin{array}{ccc}
& M \longleftrightarrow K' & \\
\nearrow \uparrow & \downarrow \searrow & \\
K_M \quad G \longleftrightarrow G' \quad K_{M'} & \\
\nwarrow \uparrow & \downarrow \nearrow & \\
& K \longleftrightarrow M' &
\end{array}
$$

In the presence of several dual pairs we will write $\theta(\psi, K, M')$ for the duality correspondence between representations of $\widetilde{K}$ and $\widetilde{M}'$ via the dual pair (K, M'), and others similarly.

Lemma 7.4 ([21], Lemma 4.1) *Let σ be a $\widetilde{K}$–type of degree d. Define:*

1. $\pi' = \theta(\psi, K, M')(\sigma)$,

2. τ' *the unique* $\widetilde{K}_{M'}$*–type of* π' *of degree* d,

3. $\tau = \theta(\psi, K_{M'}, K_M)(\tau')$.

Then

1. *Suppose τ is the $\widetilde{K}_M$–type of lowest degree in a representation π of M with $\sigma' = \theta(\psi, M, K')(\pi) \neq 0$. Then $\sigma \widetilde{\otimes} \sigma'$ occurs in $\mathcal{H}$, i.e. $\theta(\psi, \mathcal{H})(\sigma) = \sigma'$. The degrees of σ, σ', τ and τ' are all d. Furthermore all other $\widetilde{K}'$–types of τ' restricted to $\widetilde{K}'$ have degree strictly less than d.*

2. *Suppose condition (1) fails. Then σ does not occur in $\mathcal{H}$, i.e. $\theta(\psi, \sigma) = 0$, and all $\widetilde{K}'$–types of τ' restricted to $\widetilde{K}'$ have degree strictly less than d.*

This is illustrated by the following diagram.

$$
\begin{array}{ccccccccc}
K & \overset{\theta(K,M')}{\longrightarrow} & M' & \overset{\text{min. deg.}}{\longrightarrow} & K_{M'} & \overset{\theta(K_{M'},K_M)}{\longrightarrow} & K_M & \overset{\text{min. deg.}}{\longrightarrow} & M & \overset{\theta(M,K')}{\longrightarrow} & K' \\
\sigma & \longrightarrow & \pi' & \longrightarrow & \tau' & \longrightarrow & \tau & \longrightarrow & \pi & \longrightarrow & \sigma'
\end{array}
$$

The following explicit computation of the correspondence of joint harmonics is an immediate consequence of this Lemma and the formulas of Section 6. Normalizations and parameterizations of representations in terms of highest weights are as in Section 6.

Proposition 7.5 *Explicit correspondence of joint harmonics for irreducible dual pairs.*

$\theta(\psi, \mathcal{H})(\sigma) = \sigma'$ *for* σ *and* σ' *as follows.*

I. $(G, G') = (O(p, q), Sp(2n, \mathbb{R})), (K, K') = (O(p) \times O(q), U(n))$.
$$\sigma = (a_1, \ldots, a_k, 0, \ldots, 0; \delta) \check{\otimes} (b_1, \ldots, b_\ell, 0, \ldots, 0; \epsilon) \rightarrow$$
$$\sigma' = (a_1, \ldots, a_k, \overbrace{1, \ldots, 1}^{\frac{1-\delta}{2}(p-2k)}, 0, \ldots, 0,$$
$$\overbrace{-1, \ldots, -1}^{\frac{1-\epsilon}{2}(q-2\ell)}, -b_\ell, \ldots, -b_1) + (\frac{p-q}{2}, \ldots, \frac{p-q}{2}).$$
where $k + \frac{1-\delta}{2}(p - 2k) + \ell + \frac{1-\epsilon}{2}(q - 2\ell) \leq n$.

II. $(G, G') = (U(p, q), U(r, s)), (K, K') = (U(p) \times U(q), U(r) \times U(s))$.
$$\sigma = (a_1, \ldots, a_k, 0, \ldots, 0, b_1, \ldots, b_\ell; c_1, \ldots, c_m, 0, \ldots, 0, d_1, \ldots, d_n) +$$
$$\frac{1}{2}(r - s, \ldots, r - s; s - r, \ldots, s - r) \rightarrow$$
$$\sigma' = (a_1, \ldots, a_k, 0, \ldots, 0, d_1, \ldots, d_n; c_1, \ldots, c_m, 0, \ldots, 0, b_1, \ldots, b_\ell) +$$
$$\frac{1}{2}(p - q, \ldots, p - q; q - p, \ldots, q - p)$$
where the obvious inequalities hold:
$$k + \ell \leq p, \ m + n \leq q, \ k + n \leq r, \ m + \ell \leq s$$

III. $(G, G') = (Sp(p, q), O^*(2n)), (K, K') = (Sp(p) \times Sp(q), U(n))$.
$$(a_1, \ldots, a_r, 0, \ldots, 0; b_1, \ldots, b_s, 0, \ldots, 0) \rightarrow$$
$$(a_1, \ldots, a_r, 0, \ldots, 0, -b_1, \ldots, -b_s) + (p - q, \ldots, p - q) \qquad (7.6)$$
where $a_1 \geq \cdots \geq a_r > 0, b_1 \geq \cdots \geq b_s > 0, r \leq p, s \leq q, r + s \leq n$.

IV. [33] $(G, G') = (GL(m, \mathbb{R}), GL(n, \mathbb{R})), (K, K') = (O(m), O(n))$. We assume $m \leq n$.
$$(a_1, \ldots, a_k, 0, \ldots, 0; \epsilon) \rightarrow$$
$$\begin{cases} (a_1, \ldots, a_k, \overbrace{1, \ldots, 1}^{\frac{1-\epsilon}{2}(m-2k)}, 0, \ldots, 0; +) & k + \frac{1-\epsilon}{2}(m - 2k) \leq [\frac{n}{2}] \quad (7.7) \\ (a_1, \ldots, a_k, \overbrace{1, \ldots, 1}^{n-m}, 0, \ldots, 0; -) & else \end{cases}$$
where $a_1 \geq \cdots \geq a_k > 0, \ k \leq [\frac{m}{2}]$.

V. [3] $(G, G') = (GL(m, \mathbb{C}), GL(n, \mathbb{C}))$, $(K, K') = (U(m), U(n))$.

$$(a_1, \ldots, a_k, 0, \ldots, 0, b_1, \ldots, b_\ell) \rightarrow$$
$$(-b_\ell, \ldots, -b_1, 0, \ldots, 0, -a_k, \ldots, -a_1) \tag{7.8}$$

where $a_1 \geq \cdots \geq a_k > 0 > b_1 \geq \cdots \geq b_\ell$, $k + \ell \leq m, n$.

VI. [3] $(G, G') = (O(m, \mathbb{C}), Sp(2n, \mathbb{C}))$, $(K, K') = (O(m), Sp(n))$.

$$(a_1, \ldots, a_k, 0, \ldots, 0; \epsilon) \rightarrow (a_1, \ldots, a_k, \overbrace{1, \ldots, 1}^{\frac{1-\epsilon}{2}(m-2k)}, 0, \ldots, 0) \tag{7.9}$$

where $a_1 \geq \cdots \geq a_k > 0$, $k \leq [\frac{m}{2}]$, $k + \frac{1-\epsilon}{2}(m - 2k) \leq n$.

VII. [32] $(G, G') = (GL(m, \mathbb{H}), GL(n, \mathbb{H}))$, $(K, K') = (Sp(m), Sp(n))$.

$$(a_1, \ldots, a_k, 0, \ldots, 0) \rightarrow (a_1, \ldots, a_k, 0, \ldots, 0) \tag{7.10}$$

where $a_1 \geq \cdots \geq a_k > 0$ *and* $k \leq m, n$.

As an example we sketch one of the calculations involved in the proof of this Proposition.

Example. Let $(G, G') = (O(p, q), Sp(2n, \mathbb{R}))$. Then

$$(K, M') = (O(p) \times O(q), Sp(2n, \mathbb{R}) \times Sp(2n, \mathbb{R}))$$

and $(M, K') = (U(p, q), U(n))$. Finally

$$(K_M, K_{M'}) = (U(p) \times U(q), U(n) \times U(n)).$$

The embeddings $Sp(2n, \mathbb{R}) \hookrightarrow Sp(2n, \mathbb{R}) \times Sp(2n, \mathbb{R})$ and $U(n) \hookrightarrow U(n) \times U(n)$ are diagonal, and the restriction of an outer tensor product $\pi \check{\otimes} \pi'$ is the ordinary tensor product.

Let $\sigma = (a_1, \ldots, a_k, 0, \ldots, 0; \delta) \check{\otimes} (b_1, \ldots, b_\ell, 0, \ldots, 0; \epsilon)$. Then provided

$$k + \frac{1-\delta}{2}(p - 2k) \leq n\ell \quad + \frac{1-\epsilon}{2}(q - 2\ell) \leq n$$

we have

$$\tau' = (a_1 + \frac{p}{2}, \ldots, a_k + \frac{p}{2}, \overbrace{\frac{p}{2} + 1, \ldots, \frac{p}{2} + 1}^{\frac{1-\delta}{2}(p-2k)}, \frac{p}{2}, \ldots, \frac{p}{2}) \check{\otimes}$$

$$(-\frac{q}{2}, \ldots, -\frac{q}{2}, \overbrace{-\frac{q}{2} - 1, \ldots, -\frac{q}{2} - 1}^{\frac{1-\epsilon}{2}(q-2\ell)}, -b_\ell - \frac{q}{2}, \ldots, -b_1 - \frac{q}{2})$$

Then

$$\tau = (a_1 + \frac{n}{2}, \ldots, a_k + \frac{n}{2}, \overbrace{1 + \frac{n}{2}, \ldots, 1 + \frac{n}{2}}^{\frac{1-\delta}{2}(p-2k)}, \overbrace{\frac{n}{2}, \ldots, \frac{n}{2}}^{\frac{1-\epsilon}{2}(q-2\ell)}) \check{\otimes}$$

$$(-\frac{n}{2}, \ldots, -\frac{n}{2}, \overbrace{-1 - \frac{n}{2}, -1 - \frac{n}{2}}^{}, -b_\ell - \frac{n}{2}, \ldots, -b_1 - \frac{n}{2})$$

Finally σ' exists if and only if

$$k + \frac{1-\delta}{2}(p - 2k) + \ell + \frac{1-\epsilon}{2}(q - 2\ell) \leq n$$

in which case

$$\sigma' = (a_1 + \frac{p-q}{2}, \ldots, a_k + \frac{p-q}{2},$$

$$\overbrace{1 + \frac{p-q}{2}, \ldots, 1 + \frac{p-q}{2}}^{\frac{1-\delta}{2}(p-2k)}, \frac{p-q}{2}, \ldots, \frac{p-q}{2},$$

$$\overbrace{-1 + \frac{p-q}{2}, \ldots, -1 + \frac{p-q}{2}}^{\frac{1-\epsilon}{2}(q-2\ell)}, -b_\ell + \frac{p-q}{2}, \ldots, -b_1 + \frac{p-q}{2})$$

Therefore $\mathcal{H} \simeq \sum \sigma \check{\otimes} \sigma'$ where the sum runs over σ, σ' satisfying (6.9)(d).

8. Induction Principle

The discussion in this section is closely related to that of section II.6 of Roberts' second lecture [42]. The essential point is the computation of the Jacquet module of the oscillator representation with respect to the nilpotent radical of a parabolic subgroup of a dual pair. This was originally discussed in [40] in somewhat different terms, and first expressed this way in [29] for orthogonal/symplectic pairs over p–adic fields. This was generalized in [34] to all type I pairs. The case of the real dual pairs $(O(2p, 2q), Sp(2n, \mathbb{R}))$ is treated in [33], general real symplectic pairs in [4], and complex groups in [3].

We begin with a rough discussion of the principle. For the moment we ignore the covering groups, and write $G \cdot G' = G \times G'$, etc. Consider a dual pair (G, G') in some $Sp(2n, \mathbb{R})$ and parabolic subgroups $P = MN$ and $P' = M'N'$ of G and G' respectively. We assume, as is often the case, that (M, M') is itself a dual pair in some $Sp(2m, \mathbb{R})$. Let ω be an oscillator representation for $\widetilde{Sp}(2n, \mathbb{R})$ restricted to $G \times G'$, and let ω_M be an oscillator representation for $M \times M'$. The Jacquet module $\omega_{N \times N'}$ of ω is a representation of $M \times M'$, and a calculation shows it has $\omega_M \zeta$ as a quotient for some character ζ of $M \times M'$. Suppose there is a non–zero $M \times M'$ map

$$\phi : \omega_M \to \sigma \check{\otimes} \sigma'$$

for some representation $\sigma \otimes \sigma'$ of $M \times M'$. That is $\theta(\psi, M, M')(\sigma) = \sigma'$. It follows from the computation of the Jacquet module that there is a non–zero $G \times G'$–map

$$\Phi : \omega \to Ind_P^G(\sigma\xi) \check{\otimes} Ind_{P'}^{G'}(\sigma'\xi')$$

for some characters ξ, ξ'. That is $\theta(\psi, G, G')(\pi) = \pi'$ for some irreducible constituents π and π' of $Ind_P^G(\sigma\xi)$ and $Ind_{P'}^{G'}(\sigma'\xi')$ respectively.

This produces a large part of the correspondence for (G, G'). The problem, in general, is to determine the constituents of the induced modules in the image of Φ.

The same discussion holds in the p–adic case as well. In the real case, one has the extra very powerful information on K–types, and this is the primary tool for attacking the question. For example if π and a K–type σ of lowest degree in π are known, this determines a K'–type $\theta(\psi, \mathcal{H})(\sigma)$ in π', and this may be enough to describe π' explicitly.

Another important difference from the p–adic case is that the Jacquet–functor is exact over a p–adic field but not over $\mathbb{R}$, so the precise information about the filtration of the Jacquet module of [29],[34] is not available over $\mathbb{R}$. At the same time there is no notion of "supercuspidal" representation over $\mathbb{R}$, so the whole notion of "first occurrence" (cf. [42]) takes on a somewhat different form.

We turn now to a more careful description. We consider only type I dual pairs, so we are given a division algebra D of dimension $d = 1, 2$ or 4 over $\mathbb{R}$ with involution (cf. Section 5). We are also given V (resp. W) a vector space over D equipped with a non–degenerate Hermitian (resp. skew–Hermitian) form and $G = U(V)$ is the isometry group of this form (resp. $G' = U(W)$.) Then (G, G') is an irreducible reductive dual pair in $Sp(\mathbb{W}) \simeq Sp(2dn, \mathbb{R})$, where $\mathbb{W} = V \otimes_D W$ considered as a real vector space.

We consider an orthogonal direct sum

$$V = V_+ \oplus V^0 \oplus V_-$$

Here the form restricted to V^0 is non–degenerate, V_+ and V_- are isotropic, and the form defines a perfect pairing on $V_+ \times V_-$. The stabilizer $P = MN$ of V_- in G is a parabolic subgroup of G, with $M \simeq GL(D, V_+) \times U(V^0)$.

Similarly we consider

$$W = W_+ \oplus W^0 \oplus W_-$$

with $P' = M'N'$, and $M' \simeq GL(D, W_+) \times U(W^0)$.

Fix ψ and let ω be the corresponding (smooth) oscillator representation of $\widetilde{Sp}(2dn, \mathbb{R})$. We consider this as a representation of $\widetilde{G} \cdot \widetilde{G}'$ by restriction. Let $\mathcal{F}$ be the corresponding Fock model, considered as a $(\mathfrak{g} \oplus \mathfrak{g}', \widetilde{K} \cdot \widetilde{K}')$–module.

Now (M, M') is a dual pair in $Sp(\mathbb{W}_M)$ where $\mathbb{W}_M = (V_+ \otimes_D W_+) \oplus (V_- \otimes_D W_-) \oplus (V^0 \otimes W^0)$, considered as a real vector space. Let ω_M be the oscillator representation of this metaplectic group, with corresponding Fock model $\mathcal{F}_M$. A technical headache is that the inverse images of M and M' in $\widetilde{Sp}(\mathbb{W}_M)$ may not be isomorphic to $\widetilde{M}$ and $\widetilde{M}'$, so denote these covers $\overline{M}, \overline{M}'$. We consider ω_M as a representation of $\overline{M} \cdot \overline{M}'$, and $\mathcal{F}_M$ as a $(\mathfrak{m} \oplus \mathfrak{m}', \overline{K}_M \overline{K}_{M'})$–module (with the obvious notation for maximal compact subgroups of M and M' and their covers).

Let $\widetilde{P} \simeq \widetilde{M}N$ and $\widetilde{P}' \simeq \widetilde{M}'N'$ be the inverse images of P and P' in $Sp(\mathbb{W})$.

Theorem 8.1

1. There is a non–zero surjective $\widetilde{P} \cdot \widetilde{P}'$ equivariant map

$$\Phi : \omega \twoheadrightarrow \omega_M \chi$$

     where χ is some character of $\overline{M} \cdot \overline{M}'$. Here $\overline{P} \cdot \overline{P}'$ acts on the right hand side by the given action of $\overline{M} \cdot \overline{M}'$, with $N \cdot N'$ acting trivially.

2. There is a non–zero surjective $(\mathfrak{m} \oplus \mathfrak{m}', \widetilde{K}_M \cdot \widetilde{K}_{M'})$–map

$$\Phi : \mathcal{F} \twoheadrightarrow \mathcal{F}_M \chi.$$

     Here $\mathcal{F}$ may be replaced by the Lie algebra homology group $H_0(\mathfrak{n} \oplus \mathfrak{n}', \mathcal{F})$.

3. *Suppose σ (resp. σ') is an irreducible $(\mathfrak{m}, \overline{K}_M)$ (resp. $(\mathfrak{m}', \overline{K}_{M'})$) module, and σ corresponds to σ' in the duality correspondence for the dual pair (M, M'). Then there is a non–zero map*

$$\Phi : \omega \to Ind_{\widetilde{P}}^{\widetilde{G}}(\sigma\zeta)\check{\otimes}Ind_{\widetilde{P}'}^{\widetilde{G}'}(\sigma'\zeta')$$

where ζ, ζ' are certain characters of $\widetilde{M}$ and $\widetilde{M}'$.

4. *In the setting of (3), some irreducible constituents of*

$$Ind_{\widetilde{P}}^{\widetilde{G}}(\sigma\zeta)$$

and

$$Ind_{\widetilde{P}'}^{\widetilde{G}'}(\sigma'\zeta')$$

correspond via the duality correspondence for (G, G').

Note: This essentially follows from the computation of the top term of the filtration of the Jacquet module [29],[34] applied twice, once each to $P \subset G$ and $P' \subset G'$.

Note: Some covering problems are being swept under the rug. The main point is that σ and ζ are representations of $\overline{M}$, but that $\sigma\zeta$ may be identified with a representation of $\widetilde{M}$. In the orthogonal–symplectic case this is written out in in [4]. Similar but not identical covering issues are treated in the p–adic case in [34]. In [33] the same issues are addressed in a situation where the covering groups are all trivial.

Proof. The main point is to choose the proper (mixed) model of ω. For a complete polarization of $\mathbb{W}$ we take

$$X = (V \otimes_D W_+) \oplus (V_+ \otimes_D W^0) \oplus X^0$$
$$Y = (V \otimes_D W_-) \oplus (V_- \otimes_D W^0) \oplus Y^0$$

where X_0, Y_0 is an arbitrary complete polarization of $V^0 \otimes_D W^0$. Then ω is realized on the Schwarz space $\mathcal{S}(Y)$. For ω_M we take

$$X_M = (V_- \otimes_D W_+) \oplus X^0$$
$$Y_M = (V_+ \otimes_D W_-) \oplus Y^0$$

so ω_M is realized on $\mathcal{S}(Y_M)$.

An explicit calculation shows that restriction from Y to Y_M intertwines the action of $\widetilde{P}$, $\widetilde{P}'$ up to certain characters, which gives (1).

The corresponding statement for the Fock models follows from Lemma 4.1. In fact it follows that up to terms of lower degree, $\Phi : \mathcal{F} \twoheadrightarrow \mathcal{F}_M$ is given by restriction.

Statement (3) is an immediate consequence of Frobenius reciprocity [52, Proposition 6.3.5]. The final statement follows from (3) by composing Φ with the map from the image of Φ in the induced representation to an irreducible quotient of this image. $\square$

We note that in this generality there is very little that can be said about the image of Φ and an irreducible quotient of it. Even if the induced representation has a unique irreducible quotient, there is no reason *a priori* that it should be in the image of Φ.

Open problem:

Prove a version of Theorem 8.2 for cohomological induction.

It appears that some version of Theorem 8.2 should hold with parabolic induction from M replaced by cohomological induction from the Levi factor L of a theta–stable parabolic [52], [26]. This is true in many examples, and some calculations indicate it is true in some generality. The homology groups that enter into Frobenius reciprocity in this setting ([52, Proposition 6.3.2]) are not solely in degree zero. As a result in the analogue of Theorem 8.2(2) it is necessary to calculate some higher homology groups.

Such a theorem together with Theorem 8.2 would go very far towards a complete explicit understanding of the duality correspondence.

9. Examples

We give a few examples of the local theta–correspondence over $\mathbb{R}$.

Example 1. $(O(1), Sp(2n, \mathbb{R}))$.

The center of $Sp(2n, \mathbb{R})$ is $O(1) \simeq \pm 1$, and this forms a dual pair. The inverse image $\widetilde{O}(1)$ of $O(1)$ in $\widetilde{Sp}(2n, \mathbb{R})$ is isomorphic to $\mathbb{Z}/4\mathbb{Z}$ (n odd) or $\mathbb{Z}/2\mathbb{Z} \times \mathbb{Z}/2\mathbb{Z}$ (n even). This group has two genuine characters χ and χ' which correspond to the two irreducible summands of the oscillator representation ω_ψ. The labeling of χ and and χ' is a matter of convention, or equivalently of a choice of isomorphism $\widetilde{O}(1) \simeq \mathbb{Z}/4\mathbb{Z}$ or $\mathbb{Z}/2\mathbb{Z} \times \mathbb{Z}/2\mathbb{Z}$. The same two characters also give the two constituents of $\omega_{\overline{\psi}}$.

Example 2. $(U(1), U(n))$.

Here $U(1)$ is the center of the maximal compact subgroup $K = U(n)$. This dual pair describes the restriction of the oscillator representation to $\widetilde{K}$, which is quite important.

Fix ψ. With choices as in Section 6, example II (preceding Proposition 5.6) the correspondence from $U(1)$ to $U(n)$ is $(k + \frac{n}{2}) \to (k + \frac{1}{2}, \frac{1}{2}, \ldots, \frac{1}{2})$, $k = 0, 1, 2, \ldots$. The $\widetilde{K}$–types with k even (resp. odd) constitute the irreducible summand ω_ψ^+ (resp. ω_ψ^-). The lowest $\widetilde{K}$–types of these two summands (both in the sense of Vogan, and of lowest degree) are $(\frac{1}{2}, \ldots, \frac{1}{2})$ and $(\frac{3}{2}, \frac{1}{2}, \ldots, \frac{1}{2})$ respectively.

Note that the $\widetilde{K}$–types lie along a "line", i.e. their highest weights are obtained from the highest weight of the lowest $\widetilde{K}$–type by adding multiples of a single vector. This is a condition of [53]. In fact the four irreducible summands of the oscillator representations are the only non–trivial unitary representations of $\widetilde{Sp}(2n, \mathbb{R})$ ($n \geq 1$) with $\widetilde{K}$–types along a line. The oscillator representation is particularly "small", and the duality correspondence is due in part to this. It is interesting to note that it is necessary to pass to the two fold cover of $Sp(2n, \mathbb{R})$ to find these especially small representations. This is analogous to the spin representations of the two-fold cover $Spin(n)$ of $SO(n)$.

Once ψ is fixed, the dual oscillator representation $\omega_\psi^* = \omega_{\overline{\psi}}$ has $\widetilde{K}$–spectrum $(-\frac{1}{2}, \ldots, -\frac{1}{2}, -\frac{1}{2} - k)$ ($k = 0, 1, \ldots$).

Example 3. $(O(n), SL(2, \mathbb{R}))$. The duality correspondence in this case is essentially the classical theory of spherical harmonics [20],[23]. With the appropriate coordinates the Fock model is on $\mathbb{C}[z_1, \ldots, z_n]$. One of the operators coming from the (complexified) Lie algebra of $SL(2, \mathbb{R})$ is the Laplacian $\Delta = \sum_i \frac{d^2}{dz_i^2}$, and the harmonics in the sense of section 5 are the kernel of the Laplacian. Then $r^2 = \sum_i z_i^2$ and $\sum_i z_i \frac{d}{dz_i} + \frac{n}{2}$ together with Δ span the image of $\mathfrak{sl}(2, \mathbb{C})$. Theorem 6.3 (Theorem 9 of [20]) reduces to the classical statements of spherical harmonics in this case.

Example 4. $(GL(1, \mathbb{R}), GL(1, \mathbb{R}))$

This example is an illustration of the principle that no dual pair is too simple to be taken lightly. The p–adic case is treated in [34, Chapter 3, §III.7, pg. 65] and has some non–trivial analytic content. The corresponding statement over $\mathbb{R}$ for the smooth vectors is similar. Here we discuss only the Fock model.

Write the dual pair as (G_1, G_2); of course the images of G_1 and G_2 in $SL(2, \mathbb{R})$ coincide. We embed $G_1 \hookrightarrow SL(2, \mathbb{R})$ as $x \to \iota_1(x) = diag(x, x^{-1})$. Write $\widetilde{SL}(2, \mathbb{R})$ a as $\{(g, \pm 1) \,|\, g \in SL(2, \mathbb{R})\}$ with the usual cocycle [41]. Let $\widetilde{GL}(1, \mathbb{R})$ be the two–fold cover of $GL(1, \mathbb{R})$ defined by the cocycle $c(x, y) = (x, y)_\mathbb{R}$. Here $(x, y)_\mathbb{R}$ is the Hilbert symbol [46], which equals

-1 if $x, y < 0$, and 1 otherwise. This is isomorphic to the $det^{\frac{1}{2}}$ cover, or alternatively to $\mathbb{R}^{\times} \cup i\mathbb{R}^{\times} \subset \mathbb{C}^{\times}$. Then $\widetilde{\iota}_1 : (x, \epsilon) \to (\iota(x), \epsilon)$ gives an isomorphism of $\widetilde{GL}(1, \mathbb{R})$ with the inverse image of $GL(1)$ in $\widetilde{SL}(2, \mathbb{R})$. Let $\mathfrak{g}_1 \simeq \mathbb{C}$ be the complexified Lie algebra of G_1, and let $K_1 = \pm 1$ be its maximal compact subgroup, so $\widetilde{K}_1 \simeq \mathbb{Z}/4\mathbb{Z}$.

Fix ψ and the Fock model $\mathcal{F} = \mathbb{C}[z]$. We seek to show that every genuine irreducible $(\mathfrak{g}_1, \widetilde{K}_1)$–module is a quotient of $\mathcal{F}$. The action of $\widetilde{K}_1$ breaks up $\mathcal{F}$ into its two irreducible summands $\mathcal{F}^+ = \mathbb{C}[z^2]$ and $\mathcal{F}^- = z\mathbb{C}[z^2]$ for $\widetilde{SL}(2, \mathbb{R})$ (cf. Example 1). So it is enough to show any character of $\mathfrak{g}_1$ occurs as a quotient of one summand.

The action of the Lie algebra $\mathfrak{g}_1$ of $GL(1, \mathbb{R})$ is by the operator $X = z^2 - \frac{d^2}{dz^2}$ (cf. Section 4). Therefore $\mathcal{F}^{\pm}$ are each free modules for this action, and every character of $\mathfrak{g}_1$ occurs as a quotient of $\mathcal{F}^{\pm}$ in a unique way. To be explicit we consider $\mathcal{F}^+$. Fix $\lambda \in \mathbb{C} \simeq \mathfrak{g}_1^*$, with $\lambda(X) = \lambda$, and let $\mathcal{N}_{\lambda} = \mathcal{U}(\mathfrak{g}) \cdot (z^2 - \lambda)$. By induction this is a codimension one subspace of $\mathcal{F}^+$. The image $\overline{1}$ of 1 in the quotient $\mathcal{F}^+/\mathcal{N}_{\lambda}$ is non–zero, and $X \cdot \overline{1} = \overline{z^2} = \lambda\overline{1}$. The case of $\mathcal{F}^-$ is similar.

The embedding of the second copy G_2 of $GL(1)$ is $\iota_2 : x = diag(x^{-1}, x)$. The natural choice of $\widetilde{\iota}_2$ is $(x, \epsilon) \to (\iota_2(x), \epsilon))$. With this convention $\widetilde{\iota}_2^{-1} \circ \widetilde{\iota}_1 : \widetilde{G}_1 \to \widetilde{G}_2$ takes (x, ϵ) to $(x^{-1}, \epsilon) = (x, \epsilon sgn(x))^{-1}$.

With these choices, the duality correspondence is

$$\chi \to \chi^{-1} sgn$$

for any genuine $(\mathfrak{g}_1, \widetilde{K}_1)$ character χ. The sgn term comes from the twist by $sgn(x)$ in $\iota_2^{-1} \circ \iota_1$.

Of course $\widetilde{\iota}_2$ can be modified to eliminate the twist by sgn. Also the oscillator representation itself restricted to this dual pair can be normalized by tensoring with a genuine character of $\widetilde{GL}(1, \mathbb{R})$ so that the correspondence takes $\chi \to \chi^{-1}$ as χ runs over characters of $GL(1, \mathbb{R})$. This is what is normally done.

Example 5. $(GL(m, \mathbb{R}), GL(m, \mathbb{R}))$.

This generalizes Example 4. Let G be the $det^{\frac{m}{2}}$ cover of $GL(m, \mathbb{R})$. Then (G_1, G_2) is a dual pair with $G_1 \simeq G \simeq G_2$, and with the natural choice of these isomorphisms the duality correspondence takes π to $\pi^* \otimes sgn$, as π runs over all irreducible genuine representations of G. As in Example 4 the twist by sgn can be eliminated, and G replaced by $GL(m, \mathbb{R})$.

The proof is by induction on m, starting with $m = 1$ by Example 4, and the induction principle of Section 8. We normalize the correspondence to eliminate the covering groups. We write an irreducible representation π of $GL(m)$ as the unique irreducible quotient of

$$Ind_{MN}^{G_1}(\sigma \otimes 1\!\!1)$$

where $M \simeq GL(m-1) \times GL(1)$. By induction and Theorem 8.2 there is a non–zero map from ω to the tensor product of the induced module for G_1 and a similar induced modules for G_2 with σ replaced by σ^*. If these induced modules are irreducible the result is immediate. The general case follows from a deformation of parameters argument as in [3], the main point being that we have enough control over K–types to determine at least an irreducible quotient of the image of this map. The K–type information is crucial because the induced module for G_2 will have unique irreducible submodule, and not a quotient. This information is not available in the p–adic case, and these elementary techniques are not enough to determine the correspondence in this case (cf. [34]).

Example 6. $(GL(m,\mathbb{R}), GL(n,\mathbb{R}))$ Based on 5 the general case of $GL(m,\mathbb{R})$ is straightforward [33]. We normalize the correspondence and eliminate the covering groups. Suppose $m \le n$. Then every representation π of $GL(m,\mathbb{R})$ occurs in the correspondence. Let $MN \simeq GL(m) \times GL(n-m) \times N$ be the usual maximal parabolic subgroup of $GL(n)$. Then $\theta(\pi)$ is the is the unique irreducible constituent of

$$Ind_{MN}^{GL(n)}(\pi \otimes 1\!\!1 \otimes 1\!\!1)$$

containing a certain K–type of multiplicity one.

A similar result holds for $(GL(m,\mathbb{C}), GL(n,\mathbb{C}))$ [3], and also for the dual pairs $(GL(m,\mathbb{H}), GL(n,\mathbb{H}))$ [32].

Example 7. $(O(m,\mathbb{C}), Sp(2n,\mathbb{C}))$

We parameterize irreducible representations of complex groups by pairs (λ, ν) as in [35]. We fix an orthogonal group $G_1 = O(2m+\tau,\mathbb{C})$ $(\tau = 0,1)$ and consider the family of dual pairs $(G_1, G_2(n))$ with $G_2(n) = Sp(2n,\mathbb{C})$. Given an irreducible representation π_1 of G_1, there exists a non–negative integer $n(\pi_1)$ such that π occurs in the duality correspondence for $(G_1, G_2(n))$ if and only if $n \ge n(\pi_1)$. In the p–adic case this is covered in the lectures of Brooks Roberts [43], see also [33], [3].

For unexplained notation see [3].

Theorem 9.1 *Let $\pi_1 = L(\mu_1, \nu_1)$ be an irreducible representation of G_1.*

Define the integer $k = k[\mu_1]$ *by writing* $\mu_1 = (a_1, \ldots, a_k, 0, \ldots, 0; \epsilon)$ *with* $a_1 \geq a_2 \geq \cdots \geq a_k > 0$. *Write* $\nu_1 = (b_1, \ldots, b_m)$, *and define the integer* $0 \leq q = q[\mu_1, \nu_1] \leq m - k$ *to be the largest integer such that* $2q - 2 + \tau, 2q - 4 + \tau, \ldots, \tau$ *all occur (in any order) in* $\{\pm b_{k+1}, \pm b_{k+2}, \ldots, \pm b_m\}$. *After possibly conjugating by the stabilizer of* μ_1 *in* W, *we may write*

$$\mu_1 = (\overbrace{a_1, \ldots, a_k}^{k}, \overbrace{0, 0, \ldots, 0}^{m-q-k}, \overbrace{0, 0, \ldots, 0}^{q}; \epsilon)$$

$$\nu_1 = (\overbrace{b_1, \ldots, b_k}^{k}, \overbrace{b_{k+1}, \ldots, b_{m-q}}^{m-q-k}, \overbrace{2q - 2 + \tau, 2q - 4 + \tau, \ldots, \tau}^{q}).$$

Let $\mu_1' = (a_1, \ldots, a_k)$, $\nu_1' = (b_1, \ldots, b_k)$, *and* $\nu_1'' = (b_{k+1}, \ldots, b_{m-q})$.
    Then $n(\pi_1) = m - \epsilon q + \frac{1-\epsilon}{2}\tau$, *and for* $n \geq n(\pi_1)$, $\theta(\pi_1) = L(\mu_2, \nu_2)$, *where*

$$\mu_2 = (\mu_1', \overbrace{1, \ldots, 1}^{\frac{1-\epsilon}{2}(2q+\tau)}, 0, 0, \ldots, 0)$$

$$\nu_2 = (\nu_1', \overbrace{2q - 1 + \tau, 2q - 3 + \tau, \ldots, 2\epsilon q + 1 + \epsilon\tau}^{\frac{1-\epsilon}{2}(2q+\tau)}, \nu_1'',$$
$$2n - 2m - \tau, 2n - 2m - 2 - \tau, \ldots, -\epsilon(2q + \tau) + 2).$$

Example 8. $(O(p, q), Sp(2n, \mathbb{R}))$ with $p + q = 2n, 2n + 1, 2n + 2$.

    In these examples the groups are the same "size", and are of particular interest from the point of view of L–functions (cf. the lectures of Steve Kudla). This is the opposite extreme of the stable range [14]. The case $(O(2, 2), Sp(4, \mathbb{R}))$ is in [38] ($O(4, 0)$ and $O(0, 4)$ are in [25], see Section 6.). The cases $(O(p, q), Sp(2n, \mathbb{R}))$ with $p + q = 2n, 2n + 2$ and p, q even are in [33], p, q odd are only missing because of covering group technicalities. Finally $(O(p, q), Sp(2n\mathbb{R}))$ witih $p + q = 2n + 1$ is in [4], this is similar to [33] except that the covering groups are unavoidable.

    We first consider the case p, q even. In this case the covering of $Sp(2n, \mathbb{R})$ splits and the correspondence can be written in terms of the linear groups. Roughly speaking the correspondence in these cases is "functorial", and a number of nice properties hold which fail in general. In particular the minimal K–type in the sense of Vogan is always of minimal degree in this situation.

    The duality correspondence is described explicitly in terms of Langlands parameters, and these match up in a natural way. This can be expressed in terms of a homomorphism between the L–groups. The disconnectedness of $O(p,q)$ can be avoided in this range; at most one of π and $\pi \otimes sgn$ occur in the correspondence.

If $p + q = 2n + 2$ every representation of $Sp(2n, \mathbb{R})$ occurs in the correspondence with some $O(p,q)$ (perhaps more than one). If $p + q = 2n$ every representation of $Sp(2n, \mathbb{R})$ occurs with at most one $O(p,q)$ (but some may fail to occur.) This suggests that $p + q = 2n + 1$ should be particularly nice, and this is the case. Fix $\delta = \pm 1$ and consider the dual pairs $(O(p,q), Sp(2n, \mathbb{R}))$ with $(-1)^q = \delta$ and $p + q = 2n + 1$. The covering of $Sp(2n, \mathbb{R})$ is the metaplectic group, and the representations which occur are all genuine. We twist by a genuine character of the cover of $O(p,q)$ to pass to representations of the linear group.

Fix ψ. Then every genuine representation of $\widetilde{Sp}(2n, \mathbb{R})$ occurs with precisely one $O(p,q)$, and of every pair of representations π and $\pi \otimes sgn$ of $O(p,q)$ (these are not isomorphic) precisely one occurs. By restricting to $SO(p,q)$, this establishes a bijection between the set

$$\widetilde{Sp}(2n, \mathbb{R})\widehat{_{genuine}}$$

of genuine irreducible admissible representations of $\widetilde{Sp}(2n, \mathbb{R})$ and the union

$$\bigcup_{p+q=2n+1(-1)^q=\delta} SO(p,q)\widehat{}$$

of the irreducible admissible representations of the groups $SO(p,q)$.

The notion of functoriality is not well–defined for the non–linear group $\widetilde{Sp}(2n, \mathbb{R})$. Nevertheless this correspondence is "functorial" in some sense. It is naturally described in terms of Langlands parameters, and (Vogan) lowest K–types are always of lowest degree. The orbit correspondence [14] is a bijection between the regular semisimple coadjoint orbits for $Sp(2n, \mathbb{R})$ and the union of those for $O(p,q)$. At least philosophically this underlies the correspondence. Again similar but not quite so clean results hold in the case $p + q = 2n, 2n + 2$.

Example 9. Results of [33]

The dual pairs $(O(p,q), Sp(2n, \mathbb{R}))$ with p, q even are discussed extensively by Moeglin in [33]; the results on $p + q = 2n, 2n + 2$ of the preceding example are a very special case.

Roughly speaking, Moeglin first considers discrete series representations which correspond to discrete series in a dual pair with (G_1, G_2) the same size. Theorem 8.2 then produces certain constituents of some induced modules which correspond. The idea is to use the K–type information of Section 6 to determine these constituents, including their Langlands parameters. This program works provided the (Vogan) lowest K–types and K–types of lowest degree coincide for the representations in question.

The result is an explicit description of the correspondence in cases in which it is functorial. The precise conditions under which this holds are in terms of the lowest K–types; they are quite technical in general, and Example 8 is the cleanest special case. The worst case from this point of view is $(O(p,q), Sp(2n, \mathbb{R}))$ with n very small compared to p, q and $p - q$ large. In this regard we mention that the case $(O(p,q), SL(2, \mathbb{R}))$ is thoroughly described in [16], which interprets results of [50] and [39] in these terms, and also discussed in [23].

Example 10. Representations with cohomology.

Unitary representations with $(\mathfrak{g}, K)$–cohomology have been classified [54]. These have regular integral infinitesimal character, i.e. the same infinitesimal character as that of a finite–dimensional representation.

Let (G_1, G_2) be a dual pair, and let π be an irreducible representation of G_1 with non–zero $(\mathfrak{g}, K)$–cohomology (possibly with coefficients). Also assume the infinitesimal character of $\pi' = \theta(\pi)$ is regular and integral. The explicit description of the correspondence in this case is due to Jian-Shu Li [31]. It turns out that π' is a discrete series representation. The special case when (G_1, G_2) are in the stable range [14] is in [2], and in somewhat greater generality in [33].

It is known that the representations with cohomology exhaust the unitary representations with regular integral infinitesimal character [45]. Therefore Li's result implies that the theta–correspondence preserves unitarity in the case of regular integral infinitesimal character.

References

[1] J. Adams. Duality for the spin representation. www.math.umd.edu/~jda/preprints/spinduality.dvi.

[2] J. Adams. *L*-functoriality for dual pairs. *Astérisque*, (171-172):85–129, 1989. Orbites unipotentes et représentations, II.

[3] Jeffrey Adams and Dan Barbasch. Reductive dual pair correspondence for complex groups. *J. Funct. Anal.*, 132(1):1–42, 1995.

[4] Jeffrey Adams and Dan Barbasch. Genuine representations of the metaplectic group. *Compositio Math.*, 113(1):23–66, 1998.

[5] John C. Baez. The octonions. *Bull. Amer. Math. Soc. (N.S.)*, 39(2):145–205 (electronic), 2002.

[6] V. Bargmann. On a Hilbert space of analytic functions and an associated integral transform. *Comm. Pure Appl. Math.*, 14:187–214, 1961.

[7] P. Cartier. *Quantum mechanical commutation relations and θ-functions*, volume 9 of *Proc. Symp. Pure Math.* American Math. Soc., Providence, Rhode Island, 1966.

[8] Mark G. Davidson, Thomas J. Enright, and Ronald J. Stanke. Differential operators and highest weight representations. *Mem. Amer. Math. Soc.*, 94(455):iv+102, 1991.

[9] Thomas Enright, Roger Howe, and Nolan Wallach. A classification of unitary highest weight modules. In *Representation theory of reductive groups (Park City, Utah, 1982)*, volume 40 of *Progr. Math.*, pages 97–143. Birkhäuser Boston, Boston, MA, 1983.

[10] Stephen Gelbart. Holomorphic discrete series for the real symplectic group. *Invent. Math.*, 19:49–58, 1973.

[11] Stephen Gelbart. Examples of dual reductive pairs. In *Automorphic forms, representations and L-functions (Proc. Sympos. Pure Math., Oregon State Univ., Corvallis, Ore., 1977), Part 1*, Proc. Sympos. Pure Math., XXXIII, pages 287–296. Amer. Math. Soc., Providence, R.I., 1979.

[12] Kenneth I. Gross and Ray A. Kunze. Bessel functions and representation theory. II. Holomorphic discrete series and metaplectic representations. *J. Functional Analysis*, 25(1):1–49, 1977.

[13] Sigurdur Helgason. *Differential geometry, Lie groups, and symmetric spaces*, volume 34 of *Graduate Studies in Mathematics*. American Mathematical Society, Providence, RI, 2001. Corrected reprint of the 1978 original.

[14] R. Howe. L^2-duality for stable dual pairs. preprint.

[15] R. Howe. θ-series and invariant theory. In *Automorphic forms, representations and L-functions (Proc. Sympos. Pure Math., Oregon State Univ., Corvallis, Ore., 1977), Part 1*, Proc. Sympos. Pure Math., XXXIII, pages 275–285. Amer. Math. Soc., Providence, R.I., 1979.

[16] Roger Howe. On some results of Strichartz and Rallis and Schiffman. *J. Funct. Anal.*, 32(3):297–303, 1979.

[17] Roger Howe. Quantum mechanics and partial differential equations. *J. Funct. Anal.*, 38(2):188–254, 1980.

[18] Roger Howe. Dual pairs in physics: harmonic oscillators, photons, electrons, and singletons. In *Applications of group theory in physics and mathematical physics (Chicago, 1982)*, volume 21 of *Lectures in Appl. Math.*, pages 179–207. Amer. Math. Soc., Providence, RI, 1985.

[19] Roger Howe. The oscillator semigroup. In *The mathematical heritage of Hermann Weyl (Durham, NC, 1987)*, volume 48 of *Proc. Sympos. Pure Math.*, pages 61–132. Amer. Math. Soc., Providence, RI, 1988.

[20] Roger Howe. Remarks on classical invariant theory. *Trans. Amer. Math. Soc.*, 313(2):539–570, 1989.

[21] Roger Howe. Transcending classical invariant theory. *J. Amer. Math. Soc.*, 2(3):535–552, 1989.

[22] Roger Howe. Perspectives on invariant theory: Schur duality, multiplicity-free actions and beyond. In *The Schur lectures (1992) (Tel Aviv)*, volume 8 of *Israel Math. Conf. Proc.*, pages 1–182. Bar-Ilan Univ., Ramat Gan, 1995.

[23] Roger Howe and Eng-Chye Tan. *Nonabelian harmonic analysis.* Universitext. Springer-Verlag, New York, 1992. Applications of SL(2, **R**).

[24] James E. Humphreys. *Introduction to Lie algebras and representation theory*, volume 9 of *Graduate Texts in Mathematics*. Springer-Verlag, New York, 1978. Second printing, revised.

[25] M. Kashiwara and M. Vergne. On the Segal-Shale-Weil representations and harmonic polynomials. *Invent. Math.*, 44(1):1–47, 1978.

[26] Anthony W. Knapp. *Lie groups, Lie algebras, and cohomology*, volume 34 of *Mathematical Notes*. Princeton University Press, Princeton, NJ, 1988.

[27] Anthony W. Knapp. *Representation theory of semisimple groups.* Princeton Landmarks in Mathematics. Princeton University Press, Princeton, NJ, 2001. An overview based on examples, Reprint of the 1986 original.

[28] Stephen S. Kudla. Seesaw dual reductive pairs. In *Automorphic forms of several variables (Katata, 1983)*, volume 46 of *Progr. Math.*, pages 244–268. Birkhäuser Boston, Boston, MA, 1984.

[29] Stephen S. Kudla. On the local theta-correspondence. *Invent. Math.*, 83(2):229–255, 1986.

[30] Jian-Shu Li. Singular unitary representations of classical groups. *Invent. Math.*, 97(2):237–255, 1989.

[31] Jian-Shu Li. Theta lifting for unitary representations with nonzero cohomology. *Duke Math. J.*, 61(3):913–937, 1990.

[32] Jian-Shu Li, Annegret Paul, Eng-Chye Tan, and Chen-Bo Zhu. The explicit duality correspondence of $(Sp(p,q), O^*(2n))$. *J. Funct. Anal.*, 200(1):71–100, 2003.

[33] C. Mœglin. Correspondance de Howe pour les paires reductives duales: quelques calculs dans le cas archimédien. *J. Funct. Anal.*, 85(1):1–85, 1989.

[34] Colette Mœglin, Marie-France Vignéras, and Jean-Loup Waldspurger. *Correspondances de Howe sur un corps p-adique*, volume 1291 of *Lecture Notes in Mathematics*. Springer-Verlag, Berlin, 1987.

[35] K. R. Parthasarathy, R. Ranga Rao, and V. S. Varadarajan. Representations of complex semi-simple Lie groups and Lie algebras. *Ann. of Math. (2)*, 85:383–429, 1967.

[36] Dipendra Prasad. Weil representation, Howe duality, and the theta correspondence. In *Theta functions: from the classical to the modern*, volume 1 of *CRM Proc. Lecture Notes*, pages 105–127. Amer. Math. Soc., Providence, RI, 1993.

[37] Tomasz Przebinda. On Howe's duality theorem. *J. Funct. Anal.*, 81(1):160–183, 1988.

[38] Tomasz Przebinda. The oscillator duality correspondence for the pair O(2, 2), Sp(2, **R**). *Mem. Amer. Math. Soc.*, 79(403):x+105, 1989.

[39] S. Rallis and G. Schiffmann. Représentations supercuspidales du groupe métaplectique. *J. Math. Kyoto Univ.*, 17(3):567–603, 1977.

[40] Stephen Rallis. Langlands' functoriality and the Weil representation. *Amer. J. Math.*, 104(3):469–515, 1982.

[41] R. Ranga Rao. On some explicit formulas in the theory of Weil representation. *Pacific J. Math.*, 157(2):335–371, 1993.

[42] B. Roberts. The theta correspondence and witt towers. www.math.umd.edu/~jda/preprints/workshop_roberts2.dvi.

[43] B. Roberts. The weil representation and dual pairs. www.math.umd.edu/~jda/preprints/workshop_roberts1.dvi.

[44] M. Saito. Représentations unitaires des groupes symplectiques. *Math. Soc. Japan*, 24:232–251, 1972.

[45] Susana A. Salamanca-Riba. On the unitary dual of real reductive Lie groups and the $A_g(\lambda)$ modules: the strongly regular case. *Duke Math. J.*, 96(3):521–546, 1999.

[46] J.-P. Serre. *A course in arithmetic*. Springer-Verlag, New York, 1973. Translated from the French, Graduate Texts in Mathematics, No. 7.

[47] David Shale. Linear symmetries of free boson fields. *Trans. Amer. Math. Soc.*, 103:149–167, 1962.

[48] M. J. Slupinski. Dual pairs in Pin(p, q) and Howe correspondences for the spin representation. *J. Algebra*, 202(2):512–540, 1998.

[49] Shlomo Sternberg. Some recent results on the metaplectic representation. In *Group theoretical methods in physics (Sixth Internat. Colloq., Tübingen, 1977)*, volume 79 of *Lecture Notes in Phys.*, pages 117–143. Springer, Berlin, 1978.

[50] Robert S. Strichartz. Harmonic analysis on hyperboloids. *J. Functional Analysis*, 12:341–383, 1973.

[51] Hans Tilgner. Graded generalizations of Weyland- and Clifford algebras. *J. Pure Appl. Algebra*, 10(2):163–168, 1977/78.

[52] David A. Vogan, Jr. *Representations of real reductive Lie groups*, volume 15 of *Progress in Mathematics*. Birkhäuser Boston, Mass., 1981.

[53] David A. Vogan, Jr. Singular unitary representations. In *Noncommutative harmonic analysis and Lie groups (Marseille, 1980)*, volume 880 of *Lecture Notes in Math.*, pages 506–535. Springer, Berlin, 1981.

[54] David A. Vogan, Jr. and Gregg J. Zuckerman. Unitary representations with nonzero cohomology. *Compositio Math.*, 53(1):51–90, 1984.

[55] Hermann Weyl. *The classical groups*. Princeton Landmarks in Mathematics. Princeton University Press, Princeton, NJ, 1997. Their invariants and representations, Fifteenth printing, Princeton Paperbacks.

THE HEISENBERG GROUP, $SL(3, \mathbb{R})$, AND RIGIDITY

ANDREAS ČAP,* MICHAEL G. COWLING,† FILIPPO DE MARI,‡
MICHAEL EASTWOOD§ and RUPERT MCCALLUM¶

*Fakultät für Mathematik der Universität Wien
Nordbergstrasse 15, 1090 Wien, Austria
E-mail: andi@esi.ac.at

†School of Mathematics and Statistics
University of New South Wales
Sydney NSW 2052, Australia
E-mail: m.cowling@unsw.edu.au

‡DIPTEM, Universitá di Genova
Piazzale Kennedy—Padiglione D, 16129 Genova, Italy
E-mail: demari@dima.unige.it

§School of Mathematics, University of Adelaide
Adelaide SA 5005, Australia
E-mail: meastwoo@maths.adelaide.edu.au

¶School of Mathematics and Statistics
University of New South Wales
Sydney NSW 2052, Australia
E-mail: rupertmccallum@yahoo.com

Ubi materia, ibi geometria
Johannes Kepler

The second-named author of this paper discussed some of the geometric questions that underlie this paper (Darboux's theorem) with Roger Howe in the 1990s. After that, it took on a life of its own.

Čap was partially supported by projects P 15747-N05 and P19500-N13 of the Fonds zur Förderung der wissenschaftlichen Forschung. Cowling was partially supported by the Australian Research Council. De Mari was partially supported by the Progetto MIUR Cofinanziato 2005 "Analisi Armonica". Eastwood was partially supported by the Australian Research Council. McCallum was partially supported by an Australian Postgraduate Research Award and the University of New South Wales.

1. Introduction

A stratified group N is a connected, simply connected nilpotent Lie group whose Lie algebra $\mathfrak{n}$ is stratified, that is

$$n = \mathfrak{n}^1 \oplus \mathfrak{n}^2 \oplus \cdots \oplus \mathfrak{n}^s,$$

where $[\mathfrak{n}^1, \mathfrak{n}^j] = \mathfrak{n}^{j+1}$ when $j = 1, 2, \cdots, s$ (we suppose that $\mathfrak{n}^s \neq \{0\}$ and $\mathfrak{n}^{s+1} = \{0\}$), and carries an inner product for which the $\mathfrak{n}^j$ are orthogonal. The left-invariant vector fields on N that correspond to elements of $\mathfrak{n}^1$ at the identity span a subspace HN_p of the tangent space TN_p of N at each point p in N, and the corresponding distribution HN is called the horizontal subbundle of the tangent bundle TN. The commutators of length at most s of sections of HN span the tangent space at each point.

Suppose that U and V are connected open subsets of a stratified group N, and define

$$\text{Contact}(U, V) = \{f \in \text{Diffeo}(U, V) :$$
$$df(HU_p) = HV_{f(p)} \text{ for all } p \text{ in } U\}$$

(where $\text{Diffeo}(U, V)$ denotes the set of diffeomorphisms from U to a subset of V). The group is said to be rigid if $\text{Contact}(U, V)$ is finite-dimensional. If N has additional structure, one might look at spaces of maps preserving this additional structure and ask whether the space of structure preserving maps is finite-dimensional. Much fundamental work on this question has been done by N. Tanaka and his collaborators T. Morimoto and K. Yamaguchi (see, for instance, [5, 7, 10]).

Rigidity is a blessing and a curse. It makes life simpler, but it makes it harder to choose coordinates. For smooth maps, much is known about rigidity. If N is the nilradical of a parabolic subgroup of a semisimple Lie group G, then, as shown by Yamaguchi, N is rigid (and $\text{Contact}(U, V)$ is essentially a subset of G) except in a limited number of cases. If N is the free nilpotent group $N_{g,s}$ of step s on g generators, then the situation is similar: $N_{g,1}$ and $N_{2,2}$ are non-rigid, $N_{2,3}$ and $N_{g,2}$ (where $g \geq 3$) are rigid and $\text{Contact}(U, V)$ is a subset of a semisimple group (see, e.g., [10]; in this case, some contact maps are fractional linear transformations) and otherwise $N_{g,s}$ is rigid and $\text{Contact}(U, V)$ is a subset of the affine group of $N_{g,s}$, that is, the semidirect product $N_{g,s} \rtimes \text{Aut}(N_{g,s})$, where the normal factor $N_{g,s}$ gives us translations and the other factor automorphisms (see, e.g., [9]).

In a fundamental paper, P. Pansu [6] studied quasiconformal and weakly contact maps on Carnot groups (stratified groups with a natural distance function). Weakly contact maps are defined to be those that map rectifiable curves into rectifiable curves (rectifiable curves are defined using the natural distance function, but the rectifiability of a curve is independent of the distance chosen), so weakly contact maps are defined on stratified groups. Pansu showed that the weakly contact mappings of generic step 2 stratified groups are affine, and hence smooth. More recently there has been much work in quasiconformal and weakly contact maps on stratified groups, and it is a question of interest whether weakly contact maps are automatically smooth (if only because if this is true, then there is not much point in developing the theory for non-smooth maps). One result in this direction, due to L. Capogna and Cowling [1], is that 1-quasiconformal maps of Carnot groups are smooth. It then follows from work of N. Tanaka [7] that they form a finite-dimensional group, and so we have rigidity.

2. An Example

The following example is prototypical of the rigid situation, and we present a simplified version of the known proof of rigidity for smooth maps for this case. Let N be the Heisenberg group of upper diagonal 3×3 unipotent matrices

$$N = \left\{ \begin{pmatrix} 1 & x & z \\ 0 & 1 & y \\ 0 & 0 & 1 \end{pmatrix} : x, y, z \in \mathbb{R} \right\}.$$

We define the left-invariant vector fields X, Y, and Z by

$$X = \partial/\partial x, \qquad Y = \partial/\partial y + x\partial/\partial z, \qquad Z = \partial/\partial z.$$

Then $[X, Y] = Z$.

Consider $\{f \in \mathrm{Diffeo}(U, V) : df(X) = pX, \ df(Y) = qY\}$, where p and q are arbitrary functions. We might describe these as multicontact maps, as the differential preserves several subbundles of TN, not just one. A vector field M on U is called multicontact if M generates a flow of multicontact maps. This boils down to the conditions $[M, X] = p'X$ and $[M, Y] = q'Y$, where p' and q' are functions. If we write M as $aX + bY + cZ$, where a, b and c are functions, then we find the equations

$$- (Xa)X - (Xb)Y - (Xc)Z - bZ = p'X$$
$$- (Ya)X - (Yb)Y - (Yc)Z + aZ = q'Y,$$

whence

$$Xb = 0 \qquad\qquad Xc = -b$$
$$Ya = 0 \qquad\qquad Yc = a.$$

We see immediately that c determines a and b, and that

$$X^2 c = Y^2 c = 0.$$

These equations imply that c is a polynomial in X, Y and Z, and in fact that the Lie algebra of multicontact vector fields is isomorphic to $\mathfrak{sl}(3, \mathbb{R})$. It then follows that the corresponding group of smooth multicontact mappings is a finite extension of $SL(3, \mathbb{R})$. However, even if care is taken, this argument can only work for mappings that are at least twice differentiable.

Now we consider the non-smooth case. The integrated version of the multicontact equations is quite simple. We denote by $\mathcal{L}$ and $\mathcal{M}$ the sets of integral curves for X and Y, i.e.,

$$\mathcal{L} = \{l_{y,z} : y, z \in \mathbb{R}^2\}$$
$$\mathcal{M} = \{m_{x,z} : x, z \in \mathbb{R}^2\}.$$

where $l_{y,z}$ and $m_{x,z}$ are the ranges of the maps

$$s \mapsto \begin{pmatrix} 1 & s & z \\ 0 & 1 & y \\ 0 & 0 & 1 \end{pmatrix} \qquad \text{and} \qquad t \mapsto \begin{pmatrix} 1 & x & z + xt \\ 0 & 1 & t \\ 0 & 0 & 1 \end{pmatrix},$$

and for an open subset U of N, we define $\mathcal{L}_U$ to be the set of all connected components of the sets $l \cap U$ as l varies over $\mathcal{L}$, and define $\mathcal{M}_U$ similarly. We may now ask what can be said about maps that send line segments l in $\mathcal{L}_U$ into lines in $\mathcal{L}$ and line segments m in $\mathcal{M}_U$ into lines in $\mathcal{M}$.

We say that points $P_1, \ldots, P_4$ in N are in general position if neither of the sets $\cup_{i=1}^4 \{(y, z) \in \mathbb{R}^2 : P_i \in l_{y,z}\}$ and $\cup_{i=1}^4 \{(x, z) \in \mathbb{R}^2 : P_i \in m_{x,z}\}$ is contained in a line.

Theorem 2.1. *Let U be a connected open subset of N, and let φ be a map from U into N whose image contains four points in general position, and which maps connected line segments in $\mathcal{L}_U$ into lines in $\mathcal{L}$ and connected line segments in $\mathcal{M}_U$ into lines in $\mathcal{M}$. Then*

(a) *if $U = N$, then φ is an affine transformation, i.e., the composition of a translation with a dilation $(x, y, z) \mapsto (sx, ty, stz)$;*

(b) *otherwise, φ is a projective transformation.*

The proof will show what we mean by a projective transformation; essentially, N may be identified with a subset of a flag manifold, and projective transformations are those which arise from the action of $SL(3, \mathbb{R})$ on this manifold.

3. Related Questions in Two Dimensions

Before we prove Theorem 2.1, we first discuss related questions in $\mathbb{R}^2$.

Theorem 3.1. *Let U be a connected open subset of $\mathbb{R}^2$, and let φ be a map from U into $\mathbb{R}^2$ whose image contains three non-collinear points, and which maps connected line segments contained in U into lines. Then*

(a) *if $U = \mathbb{R}^2$ then φ is an affine transformation, i.e., the composition of a translation with a linear map;*

(b) *otherwise, φ is a projective transformation.*

Theorem 3.1 is a corollary of Theorem 3.2, which we state and prove shortly. But there are some interesting historical remarks to be made before we do this.

We note that (a) is a theorem of Darboux; (b) requires more work. In the special case where U and $\varphi(U)$ are both the unit disc, then (b) amounts to the identification of the geodesic-preserving maps of the hyperbolic plane in the Klein model. The paper [3] by J. Jeffers identifies the geodesic-preserving maps of the hyperbolic plane in the Poincaré model (and observes that this identification was already known). This leads to the consideration of maps of regions in the plane that preserve circular arcs, and in turn leads us to mention a related result of Carathéodory [2], which establishes that maps of regions in the plane that send circular arcs into circular arcs arise from the action of the conformal group.

It is standard that collinearity preserving mappings of the projective plane are projective, i.e., arise from the action of $PGL(3, \mathbb{R})$ on the projective space $\mathbb{P}^2(\mathbb{R})$. This may be proved by composing the collinearity preserving map with a projective map to ensure that the line at infinity in the projective plane is preserved, then applying Darboux's theorem. Conversely, if one assumes that collinearity preserving mappings of the projective plane are projective, then Darboux's theorem follows once one observes that projective maps of the projective plane that preserve the finite plane are actually affine. Thus Darboux's theorem is essentially equivalent to "the fundamental theorem of projective geometry".

We take a domain U (i.e., a connected open set) in $\mathbb{R}^2$, and we suppose that $\varphi : U \to \mathbb{R}^2$ preserves collinearity for connected line segments in an open set of directions. In the next theorem, we show that we can still control φ. Recall that three points in $\mathbb{R}^2$ are said to be in general position if they are not collinear. We call a connected line segment in U a chord.

Theorem 3.2. *Suppose that U is a domain in $\mathbb{R}^2$, and that the image of $\varphi : U \rightarrow \mathbb{R}^2$ contains three points in general position. Suppose that for each P in U, there exists a neighbourhood N_P of P and an nonempty open set of slopes S_P such that if x, y, and z in N_P lie in a line whose slope is in S_P, then $\varphi(x), \varphi(y)$ and $\varphi(z)$ are collinear. Then φ is a projective transformation.*

Proof. It suffices to show that for each P, there is a neighbourhood N_P' of P such that $\varphi\big|_{N_P'}$ is projective. For then an analytic continuation argument shows that φ is projective on U. Choose Cartesian coordinates, and take a small ball $B(P,r)$ with centre P contained in N_P.

By composing with a rotation about P, we may suppose that φ sends chords of $B(P,r)$ with slopes in $[0,\epsilon]$ into lines in $\mathbb{R}^2$, for some small ϵ. By composing with a shear transformation centred at P, we may suppose that φ sends chords of $B(P,r)$ whose slopes are not in $(-\epsilon, 0)$ into lines in $\mathbb{R}^2$, and by composing with a rotation, we may suppose that φ sends chords of $B(P,r)$ with slopes close to 0, ∞, $\pm\sqrt{3}$ or $\pm 1/\sqrt{3}$, or in the intervals $(0, 1/\sqrt{3})$ and $(\sqrt{3}, \infty)$, into lines in $\mathbb{R}^2$. By composing in the image space with a projective map, we may suppose that φ fixes the vertices and centroid P of an equilateral triangle inside $B(P,r)$ with horizontal base. This triangle will be our set N_P', as we will show that this modified φ is the identity thereon, so the original φ must have been projective.

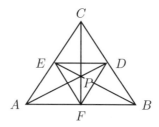

Figure 1. Subdivision.

Suppose that the triangle is as in Figure 1. Since φ sends chords with slopes 0, ∞, $\pm\sqrt{3}$ and $\pm 1/\sqrt{3}$ into lines, φ preserves the midpoints of the sides. For example, the chord CP is mapped into the line CP and the chord

AB is mapped into the line AB, and $F = CP \cap AB$, so $\varphi(F) = F$. Similarly, φ also preserves $AD \cap EF$ and $BE \cap DF$, that is, the midpoints of EF and DF, so the horizontal chord through these points is sent into the line through these points. It follows that the midpoints of AE and DB are also preserved. In this way, we deduce that not only does φ preserve the vertices of the four smaller equilateral triangles AFE, FBD, DEF and EDC, but also φ preserves the midpoints of their sides, and their centroids. We can therefore continue to subdivide and find a dense subset of the triangle of points that are fixed by φ.

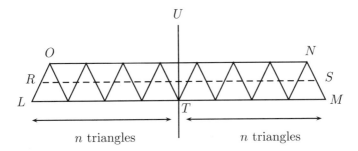

Figure 2. Horizontal chords.

We now show that all horizontal lines that meet the interior of the triangle are sent into horizontal lines, or more precisely, that a horizontal chord RS (where R and S lie on the edges of the triangle) is sent into a horizontal line $R'S'$ (where R' and S' lie on the edges of the triangle, possibly produced). If the chord is one of those that arises in the subdivision, we are done, by construction. Otherwise, for some arbitrarily large positive integer n, we can find a subdivision as shown in Figure 2, so that the chord RS passes through $2n$ congruent equilateral triangles with horizontal bases, and $2n - 1$ inverted congruent equilateral triangles with horizontal bases. If RS is as shown, then the absolute values of the slopes of LS and RM are less than $1/(2n - 1)$, so if n is large enough, the images of the chords LS and RM lie in lines. Further, no matter what slopes are involved, if R' and S' lie on AC and CB, then $R'S'$ is parallel to LM if and only if $LS' \cap R'M$ lies on the vertical line TU through the midpoint of LM. Since TU is also preserved, it follows that the image of a horizontal chord is horizontal.

We can repeat this argument to deduce that the images of chords with slopes $\pm\sqrt{3}$ that meet the interior of the triangle also have slopes of $\pm\sqrt{3}$.

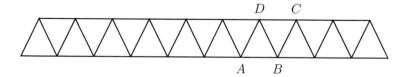

Figure 3. A rhombus.

We now show that φ is continuous. To do this, it suffices to show that each small rhombus like $ABCD$ in Figure 3 is preserved; since these can be made arbitrarily small, φ is continuous, and hence the identity. After a linear transformation centred at A, doing this is equivalent to proving the following lemma, and its proof will conclude the proof of Theorem 3.2. □

Lemma 3.3. *Let Σ denote the square $ABCD$, where the coordinates of A, B, C and D are $(0,0)$, $(1,0)$, $(1,1)$ and $(0,1)$, and suppose that $\varphi : \Sigma \to \mathbb{R}^2$ preserves the vertices of Σ, sends chords of Σ with slope 0, ± 1 and ∞ into lines of the same slope, and sends chords of Σ with slope in $(0, \infty)$ into lines. Then the image of φ is contained in the set Σ.*

Proof. Suppose that $0 < y < 1$ and take s in $(0,1)$ so that $y = s^2$. Then y is the ordinate of the point Q constructed as the intersection of AS and UV in the top left-hand diagram below, in which the ordinate of S is s, the line TS is horizontal, the line TU has slope -1, and the line UV is vertical.

The line AS has gradient in $(0,1)$, so maps into a line. Now $\varphi(A) = A$ and $\varphi(S)$ lies on BC (possibly produced), and so has coordinates $(1,t)$ for some real t. Next, the line $\varphi(T)\varphi(S)$ is horizontal and $\varphi(T)$ lies on AD (possibly produced), so $\varphi(T)$ has coordinates $(0,t)$, and the line $\varphi(T)\varphi(U)$ has gradient -1 and $\varphi(U)$ lies on AB (possibly produced), and hence $\varphi(U)$ has coordinates $(t,0)$. Now $\varphi(U)\varphi(V)$ is vertical, and so $\varphi(Q)$, which is the intersection of $\varphi(U)\varphi(V)$ and $\varphi(A)\varphi(S)$, has ordinate t^2, and lies above AB. Since φ sends horizontal chords into horizontal lines, the horizontal chord through Q maps into the horizontal line through the image of $\varphi(Q)$, and the square maps into the half-plane above the line AB.

Analogous constructions, presented in the three other diagrams in Figure 4, show that the square maps into the half planes below DC, left of AD and right of BC, together forcing the image to be in the intersection of these four half-planes, that is, Σ. □

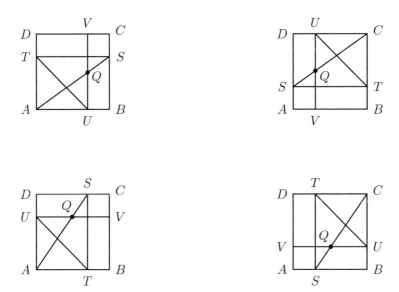

Figure 4. Constructions.

4. Proof of Theorem 2.1

Proof of (a). There is a plane at infinity, written as $\{(\infty, y, z) : y, z \in \mathbb{R}^2\}$, that compactifies the $\mathcal{L}$ lines, and a projection $\pi : (x, y, z) \mapsto (\infty, y, z)$ of N onto this plane. Any map φ of N to N that maps $\mathcal{L}$ lines into $\mathcal{L}$ lines induces a map $\dot{\varphi}$ of this plane into itself. Any non-vertical line in the plane at infinity is of the form $\pi(m)$ for some $\mathcal{M}$ line m, and $\varphi(m)$ is also an $\mathcal{M}$ line. It follows that $\dot{\varphi}$ sends non-vertical lines into non-vertical lines, and hence $\dot{\varphi}$ is a globally defined projective map, that is, an affine map, by Theorem 3.2. One can compactify the $\mathcal{M}$ lines similarly and obtain another affine map $\ddot{\varphi}$ of another plane at infinity. Unravelling things, φ must be affine.   □

A similar proof works for the case where $U \subset N$.

Theorem 2.1 (where $U = N$) has a projective analogue, which apparently goes back to E. Bertini. The projective completion of N is the flag manifold $\mathbb{F}^3(\mathbb{R})$ whose elements are full flags of subspaces of $\mathbb{R}^3$, i.e., ordered pairs

(V_1, V_2) of subspaces of $\mathbb{R}^3$ such that

$$\{0\} \subset V_1 \subset V_2 \subset \mathbb{R}^3$$

(all inclusions are proper, so $\dim(V_1) = 1$ and $\dim(V_2) = 2$). Suppose that $\varphi : \mathbb{F}^3(\mathbb{R}) \to \mathbb{F}^3(\mathbb{R})$ is a bijection. We write $\varphi(V_1, V_2)$ as $(\varphi_1(V_1, V_2), \varphi_2(V_1, V_2))$. If $\varphi_1(V_1, V_2)$ depends only on V_1, then φ induces a map $\dot\varphi$ of $\mathbb{P}^2(\mathbb{R}^3)$; similarly if $\varphi_2(V_1, V_2)$ depends only on V_2, then φ induces a map $\ddot\varphi$ of the set of planes in $\mathbb{R}^3$. If both these hypotheses hold, then the map $\dot\varphi$ of $\mathbb{P}^2(\mathbb{R})$ preserves collinearity: indeed, three collinear points in $\mathbb{P}^2(\mathbb{R})$ correspond to three one-dimensional subspaces V_1, V_1' and V_1'', of $\mathbb{P}^2(\mathbb{R})$, all contained in a common plane V_2, and by hypotheses $\varphi_1(V_1, V_2)$, $\varphi_1(V_1', V_2)$ and $\varphi_1(V_1'', V_2)$ are all subspaces of $\varphi_2(V_1, V_2)$. Conversely, any collinearity-preserving map of $\mathbb{P}^2(\mathbb{R})$ induces a map of $\mathbb{F}^3(\mathbb{R})$. From the affine theorem above, we obtain a corresponding theorem for $\mathbb{F}^3(\mathbb{R})$. Before we state this, let us agree that four points in $\mathbb{F}^3(\mathbb{R})$ are in general position if their one dimensional subspaces are in general position (no three lie in a plane) and their two dimensional subspaces are too (no three meet in a line).

Theorem 4.1. *Let U be a connected open subset of $\mathbb{F}^3(\mathbb{R})$, and suppose that φ is a map from U into $\mathbb{F}^3(\mathbb{R})$ whose image contains four points in general position. Suppose also that $\varphi_1(V_1, V_2)$ depends only on V_1 and $\varphi_2(V_1, V_2)$ depends only on V_2. Then φ is a projective transformation, i.e., φ arises from the action of (a finite extension of) $\mathrm{PGL}(3, \mathbb{R})$ on $\mathbb{F}^3(\mathbb{R})$.*

In the case where U is all $\mathbb{F}^3(\mathbb{R})$, then this is a special case of theorem of Tits [8] (also called the fundamental theorem of projective geometry). The point here is that $\mathbb{F}^3(\mathbb{R})$ may be identified with the group G/P, where G is $\mathrm{SL}(3, \mathbb{R})$ and P is the subgroup of G of lower triangular matrices. Indeed, G acts on $\mathbb{R}^3$ and P is the stabiliser of the flag $(\mathbb{R}\mathbf{e}_3, \mathbb{R}\mathbf{e}_3 + \mathbb{R}\mathbf{e}_2)$, where $\{\mathbf{e}_1, \mathbf{e}_2, \mathbf{e}_3\}$ is the standard basis of $\mathbb{R}^3$, and the map $g \mapsto (\mathbb{R}g\mathbf{e}_3, \mathbb{R}g\mathbf{e}_3 + g\mathbf{e}_2)$ identifies G/P with $\mathbb{F}^3(\mathbb{R})$. Further, we see that the stabilisers of the sets $\{(V_1, V_2) \in \mathbb{F}^3(\mathbb{R}) : V_1 = V\}$ and $\{(V_1, V_2) \in \mathbb{F}^3(\mathbb{R}) : V_2 = V\}$ are the parabolic subgroups P_1 and P_2:

$$P_1 = \left\{ \begin{pmatrix} * & * & 0 \\ * & * & 0 \\ * & * & * \end{pmatrix} \right\} \quad \text{and} \quad P_2 = \left\{ \begin{pmatrix} * & 0 & 0 \\ * & * & * \\ * & * & * \end{pmatrix} \right\}.$$

Maps φ of $\mathbb{F}^3(\mathbb{R})$ into itself with the property that $\varphi_1(V_2, V_2)$ depends only on V_1 correspond to maps of G/P that pass to maps of G/P_1, i.e., maps that preserve the fibration of G/P induced by P_1 (the fibres are the sets xP_1/P), and maps φ such that $\varphi_2(V_1, V_2)$ depends only on V_2 correspond to maps of G/P that preserve the fibration of G/P induced by P_2.

5. Final Remarks

Our original problem (or one of them) was to decide what kind of rigidity results might hold for non-smooth maps of stratified groups (or open subsets thereof). The second-named author makes the following conjecture, which holds in many examples. These include all the nilpotent groups that arise as the Iwasawa N group of a classical group of real rank 2, where the subalgebras $\mathfrak{n}_1$ and $\mathfrak{n}_2$ correspond to the simple restricted root spaces; see [4].

Conjecture 5.1. *Suppose that* $\mathfrak{n}_1$ *and* $\mathfrak{n}_2$ *are subalgebras of a stratified algebra* $\mathfrak{n}$, *such that* $\mathfrak{n}^1 = \mathfrak{n}_1^1 \oplus \mathfrak{n}_2^1$, *and for all* X *in* $\mathfrak{n}_1^1$, *there exists* Y *in* $\mathfrak{n}_2^1$ *such that* $[X, Y] \neq 0$, *and vice versa. Let* U *be a connected open subset of the corresponding group* N, *and let* $\varphi : U \to N$ *be a map that maps connected subsets of cosets of* N_i *into cosets of* N_i *when* $i = 1$ *and* 2. *Then* φ *is smooth, and the set of these maps is finite-dimensional.*

Along the lines of the previous section, Tits extended the fundamental theorem of projective geometry to semisimple Lie groups (of rank at least two). For any field (or division ring) k, the maps φ that preserve lines in k^2 are compositions of field (or ring) automorphisms with affine maps, and Tits' theorems are proved in this generality. At the affine level, these extensions include theorems such as a characterisation of maps of the plane preserving circles, a local version of which was proved by Carathéodory [2].

McCallum [4] has proved local versions of "the fundamental theorem of projective geometry" for all the classical groups, and extended these to non-discrete topological fields and division rings, and some other rings, such as the adèles.

There are a number of interesting related results about maps that preserve other geometric structures. As already mentioned, C. Carathéodory [2] studied maps of domains in the plane that send arcs of circles into arcs of circles and showed that these come from restricting the action of the conformal group, and *a fortiori* are all smooth. On the other hand, J. Jeffers [3] observes that there are bijections of the sphere that send great circles into great circles that are not even continuous.

References

[1] L. Capogna and M. G. Cowling, Conformality and Q-harmonicity in Carnot groups, *Duke Math. J.* **135** (2006), 455–479.

[2] C. Carathéodory, The most general transformations of plane regions which transform circles into circles, *Bull. Amer. Math. Soc.* **43** (1937), 573–579.

[3] J. Jeffers, Lost theorems of geometry, *Amer. Math. Manthly* **107** (2000), 800–812.

[4] R. McCallum, Generalizations of the fundamental theorem of projective geometry, Thesis, U.N.S.W., submitted (March 2007).

[5] T. Morimoto, "Lie algebras, geometric structures and differential equations on filtered manifolds", pages 205–252 in: *Lie groups, geometric structures and differential equations—one hundred years after Sophus Lie (Kyoto/Nara, 1999).* Adv. Stud. Pure Math., 37, Math. Soc. Japan, Tokyo, 2002.

[6] P. Pansu, Métriques de Carnot–Carathéodory et quasiisométries des espaces symétriques de rang un, *Ann. Math.* **129** (1989), 1–60.

[7] N. Tanaka, On differential systems, graded Lie algebras and pseudogroups, *J. Math. Kyoto Univ.* **10** (1970), 1–82.

[8] J. Tits, *Buildings of spherical type and finite BN pairs.* Lecture Notes in Math., Vol. 386. Springer-Verlag, Berlin, Heidelberg, New York, 1974.

[9] B. Warhurst, Tanaka prolongation of free Lie algebras, submitted (March 2007).

[10] K. Yamaguchi, "Differential systems associated with simple graded Lie algebras", pages 413–494 in: *Progress in differential geometry.* Adv. Stud. Pure Math., 22, Math. Soc. Japan, Tokyo, 1993.

PFAFFIANS AND STRATEGIES FOR INTEGER CHOICE GAMES

RON EVANS* and NOLAN WALLACH†

*Department of Mathematics, 0112
University of California at San Diego
La Jolla, CA 92093-0112, USA
E-mail: revans@ucsd.edu

†Department of Mathematics, 0112
University of California at San Diego
La Jolla, CA 92093-0112, USA
E-mail: nwallach@ucsd.edu

To Roger Howe on his sixtieth birthday.

The purpose of this paper is to develop optimal strategies for a simple integer choice game with a skew symmetric payoff matrix. The analysis involves the calculation of certain Pfaffians associated with these matrices.

Keywords and phrases. Pfaffian, skew-symmetric payoff matrix, matrix game, optimal strategy, three-term recurrence

1. Introduction

Alice and Bob play a game where each secretly chooses a positive integer. If both players choose the same integer then the game is a tie. Otherwise, the player that chooses the smallest integer (say Bob) wins \$1 (from Alice), unless the two integers differ by 1, in which case Alice wins w dollars from Bob. Here w is an arbitrary positive real number. This game was partially analyzed by Mendelsohn [4] about 60 years ago. It was also discussed in the book of Herstein and Kaplansky [3] and was further popularized in a book of Martin Gardner [2, Chapter 9.3].

It will be convenient to use "negative payoffs". For example, if Bob's integer is 1 less than Alice's, then Bob receives the negative payoff $v = -w$.

2000 *Mathematics Subject Classification.* Primary 15A15, 15A18, 15A57, 15A75, 91A05; Secondary 11B37, 11B83, 26C10.
Research of second author partially supported by NSF summer grant DMS-0200305.

That means that Bob pays Alice w dollars. This game yields a skew-symmetric payoff matrix $A(v)$ for "the row player" Bob, with v on the first super-diagonal and all 1's above. We denote by $A_m(v)$ the $m \times m$ submatrix of $A(v)$ consisting of the first m rows and columns; for example,

$$
A_5(v) = \begin{bmatrix}
0 & v & 1 & 1 & 1 \\
-v & 0 & v & 1 & 1 \\
-1 & -v & 0 & v & 1 \\
-1 & -1 & -v & 0 & v \\
-1 & -1 & -1 & -v & 0
\end{bmatrix}.
$$

If both players choose integers from the set $\{1, \ldots, m\}$ then $A_m(v)$ is the payoff matrix for the row player Bob. For example, if Bob plays 3 and Alice plays 1, 2, 3, 4 or 5, then Bob wins -1, $-v$, 0, v or 1, respectively.

A *strategy* for a player of this game is a list of plays each with a corresponding probability. For example, Bob could have the strategy of playing 1, 2, and 4 with probabilities $2/3$, $1/4$, and $1/12$, respectively (playing every other integer with probability 0). A *pure strategy* is a strategy where one probability is 1 and all of the rest are 0. For example, Bob's strategy is pure if he plays 4 with probability 1. An *optimal strategy* is a strategy that beats or ties any other strategy. It is easily seen that a strategy is optimal if and only if it beats or ties every pure strategy.

Mendelsohn [4] found numbers $v_4 < v_3 < v_2 < v_1 = 0$ (given below) such that: if $v_2 < v < v_1$ then the unique optimal strategy entails playing only the integers $1, 2, 3$; if $v_3 < v < v_2$ then the unique optimal strategy plays only the integers $1, 2, 3, 4, 5$; and if $v_4 < v < v_3$ then the unique optimal strategy plays only the integers $1, 2, 3, 4, 5, 6, 7$. For example, $v_2 = (-1 - \sqrt{5})/2$ and when $v_2 < v < 0$, the unique optimal strategy is to randomly choose the integers $1, 2, 3$ with respective probabilities $v/(2v - 1)$, $1/(1 - 2v)$, $v/(2v - 1)$. The aspect of this game that we find to be most striking (and our main reason for studying it) is the radical change in strategy that can be caused by a small change in the payoff v. In Theorem 1.1 below, we extend Mendelsohn's results by explicitly determining the unique optimal strategies for all $v < 0$ for which they exist. An equivalent (but less elegant) version of Theorem 1.1 was stated without proof in [1, Appendix].

For $m \geq 1$, set

$$
v_m = 1 - \frac{1}{2\left(1 - \cos \frac{\pi}{2m+1}\right)}.
$$

This is consistent with the notation v_1, v_2, v_3, v_4 in the last paragraph. We have

$$0 = v_1 > v_2 > v_3 > v_4 > \dots$$

with

$$v_m = -\left(\frac{2m+1}{\pi}\right)^2 + O(1),$$

so v_m tends to $-\infty$ quadratically.

We define a sequence of polynomials $F_m = F_m(x)$ recursively as follows: $F_{-1} = 0$, $F_0 = 1$ and

$$F_{m+1} = F_m + (x-1)F_{m-1}, \quad m \geq 0.$$

The polynomials indexed by 1, 2, 3, 4, 5, 6 are respectively

$$1, \ x, \ 2x - 1, \ x^2 + x - 1, \ 3x^2 - 2x, \ x^3 + 3x^2 - 4x + 1.$$

For $1 \leq j \leq m$ we define rational functions in v by the formula

$$p_j(m) = (-1)^{j+1} \frac{F_{j-1}(v)F_{m-j}(v)}{F_m(v)}.$$

The advertised result, which is the main content of Theorem 3.4, is

Theorem 1.1. *Suppose that there is an odd number k such that*

$$v_{k+1} < v < v_k.$$

Then $p_j(2k+1) > 0$ for $j = 1, \dots, 2k+1$, $\sum_{j=1}^{2k+1} p_j(2k+1) = 1$, and the unique optimal strategy is to choose $1, 2, \dots, 2k+1$ with respective probabilities

$$p_1(2k+1), \ p_2(2k+1), \dots, p_{2k+1}(2k+1).$$

This integer choice game can be generalized so that for each $i \geq 1$, whenever Bob chooses i and Alice chooses $i+1$, Bob's payoff is an amount $x_i < 0$ in lieu of the constant amount v. We will call this more general version the "multivariate game". The multivariate analogue of the payoff matrix $A_n(v)$ is the skew-symmetric $n \times n$ matrix

$$A_n = A_n(x_1, \dots, x_{n-1}) = \begin{bmatrix} 0 & x_1 & 1 & 1 & \dots & 1 & 1 \\ -x_1 & 0 & x_2 & 1 & \dots & 1 & 1 \\ -1 & -x_2 & 0 & x_3 & \dots & 1 & 1 \\ -1 & -1 & -x_3 & 0 & \dots & 1 & 1 \\ \vdots & \vdots & \vdots & \vdots & \ddots & \vdots & \vdots \\ -1 & -1 & -1 & -1 & \dots & 0 & x_{n-1} \\ -1 & -1 & -1 & -1 & \dots & -x_{n-1} & 0 \end{bmatrix}.$$

For indeterminates $x_0, x_1, \ldots$, we can define multivariate analogues $F_n = F_n(x_1, \ldots, x_{n-1})$ of the polynomials $F_n(x)$ recursively as follows: $F_{-1} = 0$, $F_0 = 1$ and

$$F_{m+1} = F_m + (x_m - 1)F_{m-1}, \quad m \geq 0.$$

These polynomials are intimately related to the payoff matrices A_n. For example, it will be seen in Section 5 that if n is even, $F_n(x_1, \ldots, x_{n-1})$ is the Pfaffian of $A_n(x_1, \ldots, x_{n-1})$.

Let $\mathbf{x}$ denote the infinite vector $(x_1, x_2, \ldots)$. In contrast with the single variable case, we do not know an explicit characterization of the set of all vectors $\mathbf{x}$ such that the multivariate game has a unique optimal strategy. However, in Theorem 4.2 we present the unique optimal strategy for the multivariate game in the special case that the first $2m + 1$ entries of $\mathbf{x}$ are contrained to the interior of an explicitly given unit hypercube.

We collect together results on the polynomials $F_n = F_n(x_1, \ldots, x_{n-1})$ and the payoff matrices $A_n = A_n(x_1, \ldots, x_{n-1})$ in the Appendix (Section 5). The main result in the Appendix, Theorem 5.8, shows that if no x_i equals 1, then $\ker A_{2k+1}(x_1, \ldots, x_{2k})$ is a one-dimensional space $\mathbb{R}\mathbf{t}$ where $\mathbf{t}$ is explicitly expressed in terms of the polynomials F_n as well as in terms of Pfaffians of the diagonal minors of A_{2k+1}. Theorem 5.8 is instrumental in the proof of Proposition 2.2. Propositions 2.1 and 2.2 in Section 2 give a general analysis of optimal strategies for the multivariate game. Proposition 2.2 is applied to prove the main theorems in Sections 3 and 4 (Theorems 3.4 and 4.2).

As indicated above, this paper is dedicated to Roger Howe. We hope that he enjoys it as much as we enjoyed writing it.

2. Strategies for the Multivariate Game

In this section, we provide methods for constructing optimal strategies for the multivariate game described in Section 1.

Proposition 2.1. *Suppose that there exists a vector* $\mathbf{p} = (p_1, \ldots, p_{2k+1}) \in \mathbb{R}^{2k+1}$ *such that*

$$\ker A_{2k+1}(x_1, \ldots, x_{2k}) = \mathbb{R}\mathbf{p}, \tag{2.1}$$

$$\sum_{i=1}^{2k+1} p_i = 1, \quad \text{with all } p_i > 0, \tag{2.2}$$

and

$$\sum_{i=1}^{2k} p_i + x_{2k+1}p_{2k+1} > 0. \tag{2.3}$$

Then the unique optimal strategy is to play i with probability p_i, for $1 \leq i \leq 2k + 1$. This strategy is still optimal (but not necessarily unique) if the inequalities in (2.2) *and* (2.3) *are not required to be strict.*

Proof. Let Bob play j with probability p_j for $1 \leq j \leq 2k + 1$. Suppose that Alice plays the pure strategy i. If $i \leq 2k + 1$, then Bob's payoff is

$$p_1 + \cdots + x_{i-1}p_{i-1} - x_i p_{i+1} - p_{i+2} + \cdots$$

which vanishes by (2.1). If $i > 2k + 2$, then Bob's payoff is 1. If $i = 2k + 2$, then Bob's payoff is

$$p_1 + \cdots + p_{2k} + x_{2k+1}p_{2k+1} > 0$$

by (2.3). Thus Bob beats or ties every pure strategy, so his strategy is optimal. This argument shows that Bob's strategy is still optimal if the inequalities in (2.2) and (2.3) are not required to be strict.

We now prove uniqueness. Suppose that against Bob's optimal strategy, Alice plays an optimal strategy in which she chooses i with probability r_i for $i \geq 1$. We have seen that Bob beats every pure strategy exceeding $2k + 1$, so $r_i = 0$ for every $i > 2k + 1$. For brevity, let A denote the payoff matrix in (2.1), and let $\mathbf{r}$ denote the column vector $(r_1, \ldots, r_{2k+1})$. Since Alice's strategy is optimal, all $2k + 1$ entries in the vector $A\mathbf{r}$ are ≤ 0. If at least one of these entries were strictly negative, then by (2.2), we would have $\mathbf{p}A\mathbf{r} < 0$. This is impossible, since $\mathbf{p}A = 0$ by (2.1). Thus $A\mathbf{r} = 0$. Hence by (2.1), $\mathbf{r}$ is a scalar multiple of $\mathbf{p}$. Since the sum of the entries of $\mathbf{r}$ and the sum of the entries of $\mathbf{p}$ both equal 1, we have $\mathbf{r} = \mathbf{p}$, which completes the proof of uniqueness. □

We will now apply results in the Appendix to refine Proposition 2.1.

Proposition 2.2. *Assume that $x_i < 0$ for $i \geq 1$ and that*

$$F_{2k+1}(x_1, \ldots, x_{2k}) \neq 0, \tag{2.4}$$

$$p_i := \frac{(-1)^{i+1} F_{i-1}(x_1, \ldots, x_{i-2}) F_{2k+1-i}(x_{i+1}, \ldots, x_{2k})}{F_{2k+1}(x_1, \ldots, x_{2k})} > 0, \quad 1 \leq i \leq 2k + 1, \tag{2.5}$$

and

$$\frac{F_{2k+2}(x_1, \ldots, x_{2k+1})}{F_{2k+1}(x_1, \ldots, x_{2k})} > 0. \tag{2.6}$$

Then $p_1 + \cdots + p_{2k+1} = 1$ and the unique optimal strategy is to play $i = 1, \ldots, 2k + 1$ with probabilities $p_1, \ldots, p_{2k+1}$, respectively. This strategy is still optimal if the inequalities in (2.5) *and* (2.6) *are not required to be strict.*

Proof. In the notation of (2.5), write $\mathbf{p} = (p_1, \ldots, p_{2k+1})$. We need only check that the three conditions of Proposition 2.1 hold. By (2.4) and (2.5) and Lemma 5.6, we have $p_i > 0$ and $p_1 + \cdots + p_{2k+1} = 1$. By Theorem 5.8, $\ker A_{2k+1}(x_1, \ldots, x_{2k}) = \mathbb{R}\mathbf{p}$. Thus (2.1) and (2.2) are proved, and it remains to check (2.3). The left side of (2.3) equals

$$\frac{\left(\displaystyle\sum_{i=1}^{2k}(-1)^{i+1}F_{i-1}(x_1,\ldots,x_{i-2})F_{2k+1-i}(x_{i+1},\ldots,x_{2k})\right)}{F_{2k+1}(x_1,\ldots,x_{2k})} + \frac{x_{2k+1}F_{2k}(x_1,\ldots,x_{2k-1})}{F_{2k+1}(x_1,\ldots,x_{2k})} =$$

$$\frac{\displaystyle\sum_{i=1}^{2k+1}(-1)^{i+1}F_{i-1}(x_1,\ldots,x_{i-2})F_{2k+1-i}(x_{i+1},\ldots,x_{2k}) + (x_{2k+1}-1)F_{2k}(x_1,\ldots,x_{2k-1})}{F_{2k+1}(x_1,\ldots,x_{2k})}$$

and Lemma 5.6 implies that this expression is equal to

$$\frac{F_{2k+1}(x_1,\ldots,x_{2k}) + (x_{2k+1}-1)F_{2k}(x_1,\ldots,x_{2k-1})}{F_{2k+1}(x_1,\ldots,x_{2k})}.$$

By Lemma 5.4, this in turn equals the positive expression in (2.6). This completes the proof of (2.3).   □

We remark that Proposition 2.2 can also be proved by ad hoc methods which are more elementary (but less elegant).

Consider the strategy of choosing i with probability p_i for $i \geq 1$. We say this strategy is *finite* if $p_j = 0$ for all sufficiently large j. The next result provides an example of a game with an infinite but no finite optimal strategy.

Proposition 2.3. *If $x_i = -(2^{i+1} - 3)$ for $i = 1, 2, \ldots$ then an optimal strategy is to play i with probability 2^{-i} for each $i \geq 1$. This game has no finite optimal strategy.*

Proof. Let $\mathbf{r} = (r_1, r_2, \ldots)$ with $r_i = 2^{-i}$. The strategy of playing i with probability r_i for $i \geq 1$ ties every pure strategy n, because

$$-\sum_{i\leq n-2}2^{-i} - x_{n-1}2^{1-n} + x_n2^{-n-1} + \sum_{i>n+1}2^{-i} = 0.$$

This infinite strategy is thus optimal.

Now consider another optimal strategy in which i is played with probability p_i for $i \geq 1$, where $1 = p_1 + p_2 + \cdots$. Let $\mathbf{p} = (p_1, p_2, \ldots)$, viewed as an infinite column vector. For the infinite payoff matrix A, we have $0 = \mathbf{r}A = \mathbf{r}A\mathbf{p}$. Since all entries of $A\mathbf{p}$ are ≤ 0, this implies that $A\mathbf{p} = 0$. Suppose for the purpose of contradiction that $p_i = 0$ for all $i > N$, where without loss of generality, $N = 2m$ is even. Then the submatrix

$A_{2m}(x_1, \ldots, x_{2m-1})$ of A has a nontrivial kernel, so its determinant and thus its Pfaffian vanishes. As was noted above Lemma 5.4, the Pfaffian of $A_{2m}(x_1, \ldots, x_{2m-1})$ is $F_{2m}(x_1, \ldots, x_{2m-1})$. One can show using the recurrence that

$$F_{2m}(x_1, \ldots, x_{2m-1}) = (-1)^m(2-1)(2^3-1)\cdots(2^{2m-1}-1).$$

Since this is nonzero, we have the desired contradiction to the assumption that a finite optimal strategy exists. $\square$

We remark that the optimal strategy given in Proposition 2.3 is not unique. In fact, for any a with $0 \le a \le 1/2$, it is optimal to play i with probability p_i for $i \ge 1$, where the sequence p_i is defined by the recurrence $p_1 = a$, $p_2 = (1-a)/2$, $p_3 = (1+a)/12$, and for $n \ge 4$,

$$(2^n - 2)p_n = (2^{n-1} - 3)p_{n-1} + (2^{n-1} - 3)p_{n-2} + (2 - 2^{n-2})p_{n-3}.$$

For each fixed a, we have $1 = p_1 + p_2 + \cdots$ and $p_i > 0$, except that $p_1 = 0$ in the case that $a = 0$. The case $a = 1/2$ gives the optimal strategy presented in Proposition 2.3.

3. Strategies for the Single Variable Game

In this section we will assume that all $x_i = v < 0$. We will write $A_n(v)$ for $A_n(v, v, \ldots, v)$ and $F_n(v)$ for $F_n(v, v, \ldots, v)$. Our goal is to prove Theorem 3.4.

We have the recurrence relation $F_{-1} = 0$, $F_0 = 1$, $F_1 = 1$ and for $n \ge 0$,

$$F_{n+2}(v) = F_{n+1}(v) + (v-1)F_n(v).$$

Since

$$\begin{bmatrix} 1 & v-1 \\ 1 & 0 \end{bmatrix} \begin{bmatrix} F_{n+1}(v) \\ F_n(v) \end{bmatrix} = \begin{bmatrix} F_{n+2}(v) \\ F_{n+1}(v) \end{bmatrix},$$

the standard argument implies that for $\lambda_\pm(v) = \frac{1 \pm \sqrt{4v-3}}{2}$, we have

$$F_n(v) = \frac{\lambda_+(v)^{n+1} - \lambda_-(v)^{n+1}}{\lambda_+(v) - \lambda_-(v)}.$$

One can also check directly that the right side satisfies the recurrence, using $\lambda_+\lambda_- = 1 - v$ and $\lambda_+ + \lambda_- = 1$.

Lemma 3.1. *For each $n \ge 2$, the solutions to $F_n(v) = 0$ are*

$$\xi_{n,k} = 1 - \frac{1}{2 + 2\cos(\frac{2\pi k}{n+1})}, \quad k = 1, \ldots, \left[\frac{n}{2}\right].$$

Proof. If $F_n(v) = 0$ then

$$\lambda_+(v)^{n+1} = \lambda_-(v)^{n+1}$$

so $\lambda_+(v) = \zeta \lambda_-(v)$ with $\zeta^{n+1} = 1$ and $\zeta \neq \pm 1$. Thus $\zeta \lambda_-(v) + \lambda_-(v) = 1$, so $\lambda_-(v) = \frac{1}{1+\zeta}$ and

$$1 - v = \lambda_+ \lambda_- = \frac{\zeta}{(1+\zeta)^2}.$$

Hence

$$v = \frac{1 + \zeta + \zeta^2}{1 + 2\zeta + \zeta^2} = \frac{1 + \zeta + \zeta^{-1}}{2 + \zeta + \zeta^{-1}}.$$

Now substituting $\zeta = \left(e^{\frac{2\pi i}{n+1}}\right)^k$ the lemma follows. $\square$

Note that $\xi_{2n,n}$ is the leftmost zero of F_{2n} and $\xi_{2n+1,n}$ is the leftmost zero of F_{2n+1}. The following properties of $\xi_{n,k}$ are easily checked.

Lemma 3.2. *We have*

$\xi_{n,k} < \xi_{n,l}$ *if* $k > l$,

$\xi_{2n,n} < \xi_{2m,m}$ *and* $\xi_{2n+1,n} < \xi_{2m+1,m}$ *if* $n > m$, *and*

$\xi_{2n,n} < \xi_{2n+1,n} < \xi_{2n+2,n}$.

Set $v_n = \xi_{2n,n}$ for $n \geq 1$. This definition of v_n agrees with that given in Section 1. Recall that $0 = v_1 > v_2 > \dots$.

Lemma 3.3. *If* $v_{k+1} < v < v_k$ *then*

$$(-1)^{\left[\frac{n}{2}\right]} F_n(v) > 0, \quad 0 \leq n \leq 2k+1, \tag{3.1}$$

$$(-1)^k F_{2k+2}(v) > 0. \tag{3.2}$$

Moreover, (3.1) holds for all $v < v_k$.

Proof. Let $v < v_k$. Then Lemma 3.2 implies that v is to the left of all the zeros of F_n for $n \leq 2k+1$. Since the recurrence implies that the polynomial F_n has degree $\left[\frac{n}{2}\right]$ with positive leading coefficient, (3.1) follows. When also $v > v_{k+1}$, (3.2) holds because by Lemma 3.2, v is to the right of exactly one zero of F_{2k+2}. $\square$

We are now ready to prove the main result of this section.

Theorem 3.4. *Let* $v_{k+1} \leq v < v_k$. *For* $1 \leq i \leq 2k+1$, *define*

$$p_i := \frac{(-1)^{i+1} F_{i-1}(v) F_{2k+1-i}(v)}{F_{2k+1}(v)}.$$

If $v_{k+1} < v < v_k$, *then all* $p_i > 0$ *and the unique optimal strategy is to play* i *with probability* p_i *for* $1 \leq i \leq 2k+1$. *If* $v = v_{k+1}$, *then this strategy is* *still optimal, but it is not unique, since it is also optimal to play $i + 1$ with probability p_i for $1 \leq i \leq 2k + 1$.*

Proof. By Lemma 3.3, the p_i are all well-defined positive numbers. If $v_{k+1} < v < v_k$, then appealing again to Lemma 3.3, we see that the three conditions of Proposition 2.2 are satisfied. Proposition 2.2 thus shows that playing i with probability p_i for $1 \leq i \leq 2k + 1$ is the unique optimal strategy. Now suppose that $v = v_{k+1}$. Then this strategy is still optimal, but it is not unique, since by the argument above with $k + 1$ in place of k, it is also optimal to play j with probability

$$q_j := \frac{(-1)^{j+1} F_{j-1}(v) F_{2k+3-j}(v)}{F_{2k+3}(v)}$$

for $1 \leq j \leq 2k + 3$. Observe that $q_1(v) = q_{2k+3}(v) = 0$, since $F_{2k+2}(v) = 0$. It remains to show that $p_i = q_{i+1}$. This can be proved by induction on i, using the recurrence for F_n. $\qquad\square$

4. Strategies for Some Constricted Multivariate Games

For $k \geq 1$, let V_k be the set of infinite vectors $(x_1, x_2, \dots)$ with $x_i < 0$ for all $i \geq 1$ that satisfy the three conditions of Proposition 2.2. When $(x_1, x_2, \dots) \in V_k$, Proposition 2.2 describes the unique optimal strategy for the corresponding game. The uniqueness assertion implies that $V_i \cap V_j = \emptyset$ for $i \neq j$.

Note that by Lemma 3.3 and Theorem 3.4, $(v, v, v, \dots) \in V_k$ if and only if $v_{k+1} < v < v_k$. We use this fact to give a class of multivariate games with unique optimal strategy, in Theorem 4.2.

Proposition 4.1. *Suppose that $v_{k+1} < v < v_k$. Then there exists $\varepsilon > 0$ such that if*

$$|x_i - v| < \varepsilon \text{ for } i = 1, \dots, 2k + 2$$

then $(x_1, x_2, \dots) \in V_k$.

Proof. We have $(v, v, v, \dots) \in V_k$ and V_k is open in $\mathbb{R}^\infty_{<0}$. $\qquad\square$

We next determine V_1. The conditions defining this set are $x_i < 0$ for all $i \geq 1$ and

$$\frac{x_2}{x_1 + x_2 - 1} > 0, \quad \frac{-1}{x_1 + x_2 - 1} > 0, \quad \frac{x_1}{x_1 + x_2 - 1} > 0, \quad \frac{x_1 x_3 + x_2 - 1}{x_1 + x_2 - 1} > 0.$$

All of the conditions but the last are automatic if the x_i are all negative. Thus

$$V_1 = \{(x_1, x_2, \dots) \in \mathbb{R}^\infty_{<0} \mid x_2 < \min\{0, 1 - x_1 x_3\}\}.$$

In particular, if C denotes the interior of a unit cube with vertices $-(a, b, c)$, $a, b, c \in \{0, 1\}$ then $C \times \mathbb{R}_{<0}^{\infty}$ is contained in V_1. The following theorem extends this, by giving for every $m = 1, 2, \ldots$, an open unit hypercube C_{2m+1} in $\mathbb{R}_{<0}^{2m+1}$ such that $C_{2m+1} \times \mathbb{R}_{<0}^{\infty}$ is contained in V_m. We will prove:

Theorem 4.2. *For $m = 1, 2, \ldots$, let*

$$u_m = \frac{2 \cos(\frac{\pi}{m+1})}{\cos(\frac{\pi}{m+1}) - 1}.$$

If $(x_1, x_2, \ldots) \in \mathbb{R}_{<0}^{\infty}$ satisfies

$$u_m > x_i > u_m - 1, \quad i = 1, \ldots, 2m + 1$$

then $(x_1, x_2, \ldots) \in V_m$. In particular, Proposition 2.2 describes the unique optimal strategy for the corresponding game.

The proof will occupy the rest of the section. We start with the following lemmas.

Lemma 4.3. *Let w be an indeterminate. Then fixing $w^{\frac{1}{2}}$, we have*

$$F_{2n}(w + 1, w, w + 1, \ldots, w, w + 1) =$$
$$\frac{1}{2}(w^{\frac{1}{2}})^{n-1}((w^{\frac{1}{2}} + 1)^{n+1} + (w^{\frac{1}{2}} - 1)^{n+1}), \quad n \geq 0, \quad (4.1)$$
$$F_{2n}(w, w + 1, w, \ldots, w + 1, w) =$$
$$wF_{2n-2}(w + 1, w, w + 1, \ldots, w, w + 1), \quad n \geq 1. \quad (4.2)$$

Proof. Let h_n denote the right side of (4.1). Direct calculation shows that

$$h_{n+2} = 2wh_{n+1} - w(w - 1)h_n.$$

The left side of (4.1) satisfies the same recurrence, by Lemma 5.4. Since both sides equal 1 for $n = 0$ and $1 + w$ for $n = 1$, we obtain (4.1). Each side of (4.2) also satisfies the recurrence above. Since both sides equal w for $n = 0$ and $w(w + 1)$ for $n = 1$, we obtain (4.2). $\qquad\square$

Lemma 4.4. *For fixed $m > 0$, set $u = u_m$ and $w = u_m - 1$, in the notation of Theorem 4.2. Then for $0 \leq n < m$*

$$(-1)^n F_{2n}(u, w, u, w, \ldots, u) > 0, \quad (4.3)$$

$$F_{2m}(u, w, u, w, \ldots, u) = 0, \quad (4.4)$$

$$F_{2m+2}(w, u, w, u, \ldots, w) = 0. \quad (4.5)$$

Proof. We note that if

$$t = \frac{\cos(\frac{\pi}{m+1}) + 1}{\sin(\frac{\pi}{m+1})}$$

then

$$t^2 = \frac{\cos(\frac{\pi}{m+1}) + 1}{1 - \cos(\frac{\pi}{m+1})} = -w.$$

By Lemma 4.3 with $w^{\frac{1}{2}} = it$,

$$F_{2n}(u, w, u, w, \ldots, u) = \frac{1}{2}(it)^{n-1}((it+1)^{n+1} + (it-1)^{n+1}).$$

After some simplification, the right side reduces to $\frac{1}{2}(-1)^n \frac{(\cos\frac{\pi}{m+1}+1)^{n-1}}{(\sin\frac{\pi}{m+1})^{2n}} H$, where

$$H = (1+\zeta^2)^{n+1} + (1+\zeta^{-2})^{n+1} = (\zeta+\zeta^{-1})^{n+1}(\zeta^{n+1}+\zeta^{-n-1}), \quad \zeta = e^{\frac{\pi i}{2m+2}}.$$

Thus, to prove (4.3), we must show that $H > 0$. If $n < m$, these factors in H involve positive cosines, while if $n = m$, the rightmost factor in H vanishes. This proves (4.3) and (4.4). Finally, (4.5) follows from (4.2). $\square$

We are now ready to prove Theorem 4.2. Assume that $u_m - 1 < x_i < u_m$ for $1 \le i \le 2m+1$. To satisfy the three conditions of Proposition 2.2, it suffices to prove that

$$(-1)^{[\frac{h}{2}]} F_h(x_1, \ldots, x_{h-1}) > 0, \quad 0 \le h \le 2m+1$$

and

$$(-1)^m F_{2m+2}(x_1, \ldots, x_{2m+1}) > 0.$$

It is convenient to work with $G_n(x_1, \ldots, x_{n-1}) := (-1)^{[\frac{n}{2}]} F_n(x_1, \ldots, x_{n-1})$. Lemma 5.7 implies that for $1 \le j < 2n$ we have

$$(-1)^j \frac{\partial}{\partial x_j} G_{2n}(x_1, \ldots, x_{2n-1}) = G_{j-1}(x_1, \ldots, x_{j-2}) G_{2n-j-1}(x_{j+2}, \ldots, x_{2n-1}).$$

We first use this formula to prove by induction that $G_h(x_1, \ldots, x_{h-1}) > 0$ for $0 \le h \le 2m+1$. Clearly this holds for $h = 0$ and $h = 1$. Assume that it holds for all $h < 2n$ for some n with $1 \le n \le m$. We will prove that it holds for $h = 2n$ and $h = 2n+1$. By the induction hypothesis and the derivative formula above, $G_{2n}(x_1, \ldots, x_{2n-1})$ is strictly decreasing in x_1, strictly increasing in x_2, strictly decreasing in x_3, etc. Hence

$$G_{2n}(x_1, \ldots, x_{2n-1}) > G_{2n}(u_m, u_m - 1, \ldots, u_m, u_m - 1, u_m) \ge 0$$

by (4.3) and (4.4). This proves the result for $h = 2n$. By the induction hypothesis,

$$G_{2n+1}(x_1, \ldots, x_{2n}) = G_{2n}(x_1, \ldots, x_{2n-1}) + (1 - x_{2n})G_{2n-1}(x_1, \ldots, x_{2n-2}) > 0,$$

so the result holds for $h = 2n + 1$ as well.

It remains to prove that $G_{2m+2}(x_1, \ldots, x_{2m+1}) < 0$. Applying Lemma 5.7 again we find that G_{2m+2} has the same monotonicity properties (decreasing in the odd variables, increasing in the even ones), hence

$$G_{2m+2}(x_1, \ldots, x_{2m+1}) < G_{2n}(u_m - 1, u_m, \ldots, u_m - 1, u_m, u_m - 1) = 0,$$

by (4.5). This completes the proof of Theorem 4.2.

5. Appendix: Pfaffians Associated with Payoff Matrices

In this section we will analyze the following skew-symmetric $m \times m$ matrices over a field F of characteristic 0:

$$A_m = A_m(x_1, \ldots, x_{m-1}) = \begin{bmatrix} 0 & x_1 & 1 & 1 & \ldots & 1 & 1 \\ -x_1 & 0 & x_2 & 1 & \ldots & 1 & 1 \\ -1 & -x_2 & 0 & x_3 & \ldots & 1 & 1 \\ -1 & -1 & -x_3 & 0 & \ldots & 1 & 1 \\ \vdots & \vdots & \vdots & \vdots & \ddots & \vdots & \vdots \\ -1 & -1 & -1 & -1 & \ldots & 0 & x_{m-1} \\ -1 & -1 & -1 & -1 & \ldots & -x_{m-1} & 0 \end{bmatrix}.$$

Here the superdiagonal has indeterminate entries $x_1, x_2, \ldots, x_{m-1}$ and all of the entries above the superdiagonal are 1's. In Theorem 5.8 below, we determine $\ker A_{2k+1}$ when no x_i equals 1, and we express the Pfaffians of the diagonal minors of A_{2k+1} in terms of the polynomials $F_n(x_1, \ldots, x_{n-1})$ defined in Section 1. We first need to recall some material about Grassmann algebras.

Let V be an m-dimensional vector space over F with choice of non-zero element Ω_m in the one-dimensional space $\bigwedge^m V$. With this choice there is natural isomorphism T of $\bigwedge^{m-1} V$ to the dual V^* given by the formula

$$x \wedge \eta = T(\eta)(x)\Omega_m$$

for $\eta \in \bigwedge^{m-1} V$ and $x \in V$. Let $e_1, \ldots, e_m$ be a basis of V so that $\Omega_m = e_1 \wedge \cdots \wedge e_m$. A basis of $\bigwedge^{m-1} V$ is given by the elements $e_1 \wedge \cdots \wedge \widehat{e_j} \cdots \wedge e_m$, where the circumflex indicates deletion. Thus if $x = \sum_{j=1}^{m} x_j e_j$ and if $\eta = \sum_{j=1}^{m} \eta_j (e_1 \wedge \cdots \wedge \widehat{e_j} \cdots \wedge e_m)$ then $T(\eta)(x) = \sum_{j=1}^{m} (-1)^{j-1} \eta_j x_j$.

If A is a skew-symmetric matrix of size $m \times m$ with entries a_{ij} then we define

$$\omega_A = \sum_{i<j} a_{ij} e_i \wedge e_j.$$

We note if g is an $m \times m$ matrix with transpose g^T then

$$\omega_{gAg^T} = \left(\bigwedge^2 g \right) \omega_A$$

where

$$\left(\bigwedge^k g \right) (v_1 \wedge \cdots \wedge v_k) = g v_1 \wedge \cdots \wedge g v_k.$$

If $m = 2n$ with n an integer, then the Pfaffian of A, $Pf(A)$, is given by the formula

$$\frac{\omega_A^n}{n!} = Pf(A)\Omega_{2n};$$

here the n-th power is in the Grassmann algebra. If A is a $2n+1 \times 2n+1$ skew-symmetric matrix then $\frac{\omega_A^n}{n!}$ is in $\bigwedge^{2n} F^{2n+1}$. Thus, as above, we have an element $T(\frac{\omega_A^n}{n!}) \in \left(F^{2n+1} \right)^*$. Using the standard form $(x, y) = \sum x_i y_i$, we can identify $\left(F^{2n+1} \right)^*$ with F^{2n+1}.

If A is a matrix then we denote by $A_{r,s}$ the matrix gotten by deleting the r-th row and the s-th column. Note that when A is skew-symmetric, so is A_{rr}. The following lemma is standard but not easily referenced.

Lemma 5.1. *Let A be a $2n+1 \times 2n+1$ skew symmetric matrix. Then, using the standard form to view $T(\frac{\omega_A^n}{n!})$ as an element in F^{2n+1}, we have*

$$A \text{ is of rank } 2n \text{ if and only if } \frac{\omega_A^n}{n!} \neq 0. \text{ Furthermore, } AT\left(\frac{\omega_A^n}{n!}\right) = 0. \quad (5.1)$$

Also, as an element of F^{2n+1},

$$T\left(\frac{\omega_A^n}{n!}\right) = \sum_{i=1}^{2n+1} (-1)^{i+1} Pf(A_{ii}) e_i. \quad (5.2)$$

Proof. We note that there exists $g \in GL(2n+1, F)$ such that

$$\omega_{gAg^T} = \bigwedge^2 g \; \omega_A = \sum_{i=1}^{l} e_{2i-1} \wedge e_{2i}$$

with $2l$ equal to the rank of A. We therefore see that $\frac{\omega_A^n}{n!} \neq 0$ if and only if $l = n$. To see that

$$AT\left(\frac{\omega_A^n}{n!}\right) = 0$$

it is enough to show that

$$(Ax) \wedge \frac{\omega_A^n}{n!} = 0$$

for all $x \in F^{2n+1}$. For g as above, we must show that

$$0 = \left(\bigwedge^{2n+1} g \right) (Ax) \wedge \frac{\omega_A^n}{n!} = (gAx) \wedge \frac{\omega_{gAg^T}^n}{n!}.$$

That is, we must show that for all x,

$$((gAg^T)(g^T)^{-1}x) \wedge \frac{\omega_{gAg^T}^n}{n!} = 0.$$

This follows because the image of gAg^T is contained in the span of $\{e_1, e_2, \ldots, e_{2l}\}$ and $\frac{\omega_{gAg^T}^n}{n!}$ is either zero or a nonzero scalar multiple of Ω_{2n}. This proves (5.1).

For each j we write

$$\omega_A = \omega_j + \sum_{i=1}^{2n+1} a_{ij} e_i \wedge e_j = \omega_j + \beta_j.$$

Then we note that ω_j is $\omega_{A_{jj}}$ in the basis $e_1, \ldots, e_{j-1}, e_{j+1}, \ldots, e_{2n+1}$, and $\beta_j \wedge \beta_j = 0$. Thus

$$\frac{\omega_A^n}{n!} = \frac{\omega_j^n}{n!} + n\frac{\omega_j^{n-1}}{n!}\beta_j.$$

Since the rightmost term is a multiple of e_j in the Grassmann algebra, we see that the coefficient of $e_1 \wedge \cdots \wedge \widehat{e_j} \cdots \wedge e_{2n+1}$ is $Pf(A_{jj})$. This proves (5.2). $\qquad \square$

Set

$$f_0 = 1, \quad f_n = f_n(x_1, \ldots, x_{2n-1}) = Pf(A_{2n}(x_1, \ldots, x_{2n-1})), \quad n \geq 1.$$

Lemma 5.2. *We have*

$$f_n = x_{2n-1}f_{n-1} + (x_{2n-2} - 1)f_{n-2} + (x_{2n-2} - 1)(x_{2n-4} - 1)f_{n-3} + \cdots +$$
$$(x_{2n-2} - 1)\cdots(x_4 - 1)f_1 + (x_{2n-2} - 1)\cdots(x_4 - 1)(x_2 - 1).$$

Proof. Before working with f_n, we investigate properties of the following expressions:

$$\mu_n = \sum_{i=1}^{2n-1} x_i e_i \wedge e_{i+1},$$

$$\gamma_n = \sum_{1 \leq i < j-1 < 2n} e_i \wedge e_j,$$

$$\nu_n = x_{2n-2}e_{2n-2} \wedge e_{2n-1} + x_{2n-1}e_{2n-1} \wedge e_{2n},$$

$$\xi_n = \sum_{i \leq 2n-3} e_i \wedge e_{2n-1} + \sum_{i \leq 2n-2} e_i \wedge e_{2n}$$

and
$$\delta_{j,2n} = \sum_{i \le j} e_i \wedge e_{j+1} \wedge e_{j+2} \wedge \cdots \wedge e_{2n}.$$

We have
$$\omega_{A_{2n}} = \mu_n + \gamma_n$$

with
$$\mu_n = \mu_{n-1} + \nu_n$$

and
$$\gamma_n = \gamma_{n-1} + \xi_n.$$

We will write the Grassmann multiplication of elements in the (commutative) even part of the Grassmann algebra without the wedge. We note that
$$\nu_n^2 = 0$$

and
$$\xi_n^2 = 2 \sum_{i \le 2n-3} \sum_{j \le 2n-2} e_i \wedge e_{2n-1} \wedge e_j \wedge e_{2n} =$$
$$-2 \sum_{i \le 2n-3} \sum_{j \le 2n-2} e_i \wedge e_j \wedge e_{2n-1} \wedge e_{2n} =$$
$$-2 \sum_{i \le 2n-3} \sum_{j \le 2n-3} e_i \wedge e_j \wedge e_{2n-1} \wedge e_{2n} \; - 2 \sum_{i \le 2n-3} e_i \wedge e_{2n-2} \wedge e_{2n-1} \wedge e_{2n} =$$
$$-2 \sum_{i \le 2n-3} e_i \wedge e_{2n-2} \wedge e_{2n-1} \wedge e_{2n}$$

since the first sum in the penultimate expression is 0. We write this as
$$\xi_n^2 = -2\delta_{2n-3,2n}.$$

Also
$$\nu_n \xi_n = x_{2n-2}\delta_{2n-3,2n}.$$

Similarly, for $1 < j < n$, one calculates
$$(\nu_j + \xi_j)^2 \delta_{2j-1,2n} = 0$$

and
$$(\nu_j + \xi_j)\delta_{2j-1,2n} = (x_{2j-2} - 1)\delta_{2j-3,2n}.$$

We are now ready to derive the formula for f_n. We have
$$f_n \Omega_{2n} = \frac{(\mu_n + \gamma_n)^n}{n!} =$$

$$\frac{(\mu_{n-1} + \gamma_{n-1} + \nu_n + \xi_n)^n}{n!} =$$

$$\frac{(\mu_{n-1} + \gamma_{n-1})^n}{n!} + n\frac{(\mu_{n-1} + \gamma_{n-1})^{n-1}}{n!}(\nu_n + \xi_n) +$$

$$\binom{n}{2}\frac{(\mu_{n-1} + \gamma_{n-1})^{n-2}}{n!}(\nu_n + \xi_n)^2 = C_1 + C_2 + C_3.$$

Since C_1 is of degree $2n$ in $e_1, \ldots, e_{2n-2}$ it is 0. We have

$$C_2 = f_{n-1}\Omega_{2n-2}(\nu_n + \xi_n) = x_{2n-1}f_{n-1}\Omega_{2n}.$$

We now look at C_3. Since

$$(\nu_n + \xi_n)^2 = 2\nu_n\xi_n + \xi_n^2 = 2(x_{2n-2} - 1)\delta_{2n-3,2n},$$

we have

$$C_3 = (x_{2n-2} - 1)\frac{(\mu_{n-1} + \gamma_{n-1})^{n-2}}{(n-2)!}\delta_{2n-3,2n}.$$

We now have our "bootstrap":

$$\frac{(\mu_{n-1} + \gamma_{n-1})^{n-2}}{(n-2)!}\delta_{2n-3,2n} = \frac{(\mu_{n-2} + \gamma_{n-2} + \nu_{n-1} + \xi_{n-1})^{n-2}}{(n-2)!}\delta_{2n-3,2n}$$

$$= \frac{(\mu_{n-2} + \gamma_{n-2})^{n-2}}{(n-2)!}\delta_{2n-3,2n} +$$

$$(n-2)\frac{(\mu_{n-2} + \gamma_{n-2})^{n-3}}{(n-2)!}(\nu_{n-1} + \xi_{n-1})\delta_{2n-3,2n} =$$

$$f_{n-2}\Omega_{2n} + \frac{(\mu_{n-2} + \gamma_{n-2})^{n-3}}{(n-3)!}(x_{2n-4} - 1)\delta_{2n-5,2n}.$$

Now repeat the argument on the second term, and continue in this manner, to obtain Lemma 5.2. □

The next result simplifies the recurrence relation for f_n.

Proposition 5.3. *We have*

$$f_n = (x_{2n-1} + x_{2n-2} - 1)f_{n-1} - (x_{2n-2} - 1)(x_{2n-3} - 1)f_{n-2}$$

with $f_0 = 1$ *and* $f_1 = x_1$.

Proof. The initial conditions are clear. We write the formula in Lemma 5.2 as

$$f_n = x_{2n-1}f_{n-1} + (x_{2n-2} - 1)(f_{n-2} + (x_{2n-4} - 1)f_{n-3} + \cdots +$$

$$(x_{2n-4} - 1)\cdots(x_4 - 1)f_1 + (x_{2n-4} - 1)\cdots(x_4 - 1)(x_2 - 1)).$$

This expression is (applying Lemma 5.2 with $n - 1$ replacing n)

$$f_n = x_{2n-1}f_{n-1} + (x_{2n-2} - 1)((1 - x_{2n-3})f_{n-2} + f_{n-1}) =$$

$$(x_{2n-1} + x_{2n-2} - 1)f_{n-1} - (x_{2n-2} - 1)(x_{2n-3} - 1)f_{n-2},$$

as asserted. $\qquad\qquad\qquad\qquad\qquad\qquad\qquad\qquad\qquad\qquad\qquad$ □

Define the polynomials

$$F_{2n} = F_{2n}(x_1, \ldots, x_{2n-1}) = f_n(x_1, \ldots, x_{2n-1})$$

and

$$F_{2n+1} = F_{2n+1}(x_1, \ldots, x_{2n}) = f_{n+1}(x_1, \ldots, x_{2n}, 1).$$

In particular, $F_{2m}(x_1, \ldots, x_{2m-1})$ equals the Pfaffian of $A_{2m}(x_1, \ldots, x_{2m-1})$. The following result shows that these F_m are the same multivariate polynomials that were defined by the recurrence in Section 1.

Lemma 5.4. *The polynomials $F_n(x_1, \ldots, x_{n-1})$ as defined above are the solution to the recurrence relation*

$$F_{-1} = 0, \ F_0 = F_1 = 1$$

and

$$F_{n+2} = F_{n+1} + (x_{n+1} - 1)F_n, \quad n \geq 0.$$

Furthermore,

$$F_{n+2} = (x_{n+1} + x_n - 1)F_n - (x_n - 1)(x_{n-1} - 1)F_{n-2}, \quad n \geq 1.$$

Proof. We begin by proving the first recurrence in the even case $n = 2k$. Writing $f_n = f_n(x_1, \ldots, x_{2n-1})$, we have

$$F_{2k+1}(x_1, \ldots, x_{2k}) = f_{k+1}(x_1, \ldots, x_{2k}, 1) =$$

$$x_{2k}f_k - (x_{2k} - 1)(x_{2k-1} - 1)f_{k-1} =$$

$$(x_{2k+1} + x_{2k} - 1)f_k - (x_{2k} - 1)(x_{2k-1} - 1)f_{k-1} - (x_{2k+1} - 1)f_k =$$

$$f_{k+1} - (x_{2k+1} - 1)f_k = F_{2k+2} - (x_{2k+1} - 1)F_{2k}.$$

Thus

$$F_{2k+2} = F_{2k+1} + (x_{2k+1} - 1)F_{2k}.$$

We next look at the odd case $n = 2k + 1$. Then

$$F_{2k+3}(x_1, \ldots, x_{2k+2}) = f_{k+2}(x_1, \ldots, x_{2k+2}, 1) =$$

$$x_{2k+2}f_{k+1} - (x_{2k+2} - 1)(x_{2k+1} - 1)f_k =$$

$$f_{k+1} + (x_{2k+2} - 1)(f_{k+1} - (x_{2k+1} - 1)f_k).$$

Now $f_{k+1} - (x_{2k+1} - 1)f_k = F_{2k+1}(x_1, \ldots, x_{2k})$ by the first part of this argument. Hence

$$F_{2k+3} = F_{2k+2} + (x_{2k+2} - 1)F_{2k+1}.$$

Since the initial values are obvious, this completes the proof of the first recurrence.

To prove the second recurrence, note that

$$F_{n+2} = F_{n+1} + (x_{n+1} - 1)F_n = F_n + (x_n - 1)F_{n-1} + (x_{n+1} - 1)F_n =$$

$$(x_{n+1} + x_n - 1)F_n + (x_n - 1)(-F_n + F_{n-1}) =$$

$$(x_{n+1} + x_n - 1)F_n - (x_n - 1)(x_{n-1} - 1)F_{n-2}.$$

This completes the proof of the second recurrence. $\qquad\square$

Lemma 5.5. *Let* $n \geq 2$. *For* $1 \leq i \leq n - 1$,

$$F_n(x_1, \ldots, x_{i-1}, 1, x_{i+1}, \ldots, x_{n-1}) = F_i(x_1, \ldots, x_{i-1})F_{n-i}(x_{i+1}, \ldots, x_{n-1}).$$

Proof. We prove this by induction on n. If $n = 2$, this says that $1 = F_1 F_1$, which is true. We have

$$F_n(x_1, \ldots, x_{n-2}, 1) = F_{n-1} + (1 - 1)F_{n-2} = F_{n-1}.$$

Since $F_1 = 1$ this proves the formula for $i = n - 1$. Now

$$F_n(x_1, \ldots, x_{n-3}, 1, x_{n-1}) = F_{n-1}(x_1, \ldots, x_{n-3}, 1) + (x_{n-1} - 1)F_{n-2} =$$

$$F_{n-2} + (x_{n-1} - 1)F_{n-2} = F_{n-2}F_2(x_{n-1}).$$

This proves the formula for $i = n - 2$. In particular, we now know the formula is valid for $n = 3$. Suppose that $i < n - 2$ with $n > 3$. Then

$$F_n(x_1, \ldots, x_{i-1}, 1, x_{i+1}, \ldots, x_{n-1}) = F_{n-1}(x_1, \ldots, x_{i-1}, 1, x_{i+1}, \ldots, x_{n-2})$$

$$+(x_{n-1} - 1)F_{n-2}(x_1, \ldots, x_{i-1}, 1, x_{i+1}, \ldots, x_{n-3}).$$

The induction hypothesis implies that this is equal to

$$F_i(x_1, \ldots, x_{i-1})(F_{n-1-i}(x_{i+1}, \ldots, x_{n-2})+$$

$$(x_{n-1} - 1)F_{n-2-i}(x_{i+1}, \ldots, x_{n-3}))$$

$$= F_i(x_1, \ldots, x_{i-1})F_{n-i}(x_{i+1}, \ldots, x_{n-1}).$$

$$\qquad\square$$

We now examine $F_{2k+1} = F_{2k+1}(x_1, \ldots, x_{2k})$ and $A_{2k+1} = A_{2k+1}(x_1, \ldots, x_{2k})$.

Lemma 5.6. *We have*

$$T\left(\frac{\omega_{A_{2k+1}}^k}{k!}\right) = \sum_{i=1}^{2k+1}(-1)^{i+1}F_{i-1}(x_1,\ldots,x_{i-2})F_{2k+1-i}(x_{i+1},\ldots,x_{2k})e_i.$$

Furthermore

$$F_{2k+1}(x_1,\ldots,x_{2k}) = \sum_{i=1}^{2k+1}(-1)^{i+1}F_{i-1}(x_1,\ldots,x_{i-2})F_{2k+1-i}(x_{i+1},\ldots,x_{2k}).$$

Proof. For $i = 1,\ldots,2k+1$, we consider $A_{2k+1}(x_1,\ldots,x_{2k})_{ii}$ (with notation as in Lemma 5.1). One checks that if $i = 1$ then

$$A_{2k+1}(x_1,\ldots,x_{2k})_{11} = A_{2k}(x_2,\ldots,x_{2k})$$

and if $i = 2k+1$ then

$$A_{2k+1}(x_1,\ldots,x_{2k})_{2k+1,2k+1} = A_{2k}(x_1,\ldots,x_{2k-1}).$$

For $1 < i < 2k+1$ we have

$$A_{2k+1}(x_1,\ldots,x_{2k})_{ii} = A_{2k}(x_1,\ldots,x_{i-2},1,x_{i+1},\ldots,x_{2k}).$$

The first part of the result now follows from Lemmas 5.1 and 5.5 in light of

$$Pf(A_{2k}(x_1,\ldots,x_{2k-1})) = F_{2k}(x_1,\ldots,x_{2k-1}).$$

We now turn to the formula for F_{2k+1}. This formula is easily checked for $k = 0,1,2$. Let $k \geq 3$. We will induct on k. By the second part of Lemma 5.4, we have

$$F_{2k+1}(x_1,\ldots,x_{2k}) = (x_{2k}+x_{2k-1}-1)F_{2k-1} - (x_{2k-1}-1)(x_{2k-2}-1)F_{2k-3}.$$

We now apply the induction hypothesis to F_{2k-1} and F_{2k-3} to see that F_{2k+1} equals

$$\sum_{i=1}^{2k+1}(-1)^{i+1}F_{i-1}\left((x_{2k}+x_{2k-1}-1)H_{2k-1-i} - (x_{2k-1}-1)(x_{2k-2}-1)H_{2k-3-i}\right)$$

where we ignore all terms in which negative subscripts occur, and where $H_{2k-1-i} = F_{2k-1-i}(x_{i+1},\ldots,x_{2k-2})$ and $H_{2k-3-i} = F_{2k-3-i}(x_{i+1},\ldots,x_{2k-4})$. Applying the second part of Lemma 5.4 with $n = 2k-1-i$ for each $i \leq 2k-2$, we readily complete the induction. $\square$

The following lemma will be used in the proof of Theorem 4.2.

Lemma 5.7. *For $1 \leq i \leq m$*

$$\frac{\partial}{\partial x_i}F_{m+1}(x_1,\ldots,x_m) = F_{i-1}(x_1,\ldots,x_{i-2})F_{m-i}(x_{i+2},\ldots,x_m).$$

Proof. This result is proved by essentially the same argument as in the proof of Lemma 5.5.   □

Theorem 5.8. *Suppose that $x_i \neq 1$ for all $i \geq 1$. Write $A = A_{2k+1}(x_1, \ldots, x_{2k})$. Then* $\ker A = \mathbb{R}\mathbf{t}$, *where* $\mathbf{t} = (t_1, \ldots, t_{2k+1})$ *with*

$$t_i = (-1)^{i+1} Pf(A_{ii}) = (-1)^{i+1} F_{i-1}(x_1, \ldots, x_{i-2}) F_{2k+1-i}(x_{i+1}, \ldots, x_{2k}),$$

for $i = 1, 2, \ldots, 2k + 1$.

Proof. By Lemmas 5.1 and 5.6, the vector $\mathbf{t}$ described above lies in $\ker A$, and so by Lemma 5.1, it remains to show that $\mathbf{t}$ is nonzero. Assume that $\mathbf{t}$ is zero. Then $F_{2k}(x_1, \ldots, x_{2k-1}) = t_{2k+1}$ and $F_{2k-1}(x_1, \ldots, x_{2k-2}) = t_{2k}$ both vanish. But by the first recurrence for the sequence F_n in Lemma 5.4, the vanishing of two consecutive terms of the sequence implies the vanishing of all the terms, since $x_i \neq 1$ for all i. This contradicts the fact that $F_0 = 1$.   □

We remark that Theorem 5.8 is false if one deletes the hypothesis that $x_i \neq 1$ for all $i \geq 1$. For example, $\ker A_5(0, 1, 0, 1)$ and $\ker A_5(0, 1, 1/2, 1/2)$ both have dimension 3.

References

[1] R. J. Evans and G. A. Heuer, Silverman's game on discrete sets, Linear Algebra and its Applications 166 (1992), 217-235.

[2] Martin Gardner, Time travel and other mathematical bewilderments, Freeman, N.Y., 1988.

[3] I. N. Herstein and I. Kaplansky, Matters Mathematical, 2nd ed., Chelsea, N.Y., 1978.

[4] N. S. Mendelsohn, A Psychological Game, American Mathematical Monthly 53(1946), 86-88.

WHEN IS AN *L*-FUNCTION NON-VANISHING IN PART OF THE CRITICAL STRIP?

STEPHEN GELBART

Department of Mathematics,*
Weizmann Institute of Science,
Rehovot 76100, Israel
E-mail: steve.gelbart@weizmann.ac.il

DEDICATED TO ROGER HOWE ON HIS 60TH BIRTHDAY -
WITH FRIENDSHIP AND ADMIRATION

In this paper we discuss three kinds of methods for proving that an
L-function is non-zero in a part of its critical strip. The fundamental
classical result is de la Vallée Poussin's theorem with Riemann's zeta
function. A first method of generalizing this fits in quite nicely to the
Rankin-Selberg theory of L-functions for $GL(n)$. It falls short, however, of proving all cases of our Main Theorem. A second method
of proof using Eisenstein series for $SL(2)$ was introduced by Sarnak following Selberg. This was carried out in 2002 [Sar]. In his
paper, Sarnak suggested generalizing his method using Langlands-
Shahidi theory. The result was [GLS] and [Gel-Lap]. This third
method works not only for the $GL(n)$ L-functions but for a broad
range of Langlands-Shahidi L-functions too. For a result by this
method which goes beyond "de la Vallée Poussin's method" see the
example of the ninth symmetric L-function of $GL(2)$.

Introduction

The Grand Riemann Hypothesis (GRH) states that the non-trivial zeroes of
the L-function of any automorphic cuspidal representation of $GL(n)$ have
real part equal to $1/2$. The Main Theorem discussed in this paper is a
generalization of a much weaker property of the non-vanishing of $\zeta(s)$.

In 1896, J. Hadamard and Ch. de la Vallée-Poussin proved (indepen-
dently) that $\zeta(s) \neq 0$ for $Re(s) = 1$; this non-vanishing of ζ at the edge of
the critical strip

$$0 \leq Re(s) \leq 1$$

turns out to be equivalent to the Prime Number Theorem. A little later,
de la Vallée-Poussin extended the non-vanishing of ζ to an explicit region
inside the critical strip:

---

*Nicki and J. Ira Harris Professional Chair.

Theorem 1 (1899, de la Vallée Poussin). There exists a positive constant c such that $\zeta(s)$ has no zeroes in the region

$$\sigma \geq 1 - \frac{c}{\log(|t| + 2)}.$$

Furthermore,

$$\zeta(1 + it) \gg \frac{1}{\log(t)}$$

for large t.

Remark. Of course, it is the order of magnitude of $\frac{c}{\log(|t|+2)}$ *for large* $|t|$ that is of interest, but $\log(|t| + 2)$ is used in place of $\log(|t|)$ in order to obtain a formula for all t.

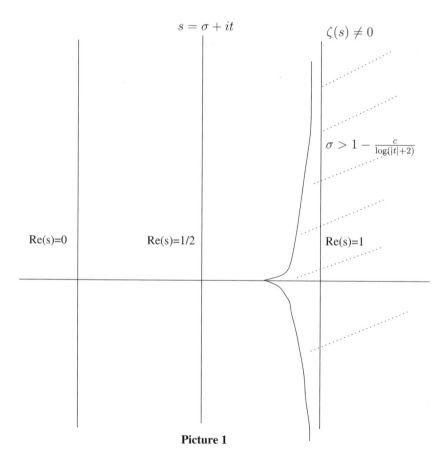

Picture 1

One could (eventually) generalize Theorem 1 by answering the following question. Let M be a connected reductive linear algebraic group, π an irreducible unitary cuspidal representation of $M(\mathbf{A})$ (of Ramanujan type), and r a finite dimensional representation of the Langlands L-group $^L M$. Let $L(s, \pi, r)$ denote the corresponding partial Langlands L-function (i.e. without archimedean factors) attached to this data. Does $L(s, \pi, r)$ then enjoy a similar zero free region?

In [Gel-Lap](see also [GLS]) we proved a much weaker result than this, both for the types of cuspidal π considered, and the size of r. To state it, let G denote a quasisplit connected reductive group.

Theorem 2 (Main Theorem). Let P be a maximal parabolic of G, and suppose $P = MU$. Let the adjoint representation r of the dual group $^L M$ on the Lie Algebra of $^L U$ be decomposed as $\bigoplus_{j=1}^{m} r_j$. Finally, suppose π is a *generic* cuspidal representation of $M(\mathbf{A})$, namely, that it admits a non-zero nondegenerate Fourier coefficient. Then there exist constants $c, n > 0$ such that

$$|L(1 + it, \pi, r_j)| \geq \frac{c}{(1 + |t|)^n},$$

$|t| \geq 1, j = 1, ..., m$.

Remarks. 1) All the concepts used in this Introduction will be explained in Section 4.

2) We note that this Theorem extends Shahidi's result that $L(1 + it, \pi, r_j) \neq 0$ (see [Shah81]).

Special Case

Take $G = SL(2), P$ equal to the Borel. Then the adjoint representation is the standard representation of $\mathbf{C}$, and for π the trivial representation of $\mathbf{A}^\times$,

$$L(s, \pi, r) = \zeta(s) = \sum_{n=1}^{\infty} \frac{1}{n^s} = \prod_p (1 - p^s)^{-1} = \zeta(\sigma + it)$$

satisfies the following: for some $n > 0$ (which we won't evaluate here)

$$|\zeta(1 + it)| \gg \frac{1}{(1 + |t|)^n}$$

for large t.

This is, of course, a weaker result than Theorem 1, but it generalizes to arbitrary Langlands-Shahidi's triples (M, G, r). Our proof of Theorem

2 leads us to the paper of [Gel-Shah]. First, we give an alternative shorter proof of their main result:

Theorem 3. Each $L(s, \pi, r_j)$ is "bounded" in vertical strips, i.e., when multiplied by the archimedeam gamma factors, the *completed* L-function is so bounded.

Then (using our results) we go on to prove our Main Theorem, which implies (by taking inverses of both sides) the Conjecture of [Gel-Shah]:

Theorem 4. Suppose π is a cuspidal generic representation of Langlands-Shahidi type. Then each $L(s, \pi, r_j)^{-1}$ is of finite order in $Re(s) \geq 1$, i.e., for some $\rho > 0$,

$$|L(s, \pi, r_j)^{-1}| \leq K e^{|s|^{\rho}}.$$

Acknowledgements. I am indebted to Erez Lapid and Peter Sarnak, collaborators in this work, for sharing with me part of their vast knowledge of L-functions. I also warmly thank Freydoon Shahidi for his help with Rankin-Selberg and Langlands-Shahidi theory, and the referee for his comments and suggestion that I attempt to write this paper as a short survey on the zero-free regions of L-functions. Last, but not least, I thank Terry Debesh for her ever cheerful help with Latex.

1. The Classical Method

Let's start by mentioning de la Vallée Poussin's Theorem as it was proved classically. Its proof was profoundly clever and simple.

The Theorem states that there exist constants $c > 0, K > 1$ such that

$$\beta < 1 - \frac{c}{\log \gamma}$$

for all roots $\rho = \beta + i\gamma$ in the range $\gamma > K$. (We are taking for granted that $Re(s) = 1$ implies $\zeta(s) \neq 0$, i.e., that $\beta < 1$.)

Taking the logarithmic derivative of Euler's identity $\zeta(s) = \prod(1 - p^{-s})^{-1}$ gives us

$$-\frac{\zeta'(s)}{\zeta(s)} = \sum \Lambda(n) n^{-s}$$

where $\Lambda(n)$ is von Mangoldt's function, i.e., 0 if n is not a power of p and $\log p$ otherwise. Therefore

$$-Re(\frac{\zeta'(s)}{\zeta(s)}) = \sum \Lambda(n) n^{-\sigma} cos(t \log n).$$

Now consider the inequality

$$3 + 4\cos\theta + \cos 2\theta = 2(1 + \cos\theta)^2 \geq 0.$$

(Hence this is the "(3,4,1) proof"....) Taking $\theta = t\log n$, and applying it to our sum, with t replaced by $0, t, 2t$ in succession, gives us

$$3[-\frac{\zeta'(\sigma)}{\zeta(\sigma)}] + 4[-Re\frac{\zeta'(\sigma + it)}{\zeta(\sigma + it)}] + [-Re\frac{\zeta'(\sigma + 2it)}{\zeta(\sigma + 2it)}] \geq 0. \tag{1}$$

The problem then is to investigate the behavior of $-\frac{\zeta'(s}{\zeta(s)}$ as $s \to 1$ for each of these three values. The end result is

$$-Re\frac{\zeta'(\sigma + it)}{\zeta(\sigma + it)} < A\log t - \frac{1}{\sigma - \beta}$$

where $1 < \sigma \leq 2$, $t \geq K > 1$, $\beta + i\gamma$ is a non-trivial root of ζ, and $A > 0$ is independent of σ and t. Solving for β, we conclude finally that - for some $c > 0$,

$$\beta < 1 - \frac{c}{\log\gamma}$$

for all roots $\rho = \beta + i\gamma$ which satisfy $\gamma \geq K$; see, for example, Section 5.2 of [Edw] for details. Hence de la Vallée Poussin's classical Theorem holds.

REMARK. It is very important in this proof that β be less than one.

2. The Rankin-Selberg Generalization of de la Vallée Poussin

There is a more modern approach to de la Vallée-Poussin's proof of Theorem 1. It's based on the construction of an auxillary L-function $D(s)$ with positive coefficients, analytic in $Re(s) > 1$, a pole at $s = 1$, and if $\zeta(\sigma + it_0) = 0$, $D(s)$ should vanish to an order at least 4 at $s = \sigma$. The one de la Vallée-Poussin took is the degree 8 L-function

$$D(s) = \zeta^3(s)\zeta^4(s + it_0)\zeta(s + 2it_0)$$

with t_0 in **R**. The (3,4,1) inequality shows that $-\frac{D'}{D}(s)$ has non-negative Dirichlet coefficients (like (1) in Section 1 holding).

The more modern way is to exploit the theory of Rankin-Selberg integrals ([JS1,2], [J-PS-S], [Shah81]) which works in a much more general setting. First consider the representation Π of $GL_3(\mathbf{A})$ induced from the 1-dimensional character $1 \times ||^{-it_0} \times ||^{it_0}$ of the Borel subgroup $B(\mathbf{A})$; this representation is automorphic but not cuspidal. Then introduce L as the Rankin-Selberg L-function attached to Π, i.e.,

$$L(s, \pi \times \tilde{\pi}) = \zeta^3(s)\zeta^2(s + it_0)\zeta^2(s - it_0)\zeta(s + 2it_0)\zeta(s - 2it_0),$$

where $\tilde{\Pi}$ denotes the contragredient of Π. Although Π is not cuspidal, it is isobaric; that is why the "multiplicativity" of the L-functions here still holds (see [JS1,2] and [Ram]). Let us again denote the Dirichlet series L by $D(s)$.

By the modern theory of Rankin-Selberg L-functions, this Dirichlet series $D(s)$ satisfies the following properties: (1) the coefficients of $D(s)$ are non-negative, and the series is absolutely convergent for $Re(s) > 1$; (2) $D(s)$ has an Euler product, so $D(s) \neq 0$ for $Re(s) > 1$; and (3) $\frac{D'(s)}{D(s)}$ real and > 1.

Now suppose $\rho = \beta + i\gamma$ is a zero of $\zeta(s)$ with $\beta \geq \frac{1}{2}$ and $\gamma \neq 0$. Take $t_0 = \gamma$. Note $D(s)$ has a pole at $s = 1$ of order ≤ 3 and a zero β of order ≥ 4. By the Lemma below, we shall show

$$\beta < 1 - \frac{c'}{log(|t_0| + 3)}. \tag{2}$$

Lemma (Special case of Goldfeld-Hoffstein-Lieman (see [HL94])). Let $D(s)$ be a Dirichlet series with non-negative coefficients, absolutely convergent for $Re(s) > 1$. Suppose also that $D(s)$ has an Euler product, so $D(s) \neq 0$ for $Re(s) > 1$, and $\frac{D'(s)}{D(s)}$ real and > 1. Let $D(s)$ have a pole of order 3 at $s = 1$ and let $\Lambda(s) = s^3(1 - s)^3 G(s)D(s)$ satisfy $\Lambda(s) = \Lambda(1 - s)$, with $\Lambda(s)$ (almost) entire or order 1. Here

$$G(s) = D^s \prod_{i=1}^{l} \Gamma(\frac{s + t_i}{2}).$$

Then there exists an absolute constant $c' > 0$ such that $D(s)$ has at most 3 (counting multiplicities) real zeroes in the interval

$$\sigma \geq 1 - \frac{c'}{log(1 + D \max |t_i|)}.$$

How does this Lemma effect the roots of our previously defined Dirichlet series $D(s)$? Because $D(s)$ has a zero at $s = \beta + it_0$ of order (*at least*) 4, and the absolute value of each t_i in its archimedean component $G(s)$ is an integral multiple of $|t_0|$, the Lemma implies

$$\beta < 1 - \frac{c'}{log(1 + D'|t_0|)}.$$

In other words, because 4 is bigger than 3, the desired inequality (2) follows! (Here we have used a slightly more general form of the Lemma, because our previously defined Dirichlet series $D(s) = L(s, \pi \times \tilde{\pi})$ does have a pole not equal to 0 or 1.)

For $GL(n)$ we now can argue similarly.

Let π denote an irreducible unitary automorphic cuspidal representation of $GL(n)$ over the rational numbers $\mathbf{Q}$. From $GL(n)$ we extend to the group $1 \otimes GL(n) \otimes GL(n)$ which we view as the Levi subgroup M of $GL(2n+1)$. Now we set Π equal to the representation of $GL(2n + 1)$ induced from $1 \times (\pi \otimes \|{}^{it_0}) \times (\tilde{\pi} \otimes \|{}^{-it_0})$ on $P = MN$. Once again, Π is automorphic but not cuspidal. Then we introduce $L(s, \Pi \times \tilde{\Pi})$ as the Rankin-Selberg L-series attached to Π, i.e.

$$L(s, \Pi \times \tilde{\Pi}) = D(s) = \zeta(s)L^2(s, \pi \times \tilde{\pi})L^2(s - it_0, \pi)$$
$$L^2(s - it_0, \tilde{\pi})L(s + 2it_0, \pi \times \pi)$$
$$L(s - 2it_0, \tilde{\pi} \times \tilde{\pi}).$$

Arguing as in the $n = 1, \pi = Id$ case we've just looked at, we conclude the result (using the Lemma):

Theorem 1'. Suppose π is an arbitrary unitary cuspidal representation of $GL(n)$. Then there exists a constant c (depending only on π) such that $L(s, \pi)$ has no zero in the region

$$\sigma \geq 1 - \frac{c}{\log(|t| + 2)}.$$

(Of course, here c means $c(\pi)$.)

Having the desired Theorem for $GL(n)$, it is natural to try to extend the result (for de la Vallée Poussin's method) for many other Rankin-Selberg L-functions. For example, in the 1980's, the Rankin-Selberg $L(s, \pi \times \pi')$ was well understood for π and π' cusp forms on $GL_n(\mathbf{A_K})$ and $GL_m(\mathbf{A_K})$ respectively, making possible a standard zero free region. For the reader interested in this attempt, see the articles of [Ogg], [Shah81], and [Mor]. In Sections 3 and 4 of this paper we describe an altogether different method which works for L-functions much greater in number (see Theorem 2 in particular).

3. An Approach Using Eisenstein Series on $SL(2, \mathbf{R})$

The non-vanishing of such a $L(s, \pi)$ for $Re(s) = 1$ was first established (before the modern Rankin-Selberg theory was developed) by Jacquet-Shalika [Ja-Sh]; they used the methods of Eisenstein series (see [Mor] for further discussion). The advantage of de la Vallée Poussin's method just discussed is that it yields a standard zero free region. Moreover, according to Langlands' general functoriality conjectures, any automorphic L-function should (someday) be a finite product of such standard L-functions. This being so, the general L-function would be non-zero in a region similar to the one for $\zeta(s)$! Of course, this is also the weakness of de la Vallée Poussin's methods: until Langlands' functoriality conjectures are known, this method is also limited to known cases of Rankin-Selberg.

In 2004, Sarnak published a completely different proof of the non-vanishing of $\zeta(s)$ by using Eisenstein series on $SL(2)$. His ingenious argument exploits the Maass-Selberg's relations and the computation of Fourier coefficients; comparing the two by Bessel's inequality gives a coarse lower bound for zeta. Here now are more details.

To define the Eisenstein series, let Γ_∞ denote the subgroup

$$\left\{ \pm \begin{pmatrix} 1 & m \\ 0 & 1 \end{pmatrix} : m \in \mathbb{Z} \right\}$$

of $SL(2, \mathbb{Z})$. Then

$$E(z, s) = \sum_{\Gamma_\infty \backslash SL(2,\mathbb{Z})} y(\gamma z)^s,$$

where $z = x + iy$, and $Re(s) > 1$. This Eisenstein series extends to a meromorphic function of s in $\mathbf{C}$, and in $Re(s) \geq 1/2$, has only a simple pole at $s = 1$. The relation with $\zeta(s)$ comes from the Fourier series expansion

$$E(z, s) = \sum_m a_m(y, s)e^{2\pi i m x}.$$

Here

$$a_0(y, s) = \int_0^1 E(z, s)dx = y^s + M(s)y^{1-s}$$

where

$$M(s) = \frac{\sqrt{\pi}\Gamma(s - \frac{1}{2})\zeta(2s - 1)}{\Gamma(s)\zeta(2s)}$$

and

$$a_m(y, s) = \int_0^1 E(z, s)e^{-2\pi i m x}dx = \frac{K_{s-\frac{1}{2}}(2\pi m y)y^{\frac{1}{2}}m^{s-\frac{1}{2}}\pi^s \sigma_{1-2s}(m)}{\Gamma(s)\zeta(2s)}$$

if $m > 0$. Recall $\sigma_{1-2s}(m) = \sum_{d|n} d^{1-2s}$ and K_s is the modified Bessel function defined by

$$K_s(z) = \frac{\pi}{2}\frac{I_{-s}(z) - I_s(z)}{sin\ \pi s}$$

with

$$I_s(z) = \sum_{m=0}^{\infty} \frac{(\frac{1}{2}z)^{s+2m}}{m!\Gamma(s+m+1)}.$$

Since $a_m(y, s)$ is analytic along the line $Re(s) = \frac{1}{2}$, the appearance of $\zeta(2s)$ in the denominator must show that it cannot be zero, i.e., we already get the wonderful relation that $\zeta(s) \neq 0$ when $Re(s) = 1$! To obtain the stronger result that Theorem 1 holds, we need to follow Sarnak's argument.

Because the new approach given in [Sar] contains entire proofs, we shall simply sketch the argument here.

Step (1). The Maass-Selberg relations express the L^2 norm of the **truncated** Eisenstein series $E_T(z, s)$(which we call the LHS) in terms of the scalar operator $M(s)$ and T (which we call the RHS):

$$\int_X |\zeta(1+2it)|^2 |E_T(z, \frac{1}{2}+it)|^2 \frac{dxdy}{y^2}$$

$$= |\zeta(1+2it)|^2 \left| 2\log T - \frac{M'}{M}(\frac{1}{2}+it) + \frac{\overline{M(\frac{1}{2}+it)}T^{2it} - M(\frac{1}{2}+it)T^{-2it}}{2it} \right|$$

where $X = SL_2(\mathbb{Z}) \setminus \mathbb{H}$, and, on the Siegel domain,

$$E_T(z, s) = \begin{cases} E(z, s) & for\ y \leq T \\ E(z, s) - y^s - M(s)y^{1-s} & for\ y > T. \end{cases}$$

Step (2). To simplify matters, take $T = 1$.

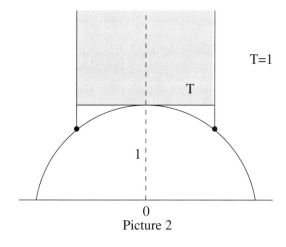

Picture 2

Using Bessel's inequality and the zeroth Fourier coefficient of $E(z,s)$, the formula in Step (1) gives us

$$LHS \gg \int_1^\infty \left| \frac{K_{it}(2\pi y)}{\Gamma(\frac{1}{2}+it)} \right|^2 \frac{dy}{y}$$

Step (3). What about the RHS? Using the expression for $M(s)$, the functional equation for ζ, and some elementary analysis, Sarnak shows that the only troublesome term,

$$\zeta(1+2it)\frac{M'}{M}(1/2+it),$$

has absolute value much less than $(\log t)^2$ when $t \geq 2$, i.e.,

$$RHS < |\zeta(1+2it)|^2 \frac{M'}{M}(1/2+it) < (\log t)^2 |\zeta(1+2it)|.$$

(The scalar operator $M(s)$ is unitary at $s = \frac{1}{2}+it$; therefore, since $T = 1$ is fixed, the two summands involving T in the RHS of Maass-Selberg are bounded by a constant, and there remains only to estimate the logarithmic derivative of M.)

Step (4). Putting the inequalities in Step (2) and Step (3) together gives

$$\int_1^\infty \left| \frac{K_{it}(2\pi y)}{\Gamma(\frac{1}{2}+it)} \right|^2 \frac{dy}{y} \ll |\zeta(1+2it)|(\log t)^2.$$

Step (5). Finally, asymptotics for K_{it} (see [Erd]) give

$$\int_1^\infty \left| \frac{K_{it}(2\pi y)}{\Gamma(\frac{1}{2}+it)} \right|^2 \frac{dy}{y} \gg \frac{1}{t},$$

so,

$$|\zeta(1+2it)| \gg \frac{1}{t(\log t)^2}.$$

Sarnak actually does better: using more than the zeroth Fourier coefficient of $E(z,s)$, the dependence of E_T on T, and some sieve theory, he arrives at

$$\zeta(1+2it) \gg \frac{1}{(\log t)^3}.$$

4. The General Method

Our approach to the general Theorem 2 roughly follows this line of thought. That is to say, "Eisenstein series" are defined in general, and Theorem 2 follows again from the Maass-Selberg formula (and parts of "Langlands-Shahidi theory"). Still, many key formulas, such as "asymtotics for K_{it}", are not known in general, and a more roundabout method is needed.

We start with F a number field, and $\mathbf{A}$ its ring of adeles. Let G denote a reductive group over F (which the reader can take to be $SL(2)$) and ^LG its Langlands dual (specifically $PGL(2, \mathbf{C})$ for $G = SL(2)$). We fix an M to be the Levi component of a maximal parabolic P of G over F, and let ^LP be the corresponding parabolic subgroup of ^LG with Levi decomposition $^LM{}^LU$. The adjoint representation of the dual ^LM on the Lie algebra of ^LU decomposes as $\oplus_{j=1}^m r_j$ where the irreducible constituents are indexed by the terms in the lower central series of ^LU. Finally, we let π be a cuspidal automorphic representation of $M(\mathbf{A})$.

In place of the Eisenstein series $E(z, s)$ of Section 3, we introduce the function

$$E(g, \varphi, s) = \sum_{\gamma \in P \backslash G} \varphi_s(\gamma g),$$

with φ_s defined in terms of s and the representation π. The exact nature of φ_s is not important to go into now; it and several other definitions in the general case are carefully defined in [Gel-Lap] but only skimmed over here. What the reader can do is follow the simple notions that come out of it for the case $G = SL(2)$. For example, for $G = SL(2)$ and $\pi = 1$, one exact choice of φ will make $E(g, \varphi, s)$ reduce (essentially) to $E(z, s)$.

One other notion we shall need is $M(s)$. Again, for $G = SL(2)$, it is a scalar intertwining operator given by multiplication. In general, by a well-known formula of Langlands ([Lan70 and 71]), $M(s)$ is given by

$$M(s)(\otimes_v \varphi_v) = \prod_{v \in S} M_v(\pi_v, s)\varphi_v \times \prod_{j=1}^m \frac{L^S(js, \tilde{\pi}, r_j)}{L^S(1 + js, \tilde{\pi}, r_j)}.$$

Here S is a sufficiently large finite set of places containing all the archimedean ones such that for all $v \notin S$ both G_v and π_v are unramified and φ_v is "the standard section". Using the operator $M(s)$, we can (and do) define the truncated Eisenstein series $\Lambda^T E(g, \phi, s)$ by generalizing the formula for $E_T(z, s)$ inside Step (1) of Section 3.

Now we shall need a new shorter proof of the main result of [Gel-Shah] which avoids the use of complex variables, notably "Matseav's Theorem". First, we prove:

Proposition. Each $L(s, \pi, r_j)$ is meromorphic of finite order (with π not necessarily a *generic* representation; see below).

This Proposition uses "only" the relation between $M(s)$ and the constant term of $E(s, \varphi, s)$ along P, and the finiteness of order of E proved by Müller ([Mul]).

Next we assume π *generic*, which means that π admits a non-zero non-degenerate Fourier coefficient ψ, i.e., a Whittaker model. This implies, in particular, that G, or equivalently M, is quasi-split over F. For our example where $G = SL(2)$ and $\pi = 1$, this is automatic.

Theorem 3′. With π generic, and any right halfplane, there exist $c, n > 0$ such that

$$|L(s, \pi, r_j)| \leq c(1 + |s|)^n.$$

This Theorem is proved *only* using facts from the theory of Eisenstein series. It implies $L(s, \pi, r_j)$ is finite of order 1. Thus, it gives (a more basic proof of) the main Theorem of [Gel-Shah]: $L(s, \pi, r_j)$ is *"bounded"in vertical strips* (taking on the Γ factor at infinity).

Now, finally, we approach a generalization of the Maass-Selberg formula and what it implies by Steps (1′) through (5′): the end result will be Theorem 2, our main result.

Step (1′). Recall that the Maass-Selberg formula allows us to write the normed square of the truncated Eisenstein series as

$$||\Lambda^T E(\cdot, \varphi, s)||_2^2 = -(M(-s)M'(s)\varphi, \varphi) + T(\varphi, \varphi)$$

$$+ \epsilon_{P, P'} \frac{(e^{2sT} M(-s)\varphi - e^{2sT} M(s)\varphi, \varphi)}{2s}$$

for $s \in i\mathbf{R}$ and $\epsilon_{P, P'}$ constants; see [Art]. Since the formula for $M(s)$ involves $|L(1 + it, \tilde{\pi}, r_i)|$, we want to find (upper and)lower bounds for $||\Lambda^T E(\cdot, \varphi, s)||_2^2$.

Step (2′). First we show that for sufficiently large T and certain g we have that the ψ coefficients of $\Lambda^T E$ and E are equal, for any ψ a non-degenerate character. Then, as in Step (2), we use the formula

$$E^\psi(g, \varphi, s) = \lambda \cdot \prod_{v \in S} \mathcal{W}_v^\psi(\varphi_v, g_v, s) \left[\prod_{j=1}^m L^S(1 + js, \tilde{\pi}, r_j) \right]^{-1}$$

discovered by Shahidi [Shah78]; here $\mathcal{W}_v$ is the local "Jacquet integral" which will be crucial in Step (5′). The end result shall be the lower bound

$$||\Lambda^T E(\cdot, \varphi, s)||_2^2 \geq |\lambda|^2 \prod_{j=1}^m |L^S(1 + js, \tilde{\pi}, r_j)|^{-2} \int_{K_S} \int_V |\mathcal{W}_S^\psi(\varphi, a_0 vk, s)|^2 dv\, dk$$

for a suitable S and neighborhoods K_S and V. Again, Bessel's inequality is used.

Step (3'). As is Step (3), we use Maass-Selberg to show that there exist c, n, N such that

$$\|\Lambda^T E(\cdot, \varphi, s)\|_2^2 \le c(1 + |s|)^n \cdot \mu_N(\varphi)^2 \cdot (1 + \sum_{j=1}^{m} |L^S(1 + js, \tilde{\pi}, r_j)|^{-1})$$

for all $s \in i\mathbb{R}$ with $|Im\, s| \ge 1$. The exact definitions are not too important; the interested reader can see [Gel-Lap]. Since T is fixed, the last two summands appearing in the formula for $\|\Lambda^T E(\cdot, \varphi, s)\|_2^2$ are bounded by a constant multiple $\|\phi\|_2^2$ for $|Im(s)| \ge 1$.

Step (4'). We now combine Step (3') (an upper bound) and Step(2') (a lower bound) for $\|\Lambda^T E(\cdot, \varphi, s)\|_2^2$. Upon multiplying the ensuing inequality by

$$|L^S(1 + js, \tilde{\pi} r_j)|^{-1} \prod_{k=1}^{m} |L^S(1 + ks, \tilde{\pi}, r_k)|^2$$

and taking into account the upper bound for the L-function on $Re(s) = 1$ (Theorem 3'- the main result of [Gel-Shah]), we obtain for any $j = 1, \ldots, m$ the inequality

$$|L^S(1 + js, \tilde{\pi}, r_j)|^{-1} \int_{K_S} \int_V |\mathcal{W}_S^\psi(\varphi, a_0 vk, s)|^2 dv\, dk \le c(1 + |s|)^n \mu_N^2(\varphi)$$

for $s \in i\mathbb{R}$ with $|Im\, s| \ge 1$ for appropriate c, n and N. This inequality is based on the theory of Eisenstein series, and in principle it applies only to K_∞-finite sections. However, both sides of the equality extend continuously to all smooth sections. Hence the inequality is valid for smooth sections as well.

Step (5'). We can get now our *lower* bound on L, i.e., our Theorem 2. Since the last inequality was good for any smooth section, in our general setup it is best to work with a special section φ having support inside the big Bruhat cell of G. This is a bit complicated, even for the case $G = SL(2)$. Suffice it to say that these local integrals for $\mathcal{W}$ are vehicles for generalizing the inequality

$$\int_1^\infty \left| \frac{K_{it}(2\pi y)}{\Gamma(\frac{1}{2} + it)} \right|^2 \frac{dy}{y} \gg \frac{1}{t},$$

and giving us

$$|\zeta(1 + 2it)| \gg \frac{1}{t(\log t)^2}.$$

The end result for a general G is

$$|L(1 + js, \pi, r_j)| \geq c(1 + |s|)^{-n},$$

exactly what Theorem 2 states.

As an application, let π denote a cuspidal representation of $GL_2(\mathbf{A})$. As in [Kim-Shah], we consider the exceptional group $G = E_8$ and construct a representation Π of the Levi factor isogenous to $GL_4 \times GL_5$ using the symmetric cube and quadruple functorial lifts of π. (If either of these representations is not cuspidal, the L-function will be a product of those for cuspidal ones for a *non*-maximal parabolic, for which Langlands-Shahidi works just as well.) The ninth symmetric power L-function of π appears as a factor in $L(s, \Pi, r_1)$, and eventually we get the following:

Corollary. There exists $c, n > 0$ such that

$$L(1 + it, \pi, Sym^9) \geq c(1 + |t|)^{-n}, t \gg 0.$$

With our current knowledge of the functorial conjectures, the lower bound here seems to be out of reach of the method of de la Vallée Poussin. (Nevertheless, Kim and Shahidi have proved the holomorphy for $Re(s) > 1$ of this ninth power L-Function in [KM2].)

References

[Art] Arthur, J., "A Trace Formula for Reductive Groups, II, Applications of a Truncation Operator", *Compositio Math.* 40, 87-121, 1980.

[Edw] Edwards, H.M., *Riemann's Zeta Function*, Dover Publications, 1974.

[Erd] Erdélyi, A. et al, *Higher Transendental Functions*, Volume 2, McGraw-Hill, 1953.

[Gel-Lap] Gelbart, S.S. and Lapid, E.M., "Lower Bounds for L-functions at the Edge of the Critical Strip", *Amer. J. of Mathematics*, 619-638, Volume 128, no. 3, 2006.

[GLS] Gelbart, S.S., Lapid, E.M., Sarnak, P., "A new method for lower bounds of L-functions", *C.R. Math Acad. Sci.* Paris 339, 91-94, 2004.

[Gel-Shah] Gelbart, S.S. and Shahidi, F. "Boundedness of Automorphic L-functions in Vertical Strips", *J.A.M.S.*, 79-107 (electronic), 2001.

[HL94] Hoffstein, J. and Lockhart, P., "Coefficients of Maass forms and the Siegel zero", *Ann. of Math. (2)*, 140 (1994), 161-181. With an appendix by Dorian Goldfield, Hoffstein and Daniel Lieman.

[Ja-Sh] Jacquet, H. and Shalika, J.A., "A non-vanishing theorem for zeta functions of GL_n", *Invent. Math*, 38 (1976/77) 1-16.

[J-PS-S] Jacquet, H., Piatetski-Shapiro, I and Shalika, J.A., "Rankin-Selberg Convolutions", *Amer. J. of Mathematics*, Volume 105, 367-464, 1983.

[JS1,2] Jacquet, H. and Shalika, J.A., "On Euler Products and the Classification of Automorphic Representations I and II", *Amer. J. of Mathematics*, Volume 103, No. 3, 499-558 and 777-815, 1981.

[Kim] Kim, H., "Functoriality for the Exterior Square of GL_4 and the Symmetric Fourth of GL_2", *J.A.M.S.* 16(1):139-183 (electronic), 2003.

[Kim-Shah] Kim, H. and Shahidi, F., "Functorial Products for $GL_2 \times GL_3$ and the Symmetric Cube for GL_2", *Ann. of M.*, (2), 155 (3):837-893, 2002.

[KM2] Kim, H. and Shahidi, F., "Holomorphy of the 9th Symmetric Power L-functions for $Re(s) > 1$", *IMRN*, Number 14, 1-7, 2006.

[Lan70,71] Langlands, R.P. "Problems in the theory of automorphic forms", *Lectures in Modern anaysis and applications, III. Lecture notes in Math.*, vol 170. Springer-Verlag. Berlin, 18-61, 1970 and *Euler Products*, Yale University Monographs 1, Yale University Press, 1971.

[Lap] Lapid, E., "A Remark on Eisenstein Series", Proceedings of the AIM Conference on Eisenstein Series, Palo Alto, 2005.

[Mor] Moreno C., "Advanced Analytic Number Theory: L-functions", *Math. Surveys and Monographs,* Vol. 115, AMS, 2005.

[Mul] Müller, W. "The trace class conjecture in the theory of automorphic forms", *Ann. of Math. (2)*, 130 (1989), 473-529.

[Ogg] Ogg, A.,"On a Convolution of L-series", *Inventiones Math.*, 7, 297-312, 1969.

[Ram] Ramakrishnan, D., "Irreducibility and Cuspidality", Representation Theory and Automorphic Forms, Edited by T. Kobayashi, W. Schmid and J. Yang, Birkhauser, Boston, September 2007, 1–27.

[Sar] Sarnak, P., "Nonvanishing of L-functions on $Re(s) = 1$", *Contributions to Automorphic Forms, Geometry and Number Theory*, 719-732, Johns Hopkins Press, 2004.

[Shah78] Shahidi, F., "Functional Equation Satisfied by Certain L-function", *Compositio Math.*, 171-207, 1978.

[Shah81] Shahidi, F., "On Certain L-functions", *Amer. J. of Mathematics*, 297-355, Vol. 103, 1981.

[Sel] Selberg, A., "Collected Papers", Springer-Verlag, 1989.

COHOMOLOGICAL AUTOMORPHIC FORMS ON UNITARY GROUPS, II: PERIOD RELATIONS AND VALUES OF L-FUNCTIONS

MICHAEL HARRIS

UFR de Mathématiques, Université Paris 7
2 Pl. Jussieu, 75251 Paris cedex 05, FRANCE
E-mail: harris@math.jussieu.fr

To Roger Howe

Introduction

This is the fourth in a series of articles devoted to the study of special values of L-functions of automorphic forms contributing to the cohomology of Shimura varieties attached to unitary groups, for the most part those attached to hermitian vector spaces over an imaginary quadratic field $\mathcal{K}$. Since these L-functions are all supposed to be motivic, all their values at integer points are conjectured to be arithmetically meaningful. For the finite set of integer points which are *critical* in the sense of Deligne, [1] the values are predicted by Deligne's conjecture [D] to be algebraic multiples of certain determinants of periods of algebraic differential forms against rational homology classes. One of the main results of [H3] is that the Deligne period of the polarized motives associated to the standard L-functions of a self-dual cohomological automorphic representation π of a unitary group G of signature $(n-1,1)$ can be expressed up to rational multiples as products of square norms of arithmetically normalized Dolbeault cohomology classes on the Shimura variety attached to G. The other main result of [H3] is an expression of the critical values *in the range of absolute convergence* in

---

Institut des Mathématiques de Jussieu, U.M.R. 7586 du CNRS. Membre, Institut Universitaire de France.

[1]The same set had been identified independently by Shimura, though without the motivic interpretation.

terms of Petersson norms of (arithmetically normalized) holomorphic automorphic forms, nearly equivalent to π but realized in general on unitary
groups of different signatures. The article [H5] removes the restriction to
the range of absolute convergence in the most important cases, corresponding to the existence of non-trivial theta lifts to larger unitary groups. [2]

Deligne's conjecture is thus reduced to a series of conjectural relations
among the Petersson norms of cohomological forms in nearly equivalent automorphic representations of unitary groups of varying signature. The article [H4], to which this article is an immediate sequel, studied the rationality
properties of the theta correspondence relating these nearly equivalent automorphic representations to one another, and to similar automorphic representations on larger unitary groups. The final section of [H4] explained how
to deduce Deligne's conjecture from a collection of hypotheses or supplementary results that had not yet been verified. The hypotheses had mostly
to do with the existence of stable base change from G to $GL(n, \mathcal{K})$, and
are reproduced below as Hypothesis 4.1.1 and 4.1.10. The supplementary
results fall into three categories:

(i) Extension of the main theorems of [H3] beyond the range of absolute
     convergence, now accomplished in [H5], under a hypothesis on the
     non-triviality of theta lifts considered in the present article;

(ii) Determination up to rational factors of a variety of archimedean
     zeta integrals, which we have not considered; and

(iii) An extension of the Rallis inner product formula, which relates
      square norms of automorphic forms to square norms of their theta
      lifts and special values of standard L-functions.

The main contributions of the present article are the the analysis of local
and global obstructions to the non-triviality of the theta correspondence,
used to verify (i) and the verification of a version of (iii). This completes the
proof of Deligne's conjecture for the L-functions considered in [H3], up to
determination of the archimedean zeta integrals, and under certain genericity hypotheses. More precise versions of our main results are provided in
§4.2.

Strictly speaking, the results of [H3] and [H5] only concern a slight weak-
ening of Deligne's conjecture, in which determination up to rational multiples is replaced by determination up to $\mathcal{K}^\times$-multiples. When Deligne's
conjecture is mentioned in the article we will understand it to mean this

---

[2]Shimura has obtained similar results in [S1] and [S2], by rather different methods.

weakened version. Moreover, to simplify some of the arguments, we make the hypothesis that $\mathcal{K}$ is unramified at the prime 2 (cf. 3.5.5 (e) and 4.1.14).

We write $U(n)$ to designate the unitary group of a hermitian vector space of dimension n over $\mathcal{K}$, regardless of signature or other local properties. The main object of study in the present article is the theta correspondence between automorphic representations of $U(n)$ and $U(n+1)$; we prove that any tempered cohomological automorphic representation π of $U(n)$ of holomorphic type admits a non-trivial lift (for some choice of splitting character, see below) to a *positive-definite* $U(n+1)$, provided π_∞ is sufficiently general (the minimal K-type of π_∞ is not too small). The condition on π_∞ is quite mild; for example, for $n = 2$, it applies to holomorphic modular forms of weight $k > 2$. Actually, the same methods work, and more easily, to show that π lifts to a positive-definite $U(n + r)$ for any $r > 0$, but the archimedean condition excludes more π as r grows, and the theta lift of π becomes more pathological; for example, for tempered π, the theta lift is only tempered for $r = 0, 1$. Nevertheless, this technique has potentially interesting applications to the construction of non-trivial extensions of Galois representations, a topic to which I will return in a subsequent paper.

We are only concerned with π for which π_∞ belongs to the discrete series. It is then conjectured that π is globally tempered; this is verified under a local hypothesis in [HL], and in general should be a consequence of the stabilization of the trace formula for $U(n)$ and the related calculation of the zeta functions of the associated Shimura varieties. We also assume π admits a base change to a cuspidal automorphic representation of $GL(n, \mathcal{K})$. There are both local and global obstructions to non-triviality of the theta lift of tempered π to $U(n + r)$, where temporarily we allow $r = 0$. We first note that the theta lift for unitary groups depends on the choice of an additive character ψ, as usual, and also on a pair of multiplicative characters (*splitting characters*) χ, χ', whose properties are recalled in §2.1. The local obstructions for $r = 0$ are studied in detail in [HKS]. For a given prime v there are two possible local $U(n)_v$, denoted $U^\pm$, where we bear in mind that what counts for the theta correspondence is not just the isomorphism class of $U(n)$ but also that of the hermitian space to which it is attached. In [HKS] we conjectured that exactly one of these $U^\pm$ admitted a non-zero local lift from a given π_v. This *dichotomy conjecture* was proved in [HKS] for most π_v and was completed by a conjecture relating the sign of the lucky $U^\pm$ to a local root number constructed from π_v as well as χ_v, χ'_v, and ψ_v. In this paper we assume (cf. (2.1.5)) that the weak Conservation Relation

proved in [KR] for orthogonal/symplectic dual reductive pairs is also valid
for unitary dual reductive pairs, and show that this easily implies the dichotomy conjecture. By making a judicious choice of splitting characters,
we can thus show that the local obstruction is trivial for $r = 0$, with no
condition on π_∞ when the latter is holomorphic. Unfortunately, the global
obstruction is the vanishing of the central value of the standard L-function.
It is expected that, up to twisting by a character of finite order, one can
guarantee that this value is non-zero, but this is not known for $n > 2$.
Thus we cannot use the theta lift from $U(n)$ to (positive-definite) $U(n)$ to
obtain period relations by means of the Rallis inner product formula. We
note, however, that there is no global obstruction for $r > 0$, because of
our hypothesis on base change and because the L-function of a cusp form
on $GL(n, \mathcal{K})$ does not vanish at integer points to the right of the critical
line. We also show that, up to replacing $U(n+r)$ by an inner form at some
set of finite primes, there is also no local obstruction for $r > 1$ when π is
tempered. Our first main technical result (Corollary 3.3) is then that for π
tempered, there is always a non-trivial theta lift to some definite $U(n+r)$
when $r \geq 2$. This allows us to apply the results of [H5] to verify (i) above
for all critical values strictly to the right of the critical line.

Our next main technical result (Theorem 3.4) considers the lifting from
$U(n)$ to (definite) $U(n+1)$. Here the global obstruction is still trivial, and
one can eliminate the local obstruction at the price of replacing the splitting
characters. This extends the application of [H5], and thus (i) above to the
critical line itself, for most character twists; this is sufficient for applications
to period relations. The results of [H5], reproduced below as Theorem 3.2,
are all based on Ichino's extension of the Siegel-Weil formula beyond the
range of absolute convergence [I1,I2]. These also suffice for application of
Theorem 3.2 to critical values at the center of the critical strip. However,
they do not suffice for the study of the value at the near central point
(corresponding in the unitary normalization to the value at $s = 1$) for π
with trivial lift to all definite $U(n+1)$; this is because, like the results of
Kudla and Rallis for orthogonal/symplectic pairs, they say nothing about
the special values of Eisenstein series that are *incoherent* (i.e., not obtained
by the Siegel-Weil formula for purely local reasons).

The final section is concerned with applications to Deligne's conjecture.
Section 4.1 reconsiders the hypotheses made in somewhat excessive generality [3] in the final section of [H4]. In light of the results of §3, most of these

---

[3]Cf. Section E.3 at the end of the introduction.

hypotheses can now be considered theorems, at least in slightly weaker versions that still suffice for the arguments of [H4]. Section 4.2 derives the applications to period relations and Deligne's conjecture on critical values. In the interest of complete disclosure, I mention that the list of hypotheses has been lengthened: in particular there is a multiplicity one hypothesis and an assumption that the automorphic representation π is tempered (cf. 4.1.4). Both of these hypotheses are surely unnecessary but their inclusion shortens some of the arguments considerably. More to the point, I have no doubt they are not only true but will be verified in the near future, so I see no reason not to use them.

Much of the local theory presented in §2 is an outgrowth of my article [HKS] with Kudla and Sweet, and I thank Steve Kudla for numerous discussions of this material over the years. Although he is not listed as an author of this paper, I would like to make it clear that I consider him equally responsible for the dichotomy theorem (Theorem 2.1.7). Matthieu Cossutta read an earlier version of the manuscript carefully and pointed out misprints and one substantial error. I also thank Jian-Shu Li for providing references for the erratum E.1, Guy Henniart and A. Minguez for informing me of the results of the latter's thesis in preparation [M], and Erez Lapid for bringing the results of [LR] to my attention. I am particularly grateful to A. Ichino for keeping me informed of his work on the extended Siegel-Weil formula, and for writing up some of his results in [I2] that were essential to the present project.

The influence of Roger Howe's approach to theta functions as a means for relating automorphic forms on different groups on every line of the present paper is too obvious to require explanation. This article, and its author, have a subtler debt to Roger as well: by pointing out the hidden premises to a question I asked him at Oberwolfach in 1985 he convinced me of the possibility of a viable theory of coherent cohomology of Shimura varieties combining arithmetic and analytic techniques; this observation structures the main calculations in [H4]. For these and many other reasons, it is a great pleasure to dedicate the present paper to Roger Howe.

Errors and Misprints in [H4]

While preparing the present article I discovered a substantial number of misprints in [H4]; these are listed at the end of this section. There are unfortunately also some mathematical errors.

E.1. The least important is explained in the following paragraph, that was inadvertently omitted from [H4] (a "Note added in proof" is announced on p. 108 but the text is missing).

Note that should have been added in proof in [H4]. The argument in IV. 5. is incomplete. Specifically, the first paragraph of the proof of Lemma IV.5.15 implicitly assumes, without proof, that small automorphic representations have multiplicity one in the automorphic spectrum. It is possible to prove this multiplicity one assertion using results of J.-S. Li on singular forms.

The results of Li to which the previous paragraph refers are contained in [L1, L2]. More precisely, the small automorphic representations π of [H4] have the property that the local components π_v are singular, and in fact of rank 1, for all places v; indeed, π is constructed as a global theta lift of a character, say η, of $U(1)$, and each π_v is thus a local theta lift from $U(1)$. Theorem A of [L2] then asserts that the multiplicity of π in the space of automorphic forms on $U(n)$ is bounded by the automorphic multiplicity of the character η on $U(1)$, which is obviously 1.

E.2. More serious is that the proof of Theorem V.1.10 of [H4], which is the main theorem, is apparently inadequate. The problem is that the oscillator representation depends on the choice of additive character, specifically at finite primes, and therefore the final rationality result needs to take this into account. I thought I had found a trick to avoid this problem (see [H4, p. 167]) but it now seems that it returns in the induction step. Specifically, Lemma IV.2.3.2 of [H4] (rationality of non-archimedean zeta integrals for $U(1)$) is expressed in terms of the rationality of local sections defining Eisenstein series, whereas Theorem V.1.10 is expressed in terms of rationality of local Schwartz-Bruhat functions, which are related to the Eisenstein series via the Siegel-Weil formula. The map denoted $\phi \mapsto \mathcal{F}_\phi$, defined in §I.2, is rational over the maximal cyclotomic extension of $\mathbb{Q}$ but not necessarily over $\mathbb{Q}$. The two kinds of rationality can be related, but (apparently) not on both sides of a seesaw, and so I can only claim the following weaker version of Theorem V.1.10:

Theorem E.2.1 (Theorem V.1.10 of [H4], corrected) *Notation is as in* [H4, §V.1]. *Let* $\alpha \in H_!^{2(r'-1)}(Sh(V^{(2)}, [W_{\Lambda(2)}])$ *be a coherent cohomology class rational over the number field* L. *Let* $\mathcal{F}$ *and* $F \in \pi(2)$ *be as in the statement of* [H4, V.1.10]. *Let* $\varphi \in \mathcal{S}(\mathbb{X}(\mathbf{A}_f))$ *be an algebraic Schwartz-Bruhat function, rational over* $\mathbb{Q}^{ab}(\pi, \chi^+, \chi'^{,+})$, *with* φ_∞ *determined by*

[H4, IV (4.6.ii)]. *Then*

$$\mathbf{p}(\pi, V, V', \chi_0)^{-1} \underline{Lift}^{-1}(\theta_{\varphi, \chi, \chi', \psi_f}(F))$$

is a $\mathbb{Q}^{ab}(\pi, \chi^+, \chi'^{,+})$-*rational class in* $H^0(Sh(V'^{,(2)}), [E_{\Lambda(2)}])$. *Moreover, for any* $\sigma \in Gal(\overline{\mathbb{Q}}/\mathcal{K})$, *we have*

$$[\mathbf{p}(\pi, V, V', \chi_0)^{-1}] \underline{Lift}^{-1}(\theta_{\sigma(\varphi), \sigma(\chi), \sigma(\chi'), \sigma(\psi_f)}(\sigma(F)))$$
$$= \sigma(\mathbf{p}(\pi, V, V', \chi_0)^{-1} \underline{Lift}^{-1}(\theta_{\varphi, \chi, \chi', \psi_f}(F))).$$

The action of σ on χ and χ' is actually defined by means of the actions on the algebraic Hecke characters χ^+ and $\chi'^{,+}$. The difference with the statement in [H4] is that the additive character has been incorporated into the notation for the theta lift, and is affected by the Galois action. This only becomes relevant in the study of non-archimedean zeta integrals, specifically in Lemma IV.2.3.2, on p. 171 (just above formula (5.5)), and again in V. (3.13); see Lemma 3.5.12, below. The map $\phi \mapsto \mathcal{F}_\phi$ is Galois equivariant in the following sense: if we incorporate ψ_f into the notation in the obvious way, then for any $\phi \in \mathcal{S}(\mathbb{X}(\mathbf{A}_f), \overline{\mathbb{Q}})$ (the Schwartz-Bruhat space, as in §I.2)

$$\sigma(\mathcal{F}_{\phi, \psi_f}) = \mathcal{F}_{\sigma(\phi), \sigma(\psi_f)}, \sigma \in Gal(\overline{\mathbb{Q}}/\mathbb{Q}). \qquad (E.2.2)$$

With this in mind, the argument of Lemma IV.2.3.2 takes care of the non-archimedean zeta integrals, and the rest of the proof is unaffected.

The notation $\sigma(\psi_f)$ is actually abusive because it is not generally the finite part of a global additive character. Thus

$$\theta_{\sigma(\varphi), \sigma(\chi), \sigma(\chi'), \sigma(\psi_f)}(\sigma(F)) \qquad (E.2.3)$$

requires explanation. In fact, the dependence on ψ_f of the theta kernel attached to φ, or to $\sigma(\varphi)$, is determined by the values of ψ_f on an open compact subgroup of $\mathbf{A}_f$; this is obvious if one rewrites the theta kernel in classical language. In other words, the theta kernel only depends on the classical additive character modulo N attached to ψ_f, where $N = N(\varphi)$ depends on φ. Call this finite additive character $\psi_{N(\varphi)}$. To define the theta kernel for $\sigma(\varphi)$ attached to $\sigma(\psi_f)$, one lets $\sigma(\psi, \phi)$ be any additive character whose associated character mod $N(\varphi)$ is $\sigma(\psi_{N(\varphi)})$. We will then have $\sigma(\psi)_\infty \neq \psi_\infty$, but this change is invisible because the archimedean Schwartz function φ_∞ (defined by [H4, IV.(4.6.ii)] – this has been added to the statement of Theorem E.2.1, and should have been recalled more clearly in [H4,§V.1]) is defined in terms of its behavior under the local theta correspondence, hence in terms of the choice of ψ_∞. In particular, changing

ψ_∞ has no effect on the local zeta integral attached to φ_∞. This correction requires us to relax our hypothesis on ψ_∞ (cf. [H4, p. 137, p. 162]).

E.2.4. *The additive character ψ_∞ is assumed to be of the form $\psi_\infty(x) = e^{2\pi i \alpha x}$ for some $\alpha \in \mathbb{Q}^\times$, $\alpha > 0$.*

I stress that, for applications to period relations, Theorem E.2.1 is just as good as the incorrect version Theorem V.1.10 stated in [H4]! The main induction step is sketched in [H4, §V.3], under a series of hypotheses, most of which are verified in the present paper. At the end, instead of the relation (3.28.2), one obtains the following version, which we write out at length:

$$A(r', \pi_0) = \frac{P^{(r')}(\pi_0)}{\prod_{j=1}^{r'} Z_{\infty,j} Q_j(\pi_0)} \in \mathbb{Q}^{ab}(\pi_0, \beta^{(2)}, \chi_0^+); \qquad (E.2.5)$$

$$\forall \sigma \in Gal(\overline{\mathbb{Q}}/\mathcal{K}), \sigma(A(r', \pi_0)) = A(r', \sigma(\pi_0)). \qquad (E.2.6)$$

Note that ψ does not appear in (E.2.6) or the left-hand side of (E.2.5) – no more than β or χ_0^+. Thus, assuming the hypotheses of [H4, §V.3], and assuming one can eliminate the $Z_{\infty,j}$, (E.2.6) suffices to imply the Deligne conjecture (up to $\mathcal{K}^\times$), as explained in [H3, H4].

E.3. Finally, and this is perhaps the main point of the present paper, for the moment some of the hypotheses used in the sketch of the proof of period relations in [H4,§V.3] can only be verified under slightly restrictive conditions. Specifically, Hypotheses (3.10)(b) and (3.20)(a) of [*loc. cit.*] assert that the relations between special values of the standard L-function of a cuspidal automorphic representation π of holomorphic type on a unitary group and Petersson norms of arithmetic holomorphic forms in π remain valid down to the center of symmetry of the functional equation; in [H3] these relations were proved in the range of absolute convergence of the integral representation of the L-function. For applications, we are most interested in the special values at the point $s = 1$ in the unitary normalization; this is a critical value under an additional hypothesis on the infinity type (corresponding to (3.20)(c) of [*loc. cit.*]). If π is an automorphic representation of some form of $U(n)$, then the results of [H5] allow us to prove this under a set of hypotheses that imply that some theta lift of π to a definite $U(n + 1)$ is non-trivial. When I wrote [H4, §V.3] I was under the impression that it was sufficient to show non-triviality of the local theta correspondences everywhere, since the Rallis inner product formula would then imply that the global theta lift is non-zero provided the L-function was non-vanishing at $s = 1$, and this was guaranteed by a standard hypothesis on base change to $GL(n)$, included here as Hypothesis 4.1.1 (in [H4] there was only a hypothesis on non-vanishing, cf. V. (3.10)(a)). However, the Rallis inner product formula only applies when the local theta correspondences patch together into a theta correspondence with a global $U(n+1)$. In general there is a parity obstruction, and the correct assertion is Theorem 3.4 of the present paper.

The relevant modifications to the hypotheses in question are explained in §4.1. It should be added that there are numerous misprints in [H4, §V.3], especially in the crucial section 3.23, almost all of which were undoubtedly in my original manuscript. Those I have found are corrected below.

Misprints and omissions in [H4].

p. 113, (1.13) The formula for the local integral in [H4] followed [HKS] in omitting a factor $\chi^{-1}(det(g'))$. This is corrected in (1.3.3) below. The missing factor plays no role in any arguments.

p. 177, (1.5): $E_{\Lambda(2)}$ should be $E_{\Lambda'(2)}$.

p. 178, Theorem 1.10: The number field L was introduced in the beginning of the statement but forgotten in the conclusion. It should be assumed that $L \supset \mathcal{K}$. On the sixth line of the statement, $\mathbb{Q}(\pi, \chi^+, \chi'^{,+})$ should be $L \cdot \mathbb{Q}(\pi, \chi^+, \chi'^{,+})$; on the next line, $Gal(\overline{\mathbb{Q}}/\mathcal{K})$ should be $Gal(\overline{\mathbb{Q}}/L)$.

p. 187, line 9: $P^{(r'),*}(\pi_0, \beta)$ should be just $P^{(r'),*}(\pi_0)$

p. 190, (3.23.1): The indices on the β_i (resp. γ_j) should run from 1 to s (resp. 1 to r).

p. 190, line -4: The condition $\mu - \frac{m-1}{2} > \frac{1}{2}$ should be $\mu - \frac{m-v}{2} \geq 0$, translating the condition from 3.20(a) that μ be strictly to the right of the center of symmetry of the L-function.

p. 191, (3.23.2): The second inequality should be $-b_s - \frac{1}{2}(n-2-\alpha(\chi')) \leq -\mu - v$.

p. 191, line 11: The congruence should be $v - \alpha(\chi) \equiv n - \alpha(\chi') - 3$ (mod 2), which implies the relation $v \equiv m$ (mod 2) asserted on the following line.

p. 191, Lemma 3.23.4. $L^{mot}(\mu, \pi', \nu^\iota, St)$ should be $L^{mot}(\mu, \pi', \check{\nu}, St)$.

p. 192, (3.23.5): The last inequality on the first line should be $\geq n - 2p_{r'} - (2k - \alpha(\chi)) + 1$.

p. 193, (3.26.2). The argument should be $\frac{1}{2}(2 - v + m - n - 1) - j$.

p. 194, (3.26.7) On the left hand side, ν^ι should be $\check{\nu}^\iota$. On the right-hand side, $P((\chi^{+,'})^\vee \cdot \nu, \iota)^s$ should be $p((\chi^{+,'})^\vee \cdot \nu, \iota)^s$.

p. 194, (3.26.8) $L^{mot,S}(\frac{1}{2}(v+n),\pi,(\check{\chi}')^+,\nu^\iota,St)$ should be $L^{mot,S}(\frac{1}{2}(n-v+1),\pi,(\check{\chi}')^+,\check{\nu}^\iota,St)$.

0. Preliminary Notation

Let E be a totally real field, $\mathcal{K}$ a totally imaginary quadratic extension of E. Let V be an n-dimensional $\mathcal{K}$-vector space, endowed with a non-degenerate hermitian form $< \bullet, \bullet >_V$, relative to the extension $\mathcal{K}/E$. We let Σ_E, resp. $\Sigma_\mathcal{K}$, denote the set of complex embeddings of E, resp. $\mathcal{K}$, and choose a CM type $\Sigma \subset \Sigma_\mathcal{K}$, i.e. a subset which upon restriction to E is identified with Σ_E. Complex conjugation in $Gal(\mathcal{K}/E)$ is denoted c.

The hermitian pairing $< \bullet, \bullet >_V$ defines an involution $\tilde{c}$ on the algebra $End(V)$ via

$$< a(v), v' >_V = < v, a^{\tilde{c}}(v') >, \qquad (0.1)$$

and this involution extends to $End(V \otimes_\mathbb{Q} R)$ for any $\mathbb{Q}$-algebra R. We define $\mathbb{Q}$-algebraic groups $U(V) = U(V, < \bullet, \bullet >_V)$ and $GU(V) = GU(V, < \bullet, \bullet >_V)$ over $\mathbb{Q}$ such that, for any $\mathbb{Q}$-algebra R,

$$U(V)(R) = \{g \in GL(V \otimes_\mathbb{Q} R) \mid g \cdot \tilde{c}(g) = 1\}; \qquad (0.2)$$

$$GU(V)(R) = \{g \in GL(V \otimes_\mathbb{Q} R) \mid g \cdot \tilde{c}(g) = \nu(g) \text{ for some } \nu(g) \in R^\times\}. \qquad (0.3)$$

Thus $GU(V)$ admits a homomorphism $\nu : GU(V) \to \mathbb{G}_m$ with kernel $U(V)$. There is an algebraic group $U_E(V)$ over E such that $U(V) \xrightarrow{\sim} R_{E/\mathbb{Q}} U_E(V)$, where $R_{E/\mathbb{Q}}$ denotes Weil's restriction of scalars functor. This isomorphism identifies automorphic representations of $U(V)$ and $U_E(V)$.

All constructions relative to hermitian vector spaces carry over without change to skew-hermitian spaces.

The quadratic Hecke character of $\mathbf{A}_E^\times$ corresponding to the extension $\mathcal{K}/E$ is denoted

$$\varepsilon_{\mathcal{K}/E} : \mathbf{A}_E^\times/E^\times N_{\mathcal{K}/E}\mathbf{A}_\mathcal{K}^\times \xrightarrow{\sim} \pm 1.$$

For any hermitian or skew-hermitian space, let

$$GU(V)(\mathbf{A})^+ = \ker \varepsilon_{\mathcal{K}/E} \circ \nu \subset GU(V)(\mathbf{A}). \qquad (0.4)$$

For any place v of E, we let $GU(V)_v^+ = GU(V)(E_v) \cap GU(V)(\mathbf{A})^+$. If v splits in $\mathcal{K}/E$, then $GU(V)_v^+ = GU(V)(E_v)$; otherwise $[GU(V)(E_v) : GU(V)_v^+] = 2$, and $GU(V)_v^+$ is the kernel of the composition of ν with the local norm residue map. We define $GU(V)^+(\mathbf{A}) = \prod_v' GU(V)_v^+$ (restricted direct product), noting the position of the superscript; we have

$$GU(V)(E) \cdot GU(V)^+(\mathbf{A}) = GU(V)(\mathbf{A})^+. \qquad (0.5)$$

1. Eisenstein Series on Unitary Similitude Groups

(1.1) Notation for Eisenstein series. The present section is largely taken from [H3, §3] and [H4, §I.1]. Let E and $\mathcal{K}$ be as in §0. Let $(W, <, >_W)$ be any hermitian space over $\mathcal{K}$ of dimension n. Define $-W$ to be the space W with hermitian form $- <, >_W$, and let $2W = W \oplus (-W)$. Set

$$W^d = \{(v, v) \mid v \in W\}, \qquad W_d = \{(v, -v) \mid v \in W\}$$

These are totally isotropic subspaces of $2W$. Let P (resp. GP) be the stablizer of W^d in $U(2W)$ (resp. $GU(2W)$). As a Levi component of P we take the subgroup $M \subset U(2W)$ which is stablizer of both W^d and W_d. Then $M \simeq GL(W^d) \xrightarrow{\sim} GL(W)$, and we let $p \mapsto A(p)$ denote the corresponding homomorphism $P \to GL(W)$. Similarly, we let $GM \subset GP$ be the stabilizer of both W^d and W_d. Then $A \times \nu : GM \to GL(W) \times \mathbb{G}_m$, with A defined as above, is an isomorphism. There is an obvious embedding

$$i_W : U(W) \times U(W) = U(W) \times U(-W) \hookrightarrow U(2W).$$

We use the same notation for the inclusion $G(U(W) \times U(-W)) \subset GU(2W)$, where as in [H3] $G(U(W) \times U(-W)) = GU(2W) \cap GU(W) \times GU(-W)$, the intersection in $GL(2W)$.

In this section we let $H = U(2W)$, viewed alternatively as an algebraic group over E or, by restriction of scalars, as an algebraic group over $\mathbb{Q}$. We choose a maximal compact subgroup $K_\infty = \prod_{v \in \Sigma_E} K_v \subset H(\mathbb{R})$; specific choices will be determined later. We also let $GH = GU(2W)$.

Let v be any place of E, $|\cdot|_v$ the corresponding absolute value on $\mathbb{Q}_v$, and let

$$\delta_v(p) = |N_{\mathcal{K}/E} \circ \det(A(p))|_v^{\frac{n}{2}} |\nu(p)|^{-\frac{1}{2}n^2}, \quad p \in GP(E_v). \qquad (1.1.1)$$

This is the local modulus character of $GP(E_v)$. The adelic modulus character of $GP(\mathbf{A})$, defined analogously, is denoted $\delta_{\mathbf{A}}$. Let χ be a Hecke character of $\mathcal{K}$. We view χ as a character of $M(\mathbb{A}_E) \xrightarrow{\sim} GL(W^d)$ via composition with det. For any complex number s, define

$$\delta^0_{P,\mathbf{A}}(p, \chi, s) = \chi(\det(A(p))) \cdot |N_{\mathcal{K}/E} \circ \det(A(p))|_v^s |\nu(p)|^{-ns}$$
$$\delta_{\mathbf{A}}(p, \chi, s) = \delta_{\mathbf{A}}(p) \delta^0_{P,\mathbf{A}}(p, \chi, s).$$

The local characters $\delta_{P,v}(\cdot, \chi, s)$ and $\delta^0_{P,v}(\cdot, \chi, s)$ are defined analogously.

Let σ be a real place of E. Then $H(E_\sigma) \xrightarrow{\sim} U(n, n)$, the unitary group of signature (n, n). As in [H97, 3.1], we identify $U(n, n)$, resp. $GU(n, n)$, with the unitary group (resp. the unitary similitude group) of the standard skew-hermitian matrix $\begin{pmatrix} 0 & I_n \\ -I_n & 0 \end{pmatrix}$. Let $K_{n,n} = U(n) \times U(n) \subset U(n,n)$ in the standard embedding, $GK_{n,n} = Z \cdot K_{n,n}$ where Z is the diagonal subgroup of $GU(n,n)$, and let $X_{n,n} = GU(n,n)/GK_{n,n}$, $X_{n,n}^+ = U(n,n)/K_{n,n}$ be the corresponding symmetric spaces. The space $X_{n,n}^+$, which can be realized as a tube domain in the space $M(n,\mathbb{C})$ of complex $n \times n$-matrices, is naturally a connected component of $X_{n,n}$; more precisely, the identity component $GU(n,n)^+$ stabilizes $X_{n,n}^+$ and identifies it with $GU(n,n)^+/GK_{n,n}$. Writing $g \in GU(n,n)$ in block matrix form

$$g = \begin{pmatrix} A & B \\ C & D \end{pmatrix}$$

with respect to bases of W_σ^d and $W_{d,\sigma}$, we identify GP with the set of $g \in GU(n,n)$ for which the block $C = 0$. In the tube domain realization, the canonical automorphy factor associated to GP and $GK_{n,n}$ is given as follows: if $\tau \in X_{n,n}$ and $g \in GU(n,n)^+$, then the triple

$$J(g,\tau) = C\tau + D, \ J'(g,\tau) = \bar{C}^t\tau + \bar{D}, \nu(g) \qquad (1.1.2)$$

defines a canonical automorphy factor with values in $GL(n,\mathbb{C}) \times GL(n,\mathbb{C}) \times GL(1,\mathbb{R})$ (note the misprint in [H97, 3.3]).

Let $\tau_0 \in X_{n,n}^+$ denote the unique fixed point of the subgroup GK_∞ and write $J(g) = J(g,\tau_0)$. Given a pair of integers (m,κ), we define a complex valued function on $GU(n,n)^+$:

$$J_{m,\kappa}(g) = \det J(g)^{-m} \cdot \det(J'(g))^{-m-\kappa} \cdot \nu(g)^{n(m+\kappa)} \qquad (1.1.3)$$

More generally, let GH^+ denote the identity component of $GH(\mathbb{R})$, and define $\mathbf{J}_{m,\kappa} \to GH^+ \to \mathbb{C}^\times$ by

$$\mathbf{J}_{m,\kappa}((g_\sigma)_{\sigma \in \Sigma_E}) = \prod_{\sigma \in \Sigma_E} J_{m,\kappa}(g_\sigma) \qquad (1.1.4)$$

We can also let m and κ denote integer valued functions on σ and define analogous automorphy factors. The subsequent theory remains valid provided the value $2m(\sigma) + \kappa(\sigma)$ is independent of σ. However, we will only treat the simpler case here.

(1.2) Formulas for the Eisenstein series. Consider the induced representation

$$I_n(s,\chi) = \mathrm{Ind}(\delta_{P,\mathbf{A}}^0(p,\chi,s)) \xrightarrow{\sim} \otimes_v I_{n,v}(\delta_{P,v}^0(p,\chi,s)), \qquad (1.2.1)$$

the induction being normalized; the local factors I_v, as v runs over places of E, are likewise defined by normalized induction. Explicitly,

$$I_n(s, \chi) = \{f : H(\mathbf{A}) \to \mathbb{C} \mid f(pg) = \delta_{P, \mathbf{A}}(p, \chi, s) f(g), p \in P(\mathbf{A}), g \in H(\mathbf{A})\}. \tag{1.2.2}$$

At archimedean places we assume our sections to be K_∞-finite. For a section $\phi(h, s; \chi) \in I_n(s, \chi)$ (cf. [H99, I.1]) we form the Eisenstein series

$$E(h, s; \phi, \chi) = \sum_{\gamma \in P(E) \backslash U(2V)(E)} \phi(\gamma h, s; \chi) \tag{1.2.3}$$

If χ is unitary, this series is absolutely convergent for $\mathrm{Re}(s) > \frac{n-1}{2}$, and it can be continued to a meromorphic function on the entire plane. *Assume henceforward* that

$$\chi \mid_{\mathbf{A}} = \varepsilon_{\mathcal{K}}^m \tag{1.2.4}$$

for a fixed positive integer m. Usually, but not invariably, we will assume $m \geq n$. The main result of [Tan] states that the possible poles of $E(h, s; \phi, \chi)$ are all simple, and can only occur at the points in the set

$$\frac{n - \delta - 2r}{2}, \ r = 0, \ldots, [\frac{n - \delta - 1}{2}], \tag{1.2.5}$$

where $\delta = 0$ if m is even and $\delta = 1$ if m is odd. We will be concerned with the residues of $E(h, s_0; \phi, \chi)$ for s_0 in the set indicated in (1.2.5), and with the values when the Eisenstein series is holomorphic at s_0.

We write $I_n(s, \chi) = I_n(s, \chi)_\infty \otimes I_n(s, \chi)_f$, the factorization over the infinite and finite primes, respectively.

We follow [H97, 3.3] and suppose the character χ has the property that

$$\chi_\sigma(z) = z^\kappa, \chi_{c\sigma}(z) = 1, \forall \sigma \in \Sigma \tag{1.2.6}$$

Then the function $\mathbf{J}_{m,\kappa}$, defined above, belongs to $I_n(m - \frac{n}{2}, \chi)_\infty$ (cf. [H97,(3.3.1)]). More generally, let

$$\mathbf{J}_{m,\kappa}(g, s + m - \frac{n}{2}) = \mathbf{J}_{m,\kappa}(g) |\det(J(g) \cdot J'(g))|^{-s} \in I_n(s, \chi)_\infty. \tag{1.2.7}$$

When $E = \mathbb{Q}$, these formulas just reduce to the formulas in [H97].

(1.3) Zeta integrals. We recall the discussion of the doubling method, applied to unitary groups; details are in [H3, (3.2)] and [H4, I.1]. Let π, π' be cuspidal automorphic representations of $GU(W) = GU(-W)$, $\phi \in \pi$, $\phi' \in \pi'$. Let $G = G(U(W) \times U(-W))$, as above, and define the Piatetski-

Shapiro-Rallis zeta integral

$$Z^+(s, \phi, \phi', \chi, \varphi)$$

$$= \int_{Z(\mathbf{A}) \cdot G(\mathbb{Q}) \backslash G(\mathbf{A})^+} E(i_W(g, g'), s, \varphi, \chi) \phi(g) \phi'(g') \chi^{-1}(g') dg dg' \quad (1.3.1)$$

with $G(\mathbf{A})^+$ as in (0.5) and the Eisenstein series as above. Let $\delta \in GU(2W)$ be a representative of the dense orbit of G on $GP \backslash GU(2W)$, cf. [KR1, $(7.2.6)$] for the analogous case of symplectic groups. The integral $(1.3.1)$ is Eulerian; if $\phi = \otimes_v \phi_v$ and $\phi' = \otimes_v \phi'_v$ are factorizable vectors then for some finite set S containing archimedean places we have

$$d_n^S(\chi, s) Z^+(s, \phi, \phi', \chi, \varphi) = \begin{cases} 0, \pi' \not\simeq \pi^\vee; \\ L^S(s + \dfrac{1}{2}, \pi, St, \chi) Z_S(s, \phi, \phi', \chi, \varphi), \end{cases} \quad (1.3.2)$$

where

$$Z_S(s, \phi, \phi', \chi, \varphi) = \int_{\prod_{v \in S} U(W_v)} (\pi(g)\phi, \phi') \chi_v^{-1}(det(g)) \cdot \varphi_v(\delta \cdot (g, 1), s) dg. \quad (1.3.3)$$

and

$$d_n^S(\chi, s) = \prod_{r=0}^{n-1} L^S(2s + n - r, \varepsilon_{\mathcal{K}}^{n-1+r}) \quad (1.3.4)$$

is a product of abelian L-functions of E with the factors at S removed. The pairing inside the integral $(1.3.3)$ is the L_2 pairing on $GU(W)(E) \backslash GU(W)(\mathbf{A})^+ / Z(\mathbf{A})$, where Z is the center of $GU(W)$. As in [H3, $(3.2.4)$, $(3.2.5)$], the factor $(1.3.3)$ can be rewritten as a product of local zeta integrals, multiplied by a global inner product. Finally, the partial L-function $L^S(s + \frac{1}{2}, \pi, St, \chi)$ is defined as in [H3, (2.7)].

Lemma 1.3.5. *The zeta integral* $Z^+(s, \phi, \phi', \chi, \varphi)$ *has at most simple poles on the set $(1.2.5)$. In particular, if $L^S(s + \frac{1}{2}, \pi, St, \chi)$ has a pole at $s = s_0$ then s_0 is in the set $(1.2.5)$, $Z_S(s, \phi, \phi', \chi, \varphi)$ is holomorphic at $s = s_0$, and*

$$res_{s=s_0} Z^+(s, \phi, \phi', \chi, \varphi)$$

$$= d_n^S(\chi, s_0)^{-1} Z_S(s_0, \phi, \phi', \chi, \varphi) \cdot res_{s=s_0} L^S(s + \frac{1}{2}, \pi, St, \chi). \quad (1.3.6)$$

Proof. The first assertion follows from the theorem of Tan quoted in (1.2). The second assertion is then an obvious consequence of the fact that all the s_0 in $(1.2.5)$ are positive and therefore in the range of absolute convergence of the Euler product $d_n^S(s)$. $\qquad \square$

The following lemma of Lapid and Rallis will be used specifically in the discussion of archimedean zeta integrals:

Lemma 1.3.6 [LR]. *Assume π is tempered. Then the zeta integral (1.3.3) converges absolutely for Re $s \geq 0$.*

Proof. If π is square integrable this is Lemma 3 of [LR]. In the general case this follows from multiplicativity of zeta integrals [LR, Proposition 2]. □

2. The Local Theta Correspondence

(2.1) Review of local theta dichotomy. In this section F is a non-archimedean local field of characteristic zero, K a quadratic extension of F, $\varepsilon_{K/F}$ the associated quadratic character. We fix an additive character $\psi : F \to \mathbb{C}^\times$. We will usually, but not always, assume the residue characteristic of F different from 2. Let W be an n-dimensional skew-hermitian space over K, relative to the extension K/F, with isometry group $G = U(W)$. Let m be a positive integer. In this section we recall the outlines of the theory [KS, HKS] of the local theta correspondence from irreducible admissible representations of G to representations of $U(V)$, where $(V, (,)_V)$ is a variable hermitian space over K of dimension m.

As in [KS] and [HKS] we choose a splitting character χ of $K^\times$ satisfying the local version of (1.2.4):

$$\chi \mid_{F^\times} = \varepsilon_{K/F}^m.$$

We also consider the doubled skew-hermitian space $2W$, as in §1. Multiplying the skew-hermitian form on W, or on $2W$, by an element $\delta \in K$ with trace zero to E transforms it into a hermitian form, without changing the isometry group. Up to isometry, the skew-hermitian space $2W$ is independent of the choice of W. The degenerate principal series $I_n(s, \chi)$ of $U(2W)$ is defined in the local setting by analogy with the global definition (1.2.2).

Define

$$\epsilon(V) = \varepsilon_{K/F}((-1)^{\frac{m(m-1)}{2}} \det V) \in \{\pm 1\} \qquad (2.1.1)$$

where $\det V = \det((x_i, x_j)_V)$, for any K-basis $\{x_1, \ldots, x_m\}$ is a well defined element of $F^\times/N_{K/F}K^\times$. Then V is determined up to isometry by the sign $\epsilon(V)$, and we let V_m^+ and V_m^- denote the corresponding pair of hermitian spaces, up to isometry.

Let $s_0 = \frac{m-n}{2}$. We refer to §4 of [HKS] for the following facts, which are proved in [KS]. Let $\mathcal{S}(V^n)$ denote the Schwartz-Bruhat space of locally constant compactly supported functions on V^n. There is a map

$$p_{V,n} : \mathcal{S}(V^n) \to I_n(s_0, \chi); \Phi \mapsto \phi_\Phi \qquad (2.1.2)$$

whose definition is recalled below, with image denoted $R_n(V, \chi)$. The following results are proved in [KS], and summarized as Proposition 4.1 of [HKS]:

Proposition (2.1.3).

(i) *If $m \le n$, then $R_n(V, \chi)$ is irreducible, and*

$$R_n(V_m^+, \chi) \oplus R_n(V_m^-, \chi)$$

is the maximal completely reducible submodule of $I_n(s_0, \chi)$.

(ii) *If $m = n$ then*

$$I_n(0, \chi) = R_n(V_n^+, \chi) \oplus R_n(V_n^-, \chi).$$

(iii) *If $n < m \le 2n$, then*

$$I_n(s_0, \chi) = R_n(V_m^+, \chi) + R_n(V_m^-, \chi)$$

and

$$Soc_{n,m}(\chi) = R_n(V_m^+, \chi) \cap R_n(V_m^-, \chi)$$

is the unique irreducible submodule of $I_n(s_0, \chi)$. Moreover, if $m < 2n$ there is a natural isomorphism

$$R_n(V_m^\pm, \chi)/Soc_{n,m}(\chi) \xrightarrow{\sim} R_n(V_{2n-m}^\pm, \chi).$$

If $m = 2n$ then $I_n(s_0, \chi) = R_n(V_{2n}^+, \chi)$ and $I_n(s_0, \chi)/R_n(V_{2n}^-, \chi)$ is the one-dimensional representation $\chi \circ \det$.

(iv) *If $m > 2n$, then $I_n(s_0, \chi) = R_n(V_m^\pm, \chi)$ is irreducible.*

The choice of splitting character χ, together with the additive character ψ, defines a Weil representation $\omega_{V,W,\chi}$ of the dual reductive pair $U(W) \times U(V)$ on an appropriate Schwartz-Bruhat space $\mathcal{S}_{V,W}$. For details, see [HKS]. When W is replaced by $2W$ (resp. V by $2V$) one obtains actions $\omega_{V,2W,\chi}$ of $U(2W) \times U(V)$ (resp. $\omega_{2V,W,\chi}$ of $U(W) \times U(2V)$) on the space $\mathcal{S}(V^n)$ introduced in (2.1.2) above, and the map $p_{V,n}$ of (2.1.2) is defined by

$$\phi_\Phi(g) = p_{V,n}(\Phi)(g) = (\omega_{V,2W,\chi}(g)\Phi)(0), g \in U(2W).$$

These actions can be chosen in such a way that

$$Res_{G \times G \times U(V)^\Delta}^{G \times U(V) \times G \times U(V)}(\omega_{V,W,\chi} \otimes \omega_{V,W,\chi}^\vee) = Res_{G \times G \times U(V)}^{U(2W) \times U(V)} \omega_{V,2W,\chi} \qquad (2.1.4)$$

(cf. [HKS,(3.5)], or [H4,I.5]). Here $U(V)^\Delta$ is the diagonal subgroup in $U(V) \times U(V)$.

2.1.5. The Conservation Relation. Let π be an irreducible admissible representation of G. We define the representation $\Theta_\chi(V, \pi)$ of $U(V)$ by

$$\Theta_\chi(V, \pi) = [\omega_{V,W,\chi} \otimes \pi]_G = Hom_G(\omega^\vee_{V,W,\chi}, \pi). \qquad (2.1.5.1)$$

Note that this is the full theta correspondence, which often gives a result larger than the Howe correspondence.

Let $m^?_\chi(\pi)$ denote the minimum integer μ such that $\Theta_\chi(\pi, V^?_\mu) \neq 0$. The analogue for unitary dual reductive pairs of the Conservation Relation, stated as Conjecture 3.6 in [KR2], is

$$m^+_\chi(\pi) + m^-_\chi(\pi) = 2n + 2. \qquad (2.1.5.2)$$

For orthogonal-symplectic pairs Kudla and Rallis prove the analogue of the weaker relation

$$m^+_\chi(\pi) + m^-_\chi(\pi) \geq 2n + 2 \qquad (2.1.5.3)$$

as Theorem 3.8.

The proof of (2.1.5.3) follows readily from the special case in which π is the trivial representation, denoted **1**:

$$m^+_\chi(\mathbf{1}) = 0, m^-_\chi(\mathbf{1}) = 2n + 2. \qquad (2.1.5.4)$$

The analogue of (2.1.5.4) is Lemma 4.2 of [KR2]. In what follows, we will admit here that the proof of this lemma, and the derivation from (2.1.5.4) of (2.1.5.3), work in the unitary case as in [KR2]. This generalization is the subject of work in progress of Z. Gong.

2.1.6. Generic behavior of the theta correspondence. Recall from [PSR, §2] the description of the set of $G \times G$-orbits on the projective variety $F(W) = P \backslash U(2W)$. The flag variety $F(W)$ is naturally the space of totally isotropic n-spaces in $2W$ for the natural hermitian form. As in [*loc. cit*], the $G \times G$-orbit of an isotropic n-plane L is completely determined by the invariant

$$d(L) = \dim(L \cap W) = \dim(L \cap (-W)).$$

Thus there are $r_0 + 1$ distinct orbits in $F(W)$, where $r_0 = [\frac{n}{2}]$ is the Witt index of W. The open orbit consists of L with $d(L) = 0$. Let $L \in F(W)$ and suppose $d(L) = r > 0$, and let $St_r = Stab(L) \subset G \times G$. If $\delta_r \in U(2W)$ is a representative of L in $F(W)$, then $\delta_r(St_r)\delta_r^{-1} \subset P$. As in [HKS, (4.10)], define a character $\xi_{r,s} : St_r \to \mathbb{C}^\times$, for any $s \in \mathbb{C}$, by

$$\xi_{r,s}(\gamma) = \chi(A(\delta_r(\gamma)\delta_r^{-1})) \cdot |\det(A(\delta_r(\gamma)\delta_r^{-1}))|^{s+\rho_n}.$$

Let $Q^r(s, \chi) = Ind^{G \times G}_{St_r} \xi_{r,s}$.

Definition 2.1.6.1. *For any $r > 0$, the irreducible representation π of G occurs in the rth boundary stratum at the point $s = s_0$ (for χ) if $Hom_{G \times G}(Q^r(s_0, \chi), \pi \otimes \chi\pi^\vee) \neq 0$. We say π does not occur in the boundary at the point $s = s_0$ (for χ) if it does not occur in the rth boundary stratum for any $r > 0$.*

We consider the principal series of G as an infinite union $\mathcal{P}$ of complex affine spaces, as in Bernstein's parametrization, of dimension equal to the split rank of G; the irreducible components of a given principal series representation are represented by the same point in $\mathcal{P}$. More precisely, let $P_0 \subset G$ be a minimal parabolic subgroup, $P_0 = M_0 \cdot A_0 \cdot N_0$, with A_0 a product of copies of $K^\times$, M_0 anisotropic semisimple, and N_0 unipotent. Let r be the split rank of G, so that $A_0 \overset{\sim}{\longrightarrow} K^{\times, r}$. Let $U(1)$ denote the kernel of the norm map from $K^\times$ to $F^\times$, viewed indifferently as a p-adic Lie group or as an anisotropic torus over F, and let $U(2)$ be the unitary group of the unique anistropic hermitian space of dimension 2 over K. We have $M_0 \overset{\sim}{\longrightarrow} 1$ if $n = 2r$ is even and G is quasi-split; otherwise $M_0 \overset{\sim}{\longrightarrow} U(n - 2r)$, where $2r = n - 1$ if n is odd, $2r = n - 2$ if n is even.

To any character $\alpha : A_0 \cdot M_0 \to \mathbb{C}^\times$ we associate the principal series

$$I(\alpha) = Ind_{P_0}^G \alpha$$

(normalized induction). These representations are generally not irreducible, and we say an irreducible admissible representation of G or G' *belongs to the principal series* if it is an irreducible constituent of some principal series representation. When $n = 2r$, so G is split, we write $\alpha = (\alpha_1, \ldots, \alpha_r)$ is an r-tuple of characters of $K^\times$; otherwise, $\alpha = (\alpha_1, \ldots, \alpha_r; \gamma)$ where the α_i are characters of $K^\times$ whereas γ is a character of the anisotropic $U(n - 2r)$. The connected components of $\mathcal{P}$ are indexed by the inertial class of α, i.e. the restrictions of the α_i to the units in $K^\times$, together with the ramified character γ when $n - 2r > 0$. The connected component corresponding to a given inertial class can be parametrized by the r-non-zero complex numbers which are values of the α_i on a fixed uniformizing parameter of K, say ϖ_K. Each connected component is thus identified with an r-tuple of elements of $\mathbb{C}^\times$, but the whole is only well-defined up to permutation of the α_i. By "almost all" principal series τ of G we mean a subset $\mathcal{P}^0$ of $\mathcal{P}$ containing an open dense subset of every inertial class. Some r-tuples correspond to reducible principal series, and therefore to several distinct irreducible representations; however almost all principal series, in the above sense, are irreducible.

Lemma 2.1.6.2. *(i) For every connected component $\mathcal{P}_\alpha$ of $\mathcal{P}$, the set of representations π of G in $\mathcal{P}_\alpha$ occurring in the boundary at any point $s = s_0$ for any fixed splitting character χ is a proper closed subset; i.e. almost all π in $\mathcal{P}$ do not occur in the boundary.*

(ii) For any irreducible principal series representation π and any fixed s_0, the set of splitting characters χ such that π occurs in the boundary at s_0 for χ is finite.

(iii) For $r > 1$, any π occurring in the rth boundary stratum is non-tempered.

Proof. The standard maximal parabolic subgroups P_r of G, $r = 0, \ldots, r_0$, are the stabilizers of isotropic subspaces of dimension r. The Levi factor L_r of P_r is isomorphic to $GL(r, K) \times U(n - 2r)$, where $U(n - 2r)$ is the unitary group of a vector space of dimension $n - 2r$ in the same Witt class as W. We claim that, for an appropriate choice of δ_r, St_r contains $GL(r, K) \times GL(r, K) \times \Delta(U(n - 2r)) \subset L_r \times L_r \subset G \times G$, where the first two factors are embedded in the obvious way and the factor $\Delta(U(n - 2r))$ is the diagonal in $U(n - 2r) \times U(n - 2r)$. Indeed, this is verified as in [KR2], §1.

Let N_r denote the unipotent radical of P_r, and let the subscript N_r denote the unnormalized Jacquet module. As in [KR2], Lemma 1.5, it follows from the above description of St_r that, if π occurs in the boundary at the point s_0 then there is an integer $r \in \{1, r_0\}$ and a character ξ'_{r,s_0} of $GL(r, K)$, derived by restriction from ξ_{r,s_0}, such that

$$Hom_{GL(r,K)}(\pi_{N_r}, \xi'_{r,s_0}) \neq 0. \tag{2.1.6.3}$$

For fixed s_0 and any r, this latter condition defines a proper closed subset of any irreducible component of $\mathcal{P}$. The remaining assertions follow immediately from the characterization (2.1.6.3). $\qquad\square$

The following theorem is joint with Steve Kudla. I recall that we are assuming the validity of the weak conservation relation (2.1.5.3) in the unitary case.

Theorem (2.1.7). *(i) For any π,*

$$Hom_{G \times G}(R_n(V, \chi), \pi \otimes (\chi \cdot \pi^\vee)) \neq 0 \Rightarrow \Theta_\chi(V, \pi) \neq 0.$$

(i') Assume the residue characteristic of F is different from 2, or π is supercuspidal, or more generally that $\Theta_\chi(V, \pi)$ admits an irreducible quotient. Then the implication in (i) is an equivalence.

(ii) Assume that π does not occur in the boundary at $s_0 = \frac{m-n}{2}$ for χ, or that $\Theta_\chi(V,\pi)$ is a non-zero irreducible representation. Then

$$d_\chi(V,\pi) = \dim Hom_{G\times G}(R_n(V,\chi), \pi \otimes (\chi \cdot \pi^\vee)) = 1.$$

(iii) If π is any admissible irreducible representation of $G = U(W)$ and $m \geq n$, then there exists V with $\dim V = m$ such that

$$\Theta_\chi(\pi, V) \neq 0 \text{ and } Hom_{G\times G}(R_n(V,\chi), \pi \otimes (\chi \cdot \pi^\vee)) \neq 0.$$

*(iv) (Theta dichotomy) With notation as in (iii), if $m \leq n$, then there is at most one such V. In particular, if $m = n$, then there is **exactly** one V of dimension n such that $\Theta_\chi(\pi, V) \neq 0$, determined by the following root number criterion:*

$$\Theta_\chi(\pi, V) \neq 0 \Leftrightarrow \varepsilon(\frac{1}{2}, \pi, \chi, \psi) = \varepsilon_{K/F}(-2)^n \varepsilon_{K/F}(\det V), \qquad (2.1.7.1)$$

where the signs ε are defined as in [HKS] §6.

Proof. Assertion (i') is [HKS, Prop. 3.1], and (iii) is contained in [HKS,Cor. 4.5 and Cor. 4.4]. Only the first part of (iii) is asserted in [loc. cit.] but the proofs are based on the second assertion. On the other hand, the proof in [loc. cit.] of the implication in (i) does not make use of the special hypotheses in (i'). If π does not occur in the boundary at s_0 for χ, then (ii) is Theorem 4.3 (ii) of [HKS]. If $\Theta_\chi(V,\pi)$ is irreducible (i) is an equivalence, by (i'). In particular, $d_\chi(V,\pi) \geq 1$ and we have to show that it is also ≤ 1 if $\Theta_\chi(V,\pi)$ is irreducible. But the dimensions of the spaces in the last three lines of (3.5) in [HKS] are all equal, in particular

$$d_\chi(V,\pi) = \dim Hom_{G\times G\times U(V)^\Delta}(\omega_{V,W,\chi} \otimes \omega_{V,W,\chi}^\vee, \pi \otimes \pi^\vee \otimes Triv)$$

where $Triv$ denotes the trivial representation of $U(V)$. The space on the right-hand side of the last formula is just

$$Hom_{U(V)^\Delta}(\Theta_\chi(V,\pi) \otimes \Theta_\chi(V,\pi^\vee), Triv)$$
$$= Hom_{U(V)^\Delta}(\Theta_\chi(V,\pi) \otimes \Theta_\chi(V,\pi)^\vee, Triv)$$
$$= Hom_{U(V)}(\Theta_\chi(V,\pi), \Theta_\chi(V,\pi))$$

where the first equality follows as in [loc. cit.] from a theorem of Mœglin-Vignéras-Waldspurger. Assertion (ii) thus follows from the irreducibility hypothesis.

Finally, when $m \leq n$, $2m < 2n + 2$, so the first part of (iv) is an immediate consequence of (2.1.5.3), and the second part then follows from (iii). The final assertion is Theorem 6.1 (ii) of [HKS], bearing in mind our sign conventions. □

Let m be as in the preceding proposition. We say π is *unambiguous* (relative to m and χ) if there is at most one V of dimension m such that $\Theta_\chi(V, \pi) \neq 0$, and *ambiguous* otherwise. For $m \leq n$, (2.1.7)(iv) asserts that every π is unambiguous. In contrast, (2.1.7)(iii) guarantees that there is always at least one V of dimension m such that $\Theta_\chi(V, \pi) \neq 0$ when $m > n$, and when $m > 2n$ (the so-called stable range) (2.1.3) (iv) implies the well-known result that any π is ambiguous. The following proposition indicates that ambiguity is the rule rather than the exception for $n < m \leq 2n$.

Proposition (2.1.8). *Let π be an irreducible admissible representation of $G = U(W)$. Let $n < m < 2n$, and assume π is unambiguous for m (and χ). Let $V_m^?$ be such that $\Theta_\chi(\pi, V_m^?) \neq 0$. Then $\Theta_\chi(\pi, V_{2n-m}^?) \neq 0$.*

Conversely, if $\Theta_\chi(\pi, V_{2n-m}^?) \neq 0$ for some choice of $? \in \{\pm 1\}$, then π is unambiguous for m, i.e. $\Theta_\chi(\pi, V_m^{-?}) = 0$.

Proof. We first note that the first assertion follows if we can construct a non-zero $G \times G$-equivariant pairing

$$I_n(s_0, \chi) \otimes [\pi \otimes (\chi \cdot \pi^\vee)]^\vee \to \mathbb{C}. \tag{2.1.8.1}$$

Indeed, such a pairing restricts to pair of $G \times G$-equivariant homomorphisms

$$\lambda^\pm : R_n(V_m^\pm, \chi) \to \pi \otimes (\chi \cdot \pi^\vee).$$

Since π is unambiguous $\lambda^\pm \neq 0$ if and only if $\pm = ?$. In particular, $f^{-?} = 0$. Since $\lambda^?$ and $\lambda^{-?}$ have the same restriction to $Soc_{n,m}(\chi)$, it follows that $\lambda^?$ factors through

$$\lambda^* \in Hom_{G \times G}(R_n(V_m^?, \chi)/Soc_{n,m}(\chi), \pi \otimes (\chi \cdot \pi^\vee))$$

$$= Hom_{G \times G}(R_n(V_{2n-m}^?, \chi), \pi \otimes (\chi \cdot \pi^\vee))$$

by (2.1.3)(iii). Then (2.1.7) (i) implies that $\Theta_\chi(\pi, V_{2n-m}^?) \neq 0$.

It thus suffices to observe [HKS, (6.27)] that the normalized zeta integral defines the required non-zero $G \times G$-equivariant pairing. For future reference, we recall the definition of the zeta integral:

$$Z(s, \varphi, \varphi', f, \chi) = \int_G f((g, 1); \chi, s)(\pi(g)\varphi, \varphi') dg$$

where $f \in I_n(s, \chi)$ is restricted as above to $G \times G$, $\varphi \in \pi$, and $\varphi' \in \chi \cdot \pi^\vee$. The normalized zeta integral is

$$\tilde{Z}(s, \varphi, \varphi', f, \chi) = d_n(s, \chi) L(s, \pi, St, \chi)^{-1} Z(s, \varphi, \varphi', f, \chi)$$

where $L(s, \pi, St, \chi)$ is the standard local L-function of π, twisted by χ (cf. [HKS,(6.20)]) and $d_n(s, \chi)$ is the normalizing factor $\prod_{r=0}^{n-1} L(2s +$

$n - r, \varepsilon_{\mathcal{K}}^{n-1+r}$), where $L(s, \bullet)$ is the local abelian Euler factor. The factor $d_n(s, \chi)$ can be ignored since it has no poles at the point s_0.

The converse follows easily from (2.1.5.3), as in [KR2]: if, say $m_\chi^+(\pi) \leq 2n - m$, then $m_\chi^- \geq m + 2$, and so the only non-vanishing lift is to V_m^+; likewise with $+$ and $-$ exchanged. □

Corollary (2.1.9). *Let π be an irreducible admissible representation of $G = U(W)$. Suppose π is ambiguous for $m > n$ and χ. Assume either that π does not occur in the boundary for $s_0 = \frac{m-n}{2}$ and χ or that $\Theta_\chi(\pi, V_m^\pm)$ is an irreducible representation of $U(V_m^\pm)$, e.g. that π is supercuspidal. Then the normalized zeta integral defines a non-zero pairing between $Soc_{n,m}(\chi)$ and $\pi \otimes (\chi \cdot \pi^\vee)$.*

Proof. The normalized zeta integral restricts to a non-zero pairing in

$$Hom_{G \times G}(R_n(V_m^?, \chi), \pi \otimes (\chi \cdot \pi^\vee))$$

for at least one ? $\in \{\pm\}$. By the multiplicity one statement in (2.1.7)(ii), this is the unique pairing for this sign. Hence if its restriction to the socle is trivial, the pairing defines a theta lift to the corresponding $V_{2n-m}^?$. But this implies π is unambiguous, which contradicts the hypothesis. □

For future reference, we describe the Galois equivariance of the spaces $R_n(V, \chi)$. Note that (1.2.4) implies that χ is a continuous character of a compact profinite group, hence takes values in $\overline{\mathbb{Q}}$ (even $\mathbb{Q}^{ab}$). The induced representation $I_n(s_0, \chi)$ in principle has a natural model, as a space of functions on $U(2W)$, defined over $\mathbb{Q}(\chi, \sqrt{p})$, where p is the residue characteristic of F; the square root may occur because of the appearance of square roots of the norm in the modulus character. We assume s_0 is chosen so that in fact these square roots only occur to even powers; this will always be the case in the applications (cf. the proof of Theorem 4.3, below). In what follows we write $R_{n,\psi}(V_m^\pm, \chi)$ for $R_n(V_m^\pm, \chi)$, including the additive character in the index. Then we have

Lemma (2.1.10). *For any $\sigma \in Gal(\overline{\mathbb{Q}}/\mathbb{Q})$,*

$$\sigma(R_{n,\psi}(V_m^\pm, \chi)) = R_{n,\sigma(\psi)}(V_m^\pm, \sigma(\chi)) \subset I_n(s_0, \sigma(\chi)).$$

Proof. The oscillator representation, split as a representation of $U(2W) \times U(V)$ is defined over $\mathbb{Q}^{ab}(\chi) = \mathbb{Q}^{ab}$. Indeed, it has a Schrödinger model over

$\mathbb{Q}^{ab}$ consisting of Schwartz-Bruhat functions with values in $\mathbb{Q}^{ab}$. The assertion follows from the explicit formulas [K] for the action on the Schrödinger model.   □

(2.2) Theta lifts of principal series. In this section K/F is an unramified quadratic extension and V and W are maximally isotropic, and $n = \dim W > 1$, $m = \dim V$. We write $G = U(W)$, as above, and $G' = U(V)$. Let $P_0 \subset G$ and $P_0' \subset G'$ be minimal parabolic subgroups, $P_0 = M_0 \cdot A_0 \cdot N_0$, $P_0' = A_0' \cdot M_0' \cdot N_0'$, with A_0, M_0, A_0', M_0' tori as in (2.1.6) and N_0, N_0' unipotent. Let $r = [\frac{n}{2}]$, so that r is the split rank over F of the torus A_0.

For $t = 1, 2, \ldots, r$ let $Q_t \subset G$ be the standard maximal parabolic stabilizing an isotropic t-plane in W. The Levi component M_t of Q_t can be identified with $GL(t, K) \times G_{n-2t}$ where $G_{n-2t} = U(W_{n-2t})$ is the unitary group of the space W_{n-2t} of dimension $n - 2t$ in the same Witt class (i.e., maximally isotropic) as W. We define $Q_t' \subset G'$ analogously, for $1 \leq t \leq [\frac{m}{2}]$. We let r_t denote the normalized Jacquet functor from representations of G to representations of M_t. For any irreducible admissible representation π of G we let $t(\pi)$ denote the maximum t such that $r_t(\pi)$ has a non-zero quotient of the form $\sigma_t \otimes \pi_t$ with σ_t a supercuspidal representation of $GL(t, K)$ and π_t a representation of G_{n-2t} ([MVW], p. 69). We also define $u(\pi)$ to be the maximum u such that $r_u(\pi)$ has a non-zero quotient of the form $\sigma_t \otimes \pi_t$ with π_t a supercuspidal representation of G_{n-2t}.

Given a character $\alpha : A_0 \cdot M_0 \to \mathbb{C}^\times$ we define the principal series $I(\alpha)$, $I(\alpha)'$ of G, G' respectively, as in (2.1.6). If π belongs to the principal series then $t(\pi) = 1$ and $u(\pi) = r$.

We choose a second splitting character χ' of $K^\times$ such that

$$\chi' \mid_{F^\times} = \varepsilon_{K/F}^n.$$

Our objective is to determine when a principal series representation of G is in the image of the theta lift from G' for some pair (χ, χ') of splitting characters. By definition when $m = 0$ G' is the trivial group and the unique theta lift to G from G' is the trivial representation. The first observation is

Lemma (2.2.1). *Suppose $b = n - m \geq 2$ and π is an irreducible admissible representation of G that belongs to the tempered principal series. Then π cannot be a theta lift from G'.*

Proof. When $n = 2$ this is true by definition, and when $n = 3$ this is due to Gelbart and Rogawski [GR]. We may thus assume $n \geq 4$, so $r \geq 2$. In general the lemma is a consequence of Kudla's induction principle [K]. This is worked out for unitary groups by Moeglin, Vignéras, and Waldspurger in [MVW, Chapter 3, IV.4], though the result there is less complete than in the cases treated in [K]. Specifically, suppose π is a theta lift of π' for some irreducible admissible representation π' of G'. Since π belongs to the principal series, the index $t(\pi)$ defined above equals 1. Then the main theorem on p. 69 of [MVW] calculates the exponents of the normalized Jacquet module $r_1(\pi)$ of π relative to the maximal parabolic P_1 of G with Levi component $K^\times \times G_{n-2}$, where G_{n-2} is an isotropic unitary group of dimension $n - 2$. In the remainder of the proof, we follow the conventions of [MVW] and consider the theta correspondence on covering groups rather than on the groups themselves using the splitting characters. Since the splitting characters are unitary they are irrelevant to determination of whether or not a representation is tempered.

Say the Jordan-Hölder constituents of $r_1(\pi)$ are of the form $\sigma_1^j \otimes \pi_{n-2}^j$ where σ_1^j are characters of $K^\times$ and π_{n-2}^j are irreducible representations of G_{n-2}. (Note that in [MVW] the groups are denoted by their Witt indices rather than by the dimensions of the corresponding hermitian spaces.) We also let P_1' denote the maximal parabolic subgroup of G' with Levi component $K^\times \times G_{m-2}'$ where G_{m-2}' is a unitary group of dimension $m - 2$ in the same Witt class as G'.

There are two choices. In case 2 (b)(i) every σ_1^j equals $| \bullet |^{\frac{b-1}{2}}$, and by the usual characterization of tempered representations by exponents this is impossible unless $b = 1$ (in particular, case 2 (b) cannot even happen when $n = m$). Indeed, when π is contained in a tempered principal series every σ_1^j and π_{n-2}^j is again tempered. In case 2 (b)(ii) π' is a quotient of the induced representation $I_{P_1'}^{G'} \sigma_1^{j,*} \times \pi_{m-2}'$ for some j where π_{m-2}' is a representation of G_{m-2}', such that π_{n-2} is a theta lift from G_{m-2} (of π_{m-2}', in fact). In this case we conclude by induction on m. (For conventions when $n \leq 1$ see the discussion following (2.2.4) below.) $\square$

Corollary (2.2.2). *Suppose π is an irreducible tempered principal series constituent. Then π is ambiguous relative to every $m > n+1$ (and every χ).*

Proof. Let $m > n + 1$ and fix χ. If $m > 2n$ then there is nothing to prove, by Proposition 2.1.3.(iv). If π were unambiguous for m and χ then π would admit a theta lift to some V_{2n-m}. If $m = 2n$ this means that π is the trivial representation. If $m < 2n$ then the previous lemma implies π is non-tempered. $\quad\Box$

The goal of this section is to prove the following proposition:

Proposition (2.2.3). *Let π be an irreducible principal series consituent. Fix $m > n$ and χ'. The set of splitting characters χ such that π is unambiguous relative to m, χ', and χ is finite.*

When $m > n + 1$ and π is tempered we have seen that this set is even empty. Henceforth we assume $m = n + b$, $H = U(V_{n-b})$ where V_{n-b} is an $n - b$-dimensional hermitian space with $\epsilon(V_{n-b}) = \epsilon(V)$. By Proposition 2.1.7 it suffices to show that the set of χ for which π is a theta lift from H with splitting character χ (and fixed χ') is finite. We first observe:

Lemma (2.2.4). *Suppose $\tau = \Theta_\chi(\pi, V_{n-b}) \neq 0$. Then τ belongs to the principal series.*

Proof. This follows by induction from the Main Theorem of [MVW], p. 69. $\quad\Box$

Lemma (2.2.4) represents π as $\Theta_\chi(\tau, W)$, and it thus remains to determine the theta lift of a principal series representation τ of H to G. The expression depends on the parity of n. We represent the contragredient principal series $\tau^\vee$ as an induced representation $I(\alpha)$ as above.

We let $|r_r(\tau^\vee)|$ denote the set of Jordan-Hölder constituents of the normalized Jacquet module $r_r(\tau^\vee)$ as above and write $|r_1(\tau^\vee)| = \{\sigma_r^j \otimes \tau_{n-b-2r}^j\}$. The σ_r^j belong to the set $\{\alpha_1, \dots, \alpha_r\}$. In order to make the induction work we specify that when $n = 3$ (resp. $n = 2$) τ_{n-2}^j (resp. σ_1^j) are representations of the trivial group which by convention are just considered trivial.

The main induction step is the following variant of the Main Theorem of [MVW], p. 69, which is itself an extension of Theorem 2.5 of [K]. In what follows τ is a principal series representation of $H = U(V_{n-b})$, corresponding to the representation π of $U(W) = U(W_n)$. In particular, the indices have changed from the beginning of the section.

Proposition (2.2.5). *(a) Under the above hypotheses, let $\alpha_i \otimes \tau_{n-b-2} \in |r_1(\tau^\vee)|$. Suppose $n > 3$. Then π is a quotient of one of the following induced representations:*

(i) $I_{P_1}^G(|\bullet|^{\frac{b-1}{2}} \cdot \chi^{-1} \otimes \pi_{n-2})$ *for some π_{n-2};*

(ii) $I_{P_1}^G(\alpha_i \cdot (\chi'/\chi) \otimes \pi_{n-2})$ *and* $\Theta_\chi(\pi_{n-2}, V_{n-b-2}) = \tau_{n-b-2}$.

(b) If $n-b = 1$, $r_1(\tau^\vee)$ is just $\tau^\vee = \gamma$. If γ is ramified then its theta lift to G is not in the principal series, hence cannot equal π. Suppose $n = 2$. Then if γ is trivial then $\pi = I(\chi'/\chi)$, where χ'/χ is viewed as a representation of the Levi component $GL(1, K)$ of the Borel subgroup of G. If $n > 2$ then we are necessarily in situation (i) of (a).

(c) If $n - b = 0$ and $n > 1$ then we are necessarily in situation (i) of (a).

Cases (i) and (ii) of (a) correspond to the two cases (i) and (ii) of 2(b) of the Main Theorem of [MVW]. Our conventions differ from those of [MVW] in that our theta lift of τ corresponds to their theta lift of $\tau^\vee$, and we have used the splitting character to obtain lifts directly from one unitary group to another. Our use of the contragredient in the definition of the theta lift explains the absence of the contragredient (σ_t^*) in the final result. The splitting character intervenes in (a) in the calculation of coinvariants of the metaplectic representation in the mixed Schrödinger model ([MVW] 3, V). Specifically, if $n = 2r$ is even and if $P \subset G$ is the stabilizer of a maximal isotropic lattice, then we have the usual Schrödinger model on the Schwartz-Bruhat space of V_{n-1}^r, with $H = U(V_{n-1})$. The action of the subgroup $GL(r) \subset P$ is multiplied by $\chi^{-1} \circ \det$ compared to the action described on p. 75 of [MVW] (cf. [K]), whereas the action of H is the linear action on the argument V_{n-1}^r, twisted by the character χ'. In contrast, if $n = 2r + 1$ is odd, and $n - b$ is even, then it is simplest to consider the lift in the opposite direction, which multiplies the action of $GL(r)$, now viewed as a Levi factor in H, by the character $\chi'^{,-1} \circ \det$, whereas the action of G is the linear action twisted by the character χ. The case where both n and $n - b$ are even is a bit more complicated to describe, since it requires a mixed model, but is treated in the same way. The signs are explained by our conventions regarding contragredients.

The special cases (b) and (c) are proved in just the same way. We mention them separately because of their role in the induction, as follows:

Proof of Proposition 2.2.3. Recall that χ' is fixed. When $n \leq 3$, (2.2.5)(a) and (b) describe π explicitly as an induced representation, depending on χ. We can thus recover χ from the Jacquet module of π with respect to the minimal parabolic P_0; in particular, χ belongs to a finite set. For general n, we apply (a). In case (i) we can already recover χ; in case (ii) we conclude by induction. We note that when $m = n$ there is no obvious restriction on the set of χ, since the twist by χ can be compensated by a twist of τ; however, there are subtle restrictions related to root numbers [HKS].   □

By induction on 2.2.5 we can calculate π explicitly in terms of τ for almost all τ, in the sense of (2.1.6). Reducible τ can be ignored in the following discussion, where by "almost all" τ such that $\Theta_\chi(\tau, W) \neq 0$ we mean the set of all τ in the intersection of the domain of the theta correspondence with $U(W)$ and some subset $\mathcal{P}^0$ as in (2.1.6). With this convention, it is possible for this intersection to be empty, hence "almost all" τ can refer to the empty set if the set of τ with $\Theta_\chi(\tau, W) \neq 0$ does not itself contain an open subset of $\mathcal{P}$.

Example 2.2.6. In the situation of Proposition 2.2.5 (a), case (ii) holds for almost all τ. Indeed, the situation in which case (i) applies is described explicitly in [MVW] in terms of the Jacquet module of τ.

The first two parts of the following corollary are essentially due to Kudla [K], and represent a special case of the analogue for unitary groups of his Theorem 2.5. Since his article does not treat unitary groups, we derive it from the results already quoted from [MVW].

Corollary 2.2.7. *(1) Suppose $n - b$ is even. For almost all τ, with $\tau^\vee = I(\alpha)$, such that $\pi = \Theta_\chi(\tau, W) \neq 0$, π is a subquotient of the principal series representation*

$$I(\alpha_1 \cdot (\chi'/\chi), \alpha_2 \cdot (\chi'/\chi), \ldots, \alpha_r(\chi'/\chi), |\bullet|^{\frac{b-1}{2}} \cdot \chi^{-1}, |\bullet|^{\frac{b-3}{2}} \cdot \chi^{-1}, \ldots, \nu),$$

where $\nu = |\bullet|^{\frac{1}{2}} \cdot \chi^{-1}$ or $\nu = \chi^{-1}$ as b is even or odd.

(2) Suppose $n - b$ is odd and γ is trivial. For almost all τ, with $\tau^\vee = I(\alpha)$, such that $\pi = \Theta_\chi(\tau, W) \neq 0$, π is a subquotient of the principal series representation

$$I(\alpha_1 \cdot (\chi'/\chi), \alpha_2 \cdot (\chi'/\chi), \ldots, \alpha_r(\chi'/\chi), |\bullet|^{\frac{b-1}{2}} \cdot \chi^{-1}, |\bullet|^{\frac{b-3}{2}} \cdot \chi^{-1}, \ldots, \nu'),$$

where $\nu' = |\bullet|^{\frac{1}{2}} \cdot \chi^{-1}$ if b is even and where $\nu' = \chi'/\chi$ if b is odd.

Proof. We may assume $n > 3$, the remaining cases being available in the literature as mentioned above. The proof then proceeds by induction on n, where at each step, until $n - b \leq 1$, we are allowed by (2.2.6) to assume we are in case (ii) of (a). At this point the proof divides into two cases, depending on whether n is even (i.e., we end up in case (c)) or n is odd (and we end up in case (b). If either b or $n - b$ is even, we are always in situation (i) of (a), and the induction proceeds as before. If b and $n - b$ are both odd, we are in situation (i) of (a) until the last step, where we are in case (b), with $H = U(1)$ and $G = U(2)$ (here the notation designates the quasi-split $U(2)$). $\square$

Remark 2.2.8. When the residue characteristic of F is odd, m is even, τ is spherical, and K/F and W are unramified, this result is obtained by Ichino [I1,Proposition 2.1] by adapting a method of Rallis. However, it does not seem that this result necessarily determines when a given τ is in the domain of the Howe correspondence. We address this question in the appendix (2.A).

2.2.9. The split case. When $K = F \oplus F$ then $G \cong GL(n,F)$, $G' \cong GL(m,F)$, and an unramified splitting character is trivial. For $m \geq n$ it is known that every irreducible τ has a non-trivial theta lift to G' [MVW, p. 64, Lemme 4, Lemme 5], which we denote π. If $m = n$, the calculation on p. 65 of [MVW] shows that $\pi = \tau^\vee$. In general, the formulas in Corollary 2.2.7 remain true in the split case; this is a special case of a result of Minguez [M, Theorem 4.1.2].

(2.3) The archimedean theta lift. We consider two unitary groups $G = U(M)$ and $G' = U(N)$ with signatures to be determined later. We will assume $N > M$ except in the statement of Theorem 2.3.5. The theta lifts of discrete series representations of $U(M)$ to cohomological representations of $U(N)$ were determined explicitly by Li in [Li90] in great generality. Certain cases of Li's results were applied in [H4].

We identify the unitary groups G, G' with the classical unitary groups of given signatures $G = U(s,r)$, $G' = U(a,b)$, with maximal compact subgroups $K = U(s) \times U(r)$, $K' = U(a) \times U(b)$, respectively. Li worked with the metaplectic representation, and considered liftings from the double cover $\tilde{G}$ of G to the double cover $\tilde{G}'$ of G'; the corresponding double covers of K and K' are denoted $\tilde{K}$ and $\tilde{K}'$, respectively. As in [KS] and [HKS], we can replace the metaplectic double covers by the more canonical $U(1)$-covering groups, which can be split over G and G' after choosing splitting characters $\chi, \chi' : \mathbb{C}^\times \to U(1)$ as in the non-archimedean case. However, Li's formulas for the theta lifts are more transparent (and easier to remember) using parameters for the double covers.

The discrete series representation $\tilde{\pi}$ of $\tilde{G}$ is determined by the highest weight $\tilde{\tau}$ of its lowest $\tilde{K}$-type, which we represent as an (s,r)-tuple of half-integers

$$\tilde{\tau} = \tilde{\tau}_+ \otimes \tilde{\tau}_-;$$
$$\tilde{\tau}_+ = (\tilde{\beta}_1 \geq \cdots \geq \tilde{\beta}_{u(+)} \geq \frac{a-b}{2} \geq \cdots \geq \frac{a-b}{2} \geq -\tilde{\gamma}_{v(+)} \geq \cdots \geq -\tilde{\gamma}_1)$$
$$\tilde{\tau}_- = (\tilde{\delta}_1 \geq \cdots \geq \tilde{\delta}_{u(-)} \geq \frac{b-a}{2} \geq \cdots \geq \frac{b-a}{2} \geq -\tilde{\alpha}_{v(-)} \geq \cdots \geq -\tilde{\alpha}_1)$$

$$(2.3.1)$$

Here $u(+) + v(+) \le s$, $u(-) + v(-) \le r$. We set

$$C_i = \tilde{\gamma}_i + \frac{a-b}{2}; \quad B_i = \tilde{\beta}_i - \frac{a-b}{2}; \quad A_i = \tilde{\alpha}_i + \frac{b-a}{2}; \quad D_i = \tilde{\delta}_i - \frac{b-a}{2}.$$

The A_i, B_i, C_i, D_i are all positive integers.

We also consider the Harish-Chandra parameter of $\tilde{\pi}$:

$$\tilde{\lambda} = (a_1, \ldots, a_{k(+)}; -b_{\ell(+)}, \ldots, -b_1) \otimes (c_1, \ldots, c_{k(-)}; -d_{\ell(-)}, \ldots, -d_1)$$

Here $k(+) + \ell(+) = s$, $k(-) + \ell(-) = r$, and all the a_i, b_i, c_i, d_i are positive integers or half-integers (depending on the parity of N as well as M). Moreover, since $\tilde{\pi}$ is in the discrete series, all of the a_i, b_i, c_i, d_i are distinct and none equals zero (this is also a consequence of the half-integral shift). The relation between $\tilde{\lambda}$ and $\tilde{\tau}$ is explicit (see [Li90, (53)] for example) and will be recalled below in a specific case of interest. The notation here is parallel but not identical to that of [Li90].

Assume

$$k(+) + \ell(-) \le a; \quad k(-) + \ell(+) \le b. \tag{2.3.2}$$

Then the theta lift $\Theta(\tilde{G} \to \tilde{G}'; \tilde{\pi}^\vee)$ to G' of the **contragredient** $\tilde{\pi}^\vee$ of $\tilde{\pi}$ is a non-trivial cohomological representation $\tilde{\pi}'$ with lowest $\tilde{K}'$-type $\tilde{\tau}' = \tilde{\tau}'_+ \otimes \tilde{\tau}'_-$, where

$$\tilde{\tau}'_+ = (B_1 + \frac{s-r}{2}, \ldots, B_{u(+)} + \frac{s-r}{2}, \frac{s-r}{2}, \ldots \frac{s-r}{2},$$
$$-A_{v(-)} + \frac{s-r}{2}, \ldots, -A_1 + \frac{s-r}{2}); \tag{2.3.2(+)}$$

$$\tilde{\tau}'_- = (D_1 + \frac{r-s}{2}, \ldots, D_{u(-)} + \frac{r-s}{2}, \frac{r-s}{2}, \ldots \frac{r-s}{2},$$
$$-C_{v(+)} + \frac{r-s}{2}, \ldots, -C_1 + \frac{r-s}{2}). \tag{2.3.2(-)}$$

In other words, the negative parameters (shifted by the difference of signatures) migrate from one part of the signature to the other, the positive parameters stay put, and zeroes are added as necessary. The representation $\tilde{\pi}'$ is identified in [Li90] explicitly as a cohomologically induced module $A_{\mathfrak{q}}(\lambda)$, determined by $\tilde{\tau}'$ and the infinitesimal character which can also be written explicitly.

On the other hand, when (2.3.2) is not satisfied, then $\Theta(\tilde{G} \to \tilde{G}'; \tilde{\pi}^\vee) = 0$.

Now choose splitting characters χ, $\chi' : \mathbb{C}^\times \to U(1)$ as in the non-archimedean case:

$$\chi \mid_{\mathbb{R}^\times} = \varepsilon_{\mathbb{C}/\mathbb{R}}^N; \quad \chi' \mid_{\mathbb{R}^\times} = \varepsilon_{\mathbb{C}/\mathbb{R}}^M. \tag{2.3.3}$$

The characters χ and χ' are determined by their restrictions to $U(1) = S^1 \subset \mathbb{C}^\times$, and can thus be indexed by integers (winding numbers) $\alpha(\chi)$, $\alpha(\chi') \in \mathbb{Z}$:

$$\chi(u) = u^{\alpha(\chi)}; \ \chi'(u) = u^{\alpha(\chi')}, u \in U(1).$$

By (2.3.3) we have $\alpha(\chi) \equiv N$ (mod 2), $\alpha(\chi') \equiv M$ (mod 2). We define

$$\tau = \tilde{\tau} + \frac{1}{2}\alpha(\chi)(1, \ldots, 1); \ \tau' = \tilde{\tau}' + \frac{1}{2}\alpha(\chi')(1, \ldots, 1).$$

We define $\tau_\pm$ and $\tau'_\pm$ likewise. Then τ (resp. τ') is an $M = (r, s)$-tuple (resp. an $N = (a, b)$-tuple) of integers defining the highest weight of a finite-dimensional representation of K (resp. K'), also denoted τ (resp. τ'), which is in turn the minimal K-type (resp. K'-type) of a representation π (resp. π') of G (resp. G'). As in [H4,(3.12)-(3.13)], we have

$$\Theta(G \to G'; \pi^\vee) = \pi' \qquad (2.3.4)$$

The following theorem is due to A. Paul [Pa].

Theorem 2.3.5. *Suppose $N \leq M$. Then there is at most one signature (a, b) such that $\Theta(G \to G' = U(a, b); \tilde{\pi}^\vee)$ is non-trivial. If $N = M$ there is exactly one such signature (a, b).*

This is an archimedean refinement of the principle of theta dichotomy studied in [HKS] for unitary groups over p-adic fields. For discrete series the signature (a, b) is determined uniquely by (2.3.2), when $N = M$. The existence and uniqueness of $U(a, b)$ when $N = M$ is stated as Theorem 0.1 of [Pa]. The uniqueness assertion when $N < M$ is not stated explicitly by Paul, but it follows immediately from her method. Indeed, suppose $\tilde{\pi}^\vee$ lifts non-trivially to $U(a, b)$ and to $U(a', b')$, with $a + b = a' + b' \leq M$. Then the proof of Theorem 2.9 of [Pa] shows that the trivial representation **1** has non-zero lift to $U(a + b', a' + b)$. But Lemma 2.10 of [Pa] shows that this is impossible unless either $a + b' = a' + b$, in which case $a = a', b = b'$, or $a + b' \geq M$ and $a' + b \geq M$, which is excluded.

On the other hand, when $N > M$ then, provided the parameters satisfy the inequalities (2.3.1) and (2.3.2), there are generally several possible signatures allowing non-trivial theta lifts. The same reasoning as in the proof of 2.3.5 (a) yields the following assertion:

Proposition 2.3.6. *Under the above hypotheses, suppose $N \geq M$. Then there are at most $N - M + 1$ signatures (a, b) such that $\Theta(G \to G' = U(a, b); \tilde{\pi}^\vee)$ is non-trivial.*

For example, if $N = M + 1$ and $b = 0$ and $b' \neq 0$ then the inequality $a' + b \geq M$ from Lemma 2.10 of [Pa] implies $(a', b') = (M, 1)$. We will be most interested in the case where $v(+) = u(-) = 0$. In this case $\tilde{\pi}^{\vee}$ is an antiholomorphic representation and the assumption that it is in fact an antiholomorphic discrete series imposes additional inequalities on the parameters.

In the present paper we are concerned with the case where $N = M + 1$, $U(M) = U(s, r)$ has signature (s, r), $U(N) = U(0, N)$ is negative-definite, and we wish to lift antiholomorphic discrete series representations of $U(M)$ – i.e., representations corresponding to antiholomorphic cohomology classes – non-trivially to (finite-dimensional irreducible) representations of $U(N)$.

Sign conventions are as in [H3] and [H4]. An antiholomorphic represen tation Π of $U(M)$ is determined by the highest weight τ of its minimal $K_{\infty} = U(s) \times U(r)$-type:

$$\tau = (-\beta_s, \ldots, -\beta_1; \alpha_1, \ldots, \alpha_r), \beta_1 \geq \beta_2 \geq \cdots \geq \beta_s \geq -\alpha_r \cdots \geq -\alpha_1.$$
$$(2.3.7)$$

All the α_i and β_j are non-negative integers and are positive when Π is in the discrete series. One checks that this implies that the dual $\tau^{\vee}$ of τ is the highest weight of an irreducible finite dimensional representation of $U(M)$ with the usual choice of positive roots. In order to have a lifting to $U(0, N)$, with respect to the splitting characters χ and χ', we need

$$-\beta_s \leq -\frac{N - \alpha(\chi')}{2} = -q_1(\chi'), \alpha_r \geq \frac{N + \alpha(\chi')}{2} = q_2(\chi') \qquad (2.3.8)$$

(In [Li90] one replaces $q_1(\chi')$ and $q_2(\chi')$ by $\frac{N}{2} = \frac{N-0}{2}$ where $(0, N)$ is the signature. Here the shift by $\frac{\alpha(\chi')}{2}$ has been incorporated, as in [H4, II.3], in order to obtain integral parameters.) For τ satisfying (2.3.8), the theta lift of the corresponding Π is the finite-dimensional representation

$$\Theta_{\chi, \chi'}(\Pi) = \Pi' = q'(\chi)(1, \ldots, 1) +$$

$$(\beta_1 - q_1(\chi'), \ldots, \beta_s - q_1(\chi'), 0, -\alpha_r + q_2(\chi'), \ldots, -\alpha_1 + q_2(\chi')) \quad (2.3.9)$$

where $q'(\chi) = \frac{r - s + \alpha(\chi)}{2}$.

One easily derives the following result from examination of the parameters above, but the methods of [Pa] provide a simpler proof:

Lemma 2.3.11. *Suppose Π is an antiholomorphic discrete series representation of $U(s, r)$ that admits a non-trivial theta lift to $U(0, r + s + 1)$ for the splitting characters χ, χ'. Then Π does not lift to $U(0, r + s - 1)$.*

Proof. We write $M = r + s$. As in the proof of Theorem 2.9 of [Pa], the assertion is reduced to the verification that, in the notation of [loc. cit.], $\theta_{M-1,M+1}(\mathbf{1}) = 0$, where $\mathbf{1}$ is the trivial representation of $U(M-1, M+1)$. But this follows again from Lemma 2.10 of [loc. cit.]   □

This has the following consequence for local zeta integrals. Let $V(a, b)$ be the hermitian space of signature (a, b), $a + b = M + 1$, and let

$$R_M(V(a, b), \chi) \subset I_M(\frac{1}{2}, \chi)_\infty$$

be the space of K_∞-finite vectors in the image of $\mathcal{S}(V(a, b)^M)_\infty$ under the analogue of the map (2.1.2).

Proposition 2.3.12. *Let Π be an antiholomorphic discrete series representation of $U(s, r)$ that admits a non-trivial theta lift to $U(0, r + s + 1)$ for the splitting characters χ, χ'. Then there is a function $f \in R_M(V(M+1, 0), \chi)$ such that, if it is extended to a section $f(h; \chi, s) \in I_M(s + \frac{1}{2}, \chi)_\infty$ (cf. (1.2.7)), then the local zeta integral $Z_\infty(s, \phi, \phi', \chi, f)$ does not vanish at $s = \frac{1}{2}$ when $\phi \otimes \phi'$ is the antiholomorphic vector in $\Pi(2)$.*

Proof. By Lemma 1.3.6, the local zeta integrals for discrete series representations converge absolutely for $Res \geq 0$, and in particular down to the unitary axis, for archimedean as well as non-archimedean local fields. The point is now the following. It follows from the results of [LZ] that

$$I_M(\frac{1}{2}, \chi)_\infty = \sum_{a+b=M+1} R_M(V(a, b))$$

(this is mainly contained in Theorem 5.5 and Proposition 5.8 of [LZ], though one has to read the notation carefully and compare it to the earlier results of Lee to derive this claim). Now the evaluation at $s = \frac{1}{2}$ of $Z_\infty(s, \phi, \phi', \chi, f)$ is non-zero for some choice of $f \in I_M(\frac{1}{2}, \chi)_\infty$. If $f \in R_M(V(a, b))$, then the non-triviality of the local zeta integral yields a non-trivial theta lift from Π to $U(a, b)$, as in the non-archimedean case (cf. the proof of Proposition 2.1.8). On the other hand, by Proposition 2.3.6, if $a + b = M + 1$, the theta lift of Π to $U(a, b)$ is zero unless $(a, b) = (M + 1, 0)$ or $(a, b) = (M, 1)$. It follows from Theorem 5.5 of [LZ] that $R_M(V(M + 1, 0)) \subset R_M(V(M, 1))$, so we can assume $f \in R_M(V(M, 1))$.

Now suppose the proposition is false. Then as in the proof of Proposition 2.1.8, the zeta integral gives a non-trivial (bilinear!) pairing

$$\Pi(2) \otimes R_M(V(M, 1))/R_M(V(M + 1, 0)) \to \mathbb{C}.$$

Again by Theorem 5.5 of [LZ], this quotient is isomorphic to $R_M(V(M - 1, 0))$. Thus (by duality) Π lifts non-trivially to $U(0, M - 1)$. But this contradicts Lemma 2.3.11. □

This result can be refined as follows. The space $R_M(V(M + 1, 0), \chi)$ is a holomorphic unitarizable representation of $U(M, M)$ (because it is a theta lift from a positive-definite unitary group). Its restriction to $U(s, r) \times U(r, s)$ decomposes discretely with multiplicity one as a sum of holomorphic representations of $U(s, r) \times U(r, s)$ – the discrete decomposition follows from unitarity, and the multiplicity one is the content of Lemma 3.3.7 of [H3].

Corollary 2.3.13. *In the statement of Proposition 2.3.12, one can take the function f to be a non-zero holomorphic vector in the unique subspace of the restriction to $U(s, r) \times U(r, s)$ of $R_M(V(M + 1, 0), \chi)$ isomorphic to the dual of $\Pi(2)$.*

Proof. Obviously only the dual $\Pi(2)^\vee$ of $\Pi(2)$ can pair non-trivially with $\Pi(2)$, and since the holomorphic K_∞-type subspace of $\Pi(2)^\vee$ is of multiplicity one, the corollary is clear. □

Appendix. Generic Calculation of the Unramified Correspondence

In this section the residue characteristic p is *odd*. A formula determining the correspondence between spherical representations of unramified unitary groups has been known for some time, but no proof seems to be available in the literature. We derive the formula in the case of odd residue characteristic from the results of Howe, Rallis, and Kudla presented in [MVW]. This is sufficient to characterize the global correspondence at almost all places.

Notation is as in §2.2; however here we do not assume $m = \dim V \leq n = \dim W$. We choose self-dual lattices $L \subset W$, $L' \subset V$, as in [MVW], Chapter 5, and let $U = Stab(L) \subset G$, $U' = Stab(L') \subset G'$. Then U and U' are hyperspecial maximal compact subgroups of the respective unitary groups. We let $\mathcal{H} = \mathcal{H}(G//U)$ denote the Hecke algebra of U-biinvariant compactly supported functions on G, $\mathcal{H}' = \mathcal{H}(G'//U')$. Both algebras are commutative, and the Satake isomorphism identifies their spectra $Z = Spec(\mathcal{H})$, $Z' = Spec(\mathcal{H}')$ with affine spaces of dimensions $[\frac{m}{2}]$, $[\frac{n}{2}]$, respectively.

Let $\mathcal{O}$ denote the ring of integers in K, $A = L \otimes_\mathcal{O} L' \subset W \otimes_K V$. The splitting characters χ, χ' are assumed to be unramified; this determines them uniquely, so they will be omitted from the notation. The additive character $\psi : F \to \mathbb{C}^\times$, implicit in all our constructions, is also unramified in the sense that ψ is trivial on the integer ring $\mathcal{O}_F$ of F but on no larger $\mathcal{O}_F$-submodule of F. Recall that W is supposed skew-hermitian and V hermitian, though this is inessential for what follows; the lattice L remains self-dual for the hermitian form obtained by means of the trace-zero element δ introduced at the beginning of §2.1, as long as the δ is a unit.

The natural skew-hermitian form on $W \otimes_K V$, composed with the trace $Tr_{K/F}$, defines a non-degenerate alternating form $< \bullet, \bullet >$ on $\mathcal{V} = R_{K/F}W \otimes_K V$. The metaplectic group $\widetilde{Sp}(\mathcal{V})$ in our setting is an extension of the symplectic group $Sp(\mathcal{V})$ by $\mathbb{C}^\times$, as in [HKS]. The oscillator representation $\omega_{V,W,\chi,\chi'}$ of $G \times G'$ is defined by means of an embedding

$$i_{V,W,\chi,\chi'} : G \times G' \hookrightarrow \widetilde{Sp}(\mathcal{V})$$

determined by the (unique) choice of splitting characters, composed with the oscillator representation of $\widetilde{Sp}(\mathcal{V})$ in one of its models. In the earlier section we worked with the Schrödinger model or mixed Schrödinger model, to which we will return below. For our present purposes, we need to use the lattice model, which is a representation of $\widetilde{Sp}(\mathcal{V})$ on the space

$$\mathcal{T} = \mathcal{T}_\psi(\mathcal{V}) = \{f \in C_c^\infty(\mathcal{V}) \mid f(a + v) = \psi(\frac{< v, a >}{2})f(v), \ a \in A, v \in \mathcal{V}\}.$$

We let $t_A \in \mathcal{T}$ denote the characteristic function of A.

Let π be an irreducible admissible representation of G. The lattice version of the theta correspondence is defined as in (2.1.5), with $\omega_{V,W,\chi}$ replaced by the isomorphic lattice model $\mathcal{T}$:

$$\Theta^T(V, \pi) = [\mathcal{T} \otimes \pi]_G.$$

We assume π is in the domain of the Howe correspondence; $\Theta^T(V, \pi) \neq 0$. Let π' be an irreducible admissible representation of G' for which there exists a surjective $G' \times G$-invariant map $p : \mathcal{T} \to \pi' \otimes \pi$; in other words, π' is the representation of G' associated to π by the Howe correspondence.

Theorem 2.A.1. *Assume π is spherical, i.e. $\pi^U \neq 0$. Then*

    (a) $(\pi')^{U'} \neq 0$;
    (b) $(\pi')^{U'} \otimes \pi^U = p(t_A)$.

Proof. The first assertion is Theorem 7.1 (b) of [Howe], proved as Theorem 5.I.10 of [MVW]. Assertion (b) is in fact a special case of Lemma 5.I.8 of [MVW], also due to Howe. □

The oscillator representation restricts to a representations of $\mathcal{H}$ and $\mathcal{H}'$ on the subspace $\mathcal{T}^{U \times U'}$. Let H, H' denote their respective images. These correspond to closed affine subschemes $Z_V = Spec(H) \subset Z$ and $Z'_W = Spec(H') \subset Z'$.

Theorem 2.A.2 (Howe).

(a) *The subalgebras H and H' of $End(\mathcal{T}^{U \times U'})$ are equal. Moreover,*

(b) $\mathcal{T}^{U \times U'} = \mathcal{H} \cdot t_A = \mathcal{H}' \cdot t_A$.

Proof. The first assertion is Theorem 7.1 (c) of [Howe], whose proof was first published as Proposition 5.I.11 of [MVW]. Part (b) is the main step of the proof of the latter proposition, on p. 107 of [MVW]. $\square$

Corollary 2.A.3. *The Howe correspondence takes spherical representations of G (in its domain) to spherical representations of G', and is defined on Satake parameters by a morphism $Z_V \to Z'$ of affine schemes. More precisely, the domain Z_V^0 of the Howe correspondence is a subset of the closed subscheme $Z_V \subset Z$, and the Howe correspondence on Satake parameters is defined by the restriction to Z_V^0 of a morphism $Z_V \to Z'$.*

Proof. Let $h : Z_V \xrightarrow{\sim} Z'_W$ be the isomorphism determined by Theorem 2.A.2. It suffices to show that $z_{\pi'} = h(z_\pi)$, whenever π and π' correspond as above, where $z_\pi \in Z_V$, $z_{\pi'} \in Z'_W \subset Z'$ are their respective Satake parameters; and that every π with $z_\pi \in Z_V$ is in the domain of the Howe correspondence. Now it follows from 2.A.1(b) and 2.A.2 (b) that the action z_π of $\mathcal{H}$ on π^U factors through the action of the quotient H on $\mathcal{T}^{U \times U'}$, and in particular that $z_\pi \in Z_V$; likewise for $z_{\pi'}$. The first claim now follows from the definition of h. $\square$

To our knowledge the exact domain Z_V^0 of the spherical Howe correspondence has not been determined in the literature. The results of [HKS] provide some qualitative information about Z_V^0. By symmetry, it suffices to assume $n \geq m$.

Proposition 2.A.5. *Suppose $n = m$. Then the domain Z_V^0 of the Howe correspondence equals Z. In particular, $Z_V = Z$.*

Proof. We apply the dichotomy criterion Theorem 6.1 (ii) of [HKS] . In our situation, both V and W are assumed to have self-dual lattices, hence $\det(V)$ and $\det(W)$ are units and so

$$\epsilon_{K/F}(\det(V)) = \epsilon_{K/F}(\det(W)) = 1.$$

Bearing in mind that K/F is unramified and p is odd, the dichotomy criterion then states

$$\Theta_\chi(\pi, V) \neq 0 \Leftrightarrow \epsilon(\frac{1}{2}, \pi, \chi, \psi) = 1.$$

The root number on the right is that defined by the Piatetski-Shapiro-Rallis doubling method, and is known to equal 1 when π, χ, and ψ are all unramified. Thus the dichotomy criterion, as strengthened by (2.1.7)(iv), implies that an unramified π belongs to Z_V^0. ☐

Corollary 2.A.6. *Suppose $n = m + 2b$, where $b \geq 0$. Then the domain Z_V^0 of the Howe correspondence equals Z.*

Proof. This follows from Kudla's persistence principle [K,MVW] and from (2.A.5). ☐

Lemma 2.A.7. *Suppose n is even and $m < n$, or n is odd and $m < n - 1$. Then the set Z_V^0 is contained in a proper Zariski closed subset of Z.*

Proof. Inspection of the parameters in Corollary 2.2.7 – where m is replaced by $n - b$ – shows that at least one parameter defining π is fixed, independently of τ, once $n > m$ or n is odd and $n > m + 1$. ☐

Remark 2.A.8. Lemma 2.A.7 remains valid whether or not V contains a self-dual lattice. Indeed, Corollary 2.2.7 calculates the image of the Howe correspondence from a unitary group of lower rank almost everywhere, avoiding the subset where the condition (a)(i) of Proposition 2.2.5 applies. However, it is obvious that the theta lifts from that subset also define a proper closed subset when $n > m$ (or $n > m + 1$ when n is odd).

Corollary 2.A.9. *Suppose n is even and $m \geq n$. Then the set Z_V^0 is a Zariski open subset of Z.*

Proof. When m is even, this is (2.A.6). By Kudla's persistence principle, it suffices to treat the case $m = n + 1$, and it suffices to show that any π outside a proper closed subset of Z is ambiguous for $n + 1$ and χ. By Proposition 2.1.8, it suffices to show that, for π outside a proper closed subset of Z, all theta lifts of π to $n - 1$-dimensional spaces vanish. But this follows from Lemma 2.A.7 and Remark 2.A.8. ☐

Corollary 2.A.10. *Suppose n is odd and $m \geq n - 1$. Then the set Z_V^0 is a Zariski open subset of Z.*

Proof. As in (2.A.9), it suffices to consider the case $m = n - 1$. Say $n = 2r + 1$. It follows from (2.A.9) that $(Z')_W^0$ is open in Z', hence is a scheme of dimension r. It then follows from Howe's theorem (2.A.2) that Z_V^0 is also of dimension r, hence is open in Z. □

Corollary 2.A.11. *Let p be an odd prime. Suppose $m \geq n$ or n is odd and $m = n - 1$. Then*

(a) The domain of the Howe correspondence contains a Zariski open subset of the space of unramified representations;

(b) On this Zariski open subset. the Howe correspondence on spherical representations is given by the formulas in Corollary 2.2.7.

Proof. Corollary 2.2.7 asserts that the Howe correspondence is given by the formulas for almost all unramified representations in the domain of the correspondence. Under the present hypotheses, (a) follows from (2.A.9) and (2.A.10). Now Corollary 2.A.3 implies that the spherical Howe correspondence is given by a morphism of schemes. The formulas in (2.2.7) are thus valid on all points of the domain of the correspondence. □

3. Applications to Special Values of L-Functions

In this section we assume $E = \mathbb{Q}$, for applications to the questions considered in [H3, H4]. We begin by introducing notation, as in [H5]. Let $G = GU(W)$, the unitary similitude group of a hermitian space $W/\mathcal{K}$ with signature (r, s) at infinity. We associate to $GU(W)$ a Shimura variety $Sh(W)$ of dimension rs, following the conventions of [H3].

Let S be a finite set of finite places of $\mathbb{Q}$. We define the motivically normalized standard L-function, with factors at S (and archimedean factors) removed, to be

$$L^{mot,S}(s, \pi \otimes \chi, St, \alpha) = L^S(s - \frac{n-1}{2}, \pi, St, \alpha). \quad (3.1)$$

The motivically normalized standard zeta integrals are defined by the corresponding shift in (2.1.7.2). A variant of the following theorem is stated as Theorem 4.3 of [H5], and generalizes Theorem 3.5.13 of [H3], to which we refer for unexplained notation.

Theorem 3.2. *Let $G = GU(W)$, a unitary group with signature (r, s) at infinity, and let π be a cuspidal automorphic representation of G. We assume $\pi \otimes \chi$ occurs in anti-holomorphic cohomology $\bar{H}^{rs}(Sh(W), E_\mu)$ where μ is the highest weight of a finite-dimensional representation of G. Let χ, α be algebraic Hecke characters of $\mathcal{K}^\times$ of type η_k and η_κ^{-1}, respectively. Let*

s_0 *be an integer which is critical for the L-function* $L^{mot,S}(s, \pi \otimes \chi, St, \alpha)$; *i.e.* s_0 *satisfies the inequalities (3.3.8.1) of [H3]:*

$$\frac{n - \kappa}{2} \leq s_0 \leq min(q_{s+1}(\mu) + k - \kappa - \mathcal{Q}(\mu), p_s(\mu - k - \mathcal{P}(\mu)), \qquad (**)$$

Define $m = 2s_0 - \kappa$. *Let* α^* *denote the unitary character* $\alpha/|\alpha|$ *and assume*

$$\alpha^* \mid_{\mathbf{A}_\mathbb{Q}^\times} = \varepsilon_\mathcal{K}^m. \qquad (3.2.1)$$

Suppose there is a positive-definite hermitian space V *of dimension* m, *a factorizable section* $\phi_f(h, s, \alpha^*) \in I_n(s, \alpha^*)_f$, *factorizable vectors* $\varphi \in \pi \otimes \chi$, $\varphi' \in \alpha^* \cdot (\pi \otimes \chi)^\vee$, *and a finite set* S *of finite primes such that*

   (a) *For every finite* v, $\phi_v \in R_n(V_v, \alpha^*)$;
   (b) *For every finite* v *in* S, π_v *does not occur in the boundary at* s_0 *for* α_v^*, *and* π_v *is ambiguous for* m *and* α^*;
   (c) *For every finite* v, $\Theta_{\alpha^*}(\pi_v \otimes \chi_v, V_v) \neq 0$;
   (d) *For every finite* v *outside* S, *all data* $(\pi_v, \chi_v, \alpha_v,$ *and the additive character* ψ_v) *are unramified.*

Then

   (i) *One can find* ϕ_f, φ, φ' *satisfying (a) and (b) such that* ϕ_f *takes values in* $(2\pi i)^{(s_0+\kappa)n} L \cdot \mathbb{Q}^{ab}$, *and such that* φ, φ' *are arithmetic over the field of definition* $E(\pi)$ *of* π_f.

   (ii) *Suppose* φ *is as in (i). Then*

$$L^{mot,S}(s_0, \pi \otimes \chi, St, \alpha) \sim_{E(\pi, \chi^{(2)} \cdot \alpha); \mathcal{K}} P(s_0, k, \kappa, \pi, \varphi, \chi, \alpha)$$

where $P(s_0, k, \kappa, \pi, \varphi, \chi, \alpha)$ *is the period*

$$(2\pi i)^{s_0 n - \frac{nw}{2} + k(r-s) + \kappa s} g(\varepsilon_\mathcal{K}^{\lceil \frac{n}{2} \rceil}) \cdot \pi^c P^{(s)}(\pi, *, \varphi) g(\alpha_0)^s p((\chi^{(2)} \cdot \alpha)^\vee, 1)^{r-s}$$

appearing in Theorem 3.5.13 of [H3].

Comments (3.2.2). (i) We have replaced the value m of [H3] by the value s_0, since m is here reserved for the dimension of an auxiliary space. The notation (r, s) for the signature of W should not interfere with the use of s as a variable.

   (ii) The hypothesis (3.2.1) guarantees that α^* can be used as a splitting character for the pair $(U(W), U(V))$, hence the condition (a) makes sense. The extension of the theta correspondence to unitary similitude groups is explained in detail in [H4,I.4] and will not be repeated here. The hypothesis should of course not be necessary, and in the cases treated by Shimura in

[S1,S2] – in which E_μ is a line bundle – there is apparently no such hypothesis. However, the cases treated here are sufficient for the applications to period relations.

(iii) In [H5] conditions (a)-(d) are replaced by conditions (a) and

(c') The normalized local zeta integrals $\tilde{Z}_v^{mot}(s, \varphi_v, \varphi'_v, \phi_v, \alpha_v^*)$ do not vanish at $s = s_0$.

We derive (c') from (a)-(d) as in [HKS]. Indeed, if the normalized local zeta integral at $s = s_0$ is non-zero then, under hypothesis (a), it defines a non-trivial element of the space

$$Hom_{U(W) \times U(W)}(R_n(V_v, \alpha^*), \pi \otimes (\chi \cdot \pi^\vee)) \qquad (3.2.3)$$

and hence the theta lift is non-trivial by Proposition 3.1 of [HKS]. Conversely, the non-triviality of the theta lift is equivalent to non-triviality of (3.2.3) [*loc. cit*]. Now the normalized local zeta integral is non-zero for some choice of data. If v is not in S, then (d) implies the nonvanishing of the normalized local zeta integral at v. If $v \in S$, then it follows from (b) and Proposition 2.1.9 that the normalized zeta integral is non-trivial for some choice of φ_v satisfying (a). Thus Theorem 3.2 does indeed follow from Theorem 4.3 of [H5].

One should not have to assume that π_v does not occur in the boundary, but for the time being we do not know how to prove that the dimension in (3.2.3) is at most one without making such a hypothesis. The following Corollary shows it is in fact unnecessary, at least for tempered π, except when $m = n + 1$:

Corollary 3.3. *Under the hypotheses of Theorem 4.3, assume π_f tempered at all but at most finitely many places. Suppose $m - n \geq 2$. Then hypotheses (a) and (c') are automatically satisfied for some choice of positive-definite V of dimension m. In particular, the conclusion of Theorem 3.2 holds unconditionally for such π and χ whenever $s_0 - \frac{n-\kappa}{2} \geq 1$.*

Proof. To simplify the exposition, we write π instead of $\pi \otimes \chi$ in what follows. Let S be the set of finite places v of $\mathbb{Q}$ that do not split in $\mathcal{K}$ and for which π_v does not belong to the principal series, together with the prime 2 and any places at which π_v is not tempered. By Proposition 2.1.6 (iii) for every $v \in S$ there exists a hermitian space V_v (up to isomorphism) such that

$$Hom_{G \times G}(R_n(V_v, \alpha^*), \pi \otimes (\alpha^* \cdot \pi^\vee)) \neq 0. \qquad (*)$$

(When v is split there is only one choice of V_v and (*) remains true, cf. 2.2.9.) Moreover, the proof of 2.1.7 (iii) in [HKS; 4.5 and p. 973] shows that the normalized local zeta integral gives a non-zero pairing in (*); hence condition (c') of (3.2.2)(iii) is satisfied. Let V be any positive-definite hermitian space of dimension m with V_v satisfying $*$ for all $v \in S$. For $v \notin S$, v not split in $\mathcal{K}$, π_v is a tempered principal series constituent, hence is ambiguous for m and α^* by Corollary 2.2.2; thus there is no local obstruction at such v. Finally, for v split in $\mathcal{K}$ there is only one V_v up to isomorphism and the analogue of $R_n(V_v, \alpha^*)$ coincides with the whole of $I_n(s_0, \alpha^*)$ [KS, Theorem 1.3]. The claim then follows from Corollary 2.1.9. $\qquad\square$

The case $m = n$ corresponds to the central critical value, where the L-function can vanish, and requires slightly different methods, as in [HK] and [KR1]. There the global induced representation breaks up as an infinite direct sum $\oplus_J I_J$ over finite sets J of finite primes. The sets J of even cardinality correspond to hermitian spaces V_J of dimension $m = n$; and the conditions (a) and (b) apply with $\phi_f \in I_J$ if and only if the theta lift to V_J is non-trivial, provided the L-function does not vanish at s_0. (If the L-function vanishes there is nothing more to prove.) On the other hand, when the cardinality of J is odd then the Eisenstein series corresponding to sections in I_J vanish. One can thus obtain an unconditional result in this case as well.

It remains to treat the case $m = n+1$, corresponding to the "near central point" of the L-function. This is more subtle, and even for tempered π we do not have unconditional results, except for "most" α:

Theorem 3.4. *Under the hypotheses of Theorem 3.2, assume π_f tempered at all but finitely many places and $m = n + 1$. Fix a finite place v that does not split in $\mathcal{K}$ such that π_v is tempered. There is a finite set of characters A_v of $K_v^\times$ satisfying (4.3.1) such that, if α is a splitting character such that $\alpha_v^* \notin A_v$, then there exists a hermitian space V of dimension $n+1$ for which hypotheses (a) and (b) of Theorem 3.2 are satisfied for π, χ, α, and s_0. In particular, for such α the conclusions of Theorem 3.2 hold unconditionally.*

Proof. We let A_v be the set of splitting characters ξ such that $\pi \otimes \chi$ is unambiguous relative to $m = n + 1$, χ', and ξ, where χ' is a fixed splitting character for $U(W)$. By Proposition 2.2.3 A_v is a finite set. Choose α such that $\alpha_v^* \notin A_v$. Now Proposition 2.1.6 implies that, for every finite place $w \neq v$ there is a hermitian space V_w (up to isomorphism) of dimension

$m + 1$ such that

$$Hom_{G \times G}(R_n(V_w, \alpha_w^*), (\pi_w \otimes \chi_w) \otimes (\alpha_w^* \cdot (\pi_w \otimes \chi_w)^\vee)) \neq 0 \qquad (3.4.1)$$

and contains the normalized zeta integral at $s = s_0$ as a non-zero element. There is exactly one global positive definite hermitian space V such that V_w is in the chosen isomorphism class at every finite prime $w \neq v$, the localization V_v at v being determined by the vanishing of the product of the local Hasse invariants. But π_v is ambiguous for m and α_v^*, hence (3.4.1) holds at v as well, and again the normalized zeta integral defines a non-zero pairing. The same is true for any Galois conjugate of α. We now conclude as in the proof of Corollary 3.3. Galois equivariance is guaranteed by Lemma 2.1.10.   □

Under additional hypotheses, the proof gives the following refinement:

Corollary 3.4.2. *Suppose π admits a global base change to a cuspidal automorphic representation Π of $GL(n, \mathcal{K})$. Then there exists a positive-definite hermitian space V over $\mathcal{K}$ of dimension $n + 1$ and a pair of splitting characters α^*, χ' such that $\Theta_{\alpha^*, \chi'}(G' \to U(V); \pi \otimes \chi) \neq 0$.*

Proof. We argue as above. If π admits a base change then so does $\pi \otimes \chi$. To simplify notation we replace π by $\pi \otimes \chi$. Under the hypothesis, we have

$$L^S(1, \pi, St, \alpha^*) = L^S(1, \Pi \otimes \alpha^* \circ \det) \neq 0; \qquad (3.4.2.1)$$

the non-vanishing of $L^S(1, \Pi \otimes \alpha^* \circ \det)$ is the result of Jacquet-Shalika.

Given a choice of local hermitian spaces V_v for all places v of E (only the non-split places matter), there exists a global hermitian space V with localizations V_v for all v if and only if the product of the Hasse invariants $\prod_v \varepsilon(V_v) = 1$. We choose χ' arbitrarily. Let S be as in the statement of Theorem 3.2, and choose α satisfying conditions (b) and (d). It follows from Lemma 2.1.6.2 and Proposition 2.2.3 that the set of such α is infinite. Let $v_0 \in S$ so that π_{v_0} is ambiguous. As above, for each non-split prime v, there exists some V_v such that $\Theta_{\chi_v, \chi_v'}(G_v' \to U(V_v); \pi_v) \neq 0$. If necessary, we switch the form V_{v_0} so that the product of the Hasse invariants is 1; then we can assume that the V_v all correspond to localizations of a global hermitian space V.

Thus the local theta lifts are non-trivial everywhere and condition (b) of [H5, Theorem 4.3] is satisfied. It remains to show that the global theta lift is non-zero. This follows from the Rallis inner product formula, which will be stated below, and (3.4.2.1). However, there is no need to quote the Rallis inner product formula *in extenso*. The nonvanishing (3.4.2.1), together with the non-vanishing of local zeta integrals, implies that the global zeta integral

$$Z^+(s, \phi, \phi', \alpha, \varphi) = \int_{Z(\mathbf{A}) \cdot G(\mathbb{Q}) \backslash G(\mathbf{A})^+} E(i_V(g, g'), \alpha, s, \varphi) \phi(g) \phi'(g') dg dg'$$

$$(3.4.2.2)$$

does not vanish. But (3.4.2.2) is the pairing of a Siegel-Weil Eisenstein series lifted from $U(V)$ against the vector $\varphi \otimes \phi' \in \pi \otimes (\alpha^* \cdot \pi^\vee)$. By seesaw duality, the theta lift of this vector to $U(V) \times U(V)$ is non-trivial. □

3.5. Extensions of the Rallis inner product formula. We recall the main theorems of [I1, I2], in the form in which they are presented in [H5]. Let $G = GU(W)$, $H = GU(2W)$, $G' = GU(V)$, with $\dim W = m$, $\dim V = n$ (note the switch!) and choose splitting characters χ, χ' as above. We consider theta lifts from G' to H. The extension to similitude groups is as in [H4, H5]: to a pair consisting of an automorphic form F on $G'(\mathbf{A})$ and a function $\phi \in \mathcal{S}(V^m)$ we obtain a function

$$\theta_{\phi, \chi, \chi'}(F) : H(E) \backslash H(\mathbf{A})^+ \to \mathbb{C}$$

where $H(\mathbf{A})^+ = GU(2W)(\mathbf{A})^+$ is as defined in (0.4).

Theorem 3.5.1 [I1]. *Suppose $n < m$, and let $s_0 = \frac{m-n}{2}$, $r_0 = 2s_0$. Let $V' = V \oplus \mathbb{H}^{r_0}$, where $\mathbb{H}$ is a hyperbolic hermitian space of dimension 2. There is an explicit real constant c_K such that, for any K_H-finite $\Phi \in \mathcal{S}((V')^n(\mathbf{A}))$,*

$$Res_{s=s_0} E(h, s, \phi_\Phi, \chi) = c_K \theta_{\pi_{V'}^V, \pi_K(\Phi), \chi, triv, \psi}(h)$$

for all $h \in H(\mathbf{A})^+$.

The index K in c_K denotes a maximal compact subgroup of $G'(\mathbf{A})$, and c_K is a factor of proportionality relating two Haar measures, calculated explicitly in [I1,§9]. We refer to [I1, §4] and Remark 3.5.15, below, for the notation $\pi_{V'}^V$ (which Ichino calls $\pi_{Q'}^Q$) and π_K. Note in particular that $I_{\chi, triv, \psi}(*)$ denotes the *regularized* theta integral, see (3.5.4.3), below. The extension to the subgroup $H(\mathbf{A})^+$ of the similitude group is as in Corollary 3.3.2 of [H5]. For $n \geq m$ Ichino has proved the following theorem; we follow the statement of [H5, Corollary 3.3.2].

Theorem 3.5.2 [I2]. *Suppose $n \geq m$ and V is positive definite. Then the Eisenstein series $E(h, s, \phi_\Phi, \chi)$ has no pole at $s = s_0$, and*

$$I_{\chi, triv, \psi}(\Phi)(h) = c \cdot E(h, s_0, \phi_\Phi, \chi)$$

for $h \in GH(\mathbf{A})^+$, where $c = 1$ if $m = n$ and $c = \frac{1}{2}$ otherwise.

In particular, we have the Rallis inner product formula: setting

$$Z^+(s, \phi, \phi', \chi, \varphi) = \int_{Z(\mathbf{A}) \cdot G(\mathbb{Q}) \backslash G(\mathbf{A})^+} E(i_W(g, g'), \chi, s, \varphi) \phi(g) \phi'(g') dg dg'$$

we have

$$Z^+(s, \phi, \phi', \chi, \varphi) = c \cdot I_{\Delta, \chi, \chi'}(\theta_{\varphi, \chi, \chi'}(\phi \otimes \phi')).$$

An almost immediate consequence of Theorem 3.5.1 is the characterization of poles of standard L-functions of automorphic representations of G. The analogous case of standard L-functions of symplectic groups has been considered by Kudla and Rallis in [KR1], and the proof goes over without change. We derive the regularized Rallis inner product formula as in [KR1,§8], and compare it to the formula proposed in [H4, V.3]. Let π be a cuspidal automorphic representation of G, χ and χ' a pair of splitting characters, $\pi(2)$ any irreducible extension to $G(G \times G^-)$ of the representation $\pi \otimes [\bar{\chi} \circ \det \otimes \bar{\pi}]$. The normalized diagonal period integrals $I_{\Delta, \chi}$ and $I_{\Delta, \chi'}$ are defined as in [loc. cit.]. Let $F \in \pi(2)$, and let $\phi(s) = \phi(h, s; \chi) \in I_m(s, \chi)$ be a section defining an Eisenstein series as in §1. The factor of the global zeta integral corresponding to the collection S of bad primes is defined, as in (1.3.3), to be

$$I_{\Delta, \chi} Z_S(\phi)(s)(F) = \int_{G_S} I_{\Delta, \chi} \pi(2)(g, 1) F \cdot \phi_S(\delta \cdot (g, 1), s) dg. \qquad (3.5.3)$$

Here $G_S = \prod_{w \in S} G(E_w)$, $\delta \in GH$ is a representative of the dense orbit of $G \times G$ on $GP \backslash GH$, and the notation $I_{\Delta, \chi}$ on the left-hand side is a reminder that we are integrating along the diagonal inside the integral over G_S.

Theorem 3.5.4. *With the above notation, let $F \in \pi(2)$ be factorizable. Assume $n < m$. Suppose $\theta_{\varphi, \chi, \chi'}(F)$ is cuspidal on $G(G' \times G'^{,-})$. Then there is a finite set S of finite places of E such that*

$$c_K I_{\Delta, \chi'}(\theta_{\varphi, \chi, \chi'}(F))$$

$$= Res_{s=s_0} d_m^S(s, \chi)^{-1} I_{\Delta, \chi}(Z_S(\Phi)(s)(F)) \cdot L^S(s + \frac{1}{2}, \pi_K \cdot \chi) \qquad (3.5.4.1)$$

where $\varphi = \pi_{V'}^V \pi_K(\Phi)$ and $d_m^S(s)$ is given by (1.3.4).

Proof. Notation is as in [H4, V.3]. We suppose we have a form $F \in \pi(2)$ of discrete series type on $G(G \times G^-)$, with $G = GU(V)$, with $\dim V = m$.

We define $\theta_{\varphi,\chi,\chi'}(F)$ on $G(G' \times G')$ as in *loc. cit.*, with $G' = GU(V')$, $\dim V' = n$; under the hypotheses of *loc. cit.*, this theta lift is holomorphic.

We recall Ichino's regularization of the theta integral in Theorem 3.5.1. We assume $n \leq m$. Let v be a finite place of E of odd residue characteristic that is inert in $\mathcal{K}$ and unramified over $\mathbb{Q}$. We assume moreover that the splitting characters χ, χ' and the additive character ψ are unramified at v, that the unitary groups G and G' are quasi-split at v, and that the function Φ is fixed by a hyperspecial maximal compact subgroup $K_v \times K'_v \subset H(E_v) \times G'(E_v)$ at v. Ichino also assumes K'_v is well placed with respect to a chosen Siegel set in $G'(\mathbf{A})$. Then [I1,§2] there exist K'_v-biinvariant functions $\alpha \in C_c^\infty(G'(E_v), \mathbb{Q}(q_v^{\frac{1}{2}}))$ and, for each such α, a non-zero constant $c_\alpha \in \mathbb{Q}(q_v^{\frac{1}{2}})^\times$ with the property that the integral defining $I_{\chi,triv,\psi}(\omega_{V,W,\chi}(\alpha)\pi_{V'}^V \pi_K(\Phi))(h)$ is absolutely convergent and, if $I_{\chi,triv,\psi}(\pi_{V'}^V \pi_K(\Phi))(h)$ is already absolutely convergent, then

$$I_{\chi,triv,\psi}(\omega_{V,W,\chi}(\omega_{V,W,\chi}(\alpha)\pi_{V'}^V \pi_K(\Phi))(h) = c_\alpha \cdot I_{\chi,triv,\psi}(\pi_{V'}^V \pi_K(\Phi))(h).$$

Then for general Φ, the regularized theta integral is defined as

$$c_\alpha^{-1} I_{\chi,triv,\psi}(\omega_{V,W,\chi}(\alpha)\pi_{V'}^V \pi_K(\Phi))(h) \tag{3.5.4.2}$$

and is independent of the choice of v and α with the required property (of killing support, cf. the proof of [I1, Prop. 2.4]). Since this property is invariant with respect to $Gal(\mathbb{Q}(q_v^{\frac{1}{2}})/\mathbb{Q})$, the regularization is rational over $\mathbb{Q}$.

By analogy with [KR1 (8.12)], if $F \in \pi(2)$ we define

$$I_{\Delta,\chi',REG}(\theta_{\varphi,\chi,\chi'}(F)) = c_\alpha^{-1} I_{\Delta,\chi'}(\theta_{\alpha*\varphi,\chi,\chi'}(F)) \tag{3.5.4.3}$$

where we write $\alpha * \varphi$ for $\omega_{V,W,\chi}(\alpha)\pi_{V'}^V \pi_K(\Phi)$, the action of α at the place v. Then the formula (3.5.4.1), with $I_{\Delta,\chi',REG}$ in the place of $I_{\Delta,\chi'}$, is precisely the analogue of Theorem 8.7 of [KR1], and is proved in the same way. Now we are assuming $\theta_{\varphi,\chi,\chi'}(F)$ is cuspidal, hence that the integral $I_{\Delta,\chi'}(\theta_{\varphi,\chi,\chi'}(F))$ over the diagonal converges absolutely. As in [KR1 (8.14)], one then has

$$I_{\Delta,\chi',REG}(\theta_{\varphi,\chi,\chi'}(F)) = I_{\Delta,\chi'}(\theta_{\varphi,\chi,\chi'}(F)) \tag{3.5.4.4}$$

which completes the proof. □

We now take $m = n + 1$ but apply Theorem 3.5.4 to the lift in the opposite direction; i.e. from a cuspidal automorphic representation $\pi'(2)$ of $G(G' \times G'^{,-})$ to $G(G \times G^-)$. We make the following hypotheses:

Hypotheses 3.5.5.

(a) *There is a cuspidal automorphic representation π of G such that $\pi' = \Theta(G \to G'; \pi^\vee)$.*

(b) *The archimedean component π_∞ of π belongs to the discrete series, and π'_∞ is of antiholomorphic type.*

(c) *The representation π occurs with multiplicity one in the space of automorphic forms on G, cuspidal or otherwise.*

(d) *The representation π admits a base change to a cuspidal automorphic representation $\pi_\mathcal{K}$ of $GL(n, \mathcal{K})$.*

(e) *At all primes w of E dividing 2, the Howe duality conjecture is valid for π_w.*

Hypothesis (a) and (b) are genuine hypotheses that define the problem under consideration. Hypotheses (c) and (e) are unnecessary but allow us to simplify the following arguments by avoiding the proof that the Petersson norm of arithmetic forms of type π'_f are well-defined up to scalars in the field of rationality of the representation. One conjectures that hypothesis (c) is always satisfied if π belongs to a stable L-packet, and this will probably be proved in the near future. If π is cuspidal and tempered then it may already be known that it is not equivalent to a non-cuspidal representation, but we have included this in Hypothesis (c). Hypothesis (d) is proved in [HL] under the assumption that the local component π_v is supercuspidal or Steinberg for some finite prime v of E that splits in $\mathcal{K}$. Hypothesis (e) is valid if all primes above 2 split in $\mathcal{K}$ (cf. [M]); it implies that the theta lift in (a) is in fact an irreducible representation of $G(\mathbf{A})$. Of course hypothesis (e) is implied by the Howe duality conjecture.

Proposition 3.5.6. *Under Hypotheses 3.5.5, the representation π' has multiplicity one in the space of cusp forms on G'. In particular, $\Theta(G' \to G; \pi'^{,\vee}) = \pi$.*

Proof. We apply an argument due to Rallis. Let π^* be a cuspidal automorphic representation of G' such that π_f^* and π'_f are isomorphic as abstract representations of $G'(\mathbf{A})$. Since π' is a theta lift from G, it follows from the parameters for the unramified correspondence determined in §2.2 (cf. [H4, I. 3.15]) that, for a sufficiently large set S of bad primes,

$$L^S(s, \pi^*, St) = L^S(s, \pi, (\chi'/\chi), St) L^S(1, \chi') \tag{3.5.6.1}$$

In particular, 3.5.5 (d) implies, as in the proof of Corollary 3.4.2, that

$$res_{s=1}L^S(s, \pi^*, (\chi')^{-1}, St) = L^S(1, \pi_\mathcal{K}, \chi^{-1})res_{s=1}\zeta_\mathcal{K}^S(s) \neq 0. \quad (3.5.6.2)$$

By the results of [I1], the arguments of [KR1,Theorem 7.2.5] apply in this situation and imply that π^* is a theta lift, with respect to the splitting character χ', from some unitary group $G^* = U(V^*)$ for some hermitian space V^* of dimension n. Now applying Proposition 2.1.7 (iv), the existence of the theta lifting from $U(V')$ to $U(V^*)$ determines the local isomorphism class of V^* at all finite primes, and Theorem 2.3.5 determines the signature of V^* at ∞. Since $\pi^* \cong \pi'$, we even know that $V_v^* \xrightarrow{\sim} V_v$ for all v. By Landherr's theorem, $V \xrightarrow{\sim} V^*$.

It follows that the theta lift, say π^{**}, of $\pi^{*,\vee}$ to G is non-trivial. Since $\pi^* \xrightarrow{\sim} \pi'$, we must have $\pi^{**} \xrightarrow{\sim} \pi$ as abstract representations. By 3.5.5 we must have $\pi^{**} = \pi$. But by adjunction the L_2 pairing on $G'(\mathbb{Q})\backslash G'(\mathbf{A})$ pairs π^* non-trivially with the theta lift of $\pi^{**} = \pi$, hence $\pi^{*,\vee}$ pairs non-trivially with π'. □

Corollary 3.5.7. *Under Hypotheses 3.5.5, let $F \in \pi'(2)$. Then the cuspidality hypothesis in Theorem 3.5.4 is automatic, and we can rewrite (3.5.4.1):*

$$c_K I_{\Delta,\chi'}(\theta_{\varphi,\chi,\chi'}(F))$$

$$= d_{n+1}^S(\frac{1}{2}, \chi)^{-1} I_{\Delta,\chi}(Z_S(\Phi)(\frac{1}{2})(F)) \cdot L^S(1, \pi_\mathcal{K}, \chi^{-1})res_{s=1}\zeta_\mathcal{K}^S(s).$$

In particular, if $E = \mathbb{Q}$, then, up to $\mathcal{K}$-rational multiples, we have

$$c_K I_{\Delta,\chi'}(\theta_{\varphi,\chi,\chi'}(F))$$

$$\sim d_{n+1}^S(\frac{1}{2}, \chi)^{-1} I_{\Delta,\chi}(Z_S(\Phi)(\frac{1}{2})(F)) \cdot L^S(1, \pi_\mathcal{K}, \chi^{-1})L(1, \varepsilon_\mathcal{K}).$$

Proof. The simplification of the residue (3.5.4.1) follows from (3.5.6.2) and (1.3.6). □

In what follows, we make use of general facts about rationality and Petersson norms developed in the following section. We work in the setting of [H4, V.1], to which we refer for notation, with $m = n + 1$. Thus $E = \mathbb{Q}$, V is of signature $(n-1, 1)$, V' of signature (r, s), $r' = r - 1$. We assume $\pi_f^\vee(2)$ occurs (with multiplicity one, by 3.5.5(c)) in $H_!^{2(r'-1)}(Sh(V^{(2)}), [W_{\Lambda(2)}])$, and that $\pi'_f(2) = \Theta(\pi(2))_f$ contributes to $H^0(Sh(V'^{,(2)}, [E_{\Lambda'(2)}]))$. Let

$$V(\pi)(2) \subset H_!^{2(r'-1)}(Sh(V^{(2)}, [W_{\Lambda(2)}]),$$

$$\text{resp. } V(\pi')(2) \subset H_!^0(Sh(V'^{,(2)}, [E_{\Lambda'(2)}]))$$

be the $\pi_f^\vee(2)$ (resp. $\pi'_f(2)$)-isotypic components. Note that we have switched π with $\pi^\vee$. The parameters Λ, Λ', etc. are defined in [H4] and do not need to be recalled here. Because local Howe duality has not been established in residue characteristic 2 we actually have to take an irreducible component $\Theta(\pi(2))_0$ in the definition of $\pi'_f(2)$ above, but this can also be ignored.

We let M be a CM field containing the fields of rationality of π, χ, and χ', over which π_f has a model. It's clear from (3.5.6.1) that π'_f ought also to have a model over M without any additional hypothesis, but we will add this to our assumptions about M. Define $V(\pi)(2)_M \subset H_!^{2(r'-1)}(Sh(V^{(2)}, [W_{\Lambda(2)}])$ and $V(\pi')(2)_M$ as in Proposition 3.6.2. We define $Q(\pi(2)), Q(\pi'(2)), Q(\pi), Q(\pi') \in \mathbb{R}^\times/M^{+,\times}$ as in Corollary 3.6.4. Define $I_{\Delta,\chi'}$ and $I_{\Delta,\chi}$ as in [H4,V.3]. The characters χ and χ are introduced only for the sake of the theta correspondence, and the function of the twisted inner products $I_{\Delta,\chi}$ and $I_{\Delta,\chi'}$ is precisely to remove the characters. In particular the rationality of elements of $V(\pi)(2)$ is independent of the choice of χ (cf. [H4, V (1.1), (1.2)]). Moreover, bearing in mind the discussion of complex conjugation in [H3,§2.5], it is clear from the definitions that, if $f \in V(\pi)(2)_M$ (resp. $F \in V(\pi')(2)_M$) then

$$I_{\Delta,\chi}(f) \sim_{M^\times} Q(\pi), \ I_{\Delta,\chi'}(F) \sim_{M^\times} Q(\pi). \tag{3.5.8}$$

Similarly, we can take

$$Q(\pi(2)) = Q(\pi)^2, \ Q(\pi'(2)) = Q(\pi')^2. \tag{3.5.9}$$

Here and in what follows, at the risk of loss of clarity (but in order to spare the reader excessively long formulas) we write $I_{\Delta,\chi}(f)$ to designate the element of $(M \otimes \mathbb{C})^\times$ obtained from the restriction of scalars $R_{M/\mathbb{Q}}\pi(2)_f$, as in [H3, (2.6.2), (2.8.1)]; likewise for the other periods.

Lemma 3.5.10. *Assume $F \in V(\pi)_M(2)$. Let Φ_S be as in Corollary 3.5.7. There is a constant $C(\Phi_S) \in \mathbb{C}^\times$ such that*

$$I_{\Delta,\chi}(Z_S(\Phi)(s_0)(F)) \in M^\times \cdot C(\Phi_S) \cdot Q(\pi).$$

Proof. The composition of the map

$$\pi \otimes \pi \to \pi \otimes \bar\pi \to \pi \otimes [\check\chi \circ \det \pi^\vee],$$

where the first arrow is complex conjugation in the second variable, with the map map $F \mapsto I_{\Delta,\chi}(Z_S(\Phi)(s_0)(F))$ defines a non-degenerate hermitian form $H_{S,\Phi}(\bullet, \bullet)$ on π_f^S. By Schur's Lemma, there is a constant $C(\Phi_S) \in \mathbb{C}^\times$ such that

$$H_{S,\Phi}(\bullet, \bullet) = C(\Phi_S) < \bullet, \bullet >_M .$$

The lemma follows immediately.   □

Since Φ and F are assumed to be factorizable, the integral (3.5.3) factors over local primes, as in (1.3.3). Let $S = S_f \cup S_\infty$ be the decomposition of S into finite and archimedean primes. Write $F = f \otimes f' \in \pi(2)$ with $f = \otimes_v f_v$, $f' = \otimes_v f'_v$. Setting

$$C(\Phi_{S_f}, F_{S_f}, \psi_{S_f}) = \prod_{v \in S_f} \int_{G_v} (\pi_v(g_v)(f_v), f'_v) \cdot \phi_S(\delta \cdot (g,1), s) dg_v$$

we have

$$I_{\Delta,\chi} Z_S(\phi)(s)(F)$$

$$= C(\Phi_{S_f}, F_{S_f}, \psi_{S_f}) \cdot \int_{G_{S_\infty}} I_{\Delta,\chi} \pi(2)(g,1) F \cdot \phi_\infty(\delta \cdot (g_\infty, 1), s) dg_\infty.$$

$$\tag{3.5.11}$$

Lemma 3.5.12. *Assume* Φ_{S_f} *and* F *are defined over* $\overline{\mathbb{Q}}$. *Then* $C(\Phi_{S_f}, F_{S_f}, \psi_{S_f})$ *is defined over* $\overline{\mathbb{Q}}$ *and, for any* $\sigma \in Gal(\overline{\mathbb{Q}}/\mathcal{K})$,

$$\sigma(C(\Phi_{S_f}, F_{S_f}, \psi_{S_f})) = C(\sigma(\Phi_{S_f}), \sigma(F_{S_f}), \sigma(\psi_{S_f})).$$

Proof. As explained in §E.2 of the introduction, this follows from [H4, Lemma IV.2.3.2], in view of the relation (E.2.2) that governs the dependence of the local zeta integrals on the choice of additive characters. There is one additional element: the local zeta integrals are defined with respect to a global Haar measure that is *rational* in the sense of [H4, p. 110].

Several identifications are implicit in the statement of the Lemma. We are assuming F to be an algebraic coherent cohomology class in $V(\pi)(2)$. If $\sigma \in Gal(\overline{\mathbb{Q}}/M)$ then F is again in $V(\pi)(2)$. If σ fixes the field of rationality of the automorphic vector bundle $[W_{\Lambda(2)}]$, then F is again a $\overline{\mathbb{Q}}$-algebraic class in $H_!^{2(r'-1)}(Sh(V^{(2)}), [W_{\Lambda(2)}])$, but in the subspace $V(\sigma(\pi)(2)$, defined in the obvious way. If σ fixes the reflex field of $Sh(V^{(2)})$, then F is a a $\overline{\mathbb{Q}}$ algebraic class in $H_!^{2(r'-1)}(Sh(V^{(2)}), \sigma[W_{\Lambda(2)}])$. Finally, we have been assuming $E = \mathbb{Q}$, so by hypothesis σ fixes the reflex field of $Sh(V^{(2)})$, which is contained in $\mathcal{K}$; however, this part of the argument works for more general totally real fields.   □

For the following corollary, we write $Z_S(\Phi, \psi_f)(s_0)(F)$ instead of $Z_S(\Phi)(s_0)(F)$ to emphasize the dependence on the additive character.

Corollary 3.5.13. *Under the hypothesis of Lemma 3.5.10, or more generally if $F \in V(\pi)_{\overline{\mathbb{Q}}}(2)$, there is a constant $C(\Phi_\infty) \in \mathbb{C}^\times$ such that,*

$$I_{\Delta,\chi}(Z_S(\Phi, \psi_f)(s_0)(F))/Q(\pi) \cdot C(\Phi_\infty) \in \overline{\mathbb{Q}}$$

and, for all $\sigma \in Gal(\overline{\mathbb{Q}}/\mathcal{K})$,

$$\sigma(I_{\Delta,\chi}(Z_S(\Phi, \psi_f)(s_0)(F))/[Q(\pi) \cdot C(\Phi_\infty))]$$
$$= I_{\Delta,\sigma(\chi)}(Z_S(\sigma(\Phi), \sigma(\psi_f)/Q(\sigma(\pi) \cdot C(\Phi_\infty)].$$

Here we need only recall that, as explained in §E.2, the archimedean local zeta integral is unaffected by the conjugation of the additive character at finite places. Also the action of σ on χ is defined in terms of the algebraic Hecke character χ^+ obtained by twisting χ by a certain power of the norm.

3.5.14. Application of Rallis' inner product formula. We continue to assume $E = \mathbb{Q}$. Start now with $f \in V(\pi)_M(2)$. By Theorem E.2.1 of the introduction (the corrected version of the main theorem of [H4]), there exists an explicit

$$\mathbf{p} = \mathbf{p}(\pi, V, V', \chi_0) \in \mathbb{C}^\times$$

defined in [H4, §V.1] such that

$$\mathbf{p}^{-1}\theta_{\phi,\chi,\chi',\psi_f} \in_{\sim_\mathcal{K}} \pi'(2)(M \cdot \mathbb{Q}^{ab}) \tag{3.5.14.1}$$

provided ϕ is M-rational, where the notation $\in_{\sim_\mathcal{K}}$ means homogeneous with respect to the $Gal(\overline{\mathbb{Q}}/\mathcal{K})$-action on χ, χ', ψ_f, etc. We abbreviate $\theta_{\phi,\chi,\chi',\psi_f}(f) = \theta(f)$, and so on, and use the notation $\sim_\mathcal{K}$ to indicate that $Gal(\overline{\mathbb{Q}}/\mathcal{K})$ is acting on the missing arguments.

Using Corollary 2.2.7 one can show that the field of rationality for $\pi'^{,S}$ is contained in $\mathbb{Q}(\pi, \chi^+, \chi'^{,+})$ for an appropriate finite set S of finite places; it suffices to take S the set of primes where the local L-packets are non-trivial (i.e., where $\pi'^{,S}$ can be completed in more than one way to an automorphic representation). Thus, just as in the proof of Lemma 3.5.12, all constructions that follow are homogeneous over $\mathbb{Q}(\pi, \chi^+, \chi'^{,+})$ and even over $\mathcal{K}$, if we allow $Gal(\overline{\mathbb{Q}}/\mathcal{K})$ to act on π, χ^+, $\chi'^{,+}$, and the automorphic vector bundle $[E_{\Lambda(2)}]$ (cf. Theorem E.2.1). Thus we can choose $M' \supset \mathbb{Q}(\pi, \chi^+, \chi'^{,+})$ to be a field of definition for π' in what follows; this choice also disappears in the end.

Let $F \in V(\pi')_{M'}(2) \subset H^0_!(Sh(V'^{,(2)}), [E_{\Lambda(2)}])$, defined by analogy with $V(\pi)_M(2)$. We know

$$I_\Delta(F) \sim_\mathcal{K} Q(\pi'), \ I_\Delta(f) \sim_\mathcal{K} Q(\pi) \qquad (3.5.14.2)$$

Combining (3.5.14.1) and (3.5.14.2) we find

$$I_\Delta(\theta(f)) \sim_\mathcal{K} \mathbf{p} \cdot Q(\pi') \qquad (3.5.14.3)$$

By Corollaries 3.5.7 and 3.5.13 we have

$$I_\Delta(\theta(F)) \sim_\mathcal{K} Q(\pi')C(\Phi_\infty)L \qquad (3.5.14.4)$$

where $L = [c_K d^S_{n+1}(\frac{1}{2}, \chi)]^{-1} L(1, \pi_\mathcal{K} \cdot \chi^{-1})$.

Now (3.5.14.1), together with Corollary 3.6.4 (applied to the doubled Shimura variety $Sh(V^{2)})$ implies

$$(\theta(f), F) \sim_\mathcal{K} \mathbf{p}Q(\pi')^2 \qquad (3.5.14.5)$$

But by the adjunction formula

$$(\theta(f), F) = (f, \theta(F)), \qquad (3.5.14.6)$$

Letting f vary among M-rational classes, it follows from (3.5.14.5), (3.5.14.6), and Corollary 3.6.5 that

Lemma 3.5.14.7. *If ϕ is M-rational, then $Q(\pi)^2[\mathbf{p} \cdot Q(\pi')^2]^{-1}\theta(F)$ is M-rational as class in $H^{2(r'-1)}(Sh(V^{(2)}), [W_{\Lambda(2)}])$. In particular*

$$I_\Delta(\theta(F)) \sim_\mathcal{K} \mathbf{p} \cdot Q(\pi')^2 Q(\pi)^{-2} \cdot Q(\pi) \sim_\mathcal{K} \mathbf{p} \cdot Q(\pi')^2 Q(\pi)^{-1}. \quad (3.5.14.8)$$

Comparing (3.5.14.8) with (3.5.14.4), we find that

$$\mathbf{p} \cdot Q(\pi') \sim_\mathcal{K} Q(\pi)C(\Phi_\infty)L. \qquad (3.5.14.9)$$

Finally, (3.5.14.9) and (3.5.14.3) yield

Corollary 3.5.14.10. *If ϕ is M-rational, then*

$$I_\Delta(\theta_{\phi,\chi,\chi'}(f)) \sim_\mathcal{K} Q(\pi)C(\Phi_\infty) \cdot [c_K d^S_{n+1}(\frac{1}{2}, \chi)]^{-1} L(1, \pi_\mathcal{K} \cdot \chi^{-1}) \cdot L(1, \varepsilon_\mathcal{K})$$

We will return to this formula in §4.

Remark 3.5.15. More on rationality of theta kernels. The results of the previous sections are based on Ichino's identification of the residue of an Eisenstein series with an explicit constant multiple of a certain theta lift. The Eisenstein series is attached to a global Schwartz-Bruhat function Φ, whereas the theta lift is attached to $\pi_{V'}^V \pi_K(\Phi)$, which is defined on a smaller space. The above discussion, especially Lemma 3.5.12, relies in an essential way on the rationality of the finite part Φ_f. However, the applications to period relations in §4 refer to rationality properties of the theta lift, and in particular to

$$[\pi_{V'}^V \pi_K(\Phi)]_f = \pi_{V'}^V \pi_K(\Phi_f),$$

where the equality indicates that π_K and $\pi_{V'}^V$ are defined by adelic integrals and thus can be factored as products of their archimedean and finite parts. They are defined respectively by

$$\pi_K(\Phi_f)(v) = \int_{K_f} \Phi_f(kv)dk, \ v \in V^n(\mathbf{A}_f); \tag{3.5.15.1}$$

$$\pi_{V'}^V \Phi_f(v') = \int_{M_{r_0,n}(\mathbf{A}_f)} \Phi \begin{pmatrix} x \\ v' \\ 0 \end{pmatrix} dx, \ v' \in V'^{,n}(\mathbf{A}_f). \tag{3.5.15.2}$$

The notation is explained in [I1,§4]. All that needs to be added here is that K_f is a maximal compact subgroup with volume 1, and the action of $k \in K_f$ on the argument preserves rationality, so (3.5.15.1) is consistent with rationality. In (3.5.15.2), the measure dx is assumed self-dual with respect to the choice of additive character ψ_f; thus (3.5.15.2) also preserves rationality, in the sense discussed in (E.2) above.

3.6. Pairings. We collect some general facts about rational structures on automorphic representations on coherent cohomology. Notation is as in [H3, H4]; we write $H_!$ as in [H4]. We let (G', X') be a Shimura datum, $Sh = Sh(G', X')$ the corresponding Shimura variety of dimension d, with reflex field $E' = E(G', X')$. In the previous section we have $G' = GU(V)$ with $\dim_K V = m$; however, the present discussion applies to general Shimura varieties. Let Π be a cuspidal automorphic representation of G' that contributes to $H_!^q(Sh, \mathcal{E})$ (cf. [H4, III.3] [4]) for some automorphic vector bundle $\mathcal{E}$. Thus $\Pi = \Pi_\infty \otimes \Pi_f$ is an irreducible admissible representation and

$$Hom_{G'(\mathbf{A}_f)}(\Pi_f, H_!^q(Sh, E)) \neq 0 \tag{3.6.1}$$

---

[4]$H_!$ denotes the image of cohomology of $\mathcal{E}^{sub}$ in $\mathcal{E}^{can}$; in [H3, 2.2.6] this is denoted $\bar{H}$.

We make no assumption regarding the dimension of the space in (3.6.1).

Proposition 3.6.2. *(i) The field of rationality of the $G'(\mathbf{A}_f)$ module Π_f, denoted $\mathbb{Q}(\Pi_f)$, viewed as a subfield of $\mathbb{C}$, is a number field contained in a CM field.*

(ii) The representation Π_f has a model over a finite extension M of $\mathbb{Q}(\Pi_f)$. More precisely, for any finite extension $L/\mathbb{Q}(\Pi_f)$, there is a finite abelian extension C of $\mathbb{Q}$, linearly disjoint from L, such that Π_f has a model over $C \cdot \mathbb{Q}(\Pi_f)$. In particular, M can also be taken to be contained in a CM field.

(iii) Let M be as in (ii) – contained in a CM field – and let $\mathcal{V}_M$ be a model of Π_f over M. Let $\mathcal{V}_M^c$ (resp. $\mathcal{V}_M^\vee$ denote the complex conjugate (resp. the contragredient) of $\mathcal{V}_M$, viewed as an M-rational $G'(\mathbf{A}_f)$-module. There is a real-valued algebraic character $\lambda : G'(\mathbf{A}_f) \to \mathbb{R}^\times$, (i.e., a character factoring through an algebraic Hecke character of the abelianization of G'), defined over $\mathbb{Q}(\Pi_f)$, such that $\mathcal{V}_M^c \xrightarrow{\sim} \mathcal{V}_M^\vee \otimes \lambda$. In other words, relative to complex conjugation on M, there is a $G'(\mathbf{A}_f)$-hermitian pairing

$$< \bullet, \bullet >_M : \mathcal{V}_M \otimes \mathcal{V}_M \to M(\lambda)$$

where $G'(\mathbf{A}_f)$ acts on $M(\lambda)$ by the character λ.

Proof. The first assertion is Theorem 4.4.1 of [BHR]. Part (ii) is presumably well-known. If $K \subset G'(\mathbf{A}_f)$ is an open compact subgroup, then the finite-dimensional $H(G'(\mathbf{A}_f), K)$-module Π^K is rational over $\mathbb{Q}(\Pi_f)$ and has a model over any finite extension $C/\mathbb{Q}(\Pi_f)$ that trivializes the Schur index; thus (ii) follows immediately for Π^K. The existence of a model of the representation Π is deduced in [V, II.4.7]. Finally, the first part of (iii) is contained in the proof of Theorem 4.4.1 of [BHR], and the second part is an immediate consequence. $\qquad\square$

Let Π and M be as in the above proposition, and let $\phi \in Hom_{G'(\mathbf{A}_f)}(\Pi_f, H_!^q(Sh, \mathcal{E}))$ be a non-zero homomorphism, $\mathcal{V}(\phi) = \phi(\Pi_f)(\mathbb{C})$. We identify the elements of $\mathcal{V}(\phi)$ with automorphic forms on $G'(\mathbb{Q})\backslash G'(\mathbf{A})$ with respect to a canonical trivialization (cf. [H3,p. 113]). The canonical trivialization depends on the choice of a CM pair $(T, x) \subset (G', X)$, and we let $E'_x = E' \cdot E(T, x)$. In the applications to unitary similitude groups we will have $E'_x = E'$, as in the proof of [H3, Lemma 2.5.12]. For any $f_1, f_2 \in \mathcal{V}(\phi)$ define the Petersson inner product

$$(f_1, f_2) = \int_{G'(\mathbb{Q})\backslash G'(\mathbf{A})/Z} f_1(g)\bar{f}_2(g)\lambda^{-1}(g)dg. \qquad (3.6.3)$$

Here $\lambda : G'(\mathbf{A}) \to \mathbb{R}^\times$ is the positive character of 3.6.2(iii) such that $f_1(g)\bar{f}_2(g)\chi^{-1}(g)$ is invariant under translation by Z.

Corollary 3.6.4. *Let* Π, *M, and ϕ be as above, and suppose the image of ϕ is a subspace of $H_!^q(Sh, \mathcal{E})$ defined over M. Let $M^+ \subset M$ be the maximal totally real subfield, and let $Z \subset G'(\mathbb{R})$ be the identity component of the center of $G'(\mathbb{R})$. There is a constant $Q(\Pi) \in \mathbb{R}^\times$, well-defined up to scalars in $M^{+,\times}$, such that, for any $f_1, f_2 \in \mathcal{V}(\phi)$ rational over M,*

$$(f_1, f_2) \in M^\times \cdot Q(\Pi).$$

Proof. The non-degenerate pairing (3.6.3) defines an isomorphism $\mathcal{V}^c \xrightarrow{\sim} \mathcal{V}^\vee \otimes \lambda$ of complex vector spaces. Schur's Lemma implies there is a complex constant such that

$$(f_1, f_2) = Q(\Pi) \cdot <f_1, f_2>_M$$

for any $f_1, f_2 \in \mathcal{V}$. The Corollary follows immediately.   □

Corollary 3.6.5. *Suppose* $\dim \mathrm{Hom}_{G'(\mathbf{A}_f)}(\Pi_f, \bar{H}^q(Sh, \mathcal{E})) = 1$, *and let $\mathcal{V}(\Pi_f)$ denote the Π_f-isotypic subspace of $H_!^q(Sh, \mathcal{E})$, $\mathcal{V}(\Pi_f)(M)$ its M-rational subspace. Let $f_2 \in \mathcal{V}(\Pi_f)$. Then $f_2 \in \mathcal{V}(\Pi_f)(M)$ if and only if, for all $f_1 \in H_!^q(Sh, \mathcal{E})$, $Q(\Pi)^{-1}(f_1, f_2) \in M$.*

Proof. Under the multiplicity one hypothesis, the Π_f-isotypic subspace of $H_!^q(Sh, \mathcal{E})$ is M-rational for any M over which Π_f has a model. Since $\mathcal{V}(\Pi_f)(M)$ pairs trivially with any subspace of $H_!^q(Sh, \mathcal{E})$ which is not Π_f-isotypic, the corollary is then obvious.   □

This can be interpreted in another way. Let $\mathcal{E}' = \Omega_{Sh}^d \otimes \mathcal{E}^\vee$. Then Serre duality defines a non-degenerate pairing

$$H_!^q(Sh, \mathcal{E}) \otimes H_!^{d-q}(Sh, \mathcal{E}') \to \mathbb{C}$$

which is rational over the reflex field E' with respect to the natural action of $Aut(\mathbb{C}/E')$ on the family of automorphic vector bundles $\mathcal{E}$. Let $\Pi'_f = \Pi^c_f \otimes \lambda^{-1}$, where λ is as in (3.6.3). Then (with respect to a a canonical trivialization) the map on cusp forms

$$f \mapsto \bar{f} \otimes \lambda^{-1} \qquad (3.6.6)$$

defines an isomorphism between the Π_f isotypic subspace of $H_!^q(Sh, \mathcal{E})$ and the Π'_f-isotypic subspace of $H_!^{d-q}(Sh, \mathcal{E}')$. In particular, if Π_f occurs with multiplicity one in $H_!^q(Sh, \mathcal{E})$, then Π'_f occurs with multiplicity one in $H_!^{d-q}(Sh, \mathcal{E}')$. Let $\Pi_{f,M}$ be the M-rational structure on Π_f induced by the rational structure of $H_!^q(Sh, \mathcal{E})$, and define $\Pi'_{f,M}$ analogously. Then the image $\bar{\Pi}_{f,M}$ of $\mathcal{V}_M$ under (3.6.6) is a new M-rational structure on the Π'_f-isotypic subspace of $H_!^{d-q}(Sh, \mathcal{E}')$, hence there is a constant $C \in \mathbb{C}^\times$ such that

$$C\Pi'_{f,M} = \bar{\Pi}_{f,M}.$$

One then sees by the arguments in [H3,(2.8)] that $C = Q(\Pi)$.

4. Applications to Period Relations

The main result of [H4] is Theorem V.1.10, corrected in the introduction to the present paper as Theorem E.2.1, which asserts that the theta lifting from $U(n)$ to $U(n+r)$, $r \geq 0$, preserves rationality up to an explicit period factor. The main application we had in mind for this theorem is to proving the period relations conjectured in [H3, (3.7.3)]. A sketch of a proof of these period relations is presented in section V.3 of [H4]. This sketch is based on a series of hypotheses. Most of these hypotheses have been verified in §3 above and in [H5]. The present section runs through the list of hypotheses and elaborates what remains to be done.

4.1. The hypotheses. Recall that $\mathcal{K}$ is an imaginary quadratic extension of $\mathbb{Q}$. As in section V.3 of [H4], we are given a cohomological automorphic representation π of a unitary similitude group G_1 of signature $(n-1,1)$ relative to $\mathcal{K}/\mathbb{Q}$. We assume π to be of the form $\pi_0 \otimes \beta$ (β replaces the character χ of [H3]) with π_0 self-dual and we are interested in special values of the standard L-function $L^{mot,S}(s, \pi_0 \otimes \beta, St, \alpha^*) = L^{mot,S}(s, \pi_{0,\mathcal{K}}, \tilde{\beta} \cdot \alpha^*)$.

Hypothesis 4.1.1. *The representation π_0 of G_1 admits a base change to a cohomological cuspidal automorphic representation $\pi_{0,\mathcal{K}}$ of $GL(n, \mathcal{K})$.*

This is slightly abusive: the base change of π_0 is to an automorphic representation $\pi_{0,\mathcal{K}} \otimes \Psi$ of $GL(n, \mathcal{K}) \times GL(1, \mathcal{K})$ (cf. [HT, Theorem VI.2.1]), and we are only retaining the first factor $\pi_{0,\mathcal{K}}$.

The article [H3] is motivated by standard conjectures concerning transfer of automorphic representations between inner forms of G_1. These will probably soon be available, at least for cohomological automorphic representations, thanks to the results of Laumon-Ngô and Waldspurger on the Fundamental Lemma for unitary groups. To avoid conjectures we nevertheless assume henceforward that π_0 satisfies the conditions of Theorem 3.1.6 of [HL], namely

Hypothesis 4.1.1 bis. *For at least two distinct places v of $\mathbb{Q}$ split in $\mathcal{K}$ the local factor $\pi_{0,v}$ is supercuspidal.*

This guarantees that π_0 satisfies Hypothesis 4.1.1 [HL,Theorem 2.2.2], hence the name. It also guarantees:

Theorem 4.1.2 [HL, Theorem 3.1.6]. *Assume Hypothesis 4.1.1 bis. Let G be an inner form of G_1. Then π_0 admits a transfer to an antiholomorphic representation π_0'' of G. More precisely, π_0'' and π_0 are nearly equivalent in the sense of being isomorphic locally at almost all places, and of course*

$$L^{mot,S}(s, \pi_0 \otimes \beta, St, \alpha^*) = L^{mot,S}(s, \pi_0'' \otimes \beta, St, \alpha^*)$$

if S is big enough.

Moreover,

Theorem 4.1.3 [HL, Theorem 3.1.5]. *Assume Hypothesis 4.1.1 bis. Then $\pi_{0,f}$ is (essentially) tempered at ∞ and at all finite places v such that either (i) v splits in $\mathcal{K}/\mathbb{Q}$ or (ii) v is unramified in $\mathcal{K}/\mathbb{Q}$ and $\pi_{0,v}$ is in the unramified principal series; the same holds for the finite part of any nearly equivalent automorphic representation on any inner form of G.*

In fact, Theorem 3.1.5 of [HL] asserts that the base change $\pi_{0,\mathcal{K}}$ of π_0 is everywhere essentially tempered, and even that there is a global character η of $\mathcal{K}_{\mathbf{A}}^\times$ such that the twist of $\pi_{0,\mathcal{K}}$ by η is tempered. This almost certainly implies formally that $\pi_{0,\mathcal{K}}$ is tempered at all non-split places, not only at those places satisfying (ii) of Theorem 4.1.3. Indeed, the standard L-function of π_0 equals the principal L-function of $\pi_{0,\mathcal{K}}$ outside a finite set of bad primes. However, the behavior of stable base change has not been worked out in general at ramified places. Thus we assume the first part of the following hypothesis in order to avoid worrying about spurious poles of local Euler factors:

Hypothesis 4.1.4. *The representation $\pi_{0,f}$ is essentially tempered. Moreover, π_0 occurs with multiplicity one in the space of automorphic forms on G_1, cuspidal or otherwise.*

Since π_0 is cuspidal, hence essentially unitary, the first part of the hypothesis means that π_0 is the twist by some global character of a tempered automorphic representation. See Remark 4.1.14 at the end of this section. The second part of the hypothesis is 3.5.5 (b) which we repeat here because it has been used in the proofs in §3.5; however, it is certainly unnecessary.

We now run through the hypotheses of [H4, V.3]. Notation is as in [*loc. cit.*].

Claim 4.1.5 (= Hypothesis (3.10)(a) of [H4,V]). $L^{mot,S}(s, \pi_0 \otimes \beta, St, \alpha^*) \neq 0$ *at* $s = \frac{m}{2}$.

When $m > n$ this is a consequence of the existence of base change, the Jacquet-Shalika theorem (cf. (3.4.2.1)), and the fact that the base change $\pi_{0,\mathcal{K}}$ is essentially tempered [HL,3.1.5].

Hypothesis 4.1.6 (= Hypothesis (3.10)(b) of [H4,V]). *Assume* $G(\mathbb{R}) = U(r,s)$, *as in the statement of Theorem 3.2, above. Theorem 3.5.13 of [H3] remains valid for the L-function* $L^{mot,S}(s - \frac{\kappa}{2}, \pi_0'' \otimes \beta, St, \alpha)$ *at the point* $s = s_0 = \frac{m+\kappa}{2}$.

It should have been pointed out in [H4] that the normalization has been changed in the course of the proof of Lemma 3.9, since α^* is a splitting character. In fact as pointed out there,

$$L^{mot,S}(s, \pi_0 \otimes \beta, St, \alpha^*) = L^{mot,S}(s - \frac{\kappa}{2}, \pi_0 \otimes \beta, St, \alpha) \qquad (4.1.7)$$

(in [H4] $\kappa = 1$ when m is odd, $\kappa = 0$ when m is even). So although Hypothesis V.3.10 (b) asserts that Theorem 3.5.13 of [H3] remains valid for $L^{mot,S}(s, \pi_0 \otimes \beta, St, \alpha^*)$ at the point $s = \frac{m}{2}$, the above formulation is actually correct.

Hypothesis 4.1.4 is that $\pi_{0,f}$ is tempered at all finite places. Thus we can apply Corollary 3.3 and Theorem 3.4. (Actually, for Theorem 3.4 the temperedness assertions in Theorem 4.1.3 are sufficient). When $m - n \geq 2$, Hypothesis 4.1.6 is true unconditionally, by Corollary 3.3. When $m = n+1$, the hypothesis is true for a large class of α; this is in fact sufficient for the applications. Thus for all practical purposes we can consider (3.10)(b) of [H4] to be valid. However, Hypothesis 4.1.6 is only true for α satisfying conditions (a)-(d) of Theorem 3.2. Here we promote the "Hypothesis" to a "Claim:

Claim 4.1.7 (= Hypothesis (3.10)(b) of [H4,V]). *For all but finitely many finite places* v *which do not split in* $\mathcal{K}$, *there is a finite set of splitting characters* A_v *of* $\mathcal{K}_v^\times$ *such that Hypothesis 4.1.6 above is valid for* α *provided* $\alpha_v^* \notin A_v$.

Claim 4.1.8 (= Hypotheses (3.10)(c) and (3.20)(d) of [H4,V]). *There exists a hermitian space* V'' *over* $\mathcal{K}$, *of any signature* (a,b), *with* $a + b = n$, *and a holomorphic automorphic representation* π'' *of* $GU(V'')$ *which is nearly equivalent to* π_0.

This follows from Theorem 4.1.2 above.

The results of §3 also apply to the hypotheses (3.20) of [H4]. We have already verified (3.20)(d) in Claim 4.1.8. Hypothesis (3.20)(c) is a mild restriction on $\pi_{0,\infty}$, to which we will return momentarily; for classical modular forms it corresponds to assuming the weight is at least 3. Given Hypothesis 4.1.1, Hypothesis (3.20)(b) is a restatement of Claim 4.1.5.

Thus it remains to verify Hypothesis (3.20)(a). This is the analogue of Hypothesis 4.1.6, with the representation $\pi_0'' \otimes \beta$ of G, an inner form of $U(n)$, replaced by the representation π' of G', an inner form of $U(n+1)$, considered in (3.5), and it is true subject to the same conditions. We let v and ν be as in [H4, 3.23].

Claim 4.1.9 (= Hypothesis (3.20)(a) of [H4,V]). *For all but finitely many finite places v which do not split in $\mathcal{K}$, there is a finite set of splitting characters A_v of $\mathcal{K}_v^\times$ such that Theorem 3.5.13 of [H3] remains valid for the standard L-function $L^{mot,S}(s, \pi', \check{\nu}, St)$ of G' at the integer point $s = \frac{n+2-v}{2}$, provided $\check{\nu}^* = \check{\nu}/|\check{\nu}| \notin A_v$.*

The discussion of [H4, 3.23] is rather complicated and it doesn't help matters that many of the formulas include misprints (corrected in the introduction to the present paper). Some remarks are nonetheless in order. The value $\frac{n+2-v}{2}$ above corresponds to the integer μ in [H4, Lemma 3.23.4], since we take $m = n + 1$. Then Theorem 3.4 applies directly to (antiholomorphic) cusp forms on G' to yield Claim 4.1.9, provided we take $\check{\nu}^*$ as the splitting character for the theta lift from G' to an appropriate definite $U(n+2)$. This does not satisfy the running assumption in [H4] that $0 \leq \alpha(\check{\nu}) \leq 1$, but in fact this assumption was only made to simplify the formulas and is irrelevant in the setting of Theorem 3.4.

Since it was not mentioned in [H4, V.3], I take this opportunity to point out that the right-hand inequalities (**) in Theorem 3.2, which are the main object of the calculation in [H4, Lemma 3.23.4], correspond to the existence of a non-trivial theta lift of π' (for the splitting characters χ' and $\check{\nu}^*$) to a definite unitary group; this is explained in [H5, Remark (4.4)(iv)]. The proof of Lemma 3.23.4 of [H4] derives these inequalities from Hypothesis (3.20)(c), which we restate as follows:

Hypothesis 4.1.10 (= Hypothesis (3.20)(c) of [H4,V]). *Let $(p_1 > p_2 > \cdots > p_n)$ be the Hodge types attached to the motive $M(\pi_0)$ in [H3,H4]:*

$$p_i = a_{0,i} + n - i, q_i = p_{n-i+1} = n - 1 - p_i,$$

where $a_{0,i}$ are defined in [H4,V, (3.2)]. We assume that for all $i = 1, 2, \ldots, n-1$,

$$p_i - p_{i+1} \geq 2.$$

In [H4,V] this hypothesis is made for $i = r'$, but for the period relations we need it to hold at every step of the induction on r'. The hypothesis guarantees that for every critical interval of $M(\pi_0)$ there is a Hecke character β in that critical interval such that the L-function of π_0 twisted by β has a critical value at $s = 1$ in the unitary normalization. This fails when $p_i - p_{i+1} = 1$ for some i; in that case, the only available critical value is at the center of symmetry, but then it is not known that one can choose β so that the L-function does not vanish, and so we are not sure to obtain non-trivial period relations.

There remain two hypotheses not presented as such in [H4]. The first is the Rallis inner product formula, stated as [H4, V (3.6)]:

$$I_{\Delta,\chi'}(\theta_{\phi,\chi,\chi'}(f)) = \delta I_{\Delta,\chi}(f) \cdot Z_\infty \cdot Z_S \cdot L^{mot,S}(\frac{m}{2}, \pi_{0,\mathcal{K}}, \tilde{\beta} \cdot \chi)/d_n(*m). \tag{4.1.11}$$

Our version of this formula is Corollary 3.5.14.10. The left-hand sides are the same, up to change of notation, and we have removed the rational factor δ and Z_S and replaced equality with $\sim_{\mathcal{K}}$. Our $Q(\pi)$ can be identified with $I_{\Delta,\chi}(f)$. Moreover, with $m = n + 1$, the point $s = \frac{m}{2}$ in the motivic normalization corresponds to the point $s = 1$ in the unitary normalization. Since π_0 is self-dual and $\tilde{\beta}$ and χ are conjugate self-dual, we can thus identify $L^{mot,S}(\frac{m}{2}, \pi_{0,\mathcal{K}}, \tilde{\beta} \cdot \chi)$ with $L(1, \pi_{\mathcal{K}} \cdot \chi^{-1})$ up to factors in $\mathbb{Q}(\pi)$ (the missing Euler factors).

Thus it remains to compare the expression

$$C(\Phi_\infty) \cdot [c_K d_{n+1}^S(\frac{1}{2}, \chi)]^{-1} \cdot L(1, \varepsilon_{\mathcal{K}}) \tag{4.1.12}$$

from Corollary 3.5.14.10. with $Z_\infty/d_n(*m)$ from [H4, V (3.6)], with $m = n + 1$. In fact, taking the normalizations into account, $d_{n+1}(s, \chi)$ is the product of $d_n(*s)$ with an additional shifted Dirichlet L-function. However, we will take the easy way out and simply define Z_∞ by the formula

$$Z_\infty = C(\Phi_\infty) \cdot \frac{d_n(*n+1)}{c_K d_{n+1}^S(\frac{1}{2}, \chi)} \cdot L(1, \varepsilon_{\mathcal{K}}). \tag{4.1.13}$$

With this definition, Corollary 3.5.14.10 can be used in place of [H4,V.3.6] in the arguments of [H4, §V.3]. Note that Z_∞ does not depend on the choice of χ but only on the restriction of χ to the idèles of $\mathbb{Q}$, which depends only on the parity of n.

Our proof of 3.5.14.10 depends on Hypotheses 3.5.5. To avoid complications, we assume

Hypothesis 4.1.14. *The prime* 2 *splits in* $\mathcal{K}$.

This guarantees 3.5.5 (e). Hypotheses 3.5.5 (c) and (d) are 4.1.4 and 4.1.3, respectively, whereas 3.5.5 (a) and (b) are running hypotheses in [H4]. The final hypothesis in [H4] is then the assertion that all the terms Z_∞ that occur as above in the various intermediate steps are actually rational numbers [H4, V.3.28.4]. We will not make this hypothesis here and so the unknown factors will remain in our final result. Note however that the situation is somewhat improved over that in [H4]; the main unknown term in our expression for Z_∞ is the local zeta integral $C(\Phi_\infty)$, which is given in terms of an integral of an automorphy factor against an (anti)-holomorphic matrix coefficient, whereas in [*loc. cit*] the matrix coefficient was obtained from a more general discrete series. Although the automorphy factor in the zeta integral is not nearly holomorphic (because no definite unitary group is involved in the theta lift), the anti-holomorphic matrix coefficients are considerably more elementary than general discrete series matrix coefficients, so there is some hope of calculating the expression Z_∞ explicitly. I hope to return to this matter in a future paper.

4.2. Period relations up to archimedean terms. Now that all the hypotheses of [H4,§V.3] have been verified, with some minor modifications, we can draw the conclusions. We let G_1 be a unitary similitude group as in 4.1.

Theorem 4.2.1 *Let* π_0 *be a self-dual cuspidal cohomological tempered automorphic representation of* G_1, *satisfying Hypotheses 4.1.1, 4.1.4, and 4.1.10. Then for any* $r = 1, 2, \ldots, n$, *there is a non-zero complex number* $Z_r(\pi_{0,\infty})$, *depending only on the infinity type of* π_0, *such that,*

$$P^{(r)}(\pi_0) \sim_\mathcal{K} Z_r(\pi_{0,\infty}) \prod_{j=1}^{r} Q_j(\pi_0).$$

This is [H4, V.3.28.2], where we write $Z_r(\pi_{0,\infty})$ for $\prod_{j=1}^{r} Z_{\infty,j}$.

Finally, we restate the last theorems of [H3], in a slightly strengthened form. We are assuming 4.1.14 (i.e. 3.5.5(e)); since the primary object of interest in [H3] is a self-dual automorphic representation of $GL(n, \mathbb{Q})$, we lose no generality by restricting attention to $\mathcal{K}$ in which 2 splits.

Theorem 4.2.2. *Notation is as in [H3], §3.7. Let* $\eta = \chi^{(2)} \cdot \alpha$, *and let* s_0 *be an integer satisfying*

$$\frac{n-\kappa}{2} \leq s_0 \leq min(p_{n-r}(\mu) - k, p_r(\mu) + k - \kappa).$$

If $s_0 \leq n - \frac{\kappa}{2}$, *we assume in addition that* α^* *is a splitting character for* n *and* s_0, *and if* $s_0 = \frac{n-\kappa+1}{2}$ *that there is a non-split finite place* v *such that* $\alpha_v^* \notin A_v$, *with* A_v *as in Theorem 3.4. Finally, if* n *is odd, we assume Hypothesis 3.7.9 on periods of abelian motives.*

Then there is a constant $Z_r(\pi_{0,\infty})$ *such that*

$$L^{mot}(s_0, \pi_0 \otimes \chi, St, \alpha) \sim_{E(\pi_0,\eta),\mathcal{K}} Z_r(\pi_{0,\infty}) \cdot c^+(M(\pi_0) \otimes RM(\eta)(s_0 + k)),$$

where c^+ *is the Deligne period.*

In other words, Deligne's conjecture is valid for the L-function $L^{mot}(s_0, \pi_0 \otimes \chi, St, \alpha)$, *up to uncontrolled factors in* $\mathcal{K}^\times$ *and the unknown factor* $Z_r(\pi_{0,\infty})$ *which depends only on the Hodge numbers of* $M(\pi_0)$. *In particular, if Deligne's conjecture is valid (up to* $\mathcal{K}^\times$*) for one* π_0 *with given infinity type, then it is valid for all such* π_0.

References

[D]   P. Deligne, Valeurs de fonctions L et périodes d'intégrales, *Proc. Symp. Pure Math.*, **XXXIII**, part 2 (1979), 313-346.

[H1]   M. Harris, Arithmetic vector bundles and automorphic forms on Shimura varieties II. *Compositio Math.*, **60** (1986), 323-378.

[H2]   M. Harris, Period invariants of Hilbert modular forms, I, *Lect. Notes Math.*, **1447** (1990), 155-202.

[H3]   M. Harris, L-functions and periods of polarized regular motives, *J.Reine Angew. Math.*, **483**, (1997) 75-161.

[H4]   M. Harris, Cohomological automorphic forms on unitary groups, I: rationality of the theta correspondence, *Proc. Symp. Pure Math*, **66.2**, (1999) 103-200.

[H5]   M. Harris, A simple proof of rationality of Siegel-Weil Eisenstein series, manuscript (2005), to appear in Proceedings of the AIM Conference on Eisenstein series and arithmetic.

[HKS]   M. Harris, S. Kudla, and W. J. Sweet, Theta dichotomy for unitary groups, *JAMS*, **9** (1996) 941-1004.

[HL]   M. Harris and J.-P. Labesse, Conditional base change for unitary groups, *Asian J. Math*, **8**, 653-684, (2004).

[Howe]   R. Howe, θ-functions and invariant theory, in *Proc. Symp. Pure Math.*, **XXXIII**, part 1 (1979), 275–285.

[HT]   M. Harris and R Taylor, *On the geometry and cohomology of some simple Shimura varieties*, Annals of Mathematics Studies, **151** (2001).

[I1]   A. Ichino, A regularized Siegel-Weil formula for unitary groups, *Math. Z.*, **247** (2004) 241-277.

[I2]   A. Ichino, On the Siegel-Weil formula for unitary groups, manuscript (2005).

[K]   S. Kudla, Splitting metaplectic covers of dual reductive pairs, *Israel J. Math*, **87** (1994) 361–401.

[KR1]   S. S. Kudla and S. Rallis, A regularized Siegel-Weil formula: the first term identity, *Ann. of Math.*, **140** (1994) 1-80.

[KR2]   S. S. Kudla and S. Rallis, On first occurrence in the local Theta correspondence, in J. Cogdell et al., eds, *Automorphic Representations, L-functions and Applications: Progress and Prospects*, Berlin: de Gruyter, 273-308 (2005).

[LR]   E. Lapid and S. Rallis, On the local factors of representations of classical groups, in J. Cogdell et al., eds, *Automorphic Representations, L-functions and Applications: Progress and Prospects*, Berlin: de Gruyter, 309-359 (2005).

[LZ]   S.-T. Lee and C.-B. Zhu, Degenerate principal series and local theta correspondence, *Trans. Am. Math. Soc.*, **350**, (1998) 5017-5046.

[L1]   J.-S. Li, On the classification of irreducible low rank unitary representations of classical groups, *Compositio Math.*, **71** (1989) 29-48.

[L2]   J.-S. Li, Automorphic forms with degenerate Fourier coefficients *Am. J. Math.*, **119** (1997) 523-578.

[M]   A. Mínguez, Correspondance de Howe ℓ-modulaire: paires duales de type II. Orsay thesis (2006).

[MVW]   C. Moeglin, M-F. Vignéras, J.-L. Waldspurger, Correspondances de Howe sur un corps p-adique, *Lecture Notes in Mathematics*, **1291** (1987).

[Pa]   A. Paul, Howe correspondence for real unitary groups, *J. Fun. Anal.* **159** (1998) 384-431.

[S1]   G. Shimura, Euler products and Eisenstein series, *CBMS Regional Conference Series in Mathematics*, **93**, Providence, R.I.: American Mathematical Society (1997).

[S2]   G. Shimura, Arithmeticity in the theory of automorphic forms, *Mathematical Series and Monographs*, **82**, Providence, R.I.: American Mathematical Society (2000).

[W]   A. Weil, Sur la formule de Siegel dans la théorie des groupes classiques, *Acta Math.*, **113** (1965) 1-87.

THE INVERSION FORMULA AND HOLOMORPHIC EXTENSION OF THE MINIMAL REPRESENTATION OF THE CONFORMAL GROUP

TOSHIYUKI KOBAYASHI* and GEN MANO†

RIMS, Kyoto University
Sakyo-ku, Kyoto, 606-8502, Japan
*E-mail: toshi@kurims.kyoto-u.ac.jp
†E-mail: gmano@kurims.kyoto-u.ac.jp

Dedicated to Roger Howe on the occasion of his 60th birthday

The minimal representation π of the indefinite orthogonal group $O(m+1,2)$ is realized on the Hilbert space of square integrable functions on $\mathbb{R}^m$ with respect to the measure $|x|^{-1}dx_1\cdots dx_m$. This article gives an explicit integral formula for the holomorphic extension of π to a holomorphic semigroup of $O(m+3,\mathbb{C})$ by means of the Bessel function. Taking its 'boundary value', we also find the integral kernel of the 'inversion operator' corresponding to the inversion element on the Minkowski space $\mathbb{R}^{m,1}$.

Mathematics Subject Classifications (2000): Primary 22E30; Secondary 22E45, 33C10, 35J10, 43A80, 43A85, 47D05, 51B20.

Keywords and phrases. minimal representation, holomorphic semigroup, Hermite operator, highest weight module, conformal group, Bessel function, Hankel transform, Schrödinger model.

Contents

*Partially supported by Grant-in-Aid for Scientific Research 18340037, Japan Society for the Promotion of Science.

1. Introduction

1.1. Semigroup generated by a differential operator D

Consider the differential operator

$$D := |x|\left(\frac{\Delta}{4} - 1\right)$$
$$= \frac{1}{4}\left(\sum_{j=1}^{m} x_j^2\right)^{\frac{1}{2}}\left(\sum_{j=1}^{m}\frac{\partial^2}{\partial x_j^2} - 4\right) \tag{1.1.1}$$

on $\mathbb{R}^m$. A distinguishing feature here is that D has the following properties (see Remark 3.4.4):

1) D extends to a self-adjoint operator on $L^2(\mathbb{R}^m, \frac{dx}{|x|})$.

2) D has only discrete spectra $\{-(j + \frac{m-1}{2}) : j = 0, 1, \cdots\}$ in $L^2(\mathbb{R}^m, \frac{dx}{|x|})$.

Therefore, one can define a continuous operator

$$e^{tD} = \sum_{k=0}^{\infty}\frac{t^k}{k!}D^k$$

for $\operatorname{Re} t \geq 0$ with the operator norm $e^{-\frac{m-1}{2}\operatorname{Re} t}$ satisfying the composition law

$$e^{t_1 D}\circ e^{t_2 D} = e^{(t_1+t_2)D}.$$

Thus, e^{tD} ($\operatorname{Re} t > 0$) forms a holomorphic one-parameter semigroup. Besides, the operator e^{tD} is self-adjoint if t is real, and is unitary if t is purely imaginary.

We ask:

Question. *Find an explicit formula of e^{tD}.*

In this context, our main results are stated as follows:

Theorem A (see Theorem 5.1.1). *The holomorphic semigroup* e^{tD} $(\operatorname{Re} t > 0)$ *is given by*

$$\left(e^{tD}u\right)(x) = \int_{\mathbb{R}^m} K^+(x, x'; t)u(x')\frac{dx'}{|x'|} \quad \text{for } u \in L^2(\mathbb{R}^m, \frac{dx}{|x|}),$$

where the integral kernel $K^+(x, x'; t)$ *is defined by*

$$K^+(x, x'; t) := \frac{2e^{-2(|x|+|x'|)\coth\frac{t}{2}}}{\pi^{\frac{m-1}{2}}\sinh^{m-1}\frac{t}{2}}\tilde{I}_{\frac{m-3}{2}}\left(\frac{\psi(x, x')}{\sinh\frac{t}{2}}\right).$$

Here, we set $\tilde{I}_\nu(z) := \left(\frac{z}{2}\right)^{-\nu}I_\nu(z)$ *(I_ν is the I-Bessel function (see (8.5.2)))* *and*

$$\psi(x, x') := 2\sqrt{2(|x||x'| + \langle x, x'\rangle)} = 4|x|^{\frac{1}{2}}|x'|^{\frac{1}{2}}\cos\frac{\theta}{2},$$

where $\theta \equiv \theta(x, x')$ *is the Euclidean angle between* x *and* x' *in* $\mathbb{R}^m$.

Particularly important is the special value at $t = \pi\sqrt{-1}$. We set

$$K^+(x, x') := \lim_{\varepsilon \downarrow 0} K^+(x, x'; \pi\sqrt{-1} + \varepsilon)$$

$$= \frac{2}{\sqrt{-1}^{m-1}\pi^{\frac{m-1}{2}}}\tilde{J}_{\frac{m-3}{2}}(\psi(x, x')).$$

Here, we set $\tilde{J}_\nu(z) := \left(\frac{z}{2}\right)^{-\nu}J_\nu(z)$ *(J_ν is the Bessel function of the first kind (see (8.5.1)))*.

Corollary B (see Theorem 6.1.1 and Corollary 6.2.1). *The unitary operator* $e^{\pi\sqrt{-1}D}$ *on* $L^2(\mathbb{R}^m, \frac{dx}{|x|})$ *is given by the Hankel-type transform:*

$$T : L^2(\mathbb{R}^m, \frac{dx}{|x|}) \to L^2(\mathbb{R}^m, \frac{dx}{|x|}), \quad u \mapsto \int_{\mathbb{R}^m} K^+(x, x')u(x')\frac{dx'}{|x'|}.$$

This transform has the following properties:

(Inversion formula) $T^{-1} = (-1)^{m+1}T$.

(Plancherel formula) $\|Tu\|_{L^2(\mathbb{R}^m, \frac{dx}{|x|})} = \|u\|_{L^2(\mathbb{R}^m, \frac{dx}{|x|})}$.

Let us explain the backgrounds and motivation of our question from three different viewpoints:

(1) Hermite semigroup and its variants (see Subsection 1.2).

(2) $\mathfrak{sl}_2$-triple of differential operators on $\mathbb{R}^m$ (see Subsection 1.3).

(3) Minimal representation of the reductive group $O(m+1, 2)$ (see Subsection 1.4).

1.2. Comparison with the Hermite operator $\mathcal{D}$

Let us compare our operator D on $L^2(\mathbb{R}^m, \frac{dx}{|x|})$ with the well-known operator $\mathcal{D}$ on $L^2(\mathbb{R}^m, dx)$ defined by

$$\mathcal{D} := \frac{1}{4}(\Delta - |x|^2). \tag{1.2.1}$$

We call this operator the *Hermite operator* following the terminology of R. Howe and E.-C. Tan [17]. Analogously to D, the Hermite operator $\mathcal{D}$ satisfies the following properties:

1) $\mathcal{D}$ extends to a self-adjoint operator on $L^2(\mathbb{R}^m, dx)$.

2) $\mathcal{D}$ has only discrete spectra $\{-\frac{1}{2}(j + \frac{m}{2}) : j = 0, 1, \cdots\}$ in $L^2(\mathbb{R}^m, dx)$.

We recall from [16, §5.3 and §6] that $\mathcal{D}$ gives rise to a holomorphic semigroup $e^{t\mathcal{D}}$ ($\operatorname{Re} t > 0$) (*Hermite semigroup*). Then the following results may be regarded as a prototype of Theorem A and Corollary B.

Fact C (see [16, §5] [31, §4.1]). *The holomorphic semigroup $e^{t\mathcal{D}}$ ($\operatorname{Re} t > 0$) is given by*

$$(e^{t\mathcal{D}}u)(x) = \int_{\mathbb{R}^m} \mathcal{K}(x, x'; t)u(x')dx',$$

where $\mathcal{K}$ is the Mehler kernel defined by

$$\mathcal{K}(x, x'; t) := \frac{1}{(2\pi \sinh \frac{t}{2})^{\frac{m}{2}}} \exp\left(-\frac{1}{2}\begin{pmatrix} x \\ x' \end{pmatrix}^t A(t) \begin{pmatrix} x \\ x' \end{pmatrix}\right). \tag{1.2.2}$$

Here, we set

$$A(t) := \frac{1}{\sinh \frac{t}{2}}\begin{pmatrix} (\cosh \frac{t}{2})I_m & -I_m \\ -I_m & (\cosh \frac{t}{2})I_m \end{pmatrix} \in GL(2m, \mathbb{R}).$$

In light of the limit formula:

$$\lim_{\varepsilon \downarrow 0} \mathcal{K}(x, x'; \pi\sqrt{-1} + \varepsilon) = \frac{1}{(2\pi\sqrt{-1})^{\frac{m}{2}}} e^{-\sqrt{-1}\langle x, x'\rangle},$$

the special value of the operator $e^{t\mathcal{D}}$ at $t = \pi\sqrt{-1}$ reduces to the (ordinary) Fourier transform:

Fact D. *The unitary operator* $e^{\pi\sqrt{-1}\mathcal{D}}$ *on* $L^2(\mathbb{R}^m, dx)$ *is nothing other than the Fourier transform* $\mathcal{F}$:

$$(\mathcal{F}f)(x) = \frac{1}{(2\pi\sqrt{-1})^{\frac{m}{2}}} \int_{\mathbb{R}^m} e^{-\sqrt{-1}\langle x,x'\rangle} f(x')dx'.$$

We shall see in Section 6 a group theoretic interpretation of the fact that T is unitary and $T^4 = \mathrm{id}$ as well as the fact that $\mathcal{F}$ is unitary and $\mathcal{F}^4 = \mathrm{id}$.

1.3. The action of $SL(2,\mathbb{R})^{\sim} \times O(m)$

The self-adjoint operator D defined by (1.1.1) arises in the context of the $\mathfrak{sl}_2$-triple of differential operators on $\mathbb{R}^m$ as follows. We define

$$\tilde{h} = 2\sum_{j=1}^{m} x_j\frac{\partial}{\partial x_j} + m - 1, \quad \tilde{e} = 2\sqrt{-1}|x|, \quad \tilde{f} = \frac{\sqrt{-1}}{2}|x|\Delta. \quad (1.3.1)$$

These operators $\tilde{h}, \tilde{e}$ and $\tilde{f}$ are skew self-adjoint operators on $L^2(\mathbb{R}^m, \frac{dx}{|x|})$, and satisfy the $\mathfrak{sl}_2$-relation:

$$[\tilde{h}, \tilde{e}] = 2\tilde{e}, \quad [\tilde{h}, \tilde{f}] = -2\tilde{f}, \quad [\tilde{e}, \tilde{f}] = \tilde{h}.$$

The operator D has the following expression

$$D = \frac{1}{2\sqrt{-1}}(-\tilde{e} + \tilde{f}), \quad (1.3.2)$$

which means that $\sqrt{-1}D$ corresponds to a generator of $\mathfrak{so}(2)$ in $\mathfrak{sl}(2,\mathbb{R})$. The $\mathfrak{sl}_2$-module $C_0^\infty(\mathbb{R}^m \setminus \{0\})$ exponentiates to a unitary representation of $SL(2,\mathbb{R})$ on $L^2(\mathbb{R}^m, \frac{dx}{|x|})$ (see Subsection 2.3 and Lemma 3.4.1).

On the other hand, there is a natural unitary representation of the orthogonal group $O(m)$ on the same space $L^2(\mathbb{R}^m, \frac{dx}{|x|})$, and the actions of $SL(2,\mathbb{R})$ and $O(m)$ mutually commute. Then we have the following discrete and multiplicity-free decomposition into irreducible representations of $SL(2,\mathbb{R}) \times O(m)$ (see [22, Theorem A]):

$$L^2(\mathbb{R}^m, \frac{dx}{|x|}) \simeq \sum_{j=0}^{\infty}{}^{\oplus} \pi_{2j+m-1}^{SL(2,\mathbb{R})} \otimes \mathcal{H}^j(\mathbb{R}^m).$$

Here, $\mathcal{H}^j(\mathbb{R}^m)$ denotes the space of harmonic polynomials on $\mathbb{R}^m$ of degree j, and $\pi_b^{SL(2,\mathbb{R})}$ stands for the irreducible unitary lowest weight representation of $SL(2,\mathbb{R})$ with minimal K-type $\mathbb{C}_b$ for $b \in \mathbb{N}_+ = \{1, 2, \cdots\}$. It is the limit of discrete series if $b = 1$, and holomorphic discrete series if $b \geq 2$.

In contrast, the Hermite operator $\mathcal{D}$ (see (1.2.1)) arises from the following $\mathfrak{sl}_2$-triple:

$$\tilde{h}' := \sum_{j=1}^{m} x_j \frac{\partial}{\partial x_j} + \frac{m}{2}, \quad \tilde{e}' := \frac{\sqrt{-1}}{2}|x|^2, \quad \tilde{f}' := \frac{\sqrt{-1}}{2}\Delta, \qquad (1.3.3)$$

where the Hermite operator $\mathcal{D}$ is given by

$$\mathcal{D} = \frac{1}{2\sqrt{-1}}(-\tilde{e}' + \tilde{f}'). \qquad (1.3.4)$$

This $\mathfrak{sl}_2$-triple also gives rise to the commutative actions of the double covering group $SL(2,\mathbb{R})^\sim$ of $SL(2,\mathbb{R})$ and $O(m)$ on $L^2(\mathbb{R}^m)$, whose irreducible decomposition amounts to (see [17, Chapter III, Theorem 2.4.4])

$$L^2(\mathbb{R}^m, dx) \simeq \sum_{j=0}^{\infty}{}^{\oplus} \pi_{j+\frac{m}{2}}^{SL(2,\mathbb{R})^\sim} \otimes \mathcal{H}^j(\mathbb{R}^m).$$

Here, $\pi_b^{SL(2,\mathbb{R})^\sim}$ stands for the irreducible unitary lowest weight representation of $SL(2,\mathbb{R})^\sim$ with minimal K-type $\mathbb{C}_b$ for $b \in \frac{1}{2}\mathbb{N}_+ = \{\frac{1}{2}, 1, \frac{3}{2}, \cdots\}$. It is the Weil representation if $b = \frac{1}{2}$, and is obtained by the representation of $SL(2,\mathbb{R})$ if $b \in \mathbb{N}_+$.

1.4. Minimal representation as hidden symmetry

The representation of $SL(2,\mathbb{R}) \times O(m)$ on $L^2(\mathbb{R}^m, \frac{dx}{|x|})$ in Subsection 1.3 extends to the irreducible unitary representation π_+ of the double covering group $G := SO_0(m+1,2)^\sim$ of the indefinite orthogonal group (see Subsections 2.3 and 3.1). If m is odd, this representation is well-defined also as a representation of $SO_0(m+1,2)$.

Similarly, the representation of $SL(2,\mathbb{R})^\sim \times O(m)$ on $L^2(\mathbb{R}^m, dx)$ extends to the unitary representation ϖ of the metaplectic group $G' = Sp(m,\mathbb{R})^\sim$.

These groups G and G' may be interpreted as *hidden symmetry* of $SL(2,\mathbb{R})^\sim \times O(m)$. Conversely, the group $SL(2,\mathbb{R})^\sim \times O(m)$ forms a 'dual pair' in each of the groups G and G'.

The unitary representations π_+ and ϖ are typical examples of 'minimal representations' of reductive Lie groups in the sense that the Gelfand–Kirillov dimension attains its minimum among infinite dimensional unitary representations or in the sense that its annihilator is the Joseph ideal in the enveloping algebra.

The unitary representation π_+ may be interpreted as the mass-zero spin-zero wave equation, or as the bound states of the Hydrogen atom (in m space dimensions), while the representation ϖ is sometimes referred to as the oscillator representation or as the (Segal–Shale–)Weil representation.

We shall review the L^2-realization of the minimal representation π_+ of $SO_0(m+1,2)^\sim$, that is, the analog of the Schrödinger model on $L^2(\mathbb{R}^m, \frac{dx}{|x|})$ in Subsection 3.1. See also [8, 16] for a nice introduction to the original Schrödinger model of the Weil representation of $Sp(m,\mathbb{R})^\sim$ on $L^2(\mathbb{R}^m)$.

To be more precise, we take

$$
e = \begin{pmatrix} 0 & 1 \\ 0 & 0 \end{pmatrix}, \quad f = \begin{pmatrix} 0 & 0 \\ 1 & 0 \end{pmatrix}, \quad h := [e,f] = \begin{pmatrix} 1 & 0 \\ 0 & -1 \end{pmatrix}
$$

to be a basis of $\mathfrak{sl}(2,\mathbb{R})$ and define injective Lie algebra homomorphisms

$$
\phi : \mathfrak{sl}(2,\mathbb{R}) \to \mathfrak{so}(m+1,2),
$$
$$
\varphi : \mathfrak{sl}(2,\mathbb{R}) \to \mathfrak{sp}(m,\mathbb{R})
$$

(see Subsection 3.3) such that the differential operators (1.3.1) and (1.3.3) are obtained via ϕ and φ, respectively, that is,

$$
d\pi_+(\phi(X)) = \tilde{X}, \quad d\varpi(\varphi(X)) = \tilde{X}'
$$

holds for $X = e, f, h$. Next we set

$$
z := \frac{1}{2\sqrt{-1}}(-e+f) = \frac{1}{2\sqrt{-1}} \begin{pmatrix} 0 & -1 \\ 1 & 0 \end{pmatrix} \in \sqrt{-1}\mathfrak{sl}(2,\mathbb{R}).
$$

Then $e^{\sqrt{-1}\mathbb{R}\phi(z)}$ is the center of the maximal compact subgroup of G for $m > 1$, while $e^{\sqrt{-1}\mathbb{R}\varphi(z)}$ is that of G' (we use the same notations ϕ and φ for their complex linear extensions).

In this context, we shall see in Lemma 3.4.3 and Remark 3.4.5 that the differential operators D and $\mathcal{D}$ are given by

$$
D = d\pi_+(\phi(z)), \quad \mathcal{D} = d\varpi(\varphi(z)). \tag{1.4.1}
$$

Thanks to these formulas, Theorem A and Corollary B are also useful in the analysis on minimal representations of $SO_0(m+1,2)^\sim$ as well as $Sp(m,\mathbb{R})^\sim$ in the following contexts:

1) The Gelfand–Gindikin program — Theorem A.

The *Gelfand–Gindikin program* asks for extending a given unitary representation of a real semisimple Lie group G to a holomorphic object of some complex submanifold in its complexification $G_{\mathbb{C}}$. Stanton and Olshanskiĭ [29, 27] independently gave a general framework of the Gelfand–Gindikin program for holomorphic discrete series. Their abstract results are enriched, for example for $G' = Sp(m, \mathbb{R})^{\sim}$, by the explicit formula of the Hermite semigroup $e^{t\mathcal{D}} = \varpi(e^{t\varphi(z)})$ for the Weil representation ϖ on $L^2(\mathbb{R}^m)$ by Howe [16]. Likewise, Theorem A gives an explicit formula of the semigroup $e^{t\mathcal{D}} = \pi_+(e^{t\phi(z)})$ for the minimal representation of $G = SO_0(m + 1, 2)^{\sim}$. Since $e^{t\phi(z)} \in G_{\mathbb{C}} \setminus G$ for $\operatorname{Re} t > 0$, Theorem A can be interpreted as a descendent of the Gelfand–Gindikin program.

2) The unitary inversion operator — Corollary B.

In the Schrödinger model of the minimal representation, G acts only on the function space $L^2(\mathbb{R}^m, \frac{dx}{|x|})$ but does not act on the underlying geometry $\mathbb{R}^m$ itself. One may observe this fact by the aforementioned formula $\tilde{f} = d\pi_+(\phi(f))$, which does not act on functions on $\mathbb{R}^m$ as a vector field but acts as a differential operator of second order on $\mathbb{R}^m$ (see (1.3.1)). To see how G acts on $L^2(\mathbb{R}^m, \frac{dx}{|x|})$, we use the facts that

1) G is generated by $\overline{P^{\max}}$ and w_0.
2) The $\overline{P^{\max}}$ action on $L^2(\mathbb{R}^m, \frac{dx}{|x|})$ is easily described (see Subsection 2.2).

Here, $\overline{P^{\max}}$ is a maximal parabolic subgroup of G (see Subsection 2.1) and $w_0 := e^{\pi\sqrt{-1}\phi(z)} \in G$ sends $\overline{P^{\max}}$ to its opposite parabolic subgroup $P^{\max}$. We set $Z := \phi(z)$. Geometrically, $\overline{P^{\max}}$ is essentially the conformal affine transformation group on the flat standard Lorentz manifold $\mathbb{R}^{m,1}$ (the *Minkowski space*), and $w_0 = e^{\pi\sqrt{-1}Z} \in G$ acts on $\mathbb{R}^{m,1}$ as the 'inversion' element (see Subsection 6.1).

Thus the representation π_+ of G on $L^2(\mathbb{R}^m, \frac{dx}{|x|})$ would be understood if we find an explicit formula for $\pi_+(w_0)$. But since the formula $w_0 = e^{\pi\sqrt{-1}\phi(z)}$ implies $\pi_+(w_0) = e^{\pi\sqrt{-1}D}$, Corollary B answers this question. This parallels the fact that the Weil representation is generated by the (natural) action of the Siegel parabolic subgroup P_{Siegel} and the Fourier transform $\mathcal{F} = e^{\pi\sqrt{-1}\mathcal{D}}$ (see Fact D).

Briefly, we pin down the analogy in the table below. Howe [16] established the left-hand side of the table for the oscillator representation ϖ of

$Sp(m, \mathbb{R})^{\sim}$, while Theorem A and Corollary B supply the right-hand side of the table for the minimal representation π_+ of $SO_0(m+1,2)^{\sim}$.

	$\mathfrak{sp}(m, \mathbb{R})$	$\mathfrak{so}(m+1,2)$
minimal representation	$(d\varpi, L^2(\mathbb{R}^m))$	$(d\pi_+, L^2(\mathbb{R}^m, \frac{dx}{|x|}))$
e	$\frac{\sqrt{-1}}{2}|x|^2$	$2\sqrt{-1}|x|$
f	$\frac{\sqrt{-1}}{2}\Delta$	$\frac{\sqrt{-1}}{2}|x|\Delta$
h	$E_x + \frac{m}{2}$	$2E_x + m - 1$
$z = \frac{-e+f}{2\sqrt{-1}}$	$\mathcal{D} := \frac{1}{4}(\Delta - |x|^2)$	$D := \frac{1}{4}|x|\Delta - |x|$
holomorphic semigroup e^{tz}	$\mathcal{K}(x, x'; t)$	$K(x, x'; t)$
inversion $e^{\pi\sqrt{-1}z}$	Fourier transform	Hankel transform
maximal parabolic subgroup	P_{Siegel}	$\overline{P^{\max}}$

Analogous results to Corollary B were previously known for some singular unitary highest weight representations. For example, see a paper [4] by Ding, Gross, Kunze and Richards for those of $U(n, n)$. Since $SU(2, 2)$ is a double covering group of $SO_0(4, 2)$, Corollary B in the case $m = 3$ essentially corresponds to [4, Corollary 7.5] in the case $(n, k) = (2, 1)$ in their notation. However, our proof based on an analytic continuation (see Theorem A) is different from theirs. Also in [20], we shall find the inversion operator $\pi(w_0)$ for the minimal representation of $O(p, q)$ for $p+q$ even, and in particular give yet another proof of Corollary B for odd m.

We also present explicit integral formulas of $\pi_+(e^{tZ})$ and $\pi_+(w_0)$ when restricted to radial functions and alike (see Theorem 4.1.1). This yields a group theoretic interpretation of some classic formulas of special functions of one variable including Weber's second exponential integral formula on Bessel function and the reciprocal and the Parseval-Plancherel formula for the Hankel transform.

This article is organized as follows. After summarizing the preliminary results on the L^2-model of the minimal representation π, we find explicitly which function arises for describing K-types in $L^2(\mathbb{R}^m, \frac{dx}{|x|})$, and define a holomorphic extension $\pi_+(e^{tZ})$ in Section 3. The integral formula of the 'radial' part of $\pi_+(e^{tZ})$ is given in Theorem 4.1.1. Theorem A is proved in Section 5 by using the result of Section 4. Taking the special value at $t = \pi\sqrt{-1}$, we obtain the integral formula of $\pi_+(w_0)$ corresponding to the inversion element w_0. This corresponds to Corollary B, and is proved in

Section 6. Our integral formula for $\pi_+(w_0)$ enables us to write explicitly the action of the whole group $G = SO_0(m+1,2)\tilde{}$ on $L^2(\mathbb{R}^m, \frac{dx}{|x|})$. This is given in Section 7. For the convenience of the reader, we collect basic formulas of special functions in a way that we use in this article.

The main results of the paper were announced in [19] with a sketch of the proof.

Notation: $\mathbb{N} = \{0,1,2,\ldots\}$, $\mathbb{N}_+ = \{1,2,3,\ldots\}$, $\mathbb{R}_{\leq 0} = \{x \in \mathbb{R} : x \leq 0\}$, $\mathbb{R}_+ = \{x \in \mathbb{R} : x > 0\}$.

2. Preliminary Results on the Minimal Representation of $O(m+1,2)$

This section gives a brief review on the known results of the L^2-model of the minimal representation of $O(m+1,2)$ in a way that we shall use later. We shall give an explicit action of the maximal parabolic subgroup $\overline{P^{\max}}$ and the Lie algebra $\mathfrak{g}$. Furthermore, we state an explicit K-type decomposition of $L^2(C_+)$ even though the action of K itself is not given explicitly here (see Section 7 for this).

2.1. Maximal parabolic subgroup of the conformal group

Let $O(m+1,2)$ be the indefinite orthogonal group which preserves the quadratic form $x_0^2 + \cdots + x_m^2 - x_{m+1}^2 - x_{m+2}^2$ of signature $(m+1,2)$. We denote by $e_0,\cdots,e_{m+2}$ the standard basis of $\mathbb{R}^{m+3}$, and by E_{ij} ($0 \leq i,j \leq m+2$) the matrix unit. We set

$$\varepsilon_j := \begin{cases} 1 & (1 \leq j \leq m), \\ -1 & (j = m+1). \end{cases}$$

We take the following elements of the Lie algebra $\mathfrak{o}(m+1,2)$:

$$\overline{N}_j := E_{j,0} + E_{j,m+2} - \varepsilon_j E_{0,j} + \varepsilon_j E_{m+2,j} \quad (1 \leq j \leq m+1),$$
$$N_j := E_{j,0} - E_{j,m+2} - \varepsilon_j E_{0,j} - \varepsilon_j E_{m+2,j} \quad (1 \leq j \leq m+1), \quad (2.1.1)$$
$$E := E_{0,m+2} + E_{m+2,0},$$

and define subalgebras of $\mathfrak{o}(m+1,2)$ by

$$\overline{\mathfrak{n}^{\max}} := \sum_{j=1}^{m+1} \mathbb{R}\overline{N}_j, \quad \mathfrak{n}^{\max} := \sum_{j=1}^{m+1} \mathbb{R}N_j, \quad \mathfrak{a} := \mathbb{R}E.$$

Then we define the following subgroups of $O(m+1,2)$:

$$M_+^{\mathrm{max}} := \{g \in O(m+1,2) : g \cdot e_0 = e_0,\ g \cdot e_{m+2} = e_{m+2}\}$$
$$\simeq O(m,1),$$
$$M^{\mathrm{max}} := M_+^{\mathrm{max}} \cup \{-I_{m+3}\} \cdot M_+^{\mathrm{max}}$$
$$\simeq O(m,1) \times \mathbb{Z}_2,$$
$$\overline{N^{\mathrm{max}}} := \exp(\overline{\mathfrak{n}^{\mathrm{max}}}),$$
$$N^{\mathrm{max}} := \exp(\mathfrak{n}^{\mathrm{max}}),$$
$$A := \exp(\mathfrak{a}).$$

For $b = (b_1, \ldots, b_{m+1}) \in \mathbb{R}^{m+1}$, we set

$$\overline{n}_b := \exp\left(\sum_{j=1}^{m+1} b_j \overline{N}_j\right) \tag{2.1.2}$$
$$= I_{m+3} + \sum_{j=1}^{m+1} b_j \overline{N}_j$$
$$+ \frac{Q(b)}{2}(-E_{0,0} - E_{0,m+1} + E_{m+1,0} + E_{m+1,m+1}),$$

where $Q(b)$ is the quadratic form of signature $(m,1)$ given by

$$Q(b) := b_1^2 + \cdots + b_m^2 - b_{m+1}^2.$$

The Lie group $\overline{N^{\mathrm{max}}}$ is abelian, and we have an isomorphism of Lie groups:

$$\mathbb{R}^{m+1} \overset{\sim}{\to} \overline{N^{\mathrm{max}}}, \quad b \mapsto \overline{n}_b.$$

It is readily seen from (2.1.2) that

$$\overline{n}_b(e_0 - e_{m+2}) = e_0 - e_{m+2}, \tag{2.1.3}$$
$$\overline{n}_b(e_0 + e_{m+2}) = {}^t(1 - Q(b), 2b_1, \ldots, 2b_{m+1}, 1 + Q(b)). \tag{2.1.4}$$

We also note

$$e^{tE}(e_0 + e_{m+2}) = e^t(e_0 + e_{m+2}). \tag{2.1.5}$$

The subgroup $M_+^{\mathrm{max}}\overline{N^{\mathrm{max}}}$ is isomorphic to the semidirect product group $O(m,1) \ltimes \mathbb{R}^{m+1}$ via the bijection $\mathbb{R}^{m+1} \overset{\sim}{\to} \overline{N^{\mathrm{max}}}$, $b \mapsto \overline{n}_b$. In this context, $M_+^{\mathrm{max}}\overline{N^{\mathrm{max}}}$ is regarded as the group of isometries of the Minkowski space $\mathbb{R}^{m,1}$, while $O(m+1,2)$ is the group of Möbius transformations on $\mathbb{R}^{m,1}$ preserving its conformal structure.

Next, we define a maximal parabolic subgroup

$$\overline{P^{\max}} := M^{\max} A \overline{N^{\max}}.$$

In our analysis of the minimal representation of $O(m+1,2)$, $\overline{P^{\max}}$ plays an analogous role to the Siegel parabolic subgroup of the metaplectic group $Sp(m, \mathbb{R})\widetilde{\ }$ for the Weil representation.

2.2. L^2-model of the minimal representation

We shall briefly review the L^2-model of the minimal representation of $O(m+1,2)$. Let $C_\pm$ be the forward and the backward light cone respectively:

$$C_\pm := \{(\zeta_1, \cdots, \zeta_{m+1}) \in \mathbb{R}^{m+1} : \pm\zeta_{m+1} > 0, \ \zeta_1^2 + \cdots + \zeta_m^2 = \zeta_{m+1}^2\},$$

and C be its disjoint union $C_+ \cup C_+$, that is, C is the conical subvariety with respect to the quadratic form of signature $(m, 1)$:

$$C = \{\zeta \in \mathbb{R}^{m+1} \setminus \{0\} : Q(\zeta) = 0\}. \tag{2.2.1}$$

Note that $M_+^{\max}$ acts on C transitively.

The measure $d\mu$ on C is naturally defined to be $\delta(Q)$, the generalized function associated to the quadratic form Q (see [11, Chapter III, §2]). Then we form a unitary representation π of $\overline{P^{\max}}$ on the Hilbert space $L^2(C, d\mu) \equiv L^2(C)$ as follows: for $\psi \in L^2(C)$,

$$(\pi(e^{tE})\psi)(\zeta) := e^{-\frac{m-1}{2}t}\psi(e^{-t}\zeta) \qquad\qquad (t \in \mathbb{R}), \tag{2.2.2}$$

$$(\pi(m)\psi)(\zeta) := \psi({}^t m\zeta) \qquad\qquad (m \in M_+^{\max}),$$

$$\tag{2.2.3}$$

$$(\pi(-I_{m+3})\psi)(\zeta) := (-1)^{\frac{m-1}{2}}\psi(\zeta), \tag{2.2.4}$$

$$(\pi(\overline{n}_b)\psi)(\zeta) := e^{2\sqrt{-1}(b_1\zeta_1 + \cdots + b_{m+1}\zeta_{m+1})}\psi(\zeta) \qquad (b \in \mathbb{R}^{m+1}). \tag{2.2.5}$$

Then π is irreducible and unitary as a $\overline{P^{\max}}$-module, and it is proved in [23, Theorem 4.9] that the $\overline{P^{\max}}$-module $(\pi, L^2(C))$ extends to an irreducible unitary representation of $O(m + 1, 2)$ if m is odd. We shall denote this representation of $O(m + 1, 2)$ by the same letter π. The direct sum decomposition

$$L^2(C) = L^2(C_+) \oplus L^2(C_-) \tag{2.2.6}$$

yields a branching law $\pi = \pi_+ \oplus \pi_-$ with respect to the restriction $O(m + 1, 2) \downarrow SO_0(m + 1, 2)$, where $SO_0(m + 1, 2)$ is the identity component of

$O(m+1,2)$. The irreducible representations π_+ and π_- of $SO_0(m+1,2)$ are contragredient to each other, one is a highest weight module, and the other is a lowest weight module.

2.3. K-type decomposition

Let $SO(2)^\sim$ be the double covering of $SO(2)$. We write η for the unique element of $SO(2)^\sim$ of order two. Then, we have an exact sequence:

$$1 \to \{1, \eta\} \to SO(2)^\sim \to SO(2) \to 1.$$

Let

$$G = SO_0(m+1,2)^\sim \tag{2.3.1}$$

be the double covering group of $SO_0(m+1,2)$ characterized as follows: a maximal compact subgroup K is of the form $K_1 \times K_2 \simeq SO(m+1) \times SO(2)^\sim$ and the kernel of the covering map $G \to SO_0(m+1,2)$ is given by $\{(1,1),(1,\eta)\}$. Likewise, the double covering group $O(m+1,2)^\sim$ of $O(m+1,2)$ is defined.

If m is odd, the irreducible representation $(\pi_\pm, L^2(C_\pm))$ defined in Subsection 2.2 extends to that of $SO_0(m+1,2)$, and therefore, also that of $G = SO_0(m+1,2)^\sim$ as we proved it more generally for $O(p,q)$ $(p+q :$ even) in [23]. We shall use the same letter π_+ to denote the extension to $SO_0(m+1,2)$ or G. If m is even, by [28], the irreducible unitary representation $(\pi_\pm, L^2(C_\pm))$ is still well-defined as a representation of G, whose Lie algebra $\mathfrak{g} = \mathfrak{so}(m+1,2)$ of G acts in the same manner as in the case of odd m (see Subsection 2.4).

The K-type formula of $(\pi_\pm, L^2(C_\pm))$ is given as follows:

$$L^2(C_\pm)_K \simeq \bigoplus_{a=0}^\infty \mathcal{H}^a(\mathbb{R}^{m+1}) \boxtimes \mathbb{C}e^{\pm(a+\frac{m-1}{2})\sqrt{-1}\theta}. \tag{2.3.2}$$

(For example, this formula can be read from [28, §1.3] by substituting $d = m-1, p = 1$ and $q = 0$.) Here, $\mathcal{H}^a(\mathbb{R}^{m+1})$ stands for the representation of $SO(m+1)$ on the space of the spherical harmonics which is irreducible if $m > 1$ (see Subsection 8.4).

Likewise, the representation $(\pi, L^2(C))$ of $O(m+1,2)^\sim$ decomposes when restricted to its maximal compact subgroup as follows (see [21, Theorem 3.6.1]):

$$L^2(C)_K \simeq \bigoplus_{a=0}^\infty \mathcal{H}^a(\mathbb{R}^{m+1}) \boxtimes \mathcal{H}^{a+\frac{m-1}{2}}(\mathbb{R}^2). \tag{2.3.3}$$

$\mathcal{H}^{a+\frac{m-1}{2}}(\mathbb{R}^2)$ decomposes into $\mathbb{C}e^{(a+\frac{m-1}{2})\sqrt{-1}\theta}\oplus\mathbb{C}e^{-(a+\frac{m-1}{2})\sqrt{-1}\theta}$ as $SO(2)$-modules (see Subsection 8.4). This corresponds to the decomposition

$$L^2(C) = L^2(C_+) \oplus L^2(C_-),$$

for which the K-type formula is given by (2.3.2).

2.4. Infinitesimal action of the minimal representation

For $\overline{N}_j, N_j (1 \leq j \leq m+1)$ and E, we define linear transformations on the space $\mathcal{S}'(\mathbb{R}^{m+1})$ of tempered distributions by

$$d\hat{\varpi}(\overline{N}_j) := 2\sqrt{-1}\zeta_j, \tag{2.4.1}$$

$$d\hat{\varpi}(N_j) := \sqrt{-1}\left(-\frac{m+3}{2}\varepsilon_j\frac{\partial}{\partial\zeta_j} - E_\zeta\varepsilon_j\frac{\partial}{\partial\zeta_j} + \frac{1}{2}\zeta_j\Box_\zeta\right), \tag{2.4.2}$$

$$d\hat{\varpi}(E) := -\frac{m+3}{2} - E_\zeta, \tag{2.4.3}$$

where we set

$$\Box_\zeta := \frac{\partial^2}{\partial\zeta_1^2} + \frac{\partial^2}{\partial\zeta_2^2} + \cdots + \frac{\partial^2}{\partial\zeta_m^2} - \frac{\partial^2}{\partial\zeta_{m+1}^2}, \quad E_\zeta := \sum_{j=1}^{m+1}\zeta_j\frac{\partial}{\partial\zeta_j}.$$

Then, we recall from [23] that this generates the infinitesimal action $d\pi$ of the Lie algebra $\mathfrak{so}(m+1,2)$, and we have the following commutative diagram for any $X \in \mathfrak{so}(m+1,2)$:

$$\begin{array}{ccc}
L^2(C)_K & \xrightarrow{\imath} & \mathcal{S}'(\mathbb{R}^{m+1}) \\
d\pi(X)\downarrow & & \downarrow d\hat{\varpi}(X) \\
L^2(C)_K & \xrightarrow{\imath} & \mathcal{S}'(\mathbb{R}^{m+1}).
\end{array} \tag{2.4.4}$$

Here, $\imath : L^2(C) \to \mathcal{S}'(\mathbb{R}^{m+1})$ is given by $u(\zeta) \mapsto u(\zeta)\delta(Q)$. This is well-defined and injective if $m > 1$ (see [23, §3.4]).

3. Branching Law of π_+

The main goal of this section is Proposition 3.2.1, which explicitly describes special functions that arise as K-types in the 'Schrödinger model' on $L^2(\mathbb{R}^m, |x|^{-1}dx)$ of the minimal representation of the double covering group G of $SO_0(m+1,2)$.

3.1. Schrödinger model of the minimal representation

We have used the variables $\zeta = (\zeta_1, \cdots, \zeta_{m+1})$ for the positive cone $C_+ \subset \mathbb{R}^{m+1}$ in Section 2, and will use the letter $x = (x_1, \cdots, x_m)$ for the coordinate of $\mathbb{R}^m$. The projection

$$p : \mathbb{R}^{m+1} \to \mathbb{R}^m, \quad (\zeta_1, \cdots, \zeta_m, \zeta_{m+1}) \mapsto (\zeta_1, \cdots, \zeta_m). \qquad (3.1.1)$$

induces a diffeomorphism from C_+ onto $\mathbb{R}^m \setminus \{0\}$, and the measure $d\mu$ on C_+ is given by $\delta(Q)$, and therefore is pushed forward to $\frac{1}{2|x|} dx = \frac{1}{2|x|} dx_1 \cdots dx_m$. Thus, we have a unitary isomorphism:

$$\sqrt{2} p^* : L^2(\mathbb{R}^m, \frac{dx}{|x|}) \xrightarrow{\sim} L^2(C_+). \qquad (3.1.2)$$

Through this isomorphism, we can realize the minimal representation of G on $L^2(\mathbb{R}^m, \frac{dx}{|x|})$ as well. Named after the Schrödinger model of the Weil representation, we say this model is the *Schrödinger model* of the minimal representation of G. We shall work with this model from now on.

3.2. K-finite functions on the forward light cone C_+

This section refines the K-type decomposition (2.3.2) by providing an explicit irreducible decomposition:

$$L^2(\mathbb{R}^m, \frac{dx}{|x|})_K = \bigoplus_{a=0}^{\infty} W_a = \bigoplus_{a=0}^{\infty} \bigoplus_{l=0}^{a} W_{a,l} \qquad (3.2.1)$$

according to the following chain of subgroups:

$$
\begin{array}{ccccc}
G & \supset & K & \supset & R := K \cap (M_+^{\max})_0 \\
\| & & \| & & \| \\
SO_0(m+1,2)^{\sim} & \supset & SO(m+1) \times SO(2)^{\sim} & \supset & SO(m).
\end{array}
$$

Here, the K-irreducible subspace W_a of $L^2(\mathbb{R}^m, \frac{dx}{|x|})$ and the R-irreducible suspace $W_{a,l}$ of W_a is characterized by

$$W_a \simeq \mathcal{H}^a(\mathbb{R}^{m+1}) \boxtimes \mathbb{C}e^{(a+\frac{m-1}{2})\theta} \quad \text{(see (2.3.2))}, \qquad (3.2.2)$$

$$W_{a,l} \simeq \mathcal{H}^l(\mathbb{R}^m), \qquad (3.2.3)$$

as K-modules and as R-modules, respectively. Here, we note that the $SO(m)$-module $\mathcal{H}^l(\mathbb{R}^m)$ occurs exactly once in the $O(m+1)$-module $\mathcal{H}^a(\mathbb{R}^{m+1})$ if $0 \le l \le a$ (see Subsection 8.4 (4)).

Proposition 3.2.1 will describe the subspace $W_{a,l}$ of $L^2(\mathbb{R}^m, \frac{dx}{|x|})$ by means of Laguerre polynomials $L_n^\alpha(x)$ (see (8.1.1) for the definition). For this, we set

$$f_{a,l}(r) := L_{a-l}^{m-2+2l}(4r)r^l e^{-2r} \quad (0 \leq l \leq a), \tag{3.2.4}$$

and define injective linear maps by

$$j_{a,l} : \mathcal{H}^l(\mathbb{R}^m) \to C^\infty(\mathbb{R}^m), \quad \phi(\omega) \mapsto (j_{a,l}\phi)(r\omega) := f_{a,l}(r)\phi(\omega).$$

Here, we have identified $\mathbb{R}^m$ with $\mathbb{R}_+ \times S^{m-1}$ by the polar coordinate

$$\mathbb{R} \times S^{m-1} \to \mathbb{R}^m, \quad (r,\omega) \mapsto r\omega. \tag{3.2.5}$$

Then, we have:

Proposition 3.2.1. 1)

$$j_{a,l}\big(\mathcal{H}^l(\mathbb{R}^m)\big) \subset L^1(\mathbb{R}^m, \frac{dx}{|x|}) \cap L^2(\mathbb{R}^m, \frac{dx}{|x|}). \tag{3.2.6}$$

2) *Furthermore, the image of $j_{a,l}$ coincides with the R-type $W_{a,l}$:*

$$j_{a,l}\big(\mathcal{H}^l(\mathbb{R}^m)\big) = W_{a,l}.$$

Remark 3.2.2. *Proposition 3.2.1 (2) asserts in particular that $j_{a,l}(\mathcal{H}^l(\mathbb{R}^m))$ and $j_{a',l}(\mathcal{H}^l(\mathbb{R}^m))$ are orthogonal to each other if $a \neq a'$ $(a, a' \geq l)$. More than this, it gives a representation theoretic proof of the fact that $\{f_{a,l}(r) : a = l, l+1, \cdots\}$ forms a complete orthogonal basis of $L^2(\mathbb{R}_+, r^{m-2}dr)$ (see Lemma 4.3.1 for the normalization).*

Remark 3.2.3. *The indefinite orthogonal group $O(p,q)$ $(p+q :$ even, $p, q \geq 2$, and $(p,q) \neq (2,2))$ has a minimal representation π whose minimal K-type is of the form $\mathcal{H}^0(\mathbb{R}^p) \otimes \mathcal{H}^{\frac{p-q}{2}}(\mathbb{R}^q)$ if $p \geq q$. In the L^2-model of π, we have proved that any $M^{\max}$-fixed vector in $\mathcal{H}^0(\mathbb{R}^p) \otimes \mathcal{H}^{\frac{p-q}{2}}(\mathbb{R}^q)$ is a scalar multiple of the function $r^{\frac{3-q}{2}}K_{\frac{q-3}{2}}(2r)$ (see [23, Theorem 5.5] for a precise statement), where $K_\nu(z)$ denotes the K-Bessel function. Since $K_{-\frac{1}{2}}(2r) = \frac{\sqrt{\pi}}{2}r^{-\frac{1}{2}}e^{-2r}$ and $L_0^\alpha(x) = 1$, we have $r^{\frac{3-q}{2}}K_{\frac{q-3}{2}}(2r) = \frac{\sqrt{\pi}}{2}e^{-2r} = \frac{\sqrt{\pi}}{2}f_{0,0}(r)$ if $q = 2$. This vector is a generator of the one dimensional vector space $W_{0,0}$.*

Remark 3.2.4 (Weil representation). *Let us compare our representation on $L^2(\mathbb{R}^m, |x|^{-1}dx)$ with the (original) Schrödinger model on $L^2(\mathbb{R}^m)$*

of the Weil representation of $G' = Sp(m, \mathbb{R})\tilde{}$. The counterpart to Proposition 3.2.1 can be stated as follows: we set for $0 \le l \le a$

$$f'_{a,l}(r) := L_{a-l}^{\frac{m-2}{2}+2l}(r^2)r^l e^{-\frac{r^2}{2}}, \qquad (3.2.7)$$

and define linear maps by

$$j'_{a,l} : \mathcal{H}^l(\mathbb{R}^m) \to C^\infty(\mathbb{R}^m), \quad \phi(\omega) \mapsto (j'_{a,l}\phi)(r\omega) = f'_{a,l}(r)\phi(\omega).$$

Then for any $\phi \in \mathcal{H}^l(\mathbb{R}^m)$, $j'_{a,l}(\phi)$ is square integrable on $\mathbb{R}^m$, and its image $j'_{a,l}(\mathcal{H}^l(\mathbb{R}^m))$ is characterized by the following properties: Let $(K', R') = (U(m)\tilde{}, O(m)\tilde{})$.

1) $j'_{a,l}(\mathcal{H}^l(\mathbb{R}^m))$ is isomorphic to $\mathcal{H}^l(\mathbb{R}^m)$ as R'-modules,

2) it is contained in the K'-type isomorphic to $S^a(\mathbb{C}) \otimes \det^{\frac{1}{2}}$.

The remaining part of this section is organized as follows. In Subsection 3.3, we shall define a central element Z of (the complexification of) the Lie algebra $\mathfrak{k}$ of K, and compute its differential action $d\pi_+(Z)$ (see Lemma 3.4.3). By using this explicit form of $d\pi_+(Z)$, we prove Proposition 3.2.1 in Subsection 3.5. Finally, in Subsection 3.6, by looking at the eigenvalues of $d\pi_+(Z)$ (see Lemma 3.5.1), we shall see that $\{\pi_+(e^{tZ}) := \exp(td\pi_+(Z)) : \operatorname{Re} t > 0\}$ forms a holomorphic semigroup of contraction operators.

In Section 5, we shall find an explicit integral kernel of this semigroup by using Proposition 3.2.1.

3.3. Description of infinitesimal generators of $\mathfrak{sl}(2, \mathbb{R})$

Let

$$e := \begin{pmatrix} 0 & 1 \\ 0 & 0 \end{pmatrix}, \quad f := \begin{pmatrix} 0 & 0 \\ 1 & 0 \end{pmatrix}, \quad h := [e, f] = \begin{pmatrix} 1 & 0 \\ 0 & -1 \end{pmatrix}$$

be the standard basis of $\mathfrak{sl}(2, \mathbb{R})$. With the notation (2.1.1), we define a Lie algebra homomorphism

$$\phi : \mathfrak{sl}(2, \mathbb{R}) \to \mathfrak{so}(m + 1, 2) \qquad (3.3.1)$$

by

$$\phi(e) = \overline{N}_{m+1}, \quad \phi(f) = N_{m+1}, \quad \phi(h) = -2E. \qquad (3.3.2)$$

In this subsection, we shall explicitly describe $d\pi_+(\phi(e)), d\pi_+(\phi(f))$, and $d\pi_+(\phi(h))$ as differential operators on $\mathbb{R}^m$.

Lemma 3.3.1. *Let* $E_x = \sum_{j=1}^{m} x_j \frac{\partial}{\partial x_j}$ *and* $\Delta = \sum_{j=1}^{m} \frac{\partial^2}{\partial x_j^2}$. *Then we have:*

$$d\pi_+(\phi(h)) = 2E_x + m - 1, \qquad (3.3.3)$$

$$d\pi_+(\phi(e)) = 2\sqrt{-1}|x|, \qquad (3.3.4)$$

$$d\pi_+(\phi(f)) = \frac{\sqrt{-1}}{2}|x|\Delta. \qquad (3.3.5)$$

Remark 3.3.2. *Lemma 3.3.1 corresponds to an analogous result for the Schrödinger model of the Weil representation* $(\varpi, L^2(\mathbb{R}^m))$ *of* $Sp(m,\mathbb{R})^\sim$ *as follows: by the matrix realization of the real symplectic Lie algebra* $\mathfrak{sp}(m,\mathbb{R})$, *we define a Lie algebra homomorphism* $\varphi : \mathfrak{sl}(2,\mathbb{R}) \to \mathfrak{sp}(m,\mathbb{R})$ *by*

$$\varphi(e) = \begin{pmatrix} 0 & I_m \\ 0 & 0 \end{pmatrix}, \quad \varphi(f) = \begin{pmatrix} 0 & 0 \\ I_m & 0 \end{pmatrix}, \quad \varphi(h) = \begin{pmatrix} I_m & 0 \\ 0 & -I_m \end{pmatrix}. \qquad (3.3.6)$$

Then, $\{d\varpi(\varphi(h)), d\varpi(\varphi(f)), d\varpi(\varphi(f))\}$ *is no other than the* $\mathfrak{sl}_2$-*triple of differential operators* $\{\tilde{h}', \tilde{e}', \tilde{f}'\}$ *on* $\mathbb{R}^m$ *given in (1.3.3).*

Proof of Lemma 3.3.1. First we compute $d\pi_+(\phi(h)) = d\pi_+(-2E)$. For this, we use the formula of $d\hat{\varpi}$ on $\mathbb{R}^{m+1}$ in Subsection 2.4, and then compute the formula of $d\pi_+$ on the positive cone C_+ (or on the coordinate space $\mathbb{R}^m$) through the embedding $\iota : L^2(C) \to \mathcal{S}'(\mathbb{R}^{m+1})$, $u(\zeta) \mapsto u(\zeta)\delta(Q)$. Since the distribution $\delta(Q)$ is homogeneous of degree -2, we note that $E_\zeta \delta(Q) = -2\delta(Q)$. Therefore,

$$\begin{aligned} (d\pi_+(-2E)u)\delta(Q) &= -2d\hat{\varpi}(E)(u\delta(Q)) && \text{by (2.4.4)} \\ &= (2E_\zeta + m + 3)(u\delta(Q)) && \text{by (2.4.3)} \\ &= (2E_\zeta + m - 1)u \cdot \delta(Q). \end{aligned}$$

Now by identifying $L^2(C_+)$ with $L^2(\mathbb{R}^m, \frac{dx}{|x|})$ by (3.1.1), we obtain (3.3.3). The second formula (3.3.4) follows immediately from (2.4.1).

We shall show the third formula (3.3.5). In light of $\phi(f) = N_{m+1}$ (see (3.3.2)), by (2.4.2), we have

$$d\hat{\varpi}(\phi(f)) = \sqrt{-1}\Big(\frac{m+3}{2}\frac{\partial}{\partial\zeta_{m+1}} + E_\zeta\frac{\partial}{\partial\zeta_{m+1}} + \frac{\zeta_{m+1}}{2}\Box_\zeta\Big). \qquad (3.3.7)$$

In order to compute the action $d\hat{\varpi}(\phi(f))$ along the cone C_+, we use the following coordinate on $\mathbb{R}^{m+1}$:

$$\mathbb{R} \times \mathbb{R}_+ \times S^{m-1} \to \mathbb{R}^{m+1}, \quad (Q, r, \omega) \mapsto (r\omega, \sqrt{r^2 - Q}). \qquad (3.3.8)$$

Claim 3.3.3. *With the above coordinate, the differential operator* $d\widehat{\varpi}(\phi(f))$ *on* $S'(\mathbb{R}^{m+1})$ *takes the form:*

$$d\widehat{\varpi}(\phi(f)) = \sqrt{-1}\sqrt{r^2 - Q}\Big(\frac{1}{2}\frac{\partial^2}{\partial r^2} + \frac{m-1}{2r}\frac{\partial}{\partial r} + \frac{\Delta_{S^{m-1}}}{2r^2} - 2Q\frac{\partial^2}{\partial Q^2} - 4\frac{\partial}{\partial Q}\Big).$$

Proof of Claim 3.3.3. We start with a new coordinate $\mathbb{R}^{m+1} = \mathbb{R}^m \oplus \mathbb{R}$ by

$$\mathbb{R}_+ \times S^{m-1} \times \mathbb{R} \to \mathbb{R}^{m+1}, \quad (R, \omega, \zeta_{m+1}) \mapsto (R\omega, \zeta_{m+1}). \qquad (3.3.9)$$

Then, clearly,

$$E_\zeta = R\frac{\partial}{\partial R} + \zeta_{m+1}\frac{\partial}{\partial \zeta_{m+1}}$$

$$\Box_\zeta = \Delta_{\mathbb{R}^m} - \frac{\partial^2}{\partial \zeta_{m+1}^2} = \Big(\frac{\partial^2}{\partial R^2} + \frac{m-1}{R}\frac{\partial}{\partial R} + \frac{1}{R^2}\Delta_{S^{m-1}}\Big) - \frac{\partial^2}{\partial \zeta_{m+1}^2}.$$

The coordinate (3.3.8) is obtained by the composition of (3.3.9) and

$$r = R, \quad Q = R^2 - \zeta_{m+1}^2.$$

In light of

$$\frac{\partial}{\partial R} = \frac{\partial}{\partial r} + 2r\frac{\partial}{\partial Q}, \quad \frac{\partial}{\partial \zeta_{m+1}} = -2\sqrt{r^2 - Q}\frac{\partial}{\partial Q}, \qquad (3.3.10)$$

we get

$$E_\zeta = r\frac{\partial}{\partial r} + 2Q\frac{\partial}{\partial Q}, \qquad (3.3.11)$$

$$\Box_\zeta = \frac{\partial^2}{\partial r^2} + 4Q\frac{\partial^2}{\partial Q^2} + 4r\frac{\partial^2}{\partial r\partial Q} + 2(m+1)\frac{\partial}{\partial Q} + \frac{m-1}{r}\frac{\partial}{\partial r} + \frac{\Delta_{S^{m-1}}}{r^2}. \qquad (3.3.12)$$

Now substituting (3.3.10), (3.3.11) and (3.3.12) into (3.3.7), we get the claim. $\qquad\Box$

Given $u \in L^2(C_+)_K$, we extend it to a distribution $\tilde{u} \equiv \tilde{u}(Q, r, \omega) \in S'(\mathbb{R}^{m+1}) \cap C^\infty(\mathbb{R}^{m+1} \setminus \{0\})$ such that

$$u\delta(Q) = \tilde{u}\delta(Q).$$

We set

$$S := 2\sqrt{r^2 - Q}\Big(Q\frac{\partial^2}{\partial Q^2} + 2\frac{\partial}{\partial Q}\Big)(\tilde{u}\delta(Q)).$$

Then, it follows from (2.4.4) and Claim 3.3.3 that

$$d\widehat{\varpi}(\phi(f))(\tilde{u}\delta(Q))$$

$$= \sqrt{-1}\Big(\sqrt{r^2 - Q}\Big(\frac{1}{2}\frac{\partial^2}{\partial r^2} + \frac{m-1}{2r}\frac{\partial}{\partial r} + \frac{\Delta_{S^{m-1}}}{2r^2}\Big)\tilde{u}\Big)\delta(Q) - S$$

$$= \sqrt{-1}\Big(1 + O\big((\tfrac{Q}{r^2})\big)\Big)\Big(\frac{r}{2}\frac{\partial^2}{\partial r^2} + \frac{m-1}{2}\frac{\partial}{\partial r} + \frac{\Delta_{S^{m-1}}}{2r}\Big)u \cdot \delta(Q) - S$$

$$= \sqrt{-1}\Big(\frac{r}{2}\frac{\partial^2}{\partial r^2} + \frac{m-1}{2}\frac{\partial}{\partial r} + \frac{\Delta_{S^{m-1}}}{2r}\Big)u \cdot \delta(Q) - S$$

$$= \frac{\sqrt{-1}}{2}r(\Delta_{\mathbb{R}^m}u)\delta(Q) - S.$$

At the second last equality, we used the formula $Q\delta(Q) = 0$. In the following claim, we shall show $S = 0$. Now, the proof of (3.3.5) is complete. Hence, we have shown Lemma 3.3.1. □

Claim 3.3.4. *For any $\tilde{u} \in C_0^\infty(\mathbb{R}^{m+1} \setminus \{0\})$, we have*

$$\Big(Q\frac{\partial^2}{\partial Q^2} + 2\frac{\partial}{\partial Q}\Big)\big(\tilde{u}\delta(Q)\big) = 0.$$

Proof of Claim 3.3.4. By the Leibniz rule, the left-hand side amounts to

$$\frac{\partial^2 \tilde{u}}{\partial Q^2}Q\delta(Q) + 2\frac{\partial \tilde{u}}{\partial Q}\Big(Q\frac{\partial}{\partial Q}\delta(Q) + \delta(Q)\Big) + \tilde{u}\Big(Q\frac{\partial^2}{\partial Q^2}\delta(Q) + 2\frac{\partial}{\partial Q}\delta(Q)\Big).$$

Hence we see that this equals 0 in light of the formulas:

$$Q\delta(Q) = 0, \quad Q\frac{\partial}{\partial Q}\delta(Q) = -\delta(Q), \quad Q\frac{\partial^2}{\partial Q^2}\delta(Q) = -2\frac{\partial}{\partial Q}\delta(Q).$$

□

3.4. Central element Z of $\mathfrak{k}_\mathbb{C}$

We extend the Lie algebra homomorphism $\phi : \mathfrak{sl}(2,\mathbb{R}) \rightarrow \mathfrak{so}(m+1,2)$ (see (3.3.2)) to the complex Lie algebra homomorphism $\phi : \mathfrak{sl}(2,\mathbb{C}) \rightarrow \mathfrak{so}(m+3,\mathbb{C})$. Consider a generator of $\mathfrak{so}(2,\mathbb{C})$ given by

$$z := \frac{\sqrt{-1}}{2}(e - f) = \frac{\sqrt{-1}}{2}\begin{pmatrix} 0 & 1 \\ -1 & 0 \end{pmatrix}. \tag{3.4.1}$$

We set

$$Z := \phi(z) \in \sqrt{-1}\mathfrak{g}. \tag{3.4.2}$$

In light of (3.3.2) and (2.1.1), we have

$$Z = \frac{\sqrt{-1}}{2}(\overline{N}_{m+1} - N_{m+1}) \tag{3.4.3}$$

$$= \sqrt{-1}(E_{m+1,m+2} - E_{m+2,m+1}). \tag{3.4.4}$$

Hence, $\sqrt{-1}Z$ is contained in the center $\mathfrak{c}(\mathfrak{k})$ of $\mathfrak{k} \simeq \mathfrak{so}(m+1) \oplus \mathfrak{so}(2)$. If $m > 1$, then $\mathfrak{c}(\mathfrak{k})$ is of one dimension, and $\sqrt{-1}Z$ generates $\mathfrak{c}(\mathfrak{k})$. By (3.4.1), we have $e^{\sqrt{-1}tZ} = I_{m+3}$ in $SO_0(m+1,2)$ if and only if $t \in 2\pi\mathbb{Z}$. Hence, $e^{\sqrt{-1}tZ} = 1$ in $G = SO_0(m+1,2)^{\sim}$ if and only if $t \in 4\pi\mathbb{Z}$ (see (2.3.1)). Therefore, we have the following lemma:

Lemma 3.4.1. *The Lie algebra homomorphism $\phi : \mathfrak{sl}(2,\mathbb{R}) \to \mathfrak{so}(m+1,2)$ (see (3.3.1)) lifts to an injective Lie group homomorphism $SL(2,\mathbb{R}) \to G$.*

By the expression (3.4.4) of Z and by the K-isomorphism (3.2.2), we have:

Lemma 3.4.2. *$d\pi_+(Z)$ acts on W_a as a scalar multiplication of $-(a + \frac{m-1}{2})$.*

Combining (3.4.1), (3.4.2), and Lemma 3.3.1, we readily get the following lemma:

Lemma 3.4.3. *The differential operator $d\pi_+(Z)$ takes the form:*

$$d\pi_+(Z) = |x|\Big(\frac{\Delta}{4} - 1\Big).$$

Remark 3.4.4. *$d\pi_+(Z)$ coincides with the operator D in Introduction. In particular, D is a self-adjoint operator on $L^2(\mathbb{R}^m, \frac{dx}{|x|})$ because $\sqrt{-1}Z \in \mathfrak{g}$ and π_+ is a unitary representation.*

Remark 3.4.5 (Weil representation). *For the Weil representation of G', the Lie algebra homomorphism $\varphi : \mathfrak{sl}(2,\mathbb{R}) \to \mathfrak{sp}(m,\mathbb{R})$ (see Remark 3.3.2) sends z to*

$$Z' := \varphi(z) = \frac{\sqrt{-1}}{2}\begin{pmatrix} 0 & I_m \\ -I_m & 0 \end{pmatrix} \in \sqrt{-1}\mathfrak{g}'.$$

Hence $\sqrt{-1}Z'$ is a central element of $\mathfrak{k}' \simeq \mathfrak{u}(m)$. The differential operator $d\varpi(Z')$ amounts to the Hermite operator (see [16, §6 (d)])

$$\mathcal{D} = \frac{1}{4}(\Delta - |x|^2).$$

See the table in Subsection 1.4 for the differential operators corresponding to e, f and h.

3.5. Proof of Proposition 3.2.1

This subsection gives a proof of Proposition 3.2.1.

1) Let us show $j_{a,l}(\phi) = f_{a,l}\phi \in L^1(\mathbb{R}^m, \frac{dx}{|x|}) \cap L^2(\mathbb{R}^m, \frac{dx}{|x|})$ for any $\phi \in \mathcal{H}^l(\mathbb{R}^m)$. By the definition (3.2.4) of $f_{a,l}$, $f_{a,l}(r)$ is regular at $r = 0$, and $f_{a,l}(r)$ decays exponentially as r tends to infinity. Therefore, $f_{a,l} \in L^1(\mathbb{R}_+, r^{m-2}dr) \cap L^2(\mathbb{R}_+, r^{m-2}dr)$. Since our measure $\frac{dx}{|x|}$ has the form

$$\frac{dx}{|x|} = r^{m-2}drd\omega, \qquad (3.5.1)$$

with respect to the polar coordinate (3.2.5), we have shown $j_{a,l}(\phi) \in L^1(\mathbb{R}^m, \frac{dx}{|x|}) \cap L^2(\mathbb{R}^m, \frac{dx}{|x|})$.

2) We set $H_{a,l} := j_{a,l}(\mathcal{H}^l(\mathbb{R}^m))$. Obviously, $H_{a,l}$ is isomorphic to $\mathcal{H}^l(\mathbb{R}^m)$ as R-modules. To see $W_{a,l} = H_{a,l}$, it is sufficient to show the following inclusion:

$$H_{a,l} \subset W_a, \qquad (3.5.2)$$

because $W_{a,l}$ is characterized as the unique subspace of W_a such that $W_{a,l} \simeq \mathcal{H}^l(\mathbb{R}^m)$ as R-modules.

To see (3.5.2), we recall from (3.2.2) that W_a is characterized as the unique subspace of $L^2(\mathbb{R}^m, \frac{dx}{|x|})$ on which $d\pi_+(Z)$ acts with eigenvalue $-(a + \frac{m-1}{2})$. Thus the inclusive relation (3.5.2) will be proved if we show the following lemma:

Lemma 3.5.1. *The operator $d\pi_+(Z)$ acts on $H_{a,l}$ as a scalar multiplication $-(a + \frac{m-1}{2})$. In other words, we have*

$$d\pi_+(Z)(f_{a,l}\phi) = -\left(a + \frac{m-1}{2}\right)f_{a,l}\phi. \qquad (3.5.3)$$

Proof of Lemma 3.5.1. Writing the differential operator $d\pi_+(Z)$ (see Lemma 3.4.3) in terms of the polar coordinate and using the definition of $f_{a,l}$ (see (3.2.4)), we see that the equation (3.5.3) is equivalent to

$$\left(\frac{r}{4}\frac{\partial^2}{\partial r^2} + \frac{m-1}{4}\frac{\partial}{\partial r} + \frac{\Delta_{S^{m-1}}}{4r} - r\right.$$
$$\left. + \left(a + \frac{m-1}{2}\right)\right)\left(\psi(4r)r^l e^{-2r}\phi(\omega)\right) = 0, \qquad (3.5.4)$$

for $\psi(r) := L_{a-l}^{m-2+2l}(r)$. The equation (3.5.4) amounts to

$$\left(4r\psi''(4r) + (m-1+2l-4r)\psi'(4r) + (a-l)\psi(4r)\right)r^l e^{-2r}\phi(\omega) = 0.$$

This is nothing but Laguerre's differential equation (8.1.2) with $n = a - l, \alpha = m - 2 + 2l$. Now the lemma follows. □

3.6. One parameter holomorphic semigroup $\pi_+(e^{tZ})$

It follows from Lemma 3.4.2 that for $t \in \mathbb{C}$ the operator

$$\pi_+(e^{tZ}) := \exp d\pi_+(tZ) = \sum_{j=0}^{\infty} \frac{1}{j!} d\pi_+(tZ)^j \qquad (3.6.1)$$

acts on $W_{a,l}$ as a scalar multiplication of $e^{-(a+\frac{m-1}{2})t}$ for any $0 \le l \le a$, of which the absolute value does not exceed 1 if $\operatorname{Re} t \ge 0$. In light of the direct sum decomposition (3.2.1), if $\operatorname{Re} t \ge 0$ then the linear map $\pi_+(e^{tZ})$: $L^2(\mathbb{R}^m, \frac{dx}{|x|})_K \to L^2(\mathbb{R}^m, \frac{dx}{|x|})_K$ extends to a continuous operator (we use the same notation $\pi_+(e^{tZ})$) on $L^2(\mathbb{R}^m, \frac{dx}{|x|})$. Furthermore, it is a contraction operator if $\operatorname{Re} t > 0$. We summarize some of basic properties of $\pi_+(e^{tZ})$:

Proposition 3.6.1. 1) *The map*

$$\{t \in \mathbb{C} : \operatorname{Re} t \ge 0\} \times L^2(\mathbb{R}^m, \frac{dx}{|x|}) \to L^2(\mathbb{R}^m, \frac{dx}{|x|}), \quad (t, f) \mapsto \pi_+(e^{tZ})f$$

$$(3.6.2)$$

is continuous.

2) *For a fixed t such that $\operatorname{Re} t \ge 0$, $\pi_+(e^{tZ})$ is characterized as the continuous operator from $L^2(\mathbb{R}^m, \frac{dx}{|x|}) \to L^2(\mathbb{R}^m, \frac{dx}{|x|})$ satisfying*

$$\pi_+(e^{tZ})u = e^{-(a+\frac{m-1}{2})t}u, \qquad (3.6.3)$$

for any $u \in W_{a,l} = \{f_{a,l}\phi : \phi \in \mathcal{H}^l(\mathbb{R}^m)\}$ (see Proposition 3.2.1) and for any $l, a \in \mathbb{N}$ such that $0 \le l \le a$.

3) *The operator norm $\|\pi_+(e^{tZ})\|$ of $\pi_+(e^{tZ})$ is $e^{-\frac{m-1}{2}\operatorname{Re} t}$.*

4) *If $\operatorname{Re} t > 0$, $\pi_+(e^{tZ})$ is a Hilbert–Schmidt operator.*

5) *If $t \in \sqrt{-1}\mathbb{R}$, then $\pi_+(e^{tZ})$ is a unitary operator.*

Remark 3.6.2. *We define a subset $\Gamma_+ := \{tZ : t > 0\}$ in $\sqrt{-1}\mathfrak{g}$. Proposition 3.6.1 indicates how the unitary representation π_+ of G extends to a holomorphic semigroup on the complex domain $G \cdot \exp\Gamma_+ \cdot G$ of $G_\mathbb{C}$. Our results may be regarded as a part of the Gelfand–Gindikin program, which tries to understand unitary representations of a real semisimple Lie group by means of holomorphic objects on an open subset of $G_\mathbb{C} \setminus G$, where $G_\mathbb{C}$ is a complexification of G (see [10, 27, 29]).*

Remark 3.6.3 (Weil representation). *In the case of the Weil representation* $(\varpi, L^2(\mathbb{R}^m))$ *of* $G' = Sp(m, \mathbb{R})^{\sim}$, *the functions* $f'_{a,l}$ *(see Remark 3.2.4) play the same role as* $f_{a,l}$ *because of the following facts:*

1) $\{f'_{a,l} : 0 \leq l \leq a\}$ *spans a complete orthogonal basis of* $L^2(\mathbb{R}_+, r^{m-1}dr)$ *(cf. Remark 3.2.2).*

2) For any $\phi \in \mathcal{H}^l(\mathbb{R}^m)$, $f'_{a,l}(r)\phi(\omega)$ $(0 \leq l \leq a)$ *are eigenfunctions of* $d\varpi(Z')$ *with negative eigenvalues* $-(\frac{m}{4} + a)$.

Owing to these facts, we obtain a holomorphic semigroup of contraction operators $\varpi(e^{tZ'}) := \exp t(d\varpi(Z'))$ $(\mathrm{Re}\, t > 0)$ *on* $L^2(\mathbb{R}^m)$. *Since* $d\varpi(Z')$ *coincides with the Hermite operator* $\mathcal{D}$ *(see Remark 3.4.5), this holomorphic semigroup is nothing but the Hermite semigroup (see [16, §5]).*

4. Radial Part of the Semigroup

This section gives an explicit integral formula for the 'radial part' of the holomorphic semigroup $\pi_+(e^{tZ})$ in the 'Schrödinger model' $L^2(\mathbb{R}^m, |x|^{-1}dx)$. The main result of Section 4 is Theorem 4.1.1. As its applications, we see that the semigroup law $\pi_+(e^{(t_1+t_2)Z}) = \pi_+(e^{t_1 Z}) \circ \pi_+(e^{t_2 Z})$ gives a simple and representation theoretic proof of the classical Weber's second exponential integral formula on Bessel functions (Corollary 4.5.1), and that taking the boundary value $\lim_{s \downarrow 0} \pi_+(e^{sZ}) = \mathrm{id}$ provides an example of a Dirac sequence (Corollary 4.6.1).

Theorem 4.1.1 will play a key role in Section 5, where we complete the proof of the main theorem of this article, namely, Theorem 5.1.1 that gives an integral formula of the holomorphic semigroup $\pi_+(e^{tZ})$ on $L^2(\mathbb{R}^m, |x|^{-1}dx)$.

4.1. Result of the section

For a complex parameter $t \in \mathbb{C}$ with $\mathrm{Re}\, t > 0$, we have defined a contraction operator $\pi_+(e^{tZ}) : L^2(\mathbb{R}^m, \frac{dx}{|x|}) \to L^2(\mathbb{R}^m, \frac{dx}{|x|})$ in Proposition 3.6.1.

We recall that Z is a central element in $\mathfrak{k}_\mathbb{C}$ (see Subsection 3.4) and that R is a subgroup of K. Therefore, $\pi_+(e^{tZ})$ intertwines with the R-action. On the other hand, the natural action of $R \simeq SO(m)$ gives a direct sum decomposition of the Hilbert space:

$$L^2(\mathbb{R}^m, \frac{dx}{|x|}) \simeq \sum_{l=0}^{\infty} {}^{\oplus} L^2(\mathbb{R}_+, r^{m-2}dr) \otimes \mathcal{H}^l(\mathbb{R}^m). \qquad (4.1.1)$$

Hence, by Schur's lemma, there exists a family of continuous operators parametrized by $l \in \mathbb{N}$:

$$\pi_{+,l}(e^{tZ}) : L^2(\mathbb{R}_+, r^{m-2}dr) \to L^2(\mathbb{R}_+, r^{m-2}dr) \qquad (4.1.2)$$

such that $\pi_+(e^{tZ})$ is diagonalized according to the direct sum decomposition (4.1.1) as follows:

$$\pi_+(e^{tZ}) = \sum_{l=0}^{\infty} {}^{\oplus} \pi_{+,l}(e^{tZ}) \otimes \mathrm{id}. \qquad (4.1.3)$$

The goal of this section is to give an explicit integral formula of $\pi_{+,l}(e^{tZ})$ on $L^2(\mathbb{R}_+, r^{m-2}dr)$ for $l \in \mathbb{N}$. We note that $\pi_{+,l}(e^{tZ})$ is a unitary operator if $\mathrm{Re}\, t = 0$ because so is $\pi_+(e^{tZ})$. Likewise, $\pi_{+,l}(e^{tZ})$ is a Hilbert–Schmidt operator if $\mathrm{Re}\, t > 0$ because so is $\pi_+(e^{tZ})$.

We now introduce the following subset of $\mathbb{C}$:

$$\Omega := \{t \in \mathbb{C} : \mathrm{Re}\, t \geq 0\} \setminus 2\pi\sqrt{-1}\mathbb{Z}, \qquad (4.1.4)$$

and define a family of analytic functions $K_l^+(r, r'; t)$ on $\mathbb{R}_+ \times \mathbb{R}_+ \times \Omega$ by the formula: for $l = 0, 1, 2, \ldots,$

$$
\begin{aligned}
K_l^+(r, r'; t) &:= \frac{2e^{-2(r+r')\coth\frac{t}{2}}}{\sinh\frac{t}{2}} (rr')^{-\frac{m-2}{2}} I_{m-2+2l}\left(\frac{4\sqrt{rr'}}{\sinh\frac{t}{2}}\right) \\
&= \frac{2^{m-1+2l}e^{-2(r+r')\coth\frac{t}{2}}(rr')^l}{(\sinh\frac{t}{2})^{m-1+2l}} \tilde{I}_{m-2+2l}\left(\frac{4\sqrt{rr'}}{\sinh\frac{t}{2}}\right). \qquad (4.1.5)
\end{aligned}
$$

Here, $\tilde{I}_\nu(z) := (\frac{z}{2})^{-\nu} I_\nu(z)$, and $I_\nu(z)$ denotes the I-Bessel function (see Subsection 8.5). We note that the denominator $\sinh\frac{t}{2}$ is nonzero for $t \in \Omega$.

We are ready to state the integral formula of $\pi_{+,l}(e^{tZ})$ for $t \in \Omega$:

Theorem 4.1.1 (Radial part of the semigroup). 1) *For* $\mathrm{Re}\, t > 0$, *the Hilbert–Schmidt operator* $\pi_{+,l}(e^{tZ})$ *on* $L^2(\mathbb{R}_+, r^{m-2}dr)$ *is given by the following integral transform:*

$$(\pi_{+,l}(e^{tZ})f)(r) = \int_0^\infty K_l^+(r, r'; t)f(r')r'^{m-2}dr'. \qquad (4.1.6)$$

The right-hand side converges absolutely for $f \in L^2(\mathbb{R}_+, r^{m-2}dr)$.

2) *If* $t \in \sqrt{-1}\mathbb{R}$ *but* $t \notin 2\pi\sqrt{-1}\mathbb{Z}$, *then the integral formula* (4.1.6) *for the unitary operator* $\pi_{+,l}(e^{tZ})$ *holds in the sense of* L^2-*convergence. Furthermore, the right-hand side converges absolutely if* f *is a finite linear combination of* $f_{a,l}$ $(a = l, l+1, \cdots)$.

Remark 4.1.2. *Let us compare Theorem 4.1.1 with the corresponding result for the Weil representation ϖ of $G' = Sp(m, \mathbb{R})^{\sim}$ realized as the Schrödinger model $L^2(\mathbb{R}^m)$. According to the direct sum decomposition of the Hilbert space:*

$$L^2(\mathbb{R}^m) \simeq \sum_{l=0}^{\infty} {}^{\oplus} L^2(\mathbb{R}_+, r^{m-1} dr) \otimes \mathcal{H}^l(\mathbb{R}^m),$$

there exists a family of continuous operators $\varpi_l(e^{tZ'})$ (Re $t > 0$) such that the holomorphic semigroup $\varpi(e^{tZ'})$ has the following decomposition:

$$\varpi(e^{tZ'}) = \sum_{l=0}^{\infty} {}^{\oplus} \varpi_l(e^{tZ'}) \otimes \mathrm{id}.$$

Then, by an analogous computation to Theorem 4.1.1, we find the kernel function of the semigroup $\varpi_l(e^{tZ'})$ is given by

$$\mathcal{K}_l(r, r'; t) := \frac{e^{-\frac{1}{2}(r^2 + r'^2) \coth \frac{t}{2}}}{\sinh \frac{t}{2}} (rr')^{-\frac{m-2}{2}} I_{\frac{m-2+2l}{2}} \left(\frac{rr'}{\sinh \frac{t}{2}} \right).$$

The relation between the kernel function $\mathcal{K}$ (Mehler kernel) of $\varpi(e^{tZ'})$ and $\mathcal{K}_l(r, r'; t)$ will be discussed in Remark 5.7.2.

This section is organized as follows. In Subsection 4.3, we give a proof of Theorem 4.1.1 for the case $\mathrm{Re}\, t > 0$, which is based on a computation of the kernel function by means of the infinite sum of the eigenfunctions. In Subsection 4.4, by taking the analytic continuation, the case $\mathrm{Re}\, t = 0$ is proved. Applications of Theorem 4.1.1 to special function theory are discussed in Subsections 4.5 and 4.6.

4.2. Upper estimate of the kernel function

In this subsection, we shall give an upper estimate of the kernel function $K_l^+(r, r'; t)$.

For $t = x + \sqrt{-1}y$, we set

$$\alpha(t) := \frac{\sinh x}{\cosh x - \cos y}, \tag{4.2.1}$$

$$\beta(t) := \frac{\cos \frac{y}{2}}{\cosh \frac{x}{2}}. \tag{4.2.2}$$

Then, an elementary computation shows

$$\operatorname{Re} \coth \frac{t}{2} = \alpha(t), \tag{4.2.3}$$

$$\operatorname{Re} \frac{1}{\sinh \frac{t}{2}} = \alpha(t)\beta(t). \tag{4.2.4}$$

For $t \in \Omega$ (see (4.1.4) for definition), we have $\cosh x - \cos y > 0$, and then,

$$\alpha(t) \geq 0 \quad \text{and} \quad |\beta(t)| < 1. \tag{4.2.5}$$

If $\operatorname{Re} t > 0$, then

$$\alpha(t) > 0. \tag{4.2.6}$$

For later purposes, we prepare:

Lemma 4.2.1. *If $y \in \mathbb{R}$ satisfies $|y| \leq 4\sqrt{rr'}$, then, for $t \in \Omega$, we have the following estimate for some constant C:*

$$\left| e^{-2(r+r')\coth \frac{t}{2}} \widetilde{I}_\nu \left(\frac{y}{\sinh \frac{t}{2}} \right) \right| \leq C e^{-2\alpha(t)(1-|\beta(t)|)(r+r')}. \tag{4.2.7}$$

Proof. Using the upper estimate of the I-Bessel function (see Lemma 8.5.1),

$$\left| \widetilde{I}_\nu \left(\frac{y}{\sinh \frac{t}{2}} \right) \right| \leq C e^{|y| \left| \operatorname{Re} \frac{1}{\sinh \frac{t}{2}} \right|},$$

we have

$$\left| e^{-2(r+r')\coth \frac{t}{2}} \widetilde{I}_\nu \left(\frac{y}{\sinh \frac{t}{2}} \right) \right| \leq C e^{-2(r+r')\operatorname{Re}\coth \frac{t}{2} + |y| \left| \operatorname{Re} \frac{1}{\sinh \frac{t}{2}} \right|}$$

$$\leq C e^{-2(r+r')\alpha(t) + 4\sqrt{rr'}\alpha(t)|\beta(t)|}$$

$$\leq C e^{-2\alpha(t)(1-|\beta(t)|)(r+r')}.$$

Here, the last inequality follows from

$$r + r' - 2|\beta(t)|\sqrt{rr'} \geq (1 - |\beta(t)|)(r + r') \tag{4.2.8}$$

for $t \in \Omega$. Thus Lemma is proved. □

Now we state a main result of this subsection:

Lemma 4.2.2. *Let* $l \in \mathbb{N}$ *and* $m \geq 2$.

1) *There exists a constant* $C > 0$ *such that*

$$|K_l^+(r, r'; t)| \leq \frac{C(rr')^l e^{-2\alpha(t)(1-|\beta(t)|)(r+r')}}{|\sinh \frac{t}{2}|^{m-1+2l}} \tag{4.2.9}$$

for any $r, r' \in \mathbb{R}_+$ *and* $t \in \Omega$.

2) *If* $\operatorname{Re} t > 0$, *then* $K_l^+(\cdot, \cdot; t) \in L^2((\mathbb{R}_+)^2, (rr')^{m-2} dr dr')$.

3) *If* $\operatorname{Re} t > 0$, *then for a fixed* $r > 0$, *we have* $K_l^+(r, \cdot; t) \in L^2(\mathbb{R}_+, r'^{m-2} dr')$.

Proof. 1) By the definition of K_l^+ (see (4.1.5)), we have

$$|K_l^+(r, r'; t)| = \frac{2^{m-1+2l}(rr')^l}{|\sinh \frac{t}{2}|^{m-1+2l}} \left| e^{-2(r+r') \coth \frac{t}{2}} \widetilde{I}_{m-2+2l}\left(\frac{4\sqrt{rr'}}{\sinh \frac{t}{2}}\right) \right|.$$

Now (4.2.9) follows from Lemma 4.2.1 by substituting $\nu = m - 2 + 2l$ and $y = 4\sqrt{rr'}$.

Since $\alpha(t) > 0$ for $\operatorname{Re} t > 0$, the statements 2) and 3) hold by 1). □

4.3. Proof of Theorem 4.1.1 (Case $\operatorname{Re} t > 0$)

We recall from Remark 3.2.2 that

$$f_{a,l}(r) = L_{a-l}^{m-2+2l}(4r) r^l e^{-2r} \quad (a = l, l+1, \ldots)$$

forms a complete orthogonal basis of $L^2(\mathbb{R}_+, r^{m-2} dr)$. Further, by the orthogonal relation of the Laguerre polynomials $L_m^\alpha(x)$ (see (8.1.3)), we have the normalization of $\{f_{a,l}\}$ as follows:

Lemma 4.3.1. *For integers* $a, b \geq l$, *we have*

$$\int_0^\infty f_{a,l}(r) f_{b,l}(r) r^{m-2} dr = \begin{cases} 0 & \text{if } a \neq b, \\ \frac{\Gamma(m-1+a+l)}{4^{m-1+2l}\Gamma(a-l+1)} & \text{if } a = b. \end{cases} \tag{4.3.1}$$

We rewrite (3.2.1) by using Proposition 3.2.1 as follows:

$$\begin{aligned}
L^2\left(\mathbb{R}^m, \frac{dx}{|x|}\right)_K &= \bigoplus_{a=0}^{\infty}\left(\bigoplus_{l=0}^{a} W_{a,l}\right) \\
&= \bigoplus_{l=0}^{\infty}\left(\bigoplus_{a=l}^{\infty} W_{a,l}\right) \\
&= \bigoplus_{l=0}^{\infty}\left(\left(\bigoplus_{a=l}^{\infty} \mathbb{C}f_{a,l}\right) \otimes \mathcal{H}^l(\mathbb{R}^m)\right).
\end{aligned} \tag{4.3.2}$$

It follows from Proposition 3.6.1 (2) and the definition (4.1.3) of $\pi_{+,l}(e^{tZ})$ that

$$\pi_{+,l}(e^{tZ})f_{a,l} = e^{-(a+\frac{m-1}{2})t}f_{a,l} \quad (a = l, l+1, \cdots).$$

Therefore, the kernel function $K_l^+(r, r'; t)$ of $\pi_{+,l}(e^{tZ})$ can be written as the infinite sum:

$$K_l^+(r, r'; t) = \sum_{a=l}^{\infty} \frac{e^{-(a+\frac{m-1}{2})t}f_{a,l}(r)\overline{f_{a,l}(r')}}{\|f_{a,l}\|_{L^2(\mathbb{R}_+, r^{m-2}dr)}^2}. \tag{4.3.3}$$

Since $\pi_{+,l}(e^{tZ})$ is a Hilbert–Schmidt operator if $\operatorname{Re} t > 0$, the right-hand side converges in $L^2((\mathbb{R}_+)^2, (rr')^{m-2}drdr')$, and therefore converges for almost all $(r, r') \in (\mathbb{R}_+)^2$.

Let us compute the infinite sum (4.3.3). For this, we set

$$\kappa(r, r'; t) := \sum_{a=l}^{\infty} \frac{\Gamma(a-l+1)}{\Gamma(m-1+a+l)}L_{a-l}^{m-2+2l}(4r)L_{a-l}^{m-2+2l}(4r')e^{-(a-l)t}.$$

Then, it follows from (4.3.1) that we have

$$K_l^+(r, r'; t) = 4^{m-1+2l}(rr')^l e^{-2(r+r')}e^{-(l+\frac{m-1}{2})t}\kappa(r, r'; t). \tag{4.3.4}$$

Now, we apply the Hille–Hardy formula (see (8.1.4)) with $\alpha = m-2+2l, n = a-l, x = 4r, y = 4r'$, and $w = e^{-t}$. We note that $|w| = e^{-\operatorname{Re} t} < 1$ by the assumption $\operatorname{Re} t > 0$. Then we have

$$\kappa(r, r'; t) = \frac{e^{\frac{(4r+4r')e^{-t}}{1-e^{-t}}}(-16rr'e^{-t})^{-\frac{m-2}{2}+l}}{1-e^{-t}}J_{m-2+2l}\left(\frac{2\sqrt{-16rr'e^{-t}}}{1-e^{-t}}\right)$$

$$= \frac{(rr')^{-\frac{m-1}{2}+l}e^{-2(r+r')\frac{2e^{-t}}{1-e^{-t}}}e^{(\frac{m-2}{2}+l)t}}{4^{m-2+2l}(1-e^{-t})}I_{m-2+2l}\left(\frac{4\sqrt{rr'}}{\sinh\frac{t}{2}}\right).$$

Hence, the formula (4.1.5) is proved.

Therefore, the right-hand side of (4.1.6) converges absolutely by the Cauchy–Schwarz inequality because $K_l^+(r, r'; t) \in L^2(\mathbb{R}_+, r'^{m-2}dr')$ for any $r > 0$ and $\operatorname{Re} t > 0$ (see Lemma 4.2.2 (3)).

Remark 4.3.2. *The special functions and related formulas that arise in the analysis of the radial part have a scheme of generalization from $SL(2, \mathbb{R})\tilde{\ }$ to $G = SO_0(m+1, 2)\tilde{\ }$ and $G' = Sp(m, \mathbb{R})\tilde{\ }$. This scheme is illustrated as* *follows:*

$$SL(2,\mathbb{R})^{\sim} \qquad \Rightarrow \qquad G \text{ or } G'$$

Segal–Shale–Weil representation $\Rightarrow$ *minimal representation*

Hermite polynomials $\Rightarrow$ *Laguerre polynomials*

Mehler's formula $\Rightarrow$ *Hille–Hardy formula*

See [17, p 116, Exercise 5 (d)] for the $SL(2,\mathbb{R})$ case. Owing to the reduction formula of Laguerre polynomials to Hermite polynomials (see (8.2.1) and (8.2.2)), the radial part $f'_{a,l}(r)$ (see Remark 3.2.4) for the Weil represen-tation of $G' = Sp(m,\mathbb{R})^{\sim}$ collapses to a constant multiple of $H_{2a-l}(r)e^{-\frac{r^2}{2}}$ ($l = 0, 1$) if $m = 1$.

Remark 4.3.3. *In [26, Chapter 2], W. Myller-Lebedeff proved the following integral formula:*

$$\int_0^\infty K_l(r,r';t)f_{a,l}(r')r'^{m-2}dr' = e^{-(a+\frac{m-1}{2})t}f_{a,l}(r) \quad \text{for } a \geq l \geq 0.$$
$$(4.3.5)$$

In view of (4.1.3) and Proposition 3.6.1 (2), the formula (4.1.6) in Theorem 4.1.1 implies (4.3.5) and vice versa. The proof of [26] is completely different from ours. Here is a brief sketch: For the partial differential operator

$$L := \frac{\partial^2}{\partial x^2} + \frac{\alpha+1}{x}\frac{\partial}{\partial x} - \frac{1}{x}\frac{\partial}{\partial t} \quad \alpha > 0,$$

one has the following identity using Green's formula,

$$\iint_D (vLu - uL^*v)dtdx = \int_{\partial D}(v\frac{\partial u}{\partial x} - u\frac{\partial v}{\partial x} + \frac{\alpha+1}{x}uv)dt + \frac{1}{x}uvdx, \quad (4.3.6)$$

for a domain $D \subset \mathbb{R}^2$. Here, L^* denotes the (formal) adjoint of L. Now, we take $u(x,t) := t^n L_n^\alpha(\frac{x}{t})$, $v(x,t) := (\frac{x}{\xi})^{\frac{\alpha}{2}}\frac{x}{\tau-t}e^{-\frac{x+\xi}{\tau-t}}I_\alpha(\frac{2\sqrt{x\xi}}{\tau-t})$, $n \in \mathbb{N}$, τ, $\xi \in \mathbb{R}$ as solutions to $Lu = 0$, $L^*v = 0$ respectively, and the domain D as a rectangular domain $D := \{(x,t) \in \mathbb{R}^2 : 0 < x < \infty, t_1 < t < t_2\}$ for some $t_1, t_2 \in \mathbb{R}$. Then by the decay properties of u and v, the integrands in the right-hand side of (4.3.6) vanish on $x = 0$ and $x = \infty$. Since the integral of the left-hand side of (4.3.6) vanishes, the integral $\int_0^\infty \frac{1}{x}u(x,t)v(x,t)dx$ becomes constant with respect to t. By taking the limit $t \to \tau$, we have $\lim_{t\to\tau}\int_0^\infty \frac{1}{x}u(x,t)v(x,t)dx = u(\xi,\tau)$ since $v(x,\tau)$ is proved to be a Dirac delta function. Hence we obtain

$$\int_0^\infty \frac{1}{x}u(x,t)v(x,t)dx = u(\xi,\tau),$$

which coincides with (4.3.5) by a suitable change of variables.

4.4. Proof of Theorem 4.1.1 (Case $\operatorname{Re} t = 0$)

Suppose $t \in \sqrt{-1}\mathbb{R}$. Then, $\pi_{+,l}(e^{tZ})$ is a unitary operator on $L^2(\mathbb{R}_+, r^{m-2}dr)$. Suppose furthermore $t \notin 2\pi\sqrt{-1}\mathbb{Z}$. For $\varepsilon > 0$ and $f \in L^2(\mathbb{R}_+, r^{m-2}dr)$, we have from Theorem 4.1.1 (1)

$$\pi_{+,l}(e^{(\varepsilon+t)Z})f = \int_0^\infty K_l^+(r, r'; \varepsilon + t)f(r')r'^{m-2}dr'.$$

By Proposition 3.6.1 (1), the left-hand side converges to $\pi_{+,l}(e^{tZ})f$ in $L^2(\mathbb{R}_+, r^{m-2}dr)$ as ε tends to 0. For the right-hand side, we have:

Claim 4.4.1. *For $t \in \sqrt{-1}\mathbb{R} \setminus 2\pi\sqrt{-1}\mathbb{Z}$,*

$$\lim_{\varepsilon \downarrow 0} \int_0^\infty K_l^+(r, r'; \varepsilon + t)f_{a,l}(r')r'^{m-2}dr'$$

$$= \int_0^\infty K_l^+(r, r'; t)f_{a,l}(r')r'^{m-2}dr'$$

and the right-hand side converges absolutely.

Proof. If $\varepsilon \in \mathbb{R}$ and $t \in \sqrt{-1}\mathbb{R}$ then

$$\left|\sinh\frac{\varepsilon+t}{2}\right|^2 = \left|\sinh\frac{\varepsilon}{2}\right|^2 + \left|\sinh\frac{t}{2}\right|^2 \geq \left|\sinh\frac{t}{2}\right|^2.$$

Therefore, it follows from (4.2.9) that

$$|K_l^+(r, r'; \varepsilon + t)| \leq \frac{C(rr')^l}{|\sinh\frac{t}{2}|^{m-1+2l}}$$

if $\varepsilon > 0$ and $t \in \sqrt{-1}\mathbb{R} \setminus 2\pi\sqrt{-1}\mathbb{Z}$, because $\varepsilon + t \in \Omega$ implies $\alpha(\varepsilon + t) \geq 0$ and $|\beta(\varepsilon + t)| < 1$. Therefore, we have

$$|K_l^+(r, r'; \varepsilon + t)f_{a,l}(r')r'^{m-2}| \leq \frac{Cr^l r'^{l+m-2}e^{-2r'}|L_{a-l}^{m-2+4l}(4r')|}{|\sinh\frac{t}{2}|^{m-1+2l}}.$$

By the Lebesgue convergence theorem, we have proved Claim. $\qquad\square$

Since linear combinations of $f_{a,l}$ span a dense subspace of $L^2(\mathbb{R}_+, r^{m-2}dr)$, 2) is proved. $\qquad\square$

4.5. Weber's second exponential integral formula

From the semigroup law:

$$\pi_+(e^{(t_1+t_2)Z}) = \pi_+(e^{t_1 Z}) \circ \pi_+(e^{t_2 Z}) \quad (\operatorname{Re} t_1, \operatorname{Re} t_2 > 0), \qquad (4.5.1)$$

we get a representation theoretic proof of classical Weber's second exponential integral for Bessel functions (see [33, §13.31 (1)]):

Corollary 4.5.1. (**Weber's second exponential integral**) *Let ν be a positive integer, and $\rho, \alpha, \beta > 0$. We have the following integral formula*

$$\int_0^\infty e^{-\rho x^2} J_\nu(\alpha x) J_\nu(\beta x) x \, dx = \frac{1}{2\rho} \exp\left(-\frac{\alpha^2 + \beta^2}{4\rho}\right) I_\nu\left(\frac{\alpha\beta}{2\rho}\right). \qquad (4.5.2)$$

Proof. It follows from the semigroup law (4.5.1) that the integral kernels for $\pi_+(e^{(t_1+t_2)Z})$ and $\pi_+(e^{t_1 Z}) \circ \pi_+(e^{t_2 Z})$ must coincide. Then, from Theorem 4.1.1, we have

$$\int_0^\infty K_l^+(r, s; t_1) K_l^+(s, r'; t_2) s^{m-2} ds = K_l^+(r, r'; t_1 + t_2). \qquad (4.5.3)$$

In view of (4.1.5), the formula (4.5.2) is obtained by (4.5.3) by the change of variables,

$$x = \sqrt{s}, \quad \alpha = \frac{4 e^{\frac{\pi\sqrt{-1}}{2}} \sqrt{r}}{\sinh \frac{t_1}{2}}, \quad \beta = \frac{4 e^{\frac{\pi\sqrt{-1}}{2}} \sqrt{r'}}{\sinh \frac{t_2}{2}}, \quad \nu = m - 2 + 2l,$$

$$\rho = 2\left(\coth \frac{t_1}{2} + \coth \frac{t_2}{2}\right).$$

$\square$

4.6. Dirac sequence operators

We shall state another corollary to Theorem 4.1.1. Let ν be a positive integer, $x, y \in \mathbb{R}, s \in \mathbb{C}$ such that $\operatorname{Re} s > 0$. For a function f on $\mathbb{R}$, let $\mathcal{T}_s$ be an operator defined by

$$\mathcal{T}_s : f(x) \mapsto \int_0^\infty A(x, y; s) f(y) dy,$$

with the kernel function

$$A(x, y; s) := (xy)^{\frac{1}{2}} \frac{e^{-\frac{1}{2}(x^2+y^2)\coth s}}{\sinh s} I_\nu\left(\frac{xy}{\sinh s}\right). \qquad (4.6.1)$$

Then we have the following corollary.

Corollary 4.6.1. 1) *The operators* $\{\mathcal{T}_s : \operatorname{Re} s > 0\}$ *form a semigroup of contraction operators on* $L^2(\mathbb{R}_+, dx)$.

2) *(Dirac sequence)* $\lim_{s \to 0} \|\mathcal{T}_s h - h\|_{L^2(\mathbb{R}_+, dx)} = 0$ *holds for all* $h \in L^2(\mathbb{R}_+, dx)$.

Remark 4.6.2. *For sufficiently small* s, *the semigroup* $\{\mathcal{T}_s : \operatorname{Re} s > 0\}$ *behaves like the Hermite semigroup (see [16]) whose kernel is given by the following Gaussian (cf. (1.2.2)):*

$$\kappa(x, y; s) = \frac{1}{\sqrt{2\pi \sinh s}} e^{-\frac{x^2}{2} \coth s + \frac{xy}{\sinh s} - \frac{y^2}{2} \coth s}$$

because $I_\nu(z) \sim \frac{1}{\sqrt{2\pi z}} e^z$ *for sufficiently large* z *(see [33, §7.23] for the asymptotic behavior of* $I_\nu(z)$*). Note that it is stated in [16, §5.5] that the Hermite semigroup forms a 'Dirac sequence'.*

Proof. Since we assume ν is a positive integer, $\nu = m - 2 + 2l$ for some $m > 3$ and $l \in \mathbb{Z}$. We change the variables $r = \frac{x^2}{4}$, $r' = \frac{y^2}{4}$, $t = 2s$ and define a unitary map

$$\Phi : L^2(\mathbb{R}_+, r^{m-2} dr) \to L^2(\mathbb{R}_+, dx), \quad (\Phi f)(x) := \left(\frac{x}{2}\right)^{\frac{2m-3}{2}} f\left(\frac{x^2}{4}\right). \quad (4.6.2)$$

Comparing (4.6.1) with (4.1.5), we have

$$\Phi^{-1} \circ \mathcal{T}_s \circ \Phi f = \pi_{+,l}(e^{2sZ}) f. \quad (4.6.3)$$

Thus by Theorem 4.1.1 (1), 1) is proved.

2) We take a limit $s \downarrow 0$ of (4.6.3). By Proposition 3.6.1 (1), the right-hand side equals f. Hence by putting $h := \Phi f$, we have $\lim_{s \downarrow 0} \mathcal{T}_s h = h$. $\square$

5. Integral Formula for the Semigroup

In this section, we shall give an explicit integral formula for the holomorphic semigroup $\exp(t d\pi_+(Z)) = \pi_+(e^{tZ})$ on $L^2(\mathbb{R}^m, \frac{dx}{|x|})$ for $\operatorname{Re} t > 0$, or more precisely, for $t \in \Omega = \{t \in \mathbb{C} : \operatorname{Re} t \geq 0\} \setminus 2\pi \sqrt{-1}\mathbb{Z}$ (see (4.1.4)). The main result of this section is Theorem 5.1.1. In particular, we give a proof of Theorem A in Introduction.

5.1. Result of the section

Let $\langle x, x' \rangle$ be the standard inner product of $\mathbb{R}^m$, $|x| := \sqrt{\langle x, x \rangle}$ be the norm. We recall the notation from Subsection 1.1:

$$\psi(x, x') := 2\sqrt{2(|x||x'| + \langle x, x' \rangle)} = 4|x|^{\frac{1}{2}} |x'|^{\frac{1}{2}} \cos\frac{\theta}{2}, \quad (5.1.1)$$

where $\theta \equiv \theta(x, x')$ is the angle between x and x' in $\mathbb{R}^m$. Let us define a kernel function $K^+(x, x'; t)$ on $\mathbb{R}^m \times \mathbb{R}^m \times \Omega$ by the following formula as in Introduction:

$$
\begin{aligned}
K^+(x, x'; t) &:= \frac{2^{\frac{m-1}{2}} e^{-2(|x|+|x'|)\coth \frac{t}{2}}}{\pi^{\frac{m-1}{2}} \sinh^{\frac{m+1}{2}} \frac{t}{2}} \psi(x, x')^{-\frac{m-3}{2}} I_{\frac{m-3}{2}}\left(\frac{\psi(x, x')}{\sinh \frac{t}{2}}\right) \\
&= \frac{2 e^{-2(|x|+|x'|)\coth \frac{t}{2}}}{\pi^{\frac{m-1}{2}} \sinh^{m-1} \frac{t}{2}} \tilde{I}_{\frac{m-3}{2}}\left(\frac{\psi(x, x')}{\sinh \frac{t}{2}}\right),
\end{aligned}
\tag{5.1.2}
$$

where $I_\nu(z)$ is the modified Bessel function of the first kind and $\tilde{I}_\nu(z) := \left(\frac{z}{2}\right)^{-\nu} I_\nu(z)$ is an entire function (see Subsection 8.5). We note that $\sinh \frac{t}{2}$ in the denominator is non-zero because $t \notin 2\pi\sqrt{-1}\mathbb{Z}$. Therefore, $K^+(x, x'; t)$ is a continuous function on $\mathbb{R}^m \times \mathbb{R}^m \times \Omega$.

We recall from Proposition 3.6.1 (3) that $\pi_+(e^{tZ})$ is a contraction operator with operator norm $\|\pi_+(e^{tZ})\| = e^{-\frac{m-1}{2}\operatorname{Re}t}$. Here is an integral formula of the holomorphic semigroup $\pi_+(e^{tZ})$:

Theorem 5.1.1 (Integral formula for the semigroup). 1) *For* $\operatorname{Re}t > 0$, $\pi_+(e^{tZ})$ *is a Hilbert–Schmidt operator on* $L^2(\mathbb{R}^m, \frac{dx}{|x|})$, *and is given by the following integral transform:*

$$
(\pi_+(e^{tZ})u)(x) = \int_{\mathbb{R}^m} K^+(x, x'; t) u(x') \frac{dx}{|x|} \quad \text{for } u \in L^2(\mathbb{R}^m, \frac{dx}{|x|}).
\tag{5.1.3}
$$

Here, the right-hand side converges absolutely.

2) *For* $t \in \sqrt{-1}\mathbb{R}$, $\pi_+(e^{tZ})$ *is a unitary operator on* $L^2(\mathbb{R}^m, \frac{dx}{|x|})$. *If* $t \in \sqrt{-1}\mathbb{R}$ *but* $t \notin 2\pi\sqrt{-1}\mathbb{Z}$, *then the right-hand side of* (5.1.3) *converges absolutely for any* $u \in L^1(\mathbb{R}^m, \frac{dx}{|x|}) \cap L^2(\mathbb{R}^m, \frac{dx}{|x|})$, *in particular, for any* K-*finite vectors in* $L^2(\mathbb{R}^m, \frac{dx}{|x|})$. *Since* K-*finite vectors span a dense subspace of* $L^2(\mathbb{R}^m, \frac{dx}{|x|})$, *the integral formula* (5.1.3) *holds in the sense of* L^2-*convergence.*

Remark 5.1.2 (Realization on the cone C_+). *Via the isomorphism* $L^2(C_+) \simeq L^2(\mathbb{R}^m, \frac{dx}{|x|})$ *(see Subsection 3.1), the above formula for* $L^2(\mathbb{R}^m, \frac{dx}{|x|})$ *can be readily transferred to the formula of the holomorphic extension* $\pi_+(e^{tZ})$ *on* $L^2(C_+)$. *For this, we define a continuous function* $\widetilde{K^+}(\zeta, \zeta'; t)$ *on* $C_+ \times C_+ \times \Omega$ *by the following formula:*

$$
\widetilde{K^+}(\zeta, \zeta'; t) := \frac{2 e^{-\sqrt{2}(|\zeta|+|\zeta'|)\coth \frac{t}{2}}}{\pi^{\frac{m-1}{2}} \sinh^{m-1} \frac{t}{2}} \tilde{I}_{\frac{m-3}{2}}\left(\frac{2\sqrt{2\langle\zeta, \zeta'\rangle}}{\sinh \frac{t}{2}}\right),
\tag{5.1.4}
$$

where $|\zeta| := \sqrt{\langle \zeta, \zeta \rangle} = (\zeta_1^2 + \cdots + \zeta_{m+1}^2)^{\frac{1}{2}}$. *Then,*

$$(\pi_+(e^{tZ})u)(\zeta) = \int_{C_+} \widetilde{K^+}(\zeta, \zeta'; t) u(\zeta') d\mu(\zeta') \quad \text{for } u \in L^2(C_+). \tag{5.1.5}$$

Remark 5.1.3 (Weil representation). *For the Weil representation ϖ of G', the corresponding semigroup of contraction operators is the Hermite semigroup $\{\varpi(e^{tZ'}) : \operatorname{Re} t > 0\}$ (see Remark 3.6.3), whose kernel function is given by the Mehler kernel $\mathcal{K}(x, x'; t)$ (see Fact C in Subsection 1.2; see also [16, §5]).*

The rest of this section is devoted to the proof of Theorem 5.1.1. Let us mention briefly a naive idea of the proof. We observe that the action of $K(= SO(m+1) \times SO(2)^\sim)$ on $L^2(\mathbb{R}^m, \frac{dx}{|x|})$ is hard to describe because K does not act on $\mathbb{R}^m$. However, the action of its subgroup $R = SO(m)$ has a simple feature, that is, we have the following direct sum decomposition:

$$L^2(\mathbb{R}^m, \frac{dx}{|x|}) \simeq \sum_{l=0}^{\infty} {}^{\oplus} L^2(\mathbb{R}_+, r^{m-2} dr) \otimes \mathcal{H}^l(\mathbb{R}^m). \tag{5.1.6}$$

We have already proved in Theorem 4.1.1 that $K_l^+(r, r'; t)$ is the kernel of $\pi_{+,l}(e^{tZ})$ which is the restriction of $\pi_+(e^{tZ})$ in each l-component of the right-hand side of (5.1.6). Theorem 5.1.1 will be proved if we decompose K^+ into K_l^+. This will be carried out in Lemma 5.6.1. An expansion formula of K^+ by K_l^+ is not used in the proof of Theorem 5.1.1, but might be of interest of its own. We shall give it in Subsection 5.7.

5.2. Upper estimates of the kernel function

In this subsection, we give an upper estimate of the kernel function $K^+(x, x'; t)$. That parallels Lemma 4.2.2.

Lemma 5.2.1. *Let $m \geq 2$. 1) There exists a constant $C > 0$ such that*

$$|K^+(r\omega, r'\omega'; t)| \leq \frac{C}{|\sinh \frac{t}{2}|^{m-1}} e^{-2\alpha(t)(1-|\beta(t)|)(r+r')}, \tag{5.2.1}$$

for any $r, r' \in \mathbb{R}_+$, $\omega, \omega' \in S^{m-1}$, and $t \in \Omega$. Here, $\alpha(t), \beta(t)$ are defined in (4.2.1) and (4.2.2), respectively.

2) *If* $\operatorname{Re} t > 0$, *then*

$$\int_{\mathbb{R}^m} \int_{\mathbb{R}^m} \left| K^+(x, x'; t) \right|^2 \frac{dx}{|x|} \frac{dx'}{|x'|} < \infty.$$

3) *If* $\operatorname{Re} t > 0$, *then for a fixed* $x \in \mathbb{R}^m$, *we have* $K^+(x, \cdot; t) \in L^2(\mathbb{R}^m, \frac{dx'}{|x'|})$.

Proof. By the definition (5.1.2) of $K^+(x, x'; t)$, we have

$$\left| K^+(r\omega, r'\omega'; t) \right| = \frac{2 e^{-2(r+r') \coth \frac{t}{2}}}{\pi^{\frac{m-1}{2}} |\sinh \frac{t}{2}|^{m-1}} \left| \tilde{I}_{\frac{m-3}{2}} \left(\frac{\psi(r\omega, r'\omega')}{\sinh \frac{t}{2}} \right) \right|.$$

By (5.1.1), we have $|\psi(r\omega, r'\omega')| \leq 4\sqrt{rr'}$. Applying Lemma 4.2.1 with $\nu = \frac{m-3}{2}$, we have

$$\left| K^+(r\omega, r'\omega'; t) \right| \leq \frac{2C e^{-\alpha(t)(1 - |\beta(t)|)(r+r')}}{\pi^{\frac{m-1}{2}} |\sinh \frac{t}{2}|^{m-1}}.$$

Replacing C with a new constant, we get (5.2.1).

The second and third statements follow from (5.2.1) because $\alpha(t) > 0$ and $|\beta(t)| < 1$ if $\operatorname{Re} t > 0$. $\qquad\Box$

5.3. Proof of Theorem 5.1.1 (Case $\operatorname{Re} t > 0$)

Suppose $\operatorname{Re} t > 0$. We set

$$(S_t u)(x) := \int_{\mathbb{R}^m} K^+(x, x'; t) u(x') \frac{dx'}{|x'|}. \tag{5.3.1}$$

By Lemma 5.2.1 (2), we observe that S_t is a Hilbert–Schmidt operator on $L^2(\mathbb{R}^m, \frac{dx}{|x|})$, and the right-hand side of (5.3.1) converges absolutely for $u \in L^2(\mathbb{R}^m, \frac{dx}{|x|})$ by the Cauchy–Schwarz inequality and by Lemma 5.2.1 (3).

The remaining assertion of Theorem 5.1.1 (1) is the equality $\pi_+(e^{tZ}) = S_t$. To see this, we observe from the definition (5.1.2) of $K^+(x, x'; t)$ that

$$K^+(kx, kx'; t) = K^+(x, x'; t) \quad \text{for all } k \in R.$$

Therefore, the operator S_t intertwines the R-action, and preserves each summand of (4.1.1). In light of the decomposition (4.1.3) of the operator $\pi_+(e^{tZ})$, the equality $\pi_+(e^{tZ}) = S_t$ will follow from:

Lemma 5.3.1. *Let* $\mathrm{Re}\,t > 0$. *For every* $l \in \mathbb{N}$, *we have*

$$\pi_{+,l}(e^{tZ}) \otimes \mathrm{id} = S_t|_{L^2(\mathbb{R}_+, r^{m-2}dr)\otimes\mathcal{H}^l(\mathbb{R}^m)}. \qquad (5.3.2)$$

We postpone the proof of Lemma 5.3.1 until Subsection 5.6.

Thus, the proof of Theorem 5.1.1 (1) is completed by admitting Lemma 5.3.1.

5.4. Proof of Theorem 5.1.1 (Case $\mathrm{Re}\,t = 0$)

Suppose $\mathrm{Re}\,t = 0$. Then, by Lemma 5.2.1 (1), we have

$$\left|K^+(x, x'; t)\right| \le \frac{C}{|\sinh\frac{t}{2}|^{m-1}}$$

because $\alpha(t) = 0$. Therefore, the right-hand side of (5.1.3) converges absolutely for any $u \in L^1(\mathbb{R}^m, \frac{dx}{|x|}) \cap L^2(\mathbb{R}^m, \frac{dx}{|x|})$, as is seen by

$$\int_{\mathbb{R}^m} |K^+(x, x'; t)u(x')| \frac{dx'}{|x'|} \le \frac{C}{|\sinh\frac{t}{2}|^{m-1}} \int_{\mathbb{R}^m} |u(x')| \frac{dx'}{|x'|} < \infty.$$

By Proposition 3.2.1 (1), we have $L^2(\mathbb{R}^m, \frac{dx}{|x|})_K \subset L^1(\mathbb{R}^m, \frac{dx}{|x|})$. Hence, the right-hand side of (5.1.3) converges absolutely, in particular, for K-finite functions.

Finally, let us show the last statement of (2). Since $W_{a,l}$ $(0 \le l \le a)$ spans $L^2(\mathbb{R}^m, \frac{dx}{|x|})_K$ (see (3.2.1)), it is sufficient to prove

$$\pi_+(e^{tZ}) = S_t \quad \text{on } W_{a,l} \ (0 \le l \le a). \qquad (5.4.1)$$

We recall from Proposition 3.2.1 that every vector $u \in W_{a,l}$ is of the form

$$u(r\omega) = f_{a,l}(r)\phi(\omega) \qquad (5.4.2)$$

for some $\phi \in \mathcal{H}^l(\mathbb{R}^m)$ (see (3.2.4) for the definition). Suppose $\varepsilon > 0$ and $t \in \sqrt{-1}\mathbb{R} \setminus 2\pi\sqrt{-1}\mathbb{Z}$. As in the proof of Claim 4.4.1, we have

$$|K^+(x, x'; \varepsilon + t)| \le \frac{C'}{|\sinh\frac{t}{2}|^{m-1}}$$

and therefore

$$|K^+(x,x';\varepsilon+t)f_{a,l}(r')\phi(\omega)r'^{m-2}| \leq \frac{Cr'^{m-2+l}e^{-2r'}|L_{a-l}^{m-2+2l}(4r')|}{|\sinh\frac{t}{2}|^{m-1}}.$$

Here, $C := C' \max_{\omega \in S^{m-1}} |\phi(\omega)|$. Hence, by the dominated convergence theorem, we have

$$\lim_{\varepsilon\downarrow0} S_{\varepsilon+t}u = S_t u$$

for any $u \in W_{a,l}$.

On the other hand, by Proposition 3.6.1 (1), we have

$$\lim_{\varepsilon\downarrow0} \pi_+(e^{(\varepsilon+t)Z})u = \pi_+(e^{tZ})u.$$

Since $\pi_+(e^{(\varepsilon+t)Z}) = S_{\varepsilon+t}$ for $\varepsilon > 0$, we have now proved (5.4.1).

5.5. Spectra of an $O(m)$-invariant operator

The rest of this section is devoted to the proof of Lemma 5.3.1.

The orthogonal group $O(m)$ acts on $L^2(S^{m-1})$ as a unitary representation, and decomposes it into irreducible representations as follows:

$$L^2(S^{m-1}) \simeq \sum_{l=0}^{\infty}{}^{\oplus} \mathcal{H}^l(\mathbb{R}^m).$$

Since this is a multiplicity-free decomposition, any $O(m)$-invariant operator S on $L^2(S^{m-1})$ acts on each irreducible component as a scalar multiplication.

The next lemma gives an explicit formula of the spectrum for an $O(m)$-invariant integral operator on $C(S^{m-1})$ in the general setting. This should be known to experts, but for the convenience of the readers, we present it in the following form:

Lemma 5.5.1. *For a continuous function h on the closed interval $[-1,1]$, we consider the following integral transform:*

$$S_h : L^2(S^{m-1}) \to L^2(S^{m-1}), \quad \phi(\omega) \mapsto \int_{S^{m-1}} h(\langle\omega,\omega'\rangle)\phi(\omega')d\omega'. \quad (5.5.1)$$

Then, S_h acts on $\mathcal{H}^l(\mathbb{R}^m)$ by a scalar multiplication of $c_{l,m}(h) \in \mathbb{C}$. The constant $c_{l,m}(h)$ is given by

$$c_{l,m}(h) = \frac{2^{m-2}\pi^{\frac{m-2}{2}}l!}{\Gamma(m-2+l)} \int_0^{\pi} h(\cos\theta)\widetilde{C}_l^{\frac{m-2}{2}}(\cos\theta)\sin^{m-2}\theta d\theta, \quad (5.5.2)$$

where $\widetilde{C}_l^{\frac{m-2}{2}}(x)$ denotes the normalized Gegenbauer polynomial (see (8.3.2)).

Example 5.5.2 (see [15, Introduction, Lemma 3.6]). For $h(x) := e^{\sqrt{-1}\lambda x}$, $c_{l,m}(h)$ amounts to

$$c_{l,m}(h) = (2\pi)^{\frac{m}{2}} e^{\frac{\sqrt{-1}}{2}\pi l} \lambda^{-\frac{m-2}{2}} J_{\frac{m-2}{2}+l}(\lambda).$$

Example 5.5.3. We set $\tilde{I}_\nu(z) = \left(\frac{z}{2}\right)^{-\nu} I_\nu(z)$ (see (8.5.6)). For $h(s) := \tilde{I}_{\frac{m-3}{2}}(\alpha\sqrt{1+s})$, we have

$$c_{l,m}(h) = 2^{\frac{3m-4}{2}} \pi^{\frac{m-1}{2}} \alpha^{-m+2} I_{m-2+2l}(\sqrt{2}\alpha) \tag{5.5.3}$$

Proof of Example 5.5.3. We apply Lemma 8.5.2 with $\nu = \frac{m-3}{2}$. Then, we have

$$\int_0^\pi \tilde{I}_{\frac{m-3}{2}}(\alpha\sqrt{1+\cos\theta})\widetilde{C}_l^{\frac{m-2}{2}}(\cos\theta)\sin^{m-2}\theta\, d\theta$$

$$= \frac{2^{\frac{m}{2}}\sqrt{\pi}\,\Gamma(m-2+l)}{\alpha^{m-2}l!} I_{m-2+2l}(\sqrt{2}\alpha).$$

Hence, (5.5.3) follows from Lemma 5.5.1. □

Proof of Lemma 5.5.1. 1) The operator S_h intertwines the $O(m)$-action because $h(\langle k\omega, k\omega'\rangle) = h(\langle \omega, \omega'\rangle)$ for $k \in O(m)$. Hence it follows from Schur's lemma that S_h acts on each irreducible $O(m)$-subspace $\mathcal{H}^l(\mathbb{R}^m)$ by the multiplication of a constant, which we shall denote by $c_{l,m}(h)$ for $l = 0, 1, 2, \ldots$. Thus, we have

$$(S_h\phi)(\omega) = c_{l,m}(h)\phi(\omega) \quad \text{for } \phi \in \mathcal{H}^l(\mathbb{R}^m). \tag{5.5.4}$$

To compute the constant $c_{l,m}(h)$, we use the following coordinate:

$$[0,\pi] \times S^{m-2} \to S^{m-1}, \quad (\theta, \eta) \mapsto \omega = (\cos\theta, \sin\theta \cdot \eta).$$

With this coordinate, we have $d\omega = \sin^{m-2}\theta d\theta d\eta$.

We set $\omega_0 = (1, 0, \cdots, 0)$. Now, we take $\phi(\omega) := \widetilde{C}_l^{\frac{m-2}{2}}(\langle\omega,\omega_0\rangle) \in \mathcal{H}^l(\mathbb{R}^m)$, which is an $O(m-1)$-invariant spherical harmonics. Then the equation (5.5.4) for $\omega = \omega_0$ amounts to

$$\frac{2\pi^{\frac{m-1}{2}}}{\Gamma(\frac{m-1}{2})} \int_0^\pi h(\cos\theta)\widetilde{C}_l^{\frac{m-2}{2}}(\cos\theta)\sin^{m-2}\theta d\theta$$

$$= \frac{\sqrt{\pi}\,\Gamma(m-2+l)}{2^{m-3}l!\,\Gamma(\frac{m-1}{2})} c_{l,m}(h),$$

because $\operatorname{vol}(S^{m-2}) = \frac{2\pi^{\frac{m-1}{2}}}{\Gamma(\frac{m-1}{2})}$ and $\phi(\omega_0) = \widetilde{C}_l^{\frac{m-2}{2}}(1) = \frac{\sqrt{\pi}\,\Gamma(m-2+l)}{2^{m-3}l!\,\Gamma(\frac{m-1}{2})}$ (see (8.3.3)). Hence, (5.5.2) is proved. $\qquad\square$

5.6. Proof of Lemma 5.3.1

This subsection gives a proof of Lemma 5.3.1.

We recall from Theorem 4.1.1 (1) that the kernel function of $\pi_{+,l}(e^{tZ})$ is given by $K_l^+(r, r'; t)$ (see (4.1.5) for definition). Therefore, the equation (5.3.2) is equivalent to the following equation between kernel functions:

Lemma 5.6.1. *For $\phi \in \mathcal{H}^l(\mathbb{R}^m)$, we have*

$$\frac{1}{2}\int_{S^{m-1}} K^+(r\omega, r'\omega'; t)\phi(\omega')d\omega' = K_l^+(r, r'; t)\phi(\omega). \tag{5.6.1}$$

Proof of Lemma 5.6.1. We set

$$h(r, r', t, s) := \frac{2e^{-2(r+r')\coth\frac{t}{2}}}{\pi^{\frac{m-1}{2}}\sinh^{m-1}\frac{t}{2}}\,\widetilde{I}_{\frac{m-3}{2}}\left(\frac{2\sqrt{2rr'(1+s)}}{\sinh\frac{t}{2}}\right). \tag{5.6.2}$$

By the definition (5.1.2) of $K^+(x, x'; t)$, we have

$$K^+(r\omega, r'\omega'; t) = h(r, r', t, \langle\omega, \omega'\rangle).$$

Consider now the integral transform $S_{h(r,r',t,\cdot)}$ on $L^2(S^{n-1})$ with kernel $h(r, r', t, \langle\omega, \omega'\rangle)$. Then it follows from Example 5.5.3 with $\alpha = \frac{2\sqrt{2rr'}}{\sinh\frac{t}{2}}$ that

$$c_{l,m}(h(r, r', t, \cdot)) = 2K_l^+(r, r'; t). \tag{5.6.3}$$

Then, (5.6.1) is a direct consequence of Lemma 5.5.1. $\qquad\square$

5.7. Expansion formulas

We recall that the kernel functions for the semigroups $\pi_+(e^{tZ})$ and $\pi_{+,l}(e^{tZ})$ are given by $K^+(r\omega, r'\omega'; t)$ and $K_l^+(r, r'; t)$. In this subsection, we shall give expansion formulas for $K^+(r\omega, r'\omega'; t)$ arising from the decomposition (see (4.1.3))

$$\pi_+(e^{tZ}) = \sum_{l=0}^{\infty}{}^{\oplus} \pi_{+,l}(e^{tZ}) \otimes \operatorname{id}.$$

Proposition 5.7.1 (Expansion formulas). *Let $m > 1$.*

1) The kernel function $K^+(x, x'; t)$ (see (5.1.2)) has the following expansion:

$$K^+(r\omega, r'\omega'; t) = \frac{1}{\pi^{\frac{m}{2}}} \sum_{l=0}^{\infty} (\frac{m-2}{2} + l) K_l^+(r, r'; t) \widetilde{C}_l^{\frac{m-2}{2}}(\langle \omega, \omega' \rangle). \quad (5.7.1)$$

2) The special value $t = \pi\sqrt{-1}$ for (5.7.1) yields the expansion formula for the Bessel function:

$$\tilde{J}_{\nu-\frac{1}{2}}(\sqrt{z}\cos\frac{\theta}{2}) = \frac{2^{4\nu}}{\sqrt{\pi}} \sum_{l=0}^{\infty} (\nu + l)(-1)^l \frac{J_{2\nu+2l}(\sqrt{z})}{z^\nu} \widetilde{C}_l^\nu(\cos\theta) \quad (5.7.2)$$

for $z \in \mathbb{R}_+, \nu \in \frac{1}{2}\mathbb{Z}$.

Remark 5.7.2 (Weil representation, Gegenbauer's expansion). *Let us compare the above result with the case of the Weil representation of G'. Then, by a similar argument to the proof of Proposition 5.7.1, we can show that the Mehler kernel $\mathcal{K}$ (see Subsection 1.2) has the following decomposition:*

$$\mathcal{K}(r\omega, r'\omega'; t) = \frac{1}{2\pi^{\frac{m}{2}}} \sum_{l=0}^{\infty} (\frac{m-2}{2} + l) \mathcal{K}_l(r, r'; t) \widetilde{C}_l^{\frac{m-2}{2}}(\langle \omega, \omega' \rangle) \quad (5.7.3)$$

if $m > 1$.

In light of the formula (see Subsection 1.2)

$$\mathcal{K}(r\omega, r'\omega'; \pi\sqrt{-1}) = \frac{1}{(2\pi\sqrt{-1})^{\frac{m}{2}}} e^{-\sqrt{-1}rr'\langle\omega,\omega'\rangle},$$

the special value at $t = \pi\sqrt{-1}$ for (5.7.3) yields the following expansion formula for the exponential function known as Gegenbauer's expansion ([9], see also [33, Chapter XI, §11.5]):

$$e^{\sqrt{-1}z\cos\phi} = 2^\nu \sum_{m=0}^{\infty} (\nu + m)\sqrt{-1}^m \frac{J_{\nu+m}(z)}{z^\nu} \widetilde{C}_m^\nu(\cos\phi) \quad (5.7.4)$$

for $z \in \mathbb{R}, \nu \in \frac{1}{2}\mathbb{Z}$. This formula corresponds to (5.7.2).

If $m = 1$, we have

$$\mathcal{K}(x, x'; t) = \frac{1}{2}\big(\mathcal{K}_0(|x|, |x'|; t)\,\mathrm{id}(x) + \mathcal{K}_1(|x|, |x'|; t)\,\mathrm{sgn}(x)\big), \quad (5.7.5)$$

which is immediately verified by the formulas for $\mathcal{K}_0, \mathcal{K}_1$ and $\mathcal{K}$ (see Remark 4.1.2).

Proof of Proposition 5.7.1. 1) First we prepare a lemma.

Lemma 5.7.3. *Let $f \in C[-1,1]$. Then f is expanded into the infinite series:*

$$f(x) = \frac{1}{2\pi^{\frac{m}{2}}} \sum_{l=0}^{\infty} (\frac{m-2}{2} + l) c_{l,m}(f) \widetilde{C}_l^{\frac{m-2}{2}}(x), \qquad (5.7.6)$$

where $c_{l,m}(f)$ is the constant defined in (5.5.2).

Proof of Lemma 5.7.3. Applying the expansion formula $f(x) = \sum_{l=0}^{\infty} \alpha_l^{\nu}(f)$ $\widetilde{C}_l^{\nu}(x)$ (see (8.3.5)) with $\nu = \frac{m-2}{2}$, we have

$$\alpha_l^{\frac{m-2}{2}}(f) = \frac{(\frac{m-2}{2} + l)}{2\pi^{\frac{m}{2}}} c_{l,m}(f), \qquad (5.7.7)$$

from (5.5.2) and (8.3.6). Hence, we get the lemma. □

Let $h(r, r', t, s)$ be as in (5.6.2). We recall from (5.6.3)

$$c_{l,m}(h(r, r', t, \cdot)) = 2K_l^+(r, r'; t).$$

Therefore, by Lemma 5.7.3, we get (5.7.1).
2) Substitute $t = \pi\sqrt{-1}$ and put

$$z := 16rr', \quad \cos\theta := \langle \omega, \omega' \rangle, \quad \nu = \frac{m-2}{2},$$

we get (5.7.2). □

6. The Unitary Inversion Operator

6.1. Result of the section

We define the 'inversion' element $w_0 \in G$ by

$$w_0 := e^{\pi\sqrt{-1}Z}.$$

Then, clearly, w_0 has the following properties:
1) w_0 is of order four.
2) w_0 normalizes $M^{\max}A$ and $\text{Ad}(w_0)\mathfrak{n}^{\max} = \overline{\mathfrak{n}^{\max}}$.
3) The group G is generated by $\overline{P^{\max}}$ and w_0.

We note that if m is odd then $e^{\pi\sqrt{-1}Z}$ is equal to $\begin{pmatrix} I_{m+1} & 0 \\ 0 & -I_2 \end{pmatrix}$ in $SO_0(m+1, 2) = G/\{1, \eta\}$.

This section gives an explicit integral formula of the unitary operator $\pi_+(w_0)$ on $L^2(\mathbb{R}^m, \frac{dx}{|x|})$. In light of $w_0 = e^{\pi\sqrt{-1}Z}$ and $\pi\sqrt{-1} \in \Omega$, we define the following kernel functions by substituting $t = \pi\sqrt{-1}$ into (5.1.2) and (4.1.5), respectively:

$$K^+(x, x') := K^+(x, x'; \pi\sqrt{-1})$$

$$= \frac{2}{e^{\frac{m-1}{2}\pi\sqrt{-1}}\pi^{\frac{m-1}{2}}} \sqrt{2\langle\zeta, \zeta'\rangle}^{-\frac{m-3}{2}} J_{\frac{m-3}{2}}(2\sqrt{2\langle\zeta, \zeta'\rangle}), \quad (6.1.1)$$

$$K_l^+(r, r') := K_l^+(r, r'; \pi\sqrt{-1})$$

$$= 2(-1)^l e^{-\frac{m-1}{2}\pi\sqrt{-1}}(rr')^{-\frac{m-2}{2}} J_{m-2+2l}(4\sqrt{rr'}). \quad (6.1.2)$$

Then, the following result is a direct consequence of Theorems 5.1.1 and 4.1.1.

Theorem 6.1.1 (Integral formula for the unitary inversion operator). *The unitary operator* $\pi_+(w_0) : L^2(\mathbb{R}^m, \frac{dx}{|x|}) \to L^2(\mathbb{R}^m, \frac{dx}{|x|})$ *is given by the following integral transform:*

$$(Tu)(x) := \int_{\mathbb{R}^m} K^+(x, x')u(x')\frac{dx}{|x|}, \quad u \in L^2(\mathbb{R}^m, \frac{dx}{|x|}). \quad (6.1.3)$$

Here, the integral (6.1.3) converges absolutely for any $u \in L^1(\mathbb{R}^m, \frac{dx}{|x|}) \cap L^2(\mathbb{R}^m, \frac{dx}{|x|})$, *in particular, for any* K*-finite vector, hence the equation (6.1.3) holds in the sense of* L^2*-convergence.*

The substitution of $t = \pi\sqrt{-1}$ into (4.1.3) gives the decomposition

$$\pi_+(w_0) = \sum_{l=0}^{\infty}{}^{\oplus} \pi_{+,l}(w_0) \otimes \mathrm{id}.$$

As $\pi_+(w_0)$ is given by the kernel $K^+(x, x')$, so is $\pi_{+,l}(w_0)$ by $K_l^+(r, r')$. Thus, we have:

Theorem 6.1.2 (Radial part of the unitary inversion operator). *The unitary operator* $\pi_{+,l}(w_0) : L^2(\mathbb{R}_+, r^{m-2}dr) \to L^2(\mathbb{R}_+, r^{m-2}dr)$ *is given by the following integral transform:*

$$(T_l f)(r) := \int_0^{\infty} K_l^+(r, r')f(r')r'^{m-2}dr'. \quad (6.1.4)$$

Here, the integral (6.1.4) converges absolutely for any $f \in L^1(\mathbb{R}_+, r^{m-2}dr) \cap L^2(\mathbb{R}_+, r^{m-2}dr)$. *In particular, the equation (6.1.4) holds in the sense of* L^2*-convergence.*

Remark 6.1.3 (Weil representation). *In the case of the Weil representation ϖ of G', the counterparts of Theorem 6.1.1 and Theorem 6.1.2 can be stated as follows: let $\omega_0 := e^{\pi\sqrt{-1}Z'}$.*
We define kernel functions $\mathcal{K}$ and $\mathcal{K}_l$ by the formulas

$$\mathcal{K}(x,x') := \mathcal{K}(x,x';\pi\sqrt{-1}) = \frac{1}{(2\pi\sqrt{-1})^{\frac{m}{2}}} e^{-\sqrt{-1}\langle x,x'\rangle},$$

$$\mathcal{K}_l(r,r') := \mathcal{K}_l(r,r';\pi\sqrt{-1}) = \sqrt{-1}^{-\frac{m-1+2l}{2}} (rr')^{-\frac{m-2}{2}} J_{\frac{m-2+2l}{2}}(rr').$$

Then,
1) The unitary operator $\varpi(w_0) : L^2(\mathbb{R}^m) \to L^2(\mathbb{R}^m)$ is given by

$$(\varpi(w_0)u)(x) = \int_{\mathbb{R}^m} \mathcal{K}(x,x')u(x')dx'.$$

Hence we see that $\varpi(w_0)$ is nothing but the Fourier transform.
2) The 'radial' part of $\varpi(w_0)$, namely, the unitary operator $\varpi_l(w_0) : L^2(\mathbb{R}_+, r^{m-1}dr) \to L^2(\mathbb{R}_+, r^{m-1}dr)$ (see Remark 4.1.2) is given by

$$(\varpi_l(w_0)f)(r) = \int_0^\infty \mathcal{K}_l(r,r')f(r')r'^{m-1}dr'.$$

6.2. Inversion and Plancherel formula

It follows from (3.6.3) that $\pi_+(w_0)^2 = \pi_+(e^{2\pi\sqrt{-1}Z}) = (-1)^{m+1}$ id. Thus, as an immediate consequence of Theorem 6.1.1, we have:

Corollary 6.2.1. *The integral transform*

$$T : u(x) \mapsto \int_{\mathbb{R}^m} K^+(x,x')\frac{u(x')}{|x'|}dx',$$

$$K^+(x,x') := \frac{2^{\frac{m-1}{2}}\psi(x,x')^{-\frac{m-3}{2}}}{e^{\frac{m-1}{2}}\pi\sqrt{-1}\pi^{\frac{m-1}{2}}} J_{\frac{m-3}{2}}(\psi(x,x'))$$

is a unitary operator on $L^2(\mathbb{R}^m, \frac{dx}{|x|})$ of order two (m: odd) and of order four (m: even), that is, we have:

(Inversion formula) $T^{-1} = (-1)^{m+1}T$,

(Plancherel formula) $\|Tu\|_{L^2(\mathbb{R}^m,\frac{dx}{|x|})} = \|u\|_{L^2(\mathbb{R}^m,\frac{dx}{|x|})}$

for all $u \in L^2(\mathbb{R}^m, \frac{dx}{|x|})$.

Remark 6.2.2 (Weil representation, see [16, Corollaries 5.7.3 and 5.7.4])). *Let us compare Corollary 6.2.1 with the corresponding result for the Schrödinger model $L^2(\mathbb{R}^m)$ of the Weil representation ϖ. The unitary operator $\varpi(w_0)$ corresponding to the "inversion element" w_0 is given by the (ordinary) Fourier transform $\mathcal{F}$ (see Fact D in Subsection 1.2). As is well-known, $\mathcal{F}$ is a unitary operator of order four. This reflects the fact that $w_0^4 = e$ in $Sp(m, \mathbb{R})^\sim$.*

6.3. The Hankel transform

Similar to Corollary 6.2.1, from the equality $\pi_+(w_0)^2 = (-1)^{m+1}\mathrm{id}$ and Theorem 6.1.2, we have:

Corollary 6.3.1. *Let ν be a positive integer, and $x, y \in \mathbb{R}$. Then the integral transform*

$$\mathcal{T}_\nu : f(x) \mapsto \int_0^\infty J_\nu(xy)f(y)\sqrt{xy}\,dy$$

is a unitary operator on $L^2((0,\infty), dx)$ of order two. Hence we have:

(Inversion formula) $\mathcal{T}_\nu^{-1} = \mathcal{T}_\nu$,

(Plancherel formula) $\|\mathcal{T}_\nu f\|_{L^2(\mathbb{R}_+,dx)} = \|f\|_{L^2(\mathbb{R}_+,dx)}$.

Remark 6.3.2. *The unitary operator $\mathcal{T}_\nu$ coincides with the Hankel transform (see [7, Chapter VIII]) and the property $\mathcal{T}_\nu^2 = \mathrm{id}$ in the corollary corresponds to its classically known reciprocal formula due to Hankel [13] (see also [7, §8.1 (1)], [33, §14.3 (3)]). The Parseval-Plancherel formula for the Hankel transform goes back to Macaulay-Owen [25].*

Proof of Corollary 6.3.1. Since we assume ν is a positive integer, $\nu = m - 2 + 2l$ for some $m > 3$ and $l \in \mathbb{Z}$. We change the variables $r = \frac{x^2}{4}$, $r' = \frac{y^2}{4}$ and define a unitary map $\Phi : L^2(\mathbb{R}_+, r^{m-2}dr) \to L^2(\mathbb{R}), (\Phi f)(x) := \left(\frac{x}{2}\right)^{\frac{2m-3}{2}} f\left(\frac{x^2}{4}\right)$. Then, we have $\mathcal{T}_\nu = e^{-(l+\frac{m-1}{2})\pi\sqrt{-1}}\Phi \circ T_l \circ \Phi^{-1}$. Since $T_l = \pi_{+,l}(w_0)$ by Theorem 6.1.2, T_l has its inverse as $T_l^{-1} = (-1)^{m+1}T_l$, $\mathcal{T}_\nu$ has its inverse given by

$$\mathcal{T}_\nu^{-1} = e^{(l+\frac{m-1}{2})\pi\sqrt{-1}}\Phi \circ (-1)^{m+1}T_l \circ \Phi^{-1} = \mathcal{T}_\nu.$$

Hence, the corollary follows.   □

6.4. Forward and backward light cones

So far, we have discussed only the irreducible unitary representation π_+ realized on $L^2(C_+)$ for the forward light cone. In this subsection, let us briefly comment on $(\pi_-, L^2(C_-))$ for the backward light cone and $(\pi, L^2(C))$ for $C = C_+ \cup C_-$.

Since $d\pi_-(Z) = -d\pi_+(Z)$ in the polar coordinate representation, we can define a semigroup of contraction operators $\{\pi_-(e^{tZ}) : \operatorname{Re} t < 0\}$ similarly on $L^2(C_-)$. All the statements about the semigroup $\pi_+(e^{tZ})$ also hold by changing the signature $t \to -t$ and replacing C_+ by C_-.

We define a function $K(\zeta, \zeta')$ on $C \times C$ by

$$K(\zeta, \zeta') := \begin{cases} K^+(\zeta, \zeta') & \text{if } \langle \zeta, \zeta' \rangle \geq 0 \\ 0 & \text{if } \langle \zeta, \zeta' \rangle < 0. \end{cases} \tag{6.4.1}$$

Here we note:

$$\{(\zeta, \zeta') \in C \times C : \langle \zeta, \zeta' \rangle \geq 0\} = (C_+ \times C_+) \cup (C_- \times C_-).$$

Originally, $K^+(\zeta, \zeta')$ was defined on $C_+ \times C_+$ (see (6.1.1)). Since $K^+(\zeta, \zeta')$ depends only on the inner product $\langle \zeta, \zeta' \rangle$, we can define $K^+(\zeta, \zeta')$ also on $C_- \times C_-$.

Corollary 6.4.1. *The unitary operator* $\pi(w_0) : L^2(C) \to L^2(C)$ *coincides with the integral transform defined by*

$$T : L^2(C) \to L^2(C), \quad u \mapsto \int_C K(\zeta, \zeta') u(\zeta') d\mu(\zeta'). \tag{6.4.2}$$

Remark 6.4.2. *We note that the kernel function* $K(\zeta, \zeta')$ *is supported on the proper subset* $\{(\zeta, \zeta') \in C \times C : \langle \zeta, \zeta' \rangle \geq 0\}$ *of* $C \times C$. *More generally, in the case of the minimal representation of* $O(p, q)$ *with* $p + q$ *: even,* ≥ 8*, we shall see in [20] that the integral kernel* $K_{p,q}(\zeta, \zeta')$ *representing the inversion element is also supported on the proper subset* $\{(\zeta, \zeta') \in C \times C : \langle \zeta, \zeta' \rangle \geq 0\}$ *if both integers* p, q *are even. This feature fails if both* p, q *are odd. In fact, the support of* $K_{p,q}$ *is the whole space* $C \times C$ *then.*

7. Explicit Actions of the Whole Group on $L^2(C)$

Building on the explicit formula of $\pi(w_0) = \pi(e^{\pi\sqrt{-1}Z})$ on $L^2(C)$ (see Corollary 6.4.1) and that of $\pi(g)$ $(g \in \overline{P^{\max}})$ (see Subsection 2.2), we can find an explicit formula of the minimal representation for the whole group G.

For simplicity, this section treats the case where m is odd. The main result of this section is Theorem 7.2.1 for the action of $O(m+1, 2)$.

7.1. Bruhat decomposition of $O(m+1, 2)$

We recall the notation in Subsection 2.1. In particular, $\overline{P^{\max}} = M^{\max}A\overline{N^{\max}}$ is a maximal parabolic subgroup of $O(m+1, 2)$. Hence, $O(m+1, 2)$ is expressed as the disjoint union:

$$O(m+1, 2) = \overline{P^{\max}} \amalg \overline{P^{\max}}w_0\overline{P^{\max}}$$
$$= \overline{P^{\max}} \amalg \overline{N^{\max}}AM^{\max}w_0\overline{N^{\max}}.$$

We begin by finding $a, b \in \mathbb{R}^{m+1}$, $t \in \mathbb{R}$ and $m \in M^{\max}$ such that

$$g = \overline{n}_b e^{tE} m w_0 \overline{n}_a \in \overline{N^{\max}}AM^{\max}w_0\overline{N^{\max}} \tag{7.1.1}$$

holds for $g \notin \overline{P^{\max}}$.

Suppose g is of the form (7.1.1), and we write $m = \delta m_+$ ($\delta = \pm 1, m_+ \in M_+^{\max}$). Then, in light of the formulas (2.1.3)–(2.1.5), we have

$$g(e_0 - e_{m+2}) = \overline{n}_b e^{tE}(\delta m_+)w_0\overline{n}_a(e_0 - e_{m+2})$$
$$= \overline{n}_b e^{tE}(\delta m_+)w_0(e_0 - e_{m+2})$$
$$= \overline{n}_b e^{tE}(\delta m_+)w_0(e_0 - e_{m+2})$$
$$= \delta e^t \overline{n}_b(e_0 + e_{m+2})$$
$$= \delta e^t \,{}^t(1 - Q(b), 2b_1, \ldots, 2b_{m+1}, 1 + Q(b)). \tag{7.1.2}$$

Now, we set

$$x \equiv {}^t(x_0, \ldots, x_{m+2}) := g(e_0 - e_{m+2}). \tag{7.1.3}$$

Then, solving (7.1.2), we have

$$x_0 + x_{m+2} = 2\delta e^t,$$
$$\frac{x_j}{x_0 + x_{m+2}} = b_j \quad (1 \le j \le m+1).$$

Likewise, if we set

$$y \equiv {}^t(y_0, \ldots, y_{m+2}) := g^{-1}(e_0 - e_{m+2}), \tag{7.1.4}$$

the following relation:

$$\frac{y_j}{y_0 + y_{m+2}} = -a_j \quad (1 \le j \le m+1)$$

must hold, because $g^{-1} = \overline{n}_{-a}e^{tE}(w_0 m^{-1}w_0^{-1})w_0\overline{n}_{-b}$.

Let $\langle \,,\, \rangle$ denote the standard positive definite inner product on $\mathbb{R}^{m+3}$. It follows from $g \in O(m+1,2)$ that

$$
\begin{aligned}
x_0 + x_{m+2} &= \langle e_0 + e_{m+2}, g(e_0 - e_{m+2}) \rangle \\
&= \langle {}^t g(e_0 + e_{m+2}), e_0 - e_{m+2} \rangle \\
&= \langle I_{m+1,2} g^{-1} I_{m+1,2}(e_0 + e_{m+2}), e_0 - e_{m+2} \rangle \\
&= \langle y, e_0 + e_{m+2} \rangle \\
&= y_0 + y_{m+2}.
\end{aligned}
$$

If $x_0 + x_{m+2} \neq 0$, we set

$$a := (a_1, \ldots, a_{m+1}), \quad a_j := \frac{-y_j}{y_0 + y_{m+2}} \quad (1 \le j \le m+1), \quad (7.1.5)\ (a)$$

$$b := (b_1, \ldots, b_{m+1}), \quad b_j := \frac{x_j}{x_0 + x_{m+2}} \quad (1 \le j \le m+1), \quad (7.1.5)\ (b)$$

$$\delta := \mathrm{sgn}(x_0 + x_{m+2}), \qquad\qquad\qquad\qquad\qquad\qquad\quad (7.1.5)\ (c)$$

$$t := \log \left| \frac{x_0 + x_{m+2}}{2} \right|, \qquad\qquad\qquad\qquad\qquad\qquad\quad (7.1.5)\ (d)$$

$$m_+ := \delta e^{-tE} \overline{n}_{-b} g \overline{n}_{-a} w_0^{-1}. \qquad\qquad\qquad\qquad\quad (7.1.5)\ (e)$$

Lemma 7.1.1. *Retain the above notation.*

1) *For $g \in O(m+1,2)$, the following three conditions are equivalent:*

   i) $g \notin \overline{P^{\max}}$,

   ii) $x_0 + x_{m+2} \neq 0$,

   iii) $y_0 + y_{m+2} \neq 0$.

2) *Suppose one of (therefore, all of) the above conditions holds. Then, the element m_+ defined by (7.1.5) (e) belongs to $M_+^{\max} \simeq O(m,1)$.*

7.2. Explicit action of the whole group

For $g \in O(m+1,2)$, we set

$$x \equiv {}^t(x_0, \ldots, x_{m+2}) := g(e_0 - e_{m+2}) \quad \text{(see (7.1.3))},$$

$$y \equiv {}^t(y_0, \ldots, y_{m+2}) := g^{-1}(e_0 - e_{m+2}) \quad \text{(see (7.1.4))}.$$

For g such that $x_0 + x_{m+2} \neq 0$, we also set $a = (a_1, \ldots, a_{m+1})$, $b = (b_1, \ldots, b_{m+1})$, $\delta \in \{\pm 1\}$, and m_+ as in (7.1.5).

If $x_0 + x_{m+2} = 0$, then $g \in \overline{P^{\max}}$ and $(\pi(g)\psi)(\zeta)$ is obtained readily by the formulas (2.2.2)–(2.2.5). For generic g such that $x_0 + x_{m+2} \neq 0$, the unitary operator $\pi(g)$ can be written by means of the kernel function $K(\zeta, \zeta')$ (see (6.4.1) for the definition), the above $x, y \in \mathbb{R}^{m+3}$ and $m_+ \in M_+^{\max} \simeq O(m, 1)$ as follows:

Theorem 7.2.1. *For $g \in O(m+1, 2)$ such that $x_0 + x_{m+2} \neq 0$, the unitary operator $\pi(g)$ is given by the following integral formula: for $\psi \in L^2(C)$,*

$$(\pi(g)\psi)(\zeta) = \left(\frac{2}{x_0 + x_{m+2}} \right)^{\frac{m-1}{2}} e^{\frac{2\sqrt{-1}(x_1\zeta_1 + \cdots + x_{m+1}\zeta_{m+1})}{x_0 + x_{m+2}}}$$

$$\int_C K \left(\frac{2\zeta}{|x_0 + x_{m+2}|}, m_+\zeta' \right) e^{\frac{-2\sqrt{-1}(y_1\zeta_1' + \cdots + y_{m+1}\zeta_{m+1}')}{y_0 + y_{m+2}}} \psi(\zeta') du(\zeta').$$

Proof. We write $g = \overline{n}_b e^t (\delta m_+) w_0 \overline{n}_a$ as in the form (7.1.1). Then, we have

$$(\pi(g)\psi)(\zeta)$$
$$= \pi(\overline{n}_b)\pi(e^{tE})\pi(\delta m_+)(\pi(w_0\overline{n}_a)\psi)(\zeta)$$
$$= e^{2\sqrt{-1}\langle b,\zeta \rangle} e^{-\frac{m-1}{2}t} \delta^{\frac{m-1}{2}} (\pi(w_0\overline{n}_a)\psi)(e^{-t\,t}m_+\zeta) \quad \text{by (2.2.2)–(2.2.5).}$$

Now, by using Corollary 6.4.1,

$$= e^{2\sqrt{-1}\langle b,\zeta \rangle} e^{-\frac{m-1}{2}t} \delta^{\frac{m-1}{2}} \int_C K(e^{-t\,t}m_+\zeta, \zeta')(\pi(\overline{n}_a)\psi)(\zeta') du(\zeta')$$

$$= e^{2\sqrt{-1}\langle b,\zeta \rangle} e^{-\frac{m-1}{2}t} \delta^{\frac{m-1}{2}} \int_C K(e^{-t}\zeta, m_+\zeta') e^{2\sqrt{-1}\langle a,\zeta' \rangle} \psi(\zeta') du(\zeta').$$

$$\square$$

8. Appendix: Special Functions

For the convenience of the reader, we collect here basic formulas of special functions in a way that we use in this article.

8.1. Laguerre polynomials

For $\alpha \in \mathbb{C}, n \in \mathbb{N}$, the *Laguerre polynomials* $L_n^\alpha(x)$ are defined by the formula (see [1, §6.2], for example):

$$L_n^\alpha(x) := \frac{x^{-\alpha} e^x}{n!} \frac{d^n}{dx^n}(e^{-x} x^{n+\alpha}) \tag{8.1.1}$$

$$= \frac{(\alpha+1)_n}{n!} \sum_{k=0}^n \frac{(-n)_k x^k}{(\alpha+1)_k k!}$$

$$= \frac{(-1)^n}{n!} x^n + \cdots + \frac{(\alpha+1)(\alpha+2)\cdots(\alpha+n)}{n!}.$$

Here, we write β_n for $\beta(\beta+1)\cdots(\beta+n-1)$. The Laguerre polynomial solves the linear ordinary differential equation of second order:

$$xu'' + (\alpha+1-x)u' + nu = 0. \tag{8.1.2}$$

Suppose $\alpha \in \mathbb{R}$ and $\alpha > -1$. Then the Laguerre polynomials $\{L_n^\alpha(x) : n = 0, 1, \cdots\}$ are complete in $L^2((0,\infty), x^\alpha e^{-x} dx)$, and satisfy the orthogonality relation (see [1, §6.5]):

$$\int_0^\infty L_m^\alpha(x) L_n^\alpha(x) x^\alpha e^{-x} dx = \frac{\Gamma(\alpha+n+1)}{n!} \delta_{mn} \quad (\alpha > -1). \tag{8.1.3}$$

The Hille–Hardy formula gives the bilinear generating function of Laguerre polynomials (see [6, §10.12 (20)], [14, §1 (3)]):

$$\sum_{n=0}^\infty \frac{\Gamma(n+1)}{\Gamma(n+\alpha+1)} L_n^\alpha(x) L_n^\alpha(y) w^n$$

$$= \frac{1}{1-w} \exp\left(-\frac{(x+y)w}{1-w}\right)(-xyw)^{-\frac{\alpha}{2}} J_\alpha\left(\frac{2\sqrt{-xyw}}{1-w}\right),$$

$$x > 0, y > 0, |w| < 1. \tag{8.1.4}$$

Here the left-hand side converges absolutely.

8.2. Hermite polynomials

Hermite polynomials $H_n(x)$ are given as special values of Laguerre polynomials. We recall from [6, II, §10.13] that

$$H_{2m}(x) = (-1)^m 2^{2m} m! L_m^{-\frac{1}{2}}(x^2), \tag{8.2.1}$$

$$H_{2m+1}(x) = (-1)^m 2^{2m+1} m! x L_m^{\frac{1}{2}}(x^2). \tag{8.2.2}$$

This reduction formula is reflected by the fact that Hermite polynomials appear in the analysis of the Weil representation of $SL(2,\mathbb{R})^\sim$, while Laguerre polynomials appear in the analysis of the minimal representation of $SO_0(m+1,2)^\sim$ and $Sp(m,\mathbb{R})^\sim$ (see Remark 4.3.2).

8.3. Gegenbauer polynomials

For $\nu \in \mathbb{C}$ and $l \in \mathbb{N}$, the Gegenbauer polynomials $C_l^\nu(x)$ are the polynomials of x of degree l given by the *Rodrigues formula* (see [1, §6.4]):

$$C_l^\nu(x) := \left(-\frac{1}{2}\right)^l \frac{(2\nu)_l}{l!\,(\nu + \frac{1}{2})_l} (1 - x^2)^{-\nu + \frac{1}{2}} \frac{d^l}{dx^l}\left((1 - x^2)^{l + \nu - \frac{1}{2}}\right). \quad (8.3.1)$$

It then follows from (8.3.1) that

$$C_l^\nu(1) = \frac{\Gamma(2\nu + l)}{l!\,\Gamma(2\nu)}.$$

We renormalize the Gegenbauer polynomial by

$$\widetilde{C}_l^\nu(x) := \Gamma(\nu) C_l^\nu(x). \quad (8.3.2)$$

Then by the duplication formula of the Gamma function:

$$\Gamma(2\nu) = \frac{2^{2\nu - 1}}{\sqrt{\pi}} \Gamma(\nu) \Gamma(n + \frac{1}{2}),$$

we have

$$\widetilde{C}_l^\nu(1) = \frac{\sqrt{\pi}\,\Gamma(2\nu + l)}{2^{2\nu - 1} l!\,\Gamma(\nu + \frac{1}{2})}. \quad (8.3.3)$$

The special value at $\nu = 0$ is given by the limit formula (see [6, **I**, §3.15.1 (14)])

$$\widetilde{C}_l^0(\cos\theta) = \lim_{\nu \to 0} \Gamma(\nu) C_l^\nu(\cos\theta) = \frac{2\cos(l\theta)}{l}.$$

Suppose $\operatorname{Re}\nu > -\frac{1}{2}$. Then, $\{\widetilde{C}_l^\nu(x) : l = 0, 1, 2, \ldots\}$ forms a complete orthogonal basis of $L^2((-1, 1), (1 - x^2)^{\nu - \frac{1}{2}} dx)$, and the norm of $\widetilde{C}_l^\nu(x)$ is given by

$$\int_{-1}^1 \widetilde{C}_l^\nu(x)^2 (1 - x^2)^{\nu - \frac{1}{2}} dx = \frac{2^{1 - 2\nu} \pi \Gamma(2\nu + l)}{l!\,(l + \nu)}, \quad (8.3.4)$$

(see [6, **I**, §3.15.1 (17)]). Therefore, $f \in L^2((-1, 1), (1 - x^2)^{\nu - \frac{1}{2}} dx)$ has the following expansion:

$$f(x) = \sum_{l=0}^\infty \alpha_l^\nu(f) \widetilde{C}_l^\nu(x), \quad (8.3.5)$$

where we set

$$\alpha_l^\nu(f) := \frac{l!\,(l + \nu)}{2^{1 - 2\nu} \pi \Gamma(2\nu + l)} \int_{-1}^1 f(x) \widetilde{C}_l^\nu(x)(1 - x^2)^{\nu - \frac{1}{2}} dx. \quad (8.3.6)$$

The following integral formula is used in the proof of Lemma 8.5.2 (see [7, §16.3 (3)]): Suppose $\operatorname{Re}\beta > -1$ and $\operatorname{Re}\nu > -\frac{1}{2}$.

$$\int_{-1}^{1}(1-x)^{\nu-\frac{1}{2}}(1+x)^{\beta}\widetilde{C}_n^{\nu}(x)dx =$$
$$\frac{2^{\beta-\nu+\frac{3}{2}}\sqrt{\pi}\,\Gamma(\beta+1)\Gamma(2\nu+n)\Gamma(\beta-\nu+\frac{3}{2})}{n!\,\Gamma(\beta-\nu-n+\frac{3}{2})\Gamma(\beta+\nu+n+\frac{3}{2})}. \qquad (8.3.7)$$

8.4. Spherical harmonics and Gegenbauer polynomials

Let $\Delta_{S^{n-1}}$ be the Laplace–Beltrami operator on the $(n-1)$-dimensional unit sphere S^{n-1}. Then the *spherical harmonics* on S^{n-1} are defined as

$$\mathcal{H}^l(\mathbb{R}^n) := \{f \in C^{\infty}(S^{n-1}) : \Delta_{S^{n-1}}f = -l(l+n-2)f\},$$
$$l = 0,1,2,\ldots.$$

The following facts are well-known:

(1) $\mathcal{H}^l(\mathbb{R}^n)$ is an irreducible representation space of $O(n)$.

(2) It is still irreducible as an $SO(n)$-module if $n \geq 3$.

(3) $\mathcal{H}^l(\mathbb{R}^2) = \mathbb{C}e^{\sqrt{-1}l\theta} \oplus \mathbb{C}e^{-\sqrt{-1}l\theta}$, $l \geq 1$ as $SO(2)$-modules, where $\theta = \tan^{-1}\frac{x_2}{x_1}$, $(x_1,x_2) \in \mathbb{R}^2$.

(4) $\mathcal{H}^l(\mathbb{R}^n)|_{O(n-1)} \simeq \bigoplus_{k=0}^{l}\mathcal{H}^k(\mathbb{R}^{n-1})$ gives an irreducible decomposition of $O(n-1)$-modules. This is also an irreducible decomposition as $SO(n-1)$-modules $(n \geq 4)$.

We regard $O(n-1)$ as the isotropy subgroup of $O(n)$ at $(1,0,\cdots,0) \in \mathbb{R}^n$. We write $(x_1,\cdots,x_n)$ for the standard coordinate of $\mathbb{R}^n$. Then, $O(n-1)$-invariant spherical harmonics are unique up to scalar, and we have:

$$\mathcal{H}^l(\mathbb{R}^n)^{O(n-1)} \simeq \mathbb{C}\widetilde{C}_l^{\frac{n-2}{2}}(x_1). \qquad (8.4.1)$$

8.5. Bessel functions

For $\nu \in \mathbb{C}, z \in \mathbb{C}\setminus\mathbb{R}_{\leq 0}$, *Bessel functions* $J_\nu(z)$ are defined by

$$J_\nu(z) := \left(\frac{z}{2}\right)^{\nu}\sum_{m=0}^{\infty}\frac{(-1)^m(\frac{z}{2})^{2m}}{m!\,\Gamma(\nu+m+1)}. \qquad (8.5.1)$$

It solves the *Bessel's differential equation* of second order:

$$u'' + \frac{1}{z}u' + \left(1 - \frac{\nu^2}{z^2}\right)u = 0.$$

The I-Bessel functions $I_\nu(z)$ (the modified Bessel functions of the first kind) are defined by

$$
I_\nu(z) := \begin{cases} e^{-\frac{\pi\nu\sqrt{-1}}{2}} J_\nu(\sqrt{-1}z) & (-\pi < \arg z \le \frac{\pi}{2}), \\ e^{\frac{3\pi\nu\sqrt{-1}}{2}} J_\nu(\sqrt{-1}z) & (\frac{\pi}{2} < \arg z < \pi). \end{cases} \tag{8.5.2}
$$

$$
= \left(\frac{z}{2}\right)^\nu \sum_{n=0}^\infty \frac{\left(\frac{z}{2}\right)^{2m}}{m!\,\Gamma(\nu+m+1)} \quad (z \in \mathbb{C} \setminus \mathbb{R}_{\le 0}). \tag{8.5.3}
$$

For a special value $\nu = \pm\frac{1}{2}$, $I_\nu(z)$ reduces to

$$
I_{\frac{1}{2}}(z) = \sqrt{\frac{2}{\pi z}}\sinh z, \quad I_{-\frac{1}{2}}(z) = \sqrt{\frac{2}{\pi z}}\cosh z. \tag{8.5.4}
$$

We set

$$
\tilde{J}_\nu(z) := \left(\frac{z}{2}\right)^{-\nu} J_\nu(z) = \sum_{m=0}^\infty \frac{(-1)^m \left(\frac{z}{2}\right)^{2m}}{m!\,\Gamma(\nu+m+1)}, \tag{8.5.5}
$$

$$
\tilde{I}_\nu(z) := \left(\frac{z}{2}\right)^{-\nu} I_\nu(z) = \sum_{m=0}^\infty \frac{\left(\frac{z}{2}\right)^{2m}}{m!\,\Gamma(\nu+m+1)}. \tag{8.5.6}
$$

We note that $\tilde{J}_\nu(z)$ and $\tilde{I}_\nu(z)$ are entire functions of z, and

$$
\tilde{J}_\nu(\sqrt{-1}z) = \tilde{I}_\nu(z).
$$

The following lemma on the estimate of I-Bessel functions are used in Subsections 4.4 and 5.2. We need an estimate of $\tilde{I}_\nu(z)$ for $\nu \ge -\frac{1}{2}$:

Lemma 8.5.1. *There exists a constant $C > 0$ such that the following estimate holds for all $z \in \mathbb{C}$:*

$$
|\tilde{I}_\nu(z)| \le C e^{|\operatorname{Re} z|}. \tag{8.5.7}
$$

Proof. First, suppose $\nu = -\frac{1}{2}$. Then $|\tilde{I}_\nu(z)| = \frac{1}{\sqrt{\pi}}|\cosh z| \le \frac{1}{\sqrt{\pi}}e^{|\operatorname{Re} z|}$.

Next, suppose $\nu > -\frac{1}{2}$. By an integral representation of the Bessel function [33, §6.15 (2)]:

$$
\tilde{I}_\nu(z) = \frac{1}{\Gamma(\nu+\frac{1}{2})\Gamma(\frac{1}{2})} \int_{-1}^1 e^{-zt}(1-t^2)^{\nu-\frac{1}{2}}\,dt,
$$

we have

$$
|\tilde{I}_\nu(z)| \leq \frac{1}{\Gamma(\nu+\frac{1}{2})\Gamma(\frac{1}{2})} \int_{-1}^{1} e^{-t\,\mathrm{Re}\,z}(1-t^2)^{\mathrm{Re}\,\nu-\frac{1}{2}}\,dt
$$

$$
\leq \frac{e^{|\,\mathrm{Re}\,z|}}{\Gamma(\nu+\frac{1}{2})\Gamma(\frac{1}{2})} \int_{-1}^{1} (1-t^2)^{\mathrm{Re}\,\nu-\frac{1}{2}}\,dt \leq C e^{|\,\mathrm{Re}\,z|}
$$

for some constant C independent of z. $\qquad\square$

The following lemma is used in Subsection 5.5, where we set $\nu = \frac{m-3}{2}, \alpha = \frac{2\sqrt{2rr'}}{\sinh\frac{t}{2}}$.

Lemma 8.5.2. *Assume* $\alpha \in \mathbb{C}, \nu \geq -\frac{1}{2}$, *and* $l \in \mathbb{N}$. *Then we have:*

$$
\int_0^\pi I_\nu\big(\alpha\sqrt{1+\cos\theta}\big)\widetilde{C}_l^{\nu+\frac{1}{2}}(\cos\theta)(1+\cos\theta)^{-\frac{\nu}{2}}\sin^{2\nu+1}\theta\,d\theta
$$

$$
= \frac{2^{\frac{3}{2}}\sqrt{\pi}\,\Gamma(2\nu+l+1)}{\alpha^{\nu+1}l!} I_{2\nu+2l+1}\big(\sqrt{2}\alpha\big). \quad (8.5.8)
$$

We could not find this formula in the literature, and so we give its proof here.

Proof. First we note that the integral (8.5.8) converges since $I_\nu\big(\alpha(\sqrt{1+\cos\theta})\big) \cdot (1+\cos\theta)^{-\frac{\nu}{2}}$ is continuous on the closed interval $[0,\pi]$. Furthermore, we have a uniformly convergent expansion

$$
I_\nu\big(\alpha(\sqrt{1+\cos\theta})\big) \cdot (1+\cos\theta)^{-\frac{\nu}{2}} = \sum_{j=0}^\infty \frac{\left(\frac{\alpha}{2}\right)^{\nu+2j}(1+\cos\theta)^j}{j!\,\Gamma(j+\nu+1)}.
$$

Now the left-hand side of (8.5.8) equals

$$
\left(\frac{\alpha}{2}\right)^\nu \sum_{j=0}^\infty \frac{\left(\frac{\alpha}{2}\right)^{2j}}{j!\,\Gamma(j+\nu+1)} \int_0^\pi (1+\cos\theta)^j \widetilde{C}_l^{\nu+\frac{1}{2}}(\cos\theta)\sin^{2\nu+1}\theta\,d\theta
$$

$$
= \left(\frac{\alpha}{2}\right)^\nu \sum_{j=0}^\infty \frac{\left(\frac{\alpha}{2}\right)^{2j}}{j!\,\Gamma(j+\nu+1)} \frac{2^{j+1}\sqrt{\pi}\,\Gamma(j+\nu+1)\Gamma(2\nu+l+1)\Gamma(j+1)}{l!\,\Gamma(j-l+1)\Gamma(2\nu+j+l+2)}
$$

$$
= \left(\frac{\alpha}{2}\right)^\nu \frac{2\sqrt{\pi}\,\Gamma(2\nu+l+1)}{l!} \sum_{j=0}^\infty \frac{\left(\frac{\alpha}{\sqrt{2}}\right)^{2j}}{\Gamma(j-l+1)\Gamma(2\nu+j+l+2)}
$$

$$
= \left(\frac{\alpha}{2}\right)^\nu \frac{2\sqrt{\pi}\,\Gamma(2\nu+l+1)}{l!} \left(\frac{\alpha}{\sqrt{2}}\right)^{2l} \sum_{n=0}^\infty \frac{\left(\frac{\alpha}{\sqrt{2}}\right)^n}{\Gamma(n+1)\Gamma(2\nu+2l+n+2)}
$$

= the right-hand side of (8.5.8).

Here the first equality follows from the formula (8.3.7). In fact, the substitution of $x = \cos\theta$ into (8.3.7) yields

$$\int_0^\pi (1 + \cos\theta)^j \widetilde{C}_l^{\nu+\frac{1}{2}} (\cos\theta) \sin^{2\nu+1}\theta d\theta$$

$$= \int_{-1}^1 (1 - x)^\nu (1 + x)^{j+\nu} \widetilde{C}_l^{\nu+\frac{1}{2}} (x) dx$$

$$= \frac{2^{j+1}\sqrt{\pi}\,\Gamma(j + \nu + 1)\Gamma(2\nu + l + 1)\Gamma(j + 1)}{l!\,\Gamma(j - l + 1)\Gamma(2\nu + j + l + 2)}.$$

Thus, the lemma has been proved. □

References

[1] G. Andrews, R. Askey and R. Roy, *Special Functions*, Cambridge, 1999.

[2] W. Beckner, Inequalities in Fourier analysis, *Ann. of Math.*, (2) **102** (1975), 159–182.

[3] B. Binegar and R. Zierau, Unitarization of a singular representation of $SO(p,q)$, *Comm. Math. Phys.*, **138** (1991), 245–258.

[4] H. Ding, K. I. Gross, R. A. Kunze, and D. St. P. Richards, Bessel functions on boundary orbits and singular holomorphic representations, *The mathematical legacy of Harish-Chandra* (Baltimore, MD, 1998), 223–254, Proc. Sympos. Pure Math., **68**, Amer. Math. Soc., Providence, RI, 2000.

[5] A. Dvorsky and S. Sahi, Explicit Hilbert spaces for certain unipotent representations II, *Invent. Math.*, **138** (1999), 203–224.

[6] A. Erdélyi et al., *Higher Transcendental Functions* **I**, **II**, McGraw-Hill, New York, 1953.

[7] A. Erdélyi et al., *Tables of Integral Transforms*, **II**, McGraw-Hill, New York, 1954.

[8] B. Folland, *Harmonic Analysis in Phase Space*, Annals of Mathematics Studies, **122**, Princeton University Press, Princeton, NJ, 1989.

[9] L. Gegenbauer, Über die Functionen X_n^m, *Wiener Sitzungsberichte*, **68** (2) (1874), 357–367.

[10] I. M. Gelfand and S. G. Gindikin, Complex manifolds whose skeltons are real semisimple groups, and the holomorphic discrete series, *Funct. Anal. Appl.*, **11** (1977), 19–27.

[11] I. M. Gelfand and G. E. Shilov, *Generalized Functions*, **I**, Academic Press, New York, 1964.

[12] V. Guillemin and S. Sternberg, Variations on a Theme by Kepler, *Amer. Math. Soc. Colloq. Publ.*, **42**, Amer. Math. Soc., Province, 1990.

[13] H. Hankel, Die Fourier'schen Reihen und Integrale für Cylinderfunctionen, *Math. Ann.*, **8** (1875), 471–494.

[14] G. H. Hardy, Summation of a series of polynomials of Laguerre, Journal of the London Mathematical Society, **7** (1932), 138–139, 192.

[15] S. Helgason, *Groups and Geometric Analysis*, Academic Press, London, 1984.

[16] R. Howe, The oscillator semigroup, Amer. Math. Soc., Proc. Symp. Pure Math., **48** (1988), 61–132.

[17] R. Howe and E. C. Tan, *Non-Abelian Harmonic Analysis*, Springer, 1992.

[18] T. Kobayashi, Conformal geometry and global solutions to the Yamabe equations on classical pseudo-Riemannian manifolds, Proceedings of the 22nd Winter School "Geometry and Physics" (Srni, 2002). Rend. Circ. Mat. Palermo (2) Suppl., **71** (2003), 15–40.

[19] T. Kobayashi and G. Mano, Integral formulas for the minimal representation of $O(p,2)$, *Acta Appl. Math.*, **86** (2005), 103–113.

[20] T. Kobayashi and G. Mano, Integral formula of the unitary inversion operator for the minimal representation of $O(p,q)$, *Proc. Japan Acad. Ser. A*, **83**, (2007), 27–31.

[21] T. Kobayashi and B. Ørsted, Analysis on the minimal representation of $O(p,q)$ I. Realization via conformal geometry, *Adv. Math.*, **180** (2003), 486–512.

[22] T. Kobayashi and B. Ørsted, Analysis on the minimal representation of $O(p,q)$ II. Branching laws, *Adv. Math.*, **180** (2003), 513–550.

[23] T. Kobayashi and B. Ørsted, Analysis on the minimal representation of $O(p,q)$ III. Ultrahyperbolic equations on $\mathbb{R}^{p-1,q-1}$, *Adv. Math.*, **180** (2003), 551–595.

[24] B. Kostant, The vanishing scalar curvature and the minimal unitary representation of $SO(4,4)$, eds. Connes et al, *Operator Algebras, Unitary Representations, Enveloping Algebras, and Invariant Theory*, Progress in Math., **92**, Birkhäuser, 1990, 85–124.

[25] P. Macaulay-Owen, Parseval's theorem for Hankel transforms, *Proc. London Math. Soc.*, (2) **45** (1939), 458–474.

[26] W. Myller-Lebedeff, Die Theorie der Integralgleichungen in Anwendung auf einige Reihenentwicklungen, *Math. Ann.*, **68** (1907), 388–416.

[27] G. I. Olshanskiĭ, Invariant cones in Lie algebras, Lie semigroups and the holomorphic discrete series, *Funct. Anal. Appl.*, **15** (1981), 275–285.

[28] S. Sahi, Explicit Hilbert spaces for certain unipotent representations, *Invent. Math.*, **110** (1992), 409–418.

[29] R. J. Stanton, Analytic extension of the holomorphic discrete series, *Amer. J. Math.*, **10** (1986), 1411–1424.

[30] G. Szegö, *Orthogonal Polynomials*, American Mathematical Society, Province, 1939.

[31] S. Thangavelu, *Lectures on Hermite and Laguerre Expansions*, Mathematical Notes **42**, Princeton University Press, Princeton, 1993.

[32] P. Torasso, Méthode des orbites de Kirillov–Duflo et représentations minimales des groupes simples sur un corps local de caractéristique nulle, *Duke Math. J.*, **90** (1997), 261–377.

[33] G. N. Watson, *A Treatise on the Theory of Bessel Functions*, Cambridge, 1922.

CLASSIFICATION DES SÉRIES DISCRÈTES POUR CERTAINS GROUPES CLASSIQUES p-ADIQUES

COLETTE MŒGLIN

Department of Mathematics
CNRS et Université de Paris 7 et Paris 6
2 Place Jussieu, F-75005 Paris, France
E-mail: moeglin@math.jussieu.fr

The goal of this paper is to prove how Arthur's results, in the case of split odd orthogonal p-adic groups, imply the Langlands' classification of discrete series. Of course this need the validity of "fundamental" lemmas which are not yet available. And this include a Langlands classification of cuspidal irreducible representations for such groups.

En l'honneur de Roger Howe pour son 60e anniversaire

L'article a pour but de montrer que seule l'hypothèse que des lemmes fondamentaux pour des paires bien précises de groupes est suffisante pour avoir les hypothèses de [10] et [11] permettant de classifier les séries discrètes des groupes symplectiques et orthogonaux "impairs" p-adiques. Le point de départ est l'idée d'Arthur exprimée pleinement sous forme de conjecture en [1] et sous forme de résultats modulo des lemmes fondamentaux en [2].

Le corps de base, noté F, est un corps p-adique; on suppose que $p \neq 2$ car nos références, surtout [19] utilisent cette hypothèse. Il n'intervient pas dans notre propre travail.

On considère les points sur F d'un groupe classique, c'est-à-dire le groupe des automorphismes d'une forme orthogonale ou symplectique sur un F-espace vectoriel, V, de dimension finie; ici on se limite aux espaces V symplectiques ou orthogonaux de dimension impaire. On note n est le rang du groupe, c'est-à-dire la partie entière, $[dim\, V/2]$ et $G(n)$ ce groupe p-adique. Si V est orthogonal, on ne suppose pas V maximalement déployée pour obtenir le résultat principal toutefois le cas non maximalement déployé s'obtient à la fin à partir du cas maximalement déployé et dans cette introduction on suppose $G(n)$ déployé. On note $G^*(n)$ le groupe complexe des automorphismes de la forme duale au sens de Langlands, c'est-à-dire si V est orthogonal de dimension impaire $G^* = Sp(2n, \mathbb{C})$, si V est symplectique de dimension paire $G^* = SO(2n+1, \mathbb{C})$. On notera n^* l'entier dimension de la représentation naturelle de G^*; c'est-à-dire que $n^* = 2n$ si $G^*(n)$ est un groupe symplectique et $2n + 1$ si $G^*(n)$ est un groupe orthogonal. Quand un groupe linéaire $GL(m, F)$ ($m \in \mathbb{N}$) est fixé, on note θ l'automorphisme de ce groupe $g \mapsto J_m^{-1}\, {}^t g^{-1} J_m$, où J_m est essentiellement la matrice antidiagonale (cf 1.1) et $\tilde{G}L(m, F)$ le produit semi-direct de $GL(m, F)$ par le groupe $\{1, \theta\}$.

L'idée de base d'Arthur pour étudier les groupes classiques est que $G(n)$ est un groupe endoscopique principal pour l'ensemble $\tilde{G}(n^*)$ qui est par définition la composante connexe de θ dans le produit semi-direct $\tilde{G}L(n^*, F)$. C'est le groupe qui permet de faire un transfert entre les objects stables pour $G(n)$ et les objets stables pour $\tilde{G}(n^*)$. Arthur a aussi expliqué comment obtenir un transfert pour les groupes orthogonaux sur un espace de dimension paire; la "seule" différence est que le transfert fait intervenir des facteurs de transfert que l'auteur ne maîtrise pas.

L'hypothèse faite dans ce papier est que le lemme fondamental pour ce couple soit vrai pour tout $m \leq n$, c'est indispensable pour l'énoncé même des résultats. Mais il faut en plus le lemme fondamental pour tous les groupes endoscopiques intervenant pour la stabilisation de $\tilde{G}(m^*)$, pour tout $m \leq n$. Et il y a un résultat d'Arthur qui doit même, je crois, supposer la validité des lemmes fondamentaux comme ci-dessus pour $m > n$; on peut sans doute borner m mais sans doute pas suffisamment pour que des cas particuliers puissent vraiment être démontrés sans que le cas général le soit. On n'a par contre pas besoin de leur analogue pondéré (sauf erreur); on utilise de façon déterminante les résultats locaux de [2] mais pas les résultats globaux.

Pour la détermination des représentations cuspidales et donc la classification complète, on a aussi besoin des lemmes fondamentaux pour tous les groupes endoscopiques elliptiques de $\tilde{G}L(n^*, F)$. Pour cette classification complète, nous nous sommes limités aux groupes G orthogonaux sur un espace de dimension impaire bien que les méthodes soient générales; la raison de cette limitation est que l'on ne veut pas ici discuter le cas des groupes orthogonaux pairs qui méritent une discussion à eux tout seuls. Pour eux, dans tous mes travaux, à la suite d'Adams, j'ai choisi de prendre comme groupe dual un groupe non connexe, le groupe orthgonal complexe;

ce n'est pas le choix d'Arthur et il y a donc à écrire les équivalences entre les 2 points de vue.

Une représentation dont la classe d'isomorphie est invariante sous l'action de θ sera dite une représentation θ-invariante. Pour tout m, on appelle série θ-dicrète une représentation irréductible tempérée de $GL(m, F)$, θ-invariante et qui n'est induite propre d'aucune représentation θ-invariante d'un sous-groupe de Levi θ-invariant de $GL(m, F)$.

Cette définition est beaucoup plus simple du côté groupe dual; on rappelle que la conjecture de Langlands locale pour les GL ([6], [7]) et les résultats de Zelevinsky ([21]) permettent d'associer à toute représentation irréductible tempérée de $GL(m, F)$ une classe de conjugaison d'un morphisme continu, borné de $W_F \times SL(2, \mathbb{C})$ dans $GL(m, \mathbb{C})$. Pour π^{GL} une représentation irréductible tempérée de $GL(m, F)$ on note $\psi_{\pi^{GL}}$, ce morphisme; le fait que π^{GL} est θ-invariante est exactement équivalent à ce que (quitte à conjuguer) $\theta \circ \psi_{\pi^{GL}} \circ \theta = \psi_{\pi^{GL}}$, où ici θ est l'automorphisme $g \mapsto {}^t g^{-1}$ de $GL(m, \mathbb{C})$. Le fait que π^{GL} est θ-discrète est équivalent à ce que le sous-groupe du centralisateur de $\psi_{\pi^{GL}}$ dans $GL(m, \mathbb{C})$ formé par les éléments θ-invariants est fini.

Soit encore m un entier et soit π^{GL} une représentation tempérée de $GL(m^*, F)$, θ-invariante; on fixe un prolongement de π^{GL} à $\tilde{GL}(m^*, F)$. Le caractère de cette représentation restreint à $\tilde{G}(n^*)$ n'est pas nécessairement stable. C'est même un résultat profond de [2] 30.1 de préciser que ce caractère est stable précisément quand le morphisme associé se factorise par $G^*(n)$; c'est pour pouvoir utiliser ce résultat que l'on a supposé la totalité des lemmes fondamentaux. Et en suivant Arthur, quand le caractère de π^{GL} vue comme distribution sur $\tilde{G}(n^*)$ est stable, on définit le paquet de π^{GL}, noté $\Pi(\pi^{GL})$ comme étant l'ensemble des représentations irréductibles, π de $G(m)$ tel que pour un bon choix de coefficients $c_\pi \in \mathbb{R}_{\geq 0}$ la distribution $tr\pi^{GL}$ de $GL(m^*, F) \times \theta$ soit un transfert de $\sum_{\pi \in \Pi(\pi^{GL})} c_\pi tr\pi$ (cf 1.3); cette condition définit uniquement les éléments de $\Pi(\pi^{GL})$. Si ψ est le morphisme correspondant à π^{GL}, on note $\Pi(\psi)$ au lieu de $\Pi(\pi^{GL})$. Comme Arthur a en vue des résultats globaux, il prolonge cette définition pour toute représentation π^{GL} composante locale d'une forme automorphe cuspidale de $GL(m^*)$; la généralisation est évidente, c'est une induction, il prolonge d'ailleurs aussi les constructions pour obtenir toutes les formes automorphes de carré intégrable mais nous n'en avons pas besoin ici. Une remarque par exemple dans [19] VI.1 (ii) est que si π^{GL} est θ-discrète alors $\Pi(\pi^{GL})$ est formé de représentations elliptiques; comme les coefficients c_π

sont positifs d'après Arthur, on peut encore remplacer elliptiques par séries
discrètes. Toutefois, on redémontrera ce dont on a besoin pour éviter cet
argument croisé.

Pour pouvoir utiliser ce qui précède, la remarque à vérifier est que si π
est une série discrète de $G(n)$ irréductible alors le caractère distribution de
cette représentation appartient à un paquet stable, c'est-à-dire qu'il existe
π^{GL}, θ-discret comme ci-dessus tel que $\pi \in \Pi(\pi^{GL})$; cela est fait en 1.3.
Ensuite, il n'est pas difficile de vérifier que cela entraîne que le support
cuspidal de π est "demi-entier" (cf. 1.4) et par voie de conséquence, cela
nécessite que les points de réductibilité des induites de cuspidales soient
demi-entiers (cf. 2). Ceci est une partie des hypothèses de [10] et [11] et
on peut aller plus loin pour avoir toutes les hypothèses. Ainsi on associe
à π le morphisme $\psi_{\pi^{GL}}$ tel que $\pi \in \Pi(\psi_{\pi^{GL}})$. A ce point, il me semble
que l'unicité de π^{GL} n'est pas complètement claire; on (re)trouvera cette
unicité de façon détournée à la fin du travail en 3.

On peut faire cette construction pour π_0 une représentation cuspidale de $G(n)$ et donc obtenir un morphisme ψ_0 de $W_F \times SL(2, \mathbb{C})$ dans
$GL(n^*, \mathbb{C})$. Ce morphisme détermine et est déterminé par les propriétés
d'irréductibilité des induites de la forme $St(\rho, a) \times \pi_0$, où ρ est une représentation irréductible, cuspidale autoduale d'un groupe $GL(d_\rho, F)$ et a est un
entier; $St(\rho, a)$ est la représentation de Steinberg généralisée de $GL(ad_\rho, F)$,
l'induite étant une représentation de $G(n + ad_\rho)$. Ce lien se fait exactement
comme conjecturé en [10] et [11]:
on voit ψ_0 comme une représentation de $W_F \times SL(2, \mathbb{C})$ dans $GL(n_0^*, \mathbb{C})$.
Toute représentation irréductible de $W_F \times SL(2, \mathbb{C})$ est le produit tensoriel
d'une représentation irréductible, notée ρ, de W_F, et d'une représentation
de dimension finie σ_a de $SL(2, \mathbb{C})$ uniquement déterminée par sa dimension
notée ici $a \in \mathbb{N}$; on note d_ρ la dimension de la représentation ρ et on
identifie ρ à une représentation cuspidale irréductible de $GL(d_\rho, F)$ grâce
à la correspondance locale de Langlands démontrée en [6], [7] et on garde
la même notation ρ. Et on montre qu'une telle représentation irréductible
$\rho \otimes \sigma_a$ de $W_F \times SL(2, \mathbb{C})$ intervient dans ψ_0 si et seulement si les 2 conditions
suivantes sont réalisées:

(1) $\rho \otimes \sigma_a$ à conjugaison près est à valeurs dans un groupe de même type
que G^*, c'est-à-dire orthogonal si ce groupe est orthogonal et symplectique
sinon;

(2) l'induite pour le groupe $G(n_0 + d_\rho)$, $St(\rho, a) \times \pi_0$ est irréductible, où $St(\rho, a)$ est la représentation de Steinberg basée sur ρ du groupe $GL(ad_\rho, F)$.

Et on montre aussi que si $\rho \otimes \sigma_a$ intervient comme sous-représentation de ψ_0, elle y intervient avec multiplicité 1.

Ces propriétés entraînent l'unicité de ψ_0 puisque l'on connaît sa décomposition en représentations irréductibles grâce à des propriétés de la représentation π_0; c'est ce que l'on a appelé la connaissance des blocs de Jordan de π_0 et de ψ_0.

Le lien avec les points de réductibilité des induites de la forme $\rho| \; |^x \times \pi_0$, où ρ est comme ci-dessus et x est un nombre réel, est le suivant. Pour ρ fixée vue comme une représentation irréductible de W_F de dimension d_ρ, on note $Jord_\rho(\psi_0)$ l'ensemble des entiers a (s'il en existe) tel que $\rho \otimes \sigma_a$ soit une sous-représentation irréductible de ψ_0. Si $Jord_\rho(\psi_0) \neq \emptyset$, on note alors a_{ρ,ψ_0} son élément maximal. Supposons maintenant que $Jord_\rho(\psi_0) = \emptyset$; si $\rho \otimes \sigma_2$ est à valeurs dans un groupe de même type que G^* (cf. ci-dessus); on pose a_{ρ,ψ_0} et sinon c'est ρ qui est à valeurs dans un groupe de même type que G^* et on pose $a_{\rho,\psi_0} := -1$.

On voit maintenant ρ comme une représentation cuspidale irréductible de $GL(d_\rho, F)$. On définit x_{ρ,π_0} comme l'unique d'après [16] réel positif ou nul tel que l'induite $\rho| \; |^{x_{\rho,\pi_0}} \times \pi_0$ soit réductible et on montre que (cf. 2, ci-dessous) que $x_{\rho,\pi_0} = (a_{\rho,\psi_0} + 1)/2$.

Au passage, on démontre que ψ_0 est sans trou c'est-à-dire que si ψ_0 contient une représentation de la forme $\rho \otimes \sigma_a$ avec $a > 2$, alors il contient aussi la représentation $\rho \otimes \sigma_{a-2}$ (ici ρ est une représentation irréductible de W_F). Pour démontrer le résultat annoncé calculant les points de réductibilité, on démontre d'abord que x_{ρ,π_0} est un demi-entier, puis on vérifie tous les cas sauf le dernier cas ci-dessus, en utilisant la compatibilité des paquets à l'induction et la restriction (cela a déjà été fait en [13]). Puis on montre que le dernier cas en résulte "par défaut".

Avec ces résultats sur les représentations cuspidales, [10] et [11] associent à toute série discrète irréductible un morphisme de $W_F \times SL(2, \mathbb{C})$ dans $GL(n^*, \mathbb{C})$, noté ψ_π, il n'y a plus qu'à démontrer la compatibilité avec les résultats d'Arthur, c'est-à-dire que $\pi \in \Pi(\psi_\pi)$; c'est fait en 3.

Dans la suite du travail, on se limite comme expliqué ci-dessus au cas où $G = SO(2n + 1, F)$ (déployé ou non). On suppose que tous les lemmes fondamentaux pour les groupes endoscopiques de $SO(2m+1, F)$ et pour $\tilde{G}L(2m, F)$ sont vérifiés pour tout m. Il y a une autre propriété dont on a besoin qui peut s'exprimer très facilement ainsi: avec la notation $\Pi(\psi)$ déjà employée, il n'y a qu'une combinaison linéaire (à homothétie près) d'éléments de $\Pi(\psi)$ qui est stable et toute combinaison linéaire stable de séries discrètes et dans l'espace vectoriel engendré par ces éléments. On discute cette propriété en 4.3 car il est vraisemblable qu'elle est dans (ou puisse être démontré par les méthodes de) [2]; mais comme elle n'est pas écrite telle quelle, on la prend comme hypothèse. Tel que nous avons écrit les arguments, nous avons besoin de cette hypothèse pour tous les morphismes ψ' à valeurs dans $Sp(2m, \mathbb{C})$ pour tout $m \leq n$. Avec les lemmes fondamentaux et cette hypothèse, on montre que l'application définie par Arthur de $\Pi(\psi)$ dans l'ensemble des caractères du groupe $Cent_{Sp(2n,\mathbb{C})}\psi/Cent(Sp(2n, \mathbb{C}))$ est une bijection; un tel résultat est propre au cas des paquets tempérés. On obtient alors la preuve de notre conjecture sur la classification des représentations cuspidales de $SO(2m+1, F)$ (pour tout $m \leq n$), à savoir cet ensemble de représentations cuspidales est en bijection avec l'ensemble des couples, ψ, ϵ, formé d'un morphisme ψ sans trou de $W_F \times SL(2, \mathbb{C})$ dans $Sp(2m, \mathbb{C})$, définissant une représentation semi-simple sans multiplicité de $W_F \times SL(2, \mathbb{C})$ et d'un caractère ϵ du centralisateur de ψ dans $Sp(2n, \mathbb{C})$, trivial sur le centre de $Sp(2n, \mathbb{C})$, qui a les propriétés suivantes:

ϵ est alterné (ou cuspidale au sens de Lusztig) c'est-à-dire: soit une sous-représentation irréductible $\rho \otimes \sigma_a$ de $W_F \times SL(2, \mathbb{C})$ de dimension ad_ρ intervenant dans ψ; elle est nécessairement à valeurs dans un groupe $Sp(ad_\rho, \mathbb{C})$ dont on note $Z_{\rho \otimes \sigma_a}$ le centre, sous-groupe à 2 éléments qui est naturellement un sous-groupe du centralisateur de ψ. Alors ϵ est alterné si pour ρ, a comme ci-dessus, la restriction de ϵ à $Z_{\rho \otimes \sigma_a} \simeq \{\pm 1\}$ est non trivial si $a = 2$ et si $a > 2$, cette restriction n'est pas le même caractère que la restriction de ϵ à $Z_{\rho \otimes \sigma_{a-2}} \simeq \{\pm 1\}$;

On remarque que pour cette conjecture précise, on n'a pas traité le cas non déployé qui devrait pourtant être analogue en remplaçant la condition sur la restriction du caractère au centre de $Sp(2n, \mathbb{C})$ par son opposé.

C'est la dernière conjecture qui manquait pour pouvoir décrire complètement les paquets de Langlands de série discrète suivant [10] et [11]; on vérifie pour finir que nos constructions sont bien compatibles avec celles d'Arthur. Il s'agit ici de vérifier que si $\pi \in \Pi(\psi)$, le caractère associé par Arthur se lit sur le module de Jacquet de π. On montre donc la propriété suivante, soit comme ci-dessus $\rho \otimes \sigma_a$ une sous-représentation irréductible de $W_F \times SL(2,\mathbb{C})$ incluse dans ψ. Supposons d'abord qu'il existe $b \in \mathbb{N}$ avec $b < a$ et $\rho \otimes [b]$ une sous-représentation de ψ; on note alors a_- le plus grand entier b vérifiant ces 2 conditions. Quand un tel b n'existe pas a_- n'est pas défini si a est impair et vaut 0 si a est pair. On note encore ρ la représentation cuspidale de $GL(d_\rho, F)$ correspondant à ρ par la correspondance de Langlands locale pour $GL(d_\rho, F)$ ([6],[7]). Soit $\pi \in \Pi(\psi)$ et $\epsilon_{\mathcal{A}}(\pi)$ le caractère du centralisateur de ψ associé par Arthur à π; on montre, avec les notations déjà introduites ci-dessus, que la restriction de $\epsilon_{\mathcal{A}}(\pi)$ à $Z_{\rho \otimes \sigma_a}$ est identique à la restriction de ce caractère à $Z_{\rho \otimes \sigma_{a_-}}$ (est triviale si $a_- = 0$) si et seulement si il existe une représentation π' du groupe $SO(2(m - d_\rho(a - a_-)/2) + 1, F)$ et une inclusion:

$$\pi \hookrightarrow \rho|\ |^{(a-1)/2} \times \cdots \times \rho|\ |^{(a_-+1)/2} \times \pi'.$$

Je remercie Laurent Clozel pour ces explications sur la globalisation de situations locales et Jean-Loup Waldspurger dont les travaux sont indispensables à cet article.

L'essentiel des idées encore développées ici et que j'utilise depuis de nombreuses années m'ont été inspirées par l'étude de la correspondance de Howe (avec l'interprétation donnée par J. Adams). C'est donc un immense plaisir pour moi que de pouvoir dédier cette article à Roger Howe pour son soixantième anniversaire. Je remercie aussi les organisateurs du congrès en son honneur pour leur accueil extrêmement chaleureux.

1. Support Cuspidal des Séries Discrètes

1.1. Notations. Dans tout l'article sauf la partie 5, si $G(n)$ est un groupe orthogonal, on le suppose déployé. On reprend les notations de l'introduction, pour tout $m \in \mathbb{N}$, $G(m), G^*(m)$, m^*. Et pour tout m, on note J la matrice

$$\begin{pmatrix} 0 & \cdots & 1 \\ \vdots & \cdots & \vdots \\ (-1)^{m-1} & \cdots & 0 \end{pmatrix}$$

de façon à pouvoir définir l'action de θ sur $GL(m, F)$ par $g \mapsto J\,{}^t g^{-1} J^{-1}$ pour tout $g \in GL(m, F)$.

1.2. Quelques rappels des résultats de [2]. On rappelle des résultats fondamentaux pour nous de [2] 30.1; soit ψ un morphisme continu borné de $W_F \times SL(2, \mathbb{C})$ dans $G^*(n)$. On le voit comme un morphisme de $W_F \times$

$SL(2,\mathbb{C}) \times SL(2,\mathbb{C})$ dans $G^*(n)$ trivial sur la deuxième copie de $SL(2,\mathbb{C})$. Arthur a défini une action de θ sur la représentation $\pi(\psi)$ de $GL(n^*, F)$ en imposant à θ d'induire l'action triviale sur l'espace des fonctionnelles de Whittaker (ceci nécessite a priori un choix d'un caractère additif mais on renvoit à [13] pour l'indépendance et une discussion plus générale); pour donner un sens précis à cela, il faut fixer, à l'aide du caractère addifif fixé, un caractère du groupe des matrices unipotentes supérieures invariant par θ; le module des co-invariants de $\pi(\psi)$ pour ce groupe unipotent et ce caractère est de dimension 1 et l'action de θ est fixée de telle sorte que θ agisse trivialement dans cet espace de co-invariants. Ceci dit, pour nous ici, la description de la normalisation n'a aucne importance.

Arthur en loc. cit. montre alors l'existence d'un paquet fini de représentations $\Pi(\psi)$ de $G(n)$ et de coefficients entiers positifs c_π tel que $tr\pi(\psi) \circ \theta$ soit un transfert stable de la distribution $\sum_{\pi \in \Pi(\psi)} c_\pi tr\, \pi$; les propriétés des coefficients sont cachés dans la définition de $\tilde{\Pi}_{fin}$ d'Arthur, ils ne joueront pas de rôle ici sauf la positivité (à un signe commun à tous les coefficients près)! Les représentations incluses dans $\Pi(\psi)$ sont uniquement déterminées grâce à l'indépendance linéaire des caractères. On dit que le morphisme ψ est θ-discret si vu comme une représentation de $W_F \times SL(2,\mathbb{C})$ dans $GL(n^*, \mathbb{C})$, il définit une représentation sans multiplicité de ce groupe; c'est exactement la même définition que celle de l'introduction.

1.3. Appartenance à un paquet stable. Soit π une série discrète irréductible de $G(n)$. Il faut savoir que la projection du caractère distribution de π sur l'ensemble des distributions stables à support dans les éléments elliptiques est non nulle. C'est un résultat local et on va en donner 2 démonstrations.

1.3.1. *argument global.* La première démonstration repose sur la globalisation suivante qui m'a été indiquée par Laurent Clozel et qui se trouve dans [5] theorem 1 B, π peut se globaliser, c'est-à-dire être considérée comme une composante locale d'une forme automorphe de carré intégrable Σ d'un groupe adélique associé à G; on peut imposer à Σ ce que l'on veut en un nombre fini de places, par exemple d'être cuspidale en une autre place et Steinberg en au moins 2 autres places. Cela assure que Σ est cuspidale et qu'elle intervient dans une formule des traces simple automatiquement stable; on peut alors d'après [2] 30.2 (b) trouver une forme automorphe $\tilde{\Sigma}$ du groupe adélique associé à $GL(n^*)$ tel que π soit dans le paquet associé par Arthur à la composante locale de $\tilde{\Sigma}$ pour notre place; mais Arthur assure même qu'en toute place la composante de Σ est dans un paquet stable dont un transfert est donnée par la composante en cette même place de Σ. On a supposé que Σ est Steinberg en au moins une place mais une telle représentation est automatiquement stable et son seul transfert possible parmi les composantes locales de forme automorphe de carré intégrable est la représentation de Steinberg; on peut le vérifier avec des arguments de module de Jacquet, ici on insiste sur le fait que le transfert se définit par des égalités sur toutes les classes de conjugaisons stables pas seulement les elliptiques. Ainsi $\tilde{\Sigma}$ est nécessairement une représentation automorphe cuspidal. Et la composante locale de $\tilde{\Sigma}$ en la place qui nous intéresse, est une représentation notée $\tilde{\pi}$ de $GL(n^*, F)$ telle que $\pi \in \Pi(\tilde{\pi})$; on sait que $\tilde{\pi}$ est θ-stable mais même plus qu'elle provient du groupe endoscopique $G(n)$, c'est un des résultats de [2] 30.2. On ne connaît pas la conjecture de Ramanujan, on écrit donc $\tilde{\pi}$ comme une induite de la forme

$$\times_{(\rho,a,x)\in\mathcal{I}} St(\rho, a)|\ |^x \times \tilde{\pi}_0 \times St(\rho, a)|\ |^{-x}, \qquad (1)$$

où $\mathcal{I}$ est un ensemble d'indices paramétrant des triplets formés d'une représentation cuspidale unitaire irréductible ρ d'un $GL(d_\rho, F)$, d'un entier a et d'un réel $x \in]0, 1/2[$ et où $\tilde{\pi}_0$ est tempérée et son paramètre est à valeurs (à conjugaison près) dans un groupe $G^*(n_0)$ pour n_0 convenable. On pose $\sigma := \times_{(\rho,a,x)\in\mathcal{I}} St(\rho, a)|\ |^x$ et la θ-stabilité entraîne que

$$\theta(\sigma) = \times_{(\rho,a,x)\in\mathcal{I}} St(\rho, a)|\ |^{-x}.$$

À $\tilde{\pi}_0$, on associe en suivant Arthur 30.1 un paquet de représentations $\Pi(\tilde{\pi}_0)$ de $G(n_0)$ où n_0 est convenable comme ci-dessus. Pour des coefficients $c_{\pi_0} \in \mathbb{R}$ positifs on a une égalité de

$$\sum_{\pi_0\in\Pi_0} c_{\pi_0} tr\, \pi_0(h') = tr\tilde{\pi}_0(g', \theta) \qquad (2)$$

pour tout h' suffisamment régulier dans $G(n_0)$, où g', θ a une classe de conjugaison stable qui correspond, dans le transfert, à la classe de conjugaison stable de h' (cf [19] I.3, III.1, III.2). Les formules explicites pour le calcul de la trace d'une induite donne ici avec les notations de (1):

$$tr(\sigma \times \tilde{\pi}_0 \times \theta(\sigma))(g, \theta) = \sum_{\pi_0\in\Pi(\tilde{\pi}_0)} c_{\pi_0} tr(\sigma \times \pi_0)(h) \qquad (3)$$

quand la classe de conjugaison stable de h et celle de (g, θ) se correspondent. Ainsi la représentation π est un sous-quotient irréductible de l'une des induites $\sigma \times \pi_0$ pour $\pi_0 \in \Pi(\pi_0)$. Remarquons aussi pour la suite que si

ψ_0 n'est pas θ discret alors il existe ψ'_0 une représentation de $W_F \times SL(2,\mathbb{C})$ dans $GL(m_0^*, \mathbb{C})$ se factorisant par $G^*(m_0)$, m_0 un entier convenable et une représentation irréductible $\rho \otimes \sigma_a$ de $W_F \times SL(2,\mathbb{C})$ tels que

$$\psi_0 = \psi'_0 \oplus (\rho \otimes \sigma_a) \oplus (\theta(\rho) \otimes \sigma_a).$$

Et, comme ci-dessus, on peut écrire de façon imagée

$$\Pi(\psi_0) = St(\rho, a) \times \Pi(\psi'_0),$$

ce qui veut dire que les éléments du membre de gauche sont tous les sous-quotients irréductibles du membre de droite, c'est-à-dire des induites $St(\rho, a) \times \pi'_0$ avec $\pi'_0 \in \Pi(\psi'_0)$. Par construction les éléments de $\Pi(\psi'_0)$ sont des représentations unitaires et les induites ci-dessus sont donc semi-simples. Dès que l'on aura démontré que σ est nécessairement inexistant, on saura que, puisque π est une série discrète, ψ_0 est θ-discret.

En particulier si π est cuspidal, alors $\psi = \psi_0$ car π ne peut être un sous-quotient de $\sigma \otimes \pi_0$ avec $\pi_0 \in \Pi(\psi_0)$ et donc ψ_0 est θ-discret.

Le défaut de cet argument est qu'il utilise de façon assez forte la partie globale des résultats d'Arthur et il n'est donc pas clair à priori, qu'il ne faille pas les lemmes fondamentaux pondérés. On va donc donné un argument local.

1.3.2. *argument d'apparence locale* . Ci-dessous tous les arguments sont locaux mais ils utilisent des résultats qui ont été obtenu avec des formules des traces simples, donc des arguments globaux. Ils m'ont été donnés par Waldspurger.

On sait que le caractère distribution d'une série discrète, π, de $G(n)$, peut se voir comme une distribution sur l'ensemble des classes de conjugaison d'éléments de $G(n)$ dont la restriction aux classes de conjugaison d'éléments elliptiques est non nulle. En évaluant sur ces intégrales orbitales, on définit une application de l'espace vectoriel engendré par ces caractères de représentations dans un espace $I_{cusp}(G(n))$, l'espace des intégrales orbitales des pseudo-coefficients, étudié par Arthur en [3]; en élargissant l'espace des caractères à l'ensemble des caractères des représentations elliptiques de $G(n)$ (définition d'Arthur) on obtient ainsi une application bijective. Arthur dans [3] a donné une décomposition en somme directe de cet espace $I_{cusp}(G(n))$, l'un des facteurs étant l'espace des distributions stables à support dans les éléments elliptiques. On a déjà véfié en [12] que pour définir la projection sur ce facteur stable, seul le lemme fondamentale stable est nécessaire. On utilise la notation $I_{cusp}^{st}(G(n))$ pour cet espace, conforme aux notations de [19] VI. 1.

Et le point est de démontrer que la projection du caractère distribution de π sur cet espace est non nulle; on le fait en localisant au voisinage de l'origine ce qui permet d'utiliser le développement asymptotique d'Harish-Chandra d'où une décomposition en somme de transformée de Fourier d'intégrales orbitales unipotentes; la dimension des orbites donnent un degré d'homogénéité pour chacun de ces coefficients. La stabilisation respecte le degré d'homogénéité. On regarde le coefficient relatif à l'orbite triviale, c'est le degré formel de la série discrète et ce coefficient est donc non nul; la transformée de Fourier de l'orbite de l'élément 0 est automatiquement stable et a donc une projection non nulle sur l'analogue de $I_{cusp}^{st}(G(n))$ dans l'algèbre de Lie. Ceci est le terme de degré 0 dans le developpement asymptotique de la projection du caractère de π sur $I_{cusp}^{st}(G(n))$ d'où la non nullité de cette projection.

On utilise ensuite [19] VI.1 proposition (b) qui démontre que le transfert induit un isomorphisme d'espace vectoriel entre $I_{cusp}^{st}(G(n))$ et son analogue stable pour $GL(n^*, F) \times \theta$; cet analogue est défini en [19] V.1 où l'existence de ce facteur direct est démontrée. On sait d'après [19] IV.5 (1) que ce dernier espace s'interprète comme combinaison linéaire de caractères de représentations tempérées θ-stable de $GL(n^*, F)$. Toutefois, [19] ne sait pas que les paramètres de ces représentations sont à valeurs dans $G^*(n)$. Pour s'en sortir il faut avoir une description de $I_{cusp}^{st}(G)$; cela est fortement suggéré par [2] paragraphe 30 mais n'est pas totalement explicite (cf. notre discussion en 4.3). Cette méthode est certainement la plus conceptuelle mais se heurte à cette difficulté. C'est pour cela que l'on a pris l'autre méthode qui nécessite de devoir montrer que $\tilde{\pi}$ est tempérée ce qui sera fait ci-dessous.

On garde les notations ψ, σ, ψ_0 introduites ci-dessus.

1.4. Support cuspidal, rappel des définitions, résultats élémentaires. Soit π une série discrète irréductible de $G(n)$. On sait définir le support cuspidal de π (comme de toute représentation irréductible de $G(n)$) comme l'union d'une représentation cuspidale irréductible, π_{cusp} d'un groupe $G(n_{cusp})$ (le support cuspidal partiel de π) et d'un ensemble de représentations cuspidales irréductibles ρ_i de groupe $GL(d_{\rho_i}, F)$ pour i parcourant un ensemble convenable, $\mathcal{I}$, d'indices. Cet ensemble est défini à permutation près et à inversion près, c'est-à-dire que l'on peut changer ρ_i en sa duale. Cet ensemble est uniquement défini par la propriété que π est un sous-quotient irréductible de l'induite

$$\times_{i \in \mathcal{I}} \rho_i \times \pi_{cusp}.$$

On note $Supp_{GL}(\pi) := \{(\rho'', x)\}$ l'ensemble des couples ρ'', x formés d'une représentation cuspidale unitaire d'un groupe linéaire et d'un réel positif ou nul tel que le support cuspidal de π soit l'union de π_{cusp} avec l'ensemble des représentations cuspidales $\rho''||^x$ pour $(\rho'', x) \in Supp_{GL}(\pi)$.

Lemme 1.4.1. *(i) Soit ρ'' une représentation cuspidale unitaire d'un groupe $GL(d_{\rho''}, F)$; on suppose qu'il existe $x'' \in \mathbb{R}$ tel que $(\rho'', x'') \in Supp_{GL}(\pi)$. Alors, il existe des réels d, f tels que $d - f + 1 \in \mathbb{N}_{\geq 1}$ et une série discrète π' tels que π soit un sous-module irréductible de l'induite*

$$< \rho''||^d, \cdots, \rho''||^f > \times \pi',$$

où la représentation entre crochet est $St(\rho'', d - f + 1)||^{(d+f)/2}$. De plus

$$Supp_{GL}(\pi) = Supp_{GL}(\pi') \cup \{(\rho, |y|); y \in [d, f]\}.$$

(ii) Pour tout $(\rho'', x) \in Supp_{GL}(\pi)$, ρ'' est autoduale et il existe un entier relatif z tel que l'induite $\rho''||^{x+z} \times \pi_{cusp}$ soit réductible.

Ces résultats ne sont pas nouveaux et bien connus des spécialistes. Par définition du support cuspidal pour tout $(\rho', x') \in Supp_{GL}(\pi)$, il existe un choix de signe $\zeta_{(\rho', x')}$ et un ordre sur $Supp_{GL}(\pi)$ tel que l'on ait une inclusion:

$$\pi \hookrightarrow \times_{(\rho', x') \in Supp_{GL}(\pi)} \rho'||^{\zeta_{\rho', x'} x'} \times \pi_{cusp} \qquad (1)$$

On fixe ρ'' comme dans l'énoncé et on prend pour f le plus petit réel de la forme $\zeta_{(\rho'', x'')} x''$ tel que $\rho''||^f$ interviennent dans (1). On peut "pousser" $\rho''||^f$ vers la gauche tout en gardant une inclusion comme dans (1). Et quand il est le plus à gauche possible, toutes les représentations intervenant avant $\rho''||^f$ sont de la forme $\rho''||^z$ avec z parcourant un segment de la forme $[d, f[$ pour d convenable vérifiant $d - f + 1 \in \mathbb{N}_{\geq 1}$. Cela entraîne qu'il existe une représentation π' irréductible et une inclusion de π dans l'induite

$$< \rho''||^d, \cdots, \rho''||^f > \times \pi'.$$

Il reste à démontrer que π' est nécessairement une série discrète. On sait que $d + f > 0$ car s'il n'en est pas ainsi, $[d, f]$ est un segment dont le milieu, $(d + f)/2$ est inférieur ou égal à 0; on pose $\delta = 0$ si $(d - f + 1)/2$ est un entier et $\delta = (d + f)/2$ sinon. Ainsi $\sum_{y \in [d, f]} y =$

$$\sum_{t \in [1, [(d-f+1)/2]]} (d - t + 1 + f + t - 1) + \delta$$
$$= [(d - f + 1)/2](d + f) + (d + f)/2 \leq 0.$$

Or le module de Jacquet de π contient la représentation $\otimes_{\ell\in[d,f]}\rho||^{\ell}\otimes\pi'$ et le critère de Casselman montre que les exposants sont une combinaison linéaire à coefficient strictement positif de toutes les racines simples. L'exposant que l'on vient de trouver ne vérifie pas cette positivité d'où notre assertion. Montrons maintenant que π' est une série discrète; il faut vérifier le critère de Casselman sur les exposants. On suppose a contrario qu'il existe un terme dans le module de Jacquet de π' de la forme $\otimes\rho'''||^{x'''}\otimes\pi_{cusp}$ qui ne satisfait pas au critère de positivité. Un tel terme donne lieu à une inclusion de π' dans l'induite $\times\rho'''||^{x'''}\times\pi_{cusp}$. Le critère de positivité de Casselman s'exprime exactement par le fait que pour tout ρ''',x''' intervenant dans cette écriture, $\sum_{(\rho,x)}x>0$ où la somme porte sur les (ρ,x) à gauche de (ρ''',x''') y compris (ρ''',x'''). Supposons que π' ne soit pas une série discrète. On peut fixer un élément de son module de Jacquet ne vérifiant pas ce critère. Et on peut considérer dans $\sum_{(\rho,x)}x$ comme précédemment, chaque somme partielle où ρ est fixé et l'une au moins est ≤ 0. Par l'argument déjà donnée on trouve une inclusion de π' dans une induite de la forme:

$$<\rho'||^{d'},\cdots,\rho'||^{f'}>\times\pi'',$$

où $[d',f']$ est un segment (i.e. $d'-f'+1\in\mathbb{N}_{\geq 1}$), ρ' est une représentation cuspidale unitaire et $\sum_{x\in[d',f']}x\leq 0$. Comme ci-dessus, ceci est équivalent à $d'+f'\leq 0$. L'induite

$$<\rho''||^{d},\cdots,\rho''||^{f}>\times<\rho'||^{d'},\cdots,\rho'||^{f'}>$$

est nécessairement irrédutible: en effet soit $\rho''\not\cong\rho'$ et le résultat est immédiat. Soit $\rho''=\rho'$, mais alors $f\leq f'$ par minimalité de f, et

$$d-d'=(d+f)-(d'+f')+(f'-f)>0,$$

d'où $[d,f]\supset[d',f']$ et le résultat. On obtient donc une inclusion de π dans l'induite:

$$<\rho''||^{d},\cdots,\rho''||^{f}>\times<\rho'||^{d'},\cdots,\rho'||^{f'}>\times\pi''\simeq$$
$$<\rho'||^{d'},\cdots,\rho'||^{f'}>\times<\rho''||^{d},\cdots,\rho''||^{f}>\times\pi''$$

et $d'+f'\leq 0$ contredit le fait que π est une série discrète. L'assertion sur le $Supp_{GL}(\pi)$ est claire par définition.

(ii) se déduit de (i) par récurrence: on reprend les notations de (i) et on admet que tous les éléments de $Supp_{GL}(\pi')$ sont de la forme (ρ,x) avec ρ autoduale. Si ρ'' n'est pas autoduale, $\rho''||^{x}\times\pi_{cusp}$ est irréductible pour tout x réel par un résultat général d'Harish-Chandra (cf. [18]) et $\rho''||^{x}$ commute avec toute représentation de la forme $\rho'||^{\pm x'}$ pour $(\rho', x') \in Supp_{GL}(\pi')$. On aurait donc un isomorphisme:

$$< \rho''||^d, \cdots, \rho''||^f > \times \pi' \simeq < (\rho'')^{*-f}, \cdots, (\rho'')^{*-d} > \times \pi'.$$

et une inclusion de π dans l'induite de droite. Or $-f - d < 0$ ce qui contredit le fait que π est une série discrète et le critère de Casselman. C'est exactement le même argument pour la deuxième partie de (ii)

1.5. Propriétés de demi-integralité du support cuspidal d'une série discrète.

Théorème 1.5.1. (i) Soit π une série discrète irréductible de $G(n)$; tout élément (ρ, x) de $Supp_{GL}(\pi)$ (notation de 1.4) est tel que ρ est autoduale et que x est demi-entier.

(ii) Soit π_0 une représentation cuspidale de $G(n)$ et ρ une représentation cuspidale irréductible autoduale de $GL(d_\rho, F)$ et $x \in \mathbb{R}$ tel que l'induite $\rho||^x \times \pi_0$ soit réductible. Alors $x \in 1/2\mathbb{Z}$.

Grâce à 1.4 (ii), pour démontrer (i), il suffit de démontrer que pour π_{cusp} une représentation cuspidale irréductible d'un groupe $G(n_{cusp})$ et pour ρ une représentation cuspidale irréductible d'un groupe $GL(d_\rho, F)$, autoduale, le réel positif ou nul, $x_{\rho, \pi_{cusp}}$ tel que l'induite $\rho||^{x_{\rho, \pi_{cusp}}} \times \pi_{cusp}$ est réductible, est un demi-entier. C'est-à-dire, en fait, (ii) et c'est donc (ii) que nous allons démontrer.

Soit donc π_{cusp} et $x_{\rho, \pi_{cusp}}$ comme ci-dessus. Si $x_{\rho, \pi_{cusp}} = 0$, l'assertion est claire. Supposons donc que $x_{\rho, \pi_{cusp}} > 0$. Alors l'induite:

$$\rho||^{x_{\rho, \pi_{cusp}}+1} \times \rho||^{x_{\rho, \pi_{cusp}}} \times \pi_{cusp}$$

a un sous-module irréductible qui est une série discrète; c'est un calcul de module de Jacquet facile on coince cette sous-représentation dans l'intersection des 2 sous-représentations suivantes: (la notation $< \sigma_1, \sigma_2 >$ représente le socle de l'induite $\sigma_1 \times \sigma_2$, c'est à dire la somme des sous-modules irréductibles, mais ici le socle est irréductible)

$$\rho||^{x_{\rho, \pi_{cusp}}+1} \times < \rho||^{x_{\rho, \pi_{cusp}}}, \pi_{cusp} >$$

$$\cap < \rho||^{x_{\rho, \pi_{cusp}}+1}, \rho||^{x_{\rho, \pi_{cusp}}} > \times \pi_{cusp}.$$

Puis on calcule les modules de Jacquet, calcul qui prouve d'abord que cette intersection est non nulle puis que c'est une série discrète.

On note π cette série discrète et on lui applique 1.3 avec les notations de ce paragraphe. En particulier, si $\psi \neq \psi_0$, $Supp_{GL}(\pi)$ contient le support cuspidal de σ (à "conjugaison" près). Mais $Supp_{GL}(\pi)$ a exactement 2 éléments $\rho||^{x_{\rho,\pi_{cusp}}+1}$ et $\rho||^{x_{\rho,\pi_{cusp}}}$; cela entraîne déjà que $\sum_{(\rho,a,x)\in\mathcal{I}} a \leq$ 2. Si $\mathcal{I}$ est réduit à un élément (ρ,a,x) avec $a=2$, les ensembles $((1+x_{\rho,\pi_{cusp}}),x_{\rho,\pi_{cusp}})$ et $((1/2+x),(-1/2+x))$ coïncident à l'ordre et au signe près. D'où par positivité, puisque $x < 1/2$

$$1 + x_{\rho,\pi_{cusp}} = 1/2 + x; \quad x_{\rho,\pi_{cusp}} = 1/2 - x.$$

Ceci est impossible. On ne pas non plus avoir $|\mathcal{I}| = 2$. On sait donc maintenant que soit $\psi = \psi_0$ soit σ est de la forme $\rho||^x$; on rappelle que $x \in]0,1/2[$ et qu'il coïncide donc avec $x_{\rho,\pi_{cusp}}$ et non avec $1 + x_{\rho,\pi_{cusp}}$. En d'autres termes soit $\pi \in \Pi(\tilde\pi)$ avec $\tilde\pi = \tilde\pi_0$ est tempérée soit un sous-quotient irréductible de l'induite $\rho||^{1+x_{\rho,\pi_{cusp}}} \times \pi_{cusp}$ est dans $\Pi(\tilde\pi_0)$. En fait l'induite est irréductible par définition de $x_{\rho,\pi_{cusp}}$ et dans ce dernier cas, ce serait toute l'induite qui serait dans $\Pi(\tilde\pi_0)$.

On montre maintenant que tout $\pi_0 \in \Pi(\tilde\pi_0)$ est tel que $Supp_{GL}(\pi_0)$ est formé de couples (ρ',x') avec x' demi-entier. Cela permettra donc de conclure grâce à ce qui précède; on aura en plus montré que σ est trivial.

On fixe ρ une représentation cuspidale unitaire irréductible d'un groupe $GL(d_\rho,F)$. Pour $z \in \mathbb{R}$, $m \in \mathbb{N}$ et π' une représentation de $G(m)$ on note $Jac_z\pi'$ l'unique élément du groupe de Grothendieck des représentations lisses de longueur finie de $G(m)$ tel que la restriction de π' au Levi $GL(d_\rho,F) \times G(m)$ le long du radical unipotent d'un parabolique de ce Levi soit de la forme $\rho||^z \otimes Jac_z\pi' \oplus \tau$ où τ est une somme de représentations irréductibles de la forme $\rho' \otimes \tau''$ avec $\rho' \not\simeq \rho||^z$. Pour $\tilde\pi$ une représentation de $GL(m^*+2d_{\rho'},F)$ on définit $Jac_z^\theta\tilde\pi$ comme l'unique élément du groupe de Grothendieck des représentations lisses de longueur finie de $GL(m,F)$ tel que la restriction de $\tilde\pi$ au Levi $GL(d_{\rho'},F) \times GL(m^*,F) \times GL(d_{\rho'},F)$ soit de la forme $\rho||^z \otimes Jac_z^\theta\tau \otimes \rho^*||^{-z} \oplus \tau'$ où τ' est une somme de représentations irréductibles de la forme $\rho' \otimes \tau'' \otimes \rho''$ avec soit $\rho' \not\simeq \rho||^z$ soit $\rho'' \not\simeq \rho^{-z}$. On a vu en [12] (mais c'est assez facile) que l'égalité des traces 1.3 (1) donne aussi une égalité (avec les mêmes coefficients):

$$\sum_{\pi\in\Pi(\tilde\pi_0)} c_{\pi_0} tr Jac_z\pi_0(h) = tr Jac_z^\theta\tilde\pi_0(g,\theta), \tag{1}$$

Le terme de droite ne peut être non nul que pour des $z \in 1/2\mathbb{Z}$. Ici on utilise la positivité des c_{π_0} annoncée par Arthur qui empêche toute simplification dans le terme de gauche. Alors pour $\pi_0 \in \Pi(\tilde\pi_0)$, $Jac_z\pi_0 \neq 0$ nécessite que $Jac_z^\theta\tilde\pi_0 \neq 0$. D'où le fait que $z \in 1/2\mathbb{Z}$. Si $Supp_{GL}(\pi_0)$ contient un élément (ρ'',x) avec $x \in \mathbb{R}$, de façon standard (en poussant vers la gauche), on montre qu'il existe x' avec $x - x' \in \mathbb{Z}$ tel que $Jac_{x'}\pi_0 \neq 0$, en faisant ici $\rho = \rho'$. Comme on vient de montrer que $x' \in 1/2\mathbb{Z}$, x aussi est un demi-entier relatif.

Cela termine la preuve du théorème.

1.6. Morphisme associé à une série discrète. Le corollaire suivant est évidemment très fortement inspiré par les travaux d'Arthur.

Corollaire 1.6.1. *Soit π une série discrète irréductible de $G(n)$, alors il existe une représentation tempérée, $\tilde{\pi}$ de $GL(n^*, F)$, θ-discrète telle que $\pi \in \Pi(\tilde{\pi})$. Ou encore, il existe un morphisme ψ de $W_F \times SL(2, \mathbb{C})$ dans $GL(n^*, \mathbb{C})$ tel que $\pi \in \Pi(\psi)$ avec ψ un morphisme θ-discret à valeurs dans $G^*(n)$.*

En tenant compte de 1.3 il faut montrer que $\tilde{\pi} = \tilde{\pi}_0$ avec les notations de ce paragraphe. Comme ci-dessus, nécessairement $Supp_{GL}(\pi)$ contient le support cuspidal de σ (à "conjugaison" près). Mais maintenant on sait que $Supp_{GL}(\pi)$ est formé de couple (ρ, x) avec x demi-entier et ce n'est pas le cas du support cuspidal de σ si σ existe vraiment. D'où le fait que $\tilde{\pi} = \tilde{\pi}_0$. Le corollaire résulte alors de 1.3.

2. Morphismes Associés aux Représentations Cuspidales de $G(n)$ et Points de Réductibilité des Induites de Cuspidales

On dit qu'un morphisme ψ de $W_F \times SL(2, \mathbb{C})$ dans un $GL(m, \mathbb{C})$ est sans trou si pour toute représentation $\rho \otimes \sigma_a$ intervenant dans ψ (notations de l'introduction) avec $a > 2$ la représentation $\rho \otimes \sigma_{a-2}$ y intervient aussi.

On précise les notations de l'introduction; soit π_0 une représentation cuspidale de $G(m)$ et soit ρ une représentation cuspidale irréductible autoduale de $GL(d_\rho, F)$ ce qui définit d_ρ. On note x_{ρ,π_0} l'unique réel positif ou nul (cf. [16]) tel que l'induite $\rho|\,|^{x_{\rho,\pi_0}} \times \pi_0$ soit réductible. Soit ψ_0 tel que $\pi_0 \in \Pi(\psi_0)$ ce qui est possible d'après le théorème précédent. Pour ρ comme ci-dessus, on pose

$Jord_\rho(\psi_0) := \{a \in \mathbb{N}$; tel que la représentation $\rho \otimes \sigma_a$ soit une sous-représentation de $\psi_0\}$.

On note a_{ρ,ψ_0} l'élément maximal de $Jord_\rho(\psi_0)$ quand cet ensemble est non vide. On dit qu'une représentation de $W_F \times SL(2, \mathbb{C})$ (ou simplement de W_F) est du type de G^* si elle est orthogonale quand V est symplectique et symplectique quand V est orthogonal. On reprend la notation $Jord(\pi_0)$ expliquée dans l'introduction

Corollaire 2.1. *Soient ρ, π_0, ψ_0 comme ci-dessus et tels que $\pi_0 \in \Pi(\psi_0)$ (cf. 1.6). Alors ψ_0 est θ discret et sans trou.*

(i) On suppose que $Jord_\rho(\psi_0) \neq \emptyset$ alors $x_{\rho,\pi_0} = (a_{\rho,\psi_0} + 1)/2$.

(ii) On suppose que $Jord_\rho(\psi_0) = \emptyset$ mais que $\rho \otimes \sigma_2$ est une représentation du type de G^*; alors $x_{\rho,\pi_0} = 1/2$.

(iii) On suppose que $Jord_\rho(\psi_0) = \emptyset$ et que ρ vue comme représentation de W_F est de même type que G^*. Alors $x_{\rho,\pi_0} = 0$.

(iv) $Jord(\pi_0)$ coïncide avec $Jord(\psi_0)$. En particulier ψ_0 est uniquement déterminé par π_0.

Le fait que ψ_0 est θ discret est dans 1.3. Le fait que ψ_0 soit sans trou et les assertions (i) et (ii) sont démontrées en [13]. On va redémontrer l'essentielle de ces assertions pour la commodité du lecteur; on ne redonne pas la démonstration de ce que si $Jord_\rho(\psi_0) \neq \emptyset$ alors $x_{\rho,\pi_0} = (a_{\rho,\psi_0}+1)/2$; la démonstration est plus simple mais de même nature que celle de (ii) que l'on va redonner.

En [9], cf. l'introduction, on a montré l'inégalité, où la somme porte sur toutes les représentations cuspidales irréductibles autoduales d'un $GL(d,F)$:

$$\sum_\rho \sum_{\ell \in [1,[x_{\rho,\pi_0}]]} d_\rho(2x_\rho - 2\ell + 1) \leq m^*.$$

Cette somme ne voit que les ρ tel que $x_\rho > 1/2$. Or si on se limite dans le terme de gauche aux représentations ρ telles que $Jord_\rho(\psi_0) \neq \emptyset$, ce terme est supérieur ou égal à $\sum_{(\rho,a) \in Jord(\psi)} ad_\rho$ avec égalité si et seulement si $Jord(\psi_0)$ est sans trou. Or cette somme vaut m^* car c'est exactement la somme des dimensions des sous-représentations de ψ_0. Et on a donc démontré que ψ_0 est sans trou et que pour tout ρ tel que $Jord_\rho(\psi_0) = \emptyset$, $x_{\rho,\pi_0} \in [0, 1/2]$. Comme on sait déjà que x_{ρ,π_0} est un demi-entier, il reste les possibilités 0 ou 1/2. On remarque ici que le fait que x_{ρ,π_0} soit entier ou non résulte de la parité de a_{ρ,ψ_0} et donc de savoir si le morphisme de W_F dans $GL(d_\rho, \mathbb{C})$ associé à ρ est du type de G^* ou non. Ce sont les conditions de parités voulues dans la définition des blocs de Jordan.

Avec cela (iv) résulte de [9] 4.3.

Il reste donc à montrer, pour toute représentation autoduale cuspidale irréductible, ρ, telle que $Jord_\rho(\psi_0) = \emptyset$, l'équivalence entre le fait que ρ est telle que $x_{\rho,\pi_0} = 1/2$ et le fait que $\rho \otimes \sigma_2$ est de même type que G^*. On rappelle que σ_2 est une représentation symplectique et donc que $\rho \otimes \sigma_2$ est une représentation symplectique si ρ est une représentation orthogonale et est une représentation orthogonale si ρ est une représentation symplectique.

Soit donc ρ tel que $x_{\rho,\pi_0} = 1/2$; on note ici π_d l'unique sous-module irréductible de l'induite $\rho||^{1/2} \times \pi_0$; c'est une série discrète par hypothèse de réductibilité. On considère une représentation tempérée, θ-discrète, $\tilde{\pi}_d$ telle que $\pi \in \Pi(\tilde{\pi}_d)$ et on écrit avec des coefficients $c_{\pi'}$ positifs, comme en 1.3

$$\sum_{\pi' \in \Pi(\tilde{\pi}_d)} c_{\pi'} tr\pi'(h) = tr\tilde{\pi}_d(g, \theta).$$

Comme $\tilde{\pi}_d$ est une représentation tempérée, on l'écrit comme une induite de représentations de Steinberg généralisées, $\times_{(\rho',a) \in \mathcal{I}} St(\rho', a)$ où $\mathcal{I}$ est un ensemble d'indices paramétrant des couples (ρ', a) formés d'une représentation cuspidale unitaire ρ' et d'un entier a; on sait que $\mathcal{I}$ est sans multiplicité puisque $\tilde{\pi}_d$ est θ-discret. Avec les notations Jac_z et Jac_z^θ de la preuve de 1.6 pour ρ fixé comme ici, on a encore que $Jac_{1/2}^\theta \tilde{\pi}_d$ est un transfert de

$$\sum_{\pi' \in \Pi(\tilde{\pi})} c_{\pi'} tr Jac_{1/2}\pi'.$$

Or $Jac_{1/2}\tilde{\pi}_d = \pi_0$, ainsi $Jac_{1/2}^\theta \tilde{\pi}_d \neq 0$. Cela nécessite que $\mathcal{I}$ contienne $(\rho, 2)$; il le contient avec multiplicité exactement 1. On obtient alors:

$$Jac_{1/2}^\theta \tilde{\pi}_d = \times_{(\rho',a) \in \mathcal{I} - \{(\rho,2)\}} St(\rho', a) =: \tilde{\pi}_0$$

et $\pi_0 \in \Pi(\tilde{\pi}_0)$. On note encore ψ_0 le morphisme associé à π_0 et le paramètre de $\tilde{\pi}_d$ est donc la somme de ψ_0 et de $\rho \otimes \sigma_2$ (ici ρ est la représentation irréductible de W_F associée à ρ via la correspondance de Langlands et σ_2 est la représentation irréductible de dimension 2 de $SL(2, \mathbb{C})$). D'après la description d'Arthur (formule (30.15) de 30.2) ce paquet doit être de même type que G^*; comme c'est déjà le cas de ψ_0, il en est de même de $\rho \otimes \sigma_2$.

On suppose maintenant que ρ est tel que $\rho \otimes \sigma_2$ est de même type que G^* et que $Jord_\rho(\psi) = \emptyset$; on doit montrer que $x_{\rho,\pi_0} = 1/2$. On note $\psi_d := \rho \otimes \sigma_2 \oplus \psi_0$ le morphisme de $W_F \times SL(2, \mathbb{C})$ dans $GL(m^* + 2d_\rho, \mathbb{C})$ et $\tilde{\pi}_d$ la représentation tempérée de $GL(m^* + 2d_\rho, F)$ lui correspondant. Le morphisme ψ_d est à valeurs dans $G^*(m + d_\rho)$. On a donc encore un transfert, pour des bons coefficients positifs ([2] 30.1)

$$\sum_{\pi' \in \Pi(\psi_d)} c_{\pi'} tr\, \pi'(h) = tr\tilde{\pi}_d(g, \theta). \tag{1}$$

Et avec les notations Jac_z et Jac_z^θ déjà introduite pour ρ fixé et $z \in \mathbb{R}$:

$$\sum_{\pi' \in \Pi(\psi_d)} c_{\pi'} Jac_z tr\, \pi'(h') = Jac_z^\theta tr\tilde{\pi}_d(g', \theta). \tag{2}$$

On l'applique d'abord avec $z = 1/2$; le terme de droite n'est pas nul et il contient $\tilde{\pi}_0$; il est même réduit à $\tilde{\pi}_0$ car par hypothèse $Jord_\rho(\pi_0) = 0$ et on a déjà identifié $Jord_\rho(\pi_0)$ avec les entiers a tel que $\rho \otimes \sigma_a$ soit une sous-représentation de ψ_0. Donc ψ_0 ne contient pas $\rho \otimes \sigma_2$. Ainsi le terme de gauche de (2) est le paquet associé à ψ_0 et il contient donc π_0. Ainsi il existe $\pi' \in \Pi(\psi_d)$ tel que $Jac_{1/2}\pi'$ contienne π_0 ou encore que π' est un sous-quotient irréductible de $\rho||^{1/2} \times \pi_0$. Comme $Jac_{1/2}\pi' \neq 0$, π' est un sous-module irréductible de l'induite $\rho||^{1/2} \times \pi_0$. Cette induite est nécessairement réductible car sinon $Jac_{-1/2}\pi' \neq 0$ alors que pour $z = -1/2$ le terme de droite de (2) est nul. Ainsi $x_{\rho,\pi_0} = 1/2$ comme cherché et cela termine la démonstration.

3. Classification et Paquet de Langlands

Soit π une série discrète irréductible de $G(n)$; nous lui avons associé un morphisme de $W_F \times SL(2,\mathbb{C})$ dans $GL(n^*,\mathbb{C})$, à l'aide de ses blocs de Jordan. On note ψ_π ce morphisme. En suivant Arthur, on lui a aussi associé un morphisme en 1.6, le théorème ci-dessous dit que ces morphismes sont conjugués:

Théorème 3.1. *On a $\pi \in \Pi(\psi_\pi)$.*

Ce théorème montre que nos constructions sont compatibles avec celles d'Arthur et que le morphisme associé à une série discrète à l'aide de ses blocs de Jordan est bien celui conjecturé par Langlands. Toutefois, il n'est pas démontré que les coefficients permettant de construire la distribution stable associée au paquet sont égaux à 1; ce problème est résolu par Waldspurger en [20] pour certaines représentations venant des constructions de Lusztig ([8]) et pour $G(n) = SO(2n + 1, F)$. Et il n'est pas non plus démontré que les représentations à l'intérieur du paquet sont classifiées par les caractères du centralisateur; ce point semble moins sérieux et nous règlerons le cas des groupes orthogonaux déployés ici avec une méthode tout à fait générale.

Pour éviter les confusions, on note ψ plutôt que ψ_π le morphisme associé à π par [10] et [11] à l'aide des blocs de Jordan; on rappellera les propriétés qui le caractérise ci-dessous. Et on note $\tilde{\psi}$ celui associé essentiellement par Arthur en 1.6 et $\tilde{\pi}$ la représentation tempérée de $GL(n^*, F)$ associé à $\tilde{\psi}$. Il faut démontrer que ψ et $\tilde{\psi}$ sont conjugués. On réutilise les notations Jac_z et Jac_z^θ de 1.5. On vérifie d'abord que $\tilde{\psi}$ est θ-discret. Sinon π serait un sous-quotient d'une induite de la forme $St(\rho, a) \times \pi'$ avec π' une représentation au moins unitaire; l'induite est alors semi-simple et π est tempéré mais non discret. Ainsi pour tout $z \in \mathbb{R}$, $Jac_z^\theta\tilde{\pi}$ est 0 ou est une représentation tempérée irréductible. On a vu en loc. cit. que pour tout $z \in \mathbb{R}$,

$$Jac_z\pi \in \Pi(Jac_z^\theta\tilde{\pi}).$$

Si π est une représentation cuspidale, le fait que ψ et $\tilde{\psi}$ sont conjugués a été vu en 2 (iv). On suppose donc que π n'est pas cuspidale. D'après la construction de [10], 2 cas sont à distinguer. Dans ce qui suit ρ est une représentation cuspidale autoduale irréductible d'un $GL(d_\rho, F)$.

1e cas: il existe x_0 de la forme $(a-1)/2$ avec $a \in \mathbb{N}_{>1}$ et une série discrète π' tel que π soit l'unique sous-module irréductible de l'induite $\rho| \, |^{x_0} \times \pi'$. Dans ce cas, si on note ψ' le morphisme de $W_F \times SL(2, \mathbb{C})$ dans $G^{n^*-d_\rho}$, alors ψ' contient $(\rho, a-2)$ comme bloc de Jordan ($a = 2$ est accepté) et ψ s'obtient en remplaçant ce bloc de Jordan par (ρ, a).

Concluons dans ce cas; comme $Jac^\theta_{x=(a-1)/2}\tilde{\pi}_v \neq 0$ et est irréductible, cela veut dire que la représentation $\rho \otimes \sigma_a$ intervient avec multiplicité exactement 1 dans $\tilde{\psi}$ et que la représentation obtenue s'obtient simplement en remplaçant $\tilde{\psi}$ par un morphisme $\tilde{\psi}'$ ayant les mêmes sous-représentations irréductibles sauf cette représentation qui devient $\rho \otimes \sigma_{a-2}$. Il suffit d'appliquer par exemple une hypothèse de récurrence pour savoir que $\tilde{\psi}'$ et ψ' sont conjugués pour obtenir la même assertion pour $\tilde{\psi}$ et ψ.

2e cas: il existe x_0 de la forme $(a + 1)/2$ avec $a \in \mathbb{N}_{\geq 1}$ et une série discrète π' tel que π soit l'un des 2 sous-modules irréductibles de l'induite:

$$< \rho| \, |^{(a+1)/2}, St(\rho, a) > \times \pi'.$$

On note ψ' le morphisme associé à π' par nos construction et on admet encore par récurrence que $\pi' \in \Pi(\psi')$. La différence avec le cas 1 est que $Jac_{x_0}\pi$ est une représentation π_1 qui est une représentation tempérée et non plus une série discrète. Toutefois, nous avons déjà vérifié que puisque $\pi' \in \Pi(\psi')$, alors $\pi_1 \in \Pi(\psi' \oplus (\rho \otimes \sigma_a \oplus \rho \otimes \sigma_a))$ et comme ci-dessus,

$$\tilde{\psi} = \psi' \oplus \rho \otimes \sigma_{a+2} \oplus \rho \otimes \sigma_a.$$

Mais d'après nos constructions, on a bien que $Jord(\psi)$ se déduit de $Jord(\psi')$ en ajoutant les 2 blocs (ρ, a) et $(\rho, a + 2)$ ce qui est exactement la même chose. Ceci prouve le théorème.

4. Classification à la Langlands des Séries Discrètes de $SO(2n + 1, F)$

Dans cette partie on suppose que $G(n)$ est la forme déployée du groupe $SO(2n + 1, F)$. On a en vue la classification de Langlands des séries discrètes. Soit ψ un morphisme θ-discret de $W_F \times SL(2, \mathbb{C})$ dans $GL(2n, \mathbb{C})$, on dit qu'il est θ-stable s'il se factorise par $Sp(2n, \mathbb{C})$.

4.1. Définition de l'espace $I_{cusp}(\ )$. On note $C_{cusp}(G)$ l'ensemble des fonctions lisses cuspidales de G, c'est-à-dire celles dont les intégrales orbitales sur les éléments non elliptiques sont nulles et $I_{cusp}(G)$ est l'image de cet espace vectoriel de fonctions modulo le sous-espace des fonctions lisses dont toutes les intégrales orbitales sont nulles. L'article d'Arthur [3] montre toute l'importance de cet espace; dans cet article le groupe est supposé connexe, ce qui est le cas de G. On définit $I_{cusp}^{st}(G)$ comme l'ensemble des éléments qui ont des intégrales orbitales constantes sur les classes de conjugaison stable. On appelle $I_{cusp}^{ns}(G)$ l'ensemble des éléments dont la somme des intégrales orbitales sur toute classe de conjugaison stable est nulle. Waldspurger a remarqué (cf. [12] 4.5) que sans aucun lemme fondamental, on peut montrer la décomposition:

$$I_{cusp}(G) = I_{cusp}^{st}(G) \oplus I_{cusp}^{ns}(G). \tag{1}$$

Le point est que la stabilité commute à la transformation de Fourier [17]. Pour aller plus loin, on suppose la validité des lemmes fondamentaux pour les groupes endoscopiques de G les lemmes fondamentaux et pour les algèbres de Lie, introduits par Waldspurger. Pour toute donnée endoscopique, noté abusivement H de G, on peut d'une part définir $I_{cusp}^{st}(H)$ (en remplaçant G par H dans les notations ci-dessus) et un transfert de $I_{cusp}^{st}(H)$ dans $I_{cusp}(G)$; on note $I_{cusp}^{H-st}(G)$ l'image; cette image peut se définir directement. Et le résultat principal de [3] est de prouver l'égalité:

$$I_{cusp}^{ns}(G) = \oplus_H I_{cusp}^{H-st}(G). \tag{2}$$

Waldspurger définit et étudie l'espace analogue à $I_{cusp}(G)$ pour les groupes non connexes de la forme $\tilde{G}L(m', F)$ et montre en [19] VI.1 que sous l'hypothèse d'un lemme fondamental convenable, le transfert induit un isomorphisme de $I_{cusp}^{st}(G)$ sur $I_{cusp}^{st}(\tilde{G}L(2n, F))$. On peut ainsi récrire (1) et (2) en

$$I_{cusp}(G) = \oplus_H I_{cusp}^{H-st}(G), \tag{3}$$

où ici H parcourt toutes les données endoscopiques en incluant pour la partie stable $H = \tilde{G}L(2n, F)$.

4.2. Représentations elliptiques et I_{cusp}. Soit π une série discrète; on sait qu'elle possède un pseudo-coefficient et on peut définir la projection de ce pseudo coefficient sur l'ensemble des fonctions cuspidales; pour le cas le plus nouveau celui de $\tilde{G}L(2m, F)$ ceci est expliqué en [19] II. Ceci s'étend

aux représentations elliptiques et permet de définir pour toute représentation elliptique un élément de $I_{cusp}(G)$; $I_{cusp}(G)$ est engendré comme espace
vectoriel par ces images.

Soit maintenant ψ un morphisme de $W_F \times SL(2,\mathbb{C})$ dans $Sp(2n,\mathbb{C})$;
on supppose que ψ est θ-discret, c'est-à-dire que la représentation définie
quand on inclut $Sp(2n,\mathbb{C})$ dans $GL(2n,\mathbb{C})$ est sans multiplicité; pour aller
plus vite on appelle un tel morphisme θ-stable (car il est à image dans
$Sp(2n,\mathbb{C})$) et θ-discret. On a alors vérifié que les éléments de $\Pi(\psi)$ sont
des séries discrètes.

On suppose maintenant que ψ n'est pas θ-discret; $\Pi(\psi)$ est toujours
défini uniquement par la propriété de transfert stable. Puisque ψ n'est pas
θ-discret, ψ est à valeurs dans un sous-groupe de Levi d'un parabolique
θ-stable et les éléments de $\Pi(\psi)$ sont les sous-modules irréductibles d'une
induite de représentations tempérées; en particulier $\Pi(\psi)$ ne contient pas
de séries discrètes.

On note $I_{cusp}(G)[\psi]$ le sous-espace vectoriel de $I_{cusp}(G)$ engendré par
l'image des représentations elliptiques combinaisons linéaires de représentations incluses dans $\Pi(\psi)$; cet espace peut-être 0. Mais si ψ est θ-discret
on a:

$$dim\, I_{cusp}(G)[\psi] = |\Pi(\psi)|. \tag{1}$$

On a vérifié en 3 que pour π une série discrète, un morphisme ψ tel que
$\pi \in \Pi(\psi)$ est uniquement déterminé par les blocs de Jordan de π; en
particulier si $\Pi(\psi)$ contient une série discrète alors pour tout morphisme
ψ' θ-stable, $\Pi(\psi) \cap \Pi(\psi') = \emptyset$.

On pose $I_{cusp}(G)[nd]$ la somme des espaces $I_{cusp}(G)[\psi]$ pour tous les
morphismes ψ non θ-discret. On a donc la décomposition en somme directe:

$$I_{cusp}(G) = \oplus_\psi I_{cusp}(G)[\psi] \oplus I_{cusp}(G)[nd]. \tag{2}$$

4.3. Description de $I_{cusp}^{st}(G)$. Pour tout morphisme ψ θ-stable et θ-
discret, en [2] 30.1 (30.14) (pour $s = 1$), il est construit un élément de
$I_{cusp}^{st}(G)[\psi]$. On a besoin d'avoir que $I_{cusp}^{st}(G)$ est engendré par ces éléments.
Ceci n'est pas écrit dans [2] mais est très fortement suggéré par les résultats
de loc.cit. pour les raisons suivantes.

Pour tout morphisme ψ, θ-discret mais non nécessairement θ-stable, on
note $I_{cusp}(\tilde{G})[\psi]$ le sous-espace vectoriel de $I_{cusp}(\tilde{GL}(2n,F))$ engendré par
l'image d'un prolongement à $\tilde{GL}(2n,F)$ de la représentation tempérée de
$GL(2n,F)$ définie par ψ; il est de dimension 1. Waldspurger a montré que
$I_{cusp}(\tilde{GL}(2n,F))$ est engendré par ces sous-espaces vectoriels. Le problème

est donc de savoir si $I^{st}_{cusp}(\tilde{G}L(2n,F))$ est engendré par ceux de ces éléments qui correspondent aux ψ, θ-stable. A ψ simplement θ-discret, Arthur lui-même associe un groupe endoscopique elliptique de $\tilde{G}L(2m,F)$ en [2] pages 235 et 236: on décompose ψ en la somme de 2 morphismes $\psi_s \oplus \psi_o$ où ψ_s est à valeurs dans un groupe symplectique, $Sp(2m_s,\mathbb{C})$ alors que ψ_o est à valeurs dans un groupe orthogonal, $O(2m_o,\mathbb{C})$. Le déterminant de la restriction de ψ_o à W_F donne un caractère quadratique η du corps de base (par réciprocité) et le groupe endoscopique est déterminé par m_s, m_o et η. On note H_ψ ce groupe endoscopique, c'est le groupe $Sp(2m_s,F) \times O(2m_o,F)$ où le groupe orthogonal est le groupe de la forme orthogonale de dimension $2m_o$ de discriminant (normalisé) η et d'invariant de Hasse est $+1$. Par 30.1, Arthur associe à ψ un élément stable de $I_{cusp}(H_\psi)$ puisque l'on est dans une situation produit mais il ne transfert pas cet élément en un élément de $I^{H_\psi-st}_{cusp}(\tilde{G}L(2m,F))$. Comme on admet les lemmes fondamentaux on peut bien faire ce transfert et le point est de montrer que ce transfert coïncide avec l'image de $\pi(\psi) \circ \theta$ dans $I_{cusp}(\tilde{G}_m)$. On admet donc pour la suite de cette partie l'hypothèse suivante:

$I^{st}_{cusp}(\tilde{G}L(2n,F))$ *est la somme des espaces* $I_{cusp}(\tilde{G})[\psi]$, *où* ψ *parcourt l'ensemble des morphismes* θ*-stables et* θ*-discrets.*

La conséquence de cette hypothèse est que la décomposition 4.2 (2) est compatible à la projection sur $I^{st}_{cusp}(G)$ et précisément que l'on a:

$$I^{st}_{cusp}(G) = \oplus_\psi \left(I_{cusp}(G)[\psi] \cap I^{st}_{cusp}(G) \right), \tag{1}$$

où ψ parcourt l'ensemble des classes de conjugaison de morphismes θ-stables et θ-discrets.

Montrons que cette hypothèse entraîne aussi la décomposition en somme directe pour tout ψ morphisme θ-stable et θ-discret:

$$I_{cusp}(G)[\psi] = \oplus_H \left(I^{H-st}_{cusp}(G) \cap I_{cusp}(G)[\psi] \right), \tag{2}$$

où H parcourt le même ensemble qu'en 4.1 (3).

Ceci n'étant pas directement dans [2] 30.1, il faut le vérifier. Et il suffit de prouver l'analogue de (1) pour H un groupe endoscopique de $Sp(2m,F)$; pour un tel groupe il faut aussi tenir compte de $I_{cusp}(G)[nd]$. Soit H un produit de groupes orthogonaux de la forme $SO(2m_1+1,F) \times SO(2m_2+1,F)$

avec $m = m_1 + m_2$; on applique la propriété (1) à chaque facteur pour décrire $I_{cusp}^{st}(H)$ comme somme directe des espaces vectoriels de dimension 1, $I_{cusp}^{st}(H)[\psi_1 \times \psi_2]$, où ψ_i, pour $i = 1, 2$ est un morphisme θ-stable et θ-discret à valeurs dans $Sp(2m_i, \mathbb{C})$. On sait transférer un tel espace vectoriel dans $I_{cusp}(G)$ grâce à [2] 30.1 (30.14); le transfert est à valeurs dans $I_{cusp}(G)(\psi_{1,2})$ où $\psi_{1,2}$ est le morphisme de $W_F \times SL(2, \mathbb{C})$ dans $Sp(2m, \mathbb{C})$ obtenu en composant $\psi_1 \times \psi_2$ avec l'inclusion naturelle de $Sp(2m_1, \mathbb{C}) \times Sp(2m_2, \mathbb{C})$ dans $Sp(2m, \mathbb{C})$. Cela donne exactement la décomposition:

$$I_{cusp}^{H-st}(G)$$

$$= \oplus_\psi \left(I_{cusp}(G)[\psi] \cap I_{cusp}^{H-st}(G) \right) \oplus \left(I_{cusp}(G)[nd] \cap I_{cusp}^{H-st}(G) \right).$$

On en déduit la double somme:

$$I_{cusp}(G) = \oplus_H \left(\oplus_\psi (I_{cusp}(G)[\psi] \cap I_{cusp}^{H-st}(G)) \right.$$

$$\left. \oplus I_{cusp}(G)[nd] \cap I_{cusp}^{H-st}(G) \right).$$

Et on peut inverser les sommations pour obtenir le résultat cherché (2).

4.4. Classification des séries discrètes. On rappelle que l'on a admis tous les lemmes fondamentaux de notre situation et l'hypothèse de 4.3.

Théorème 4.4.1. *Soit ψ un morphisme θ-stable et θ-discret que l'on voit comme une représentation de $W_F \times SL(2, \mathbb{C})$ dans $GL(2m, \mathbb{C})$. On note ℓ_ψ le nombre de sous-représentations irréductibles incluses dans ψ. Alors:*

$$|\Pi(\psi)| = 2^{\ell_\psi - 1}.$$

Et l'application définie par Arthur qui associe un caractère du centralisateur de ψ dans $Sp(2n, \mathbb{C})$ est une bijection sur l'ensemble des caractères de restriction triviale au centre de $Sp(2n, \mathbb{C})$.

On sait déjà que $I_{cusp}^{st}(G)[\psi]$ est de dimension 1. On fixe $H = SO(2m_1 + 1, F) \times SO(2m_2 + 1, F)$ et on doit calculer le nombre de couples ψ_1, ψ_2 tels que ψ_i est θ-stable et θ-discret pour le groupe $SO(2m_i+1, F)$ (pour $i = 1, 2$) et tels ψ soit conjugué de $\psi_1 \times \psi_2$ puisque chacun de ces couples donnent un élément de $I_{cusp}^{H-st}(G)[\psi]$ et que cela l'engendre complètement. Il est plus simple de faire ce calcul en laissant varier la décomposition de m en $m_1 + m_2$; ensuite il faut diviser le résultat par 2 à cause des isomorphismes entre groupes endoscopiques. Ici on autorise $m_1 m_2 = 0$, c'est-à-dire que l'on retrouve 2 fois la partie correspondant à $I^{st}_{cusp}(G)[\psi]$. On note $Jord(\psi)$ l'ensemble des sous-représentations irréductibles de $W_F \times SL(2,\mathbb{C})$ incluses dans ψ; on rappelle que c'est un ensemble sans multiplicité. Le nombre de décomposition de ψ en $\psi_1 \times \psi_2$ est précisément le nombre de décomposition de $Jord(\psi)$ en 2 sous-ensembles, c'est à dire 2^{ℓ_ψ} puisque ℓ_ψ est le nombre d'éléments de $Jord(\psi)$ par définition. On obtient le nombre d'élément de $\Pi(\psi)$ quand on a divisé par 2.

On vérifie aisément que le cardinal du centralisateur de ψ dans $Sp(2m,\mathbb{C})$ est 2^{ℓ_ψ} et que le centre est de cardinal 2. Ainsi le groupe des caractères du centralisateur de ψ dans $Sp(2m,\mathbb{C})$ triviaux sur le centre de $Sp(2m,\mathbb{C})$ est aussi de cardinal $2^{\ell_\psi-1}$. Pour démontrer la deuxième assertion du théorème, il suffit donc de montrer que l'application définie par Arthur est surjective. On note $\epsilon_\mathcal{A}(\pi)$ cette application. Ce que l'on connaît est le rang de la matrice dont les lignes sont indéxées par les éléments de $Centr_{Sp(2m,\mathbb{C})}\psi/Cent(Sp(2m,\mathbb{C}))$ et les colonnes par les éléments de $\Pi(\psi)$, les coefficients de la matrice étant

$$\epsilon_\mathcal{A}(\pi)(s); s \in Centr_{Sp(2m,\mathbb{C})}\psi/Cent(Sp(2m,\mathbb{C})), \pi \in \Pi(\psi).$$

C'est une matrice carré. Vérifions que le rang de cette matrice est le nombre de ses lignes:

fixons une ligne donc $s \in Centr_{Sp(2m,\mathbb{C})}\psi/Cent(Sp(2m,\mathbb{C}))$. Un tel élément fixe une décomposition de ψ en $\psi_1 \times \psi_2$, où pour $i = 1,2$, ψ_i est inclus dans l'espace propre pour l'une des 2 valeurs propres de s; comme s n'est défini que modulo le centre de $Sp(2m,\mathbb{C})$, cette décomposition n'est définie qu'à l'échange près des 2 facteurs. L'élément s définit aussi un groupe endoscopique, H_s, à isomorphisme près (qui est le groupe $\tilde{GL}(2m,F)$ si s est central). Et l'élément

$$\sum_\pi \epsilon_\mathcal{A}(\pi)(s)\,\pi \qquad\qquad (*)_s$$

a pour image dans $I_{cusp}(G)$ l'élément [2] (30.14) correspondant à H_s et à la décomposition $\psi = \psi_1 \times \psi_2$ de ψ. Réciproquement à toute décomposition de ψ en $\psi_1 \times \psi_2$, on associe un élément s simplement en donnant ses 2 espaces propres comme ci-dessus. Ainsi les éléments $(*)_s$ donne la base de $I_{cusp}(G)[\psi]$ déjà considérée. Ils sont linéairement indépendants et forment un ensemble de cardinal $2^{\ell_\psi-1}$ puisque ce nombre est la dimension de l'espace vectoriel $I_{cusp}(G)[\psi]$. Cela prouve notre assertion sur le rang de la matrice et termine la preuve.

4.5. Caractère et module de Jacquet. On fixe ψ un morphisme θ-stable et θ-discret. On reprend les notations de l'introduction; soit $\rho \otimes \sigma_a$ une sous-représentation irréductible de ψ vu comme représentation de $W_F \times SL(2,\mathbb{C})$. Dans l'introduction on a défini l'entier a_- dans les 2 cas suivants:

soit il existe $b \in \mathbb{N}$ tel que $\rho \otimes \sigma_b$ soit une sous-représentation irréductible de ψ avec $b < a_-$, auquel cas on pose a_- le plus grand des entiers b avec ces propriétés

soit il n'existe pas d'entier b comme ci-dessus mais a est pair et on pose $a_- = 0$.

On a noté $Z_{\rho \otimes \sigma_a}$ le sous-groupe à 2 éléments du centralisateur de ψ dans $Sp(2n,\mathbb{C})$ qui correspond à cette sous-représentation $\rho \otimes \sigma_a$; précisément cette représentation est à valeurs dans un sous-groupe symplectique de $Sp(2n,\mathbb{C})$ et $Z_{\rho \otimes \sigma_a}$ est le centre de ce sous-groupe.

Soit $\pi \in \Pi(\psi)$ et notons $\epsilon_\mathcal{A}(\pi)$ le caractère du centralisateur de ψ qu'Arthur associe à π; $\epsilon_\mathcal{A}(\pi)$ est connu quand on connaît toutes ses restrictions aux sous-groupes $Z_{\rho \otimes \sigma_a}$. On avait remarqué en [10] que ce caractère devait être lié aux propriétés de modules de Jacquet de π. C'était le point de départ des classifications de [10] et [11] et on va montrer que $\epsilon_\mathcal{A}(\pi)$ a bien la propriété qui permet de définir les caractères en loc. cit. ou encore que $\epsilon_\mathcal{A}(\pi)$ coïncide avec le caractère associé à π par [10] là où nous l'avions défini.

Pour unifier le théorème ci-dessous, on dit, par convention que la restriction d'un caractère à $Z_{\rho \otimes \sigma_{a_-}}$ est le caractère trivial si $a_- = 0$, cas où $Z_{\rho \otimes \sigma_{a_-}}$ n'est pas le groupe $\{\pm 1\}$. On note d_ρ la dimension de la représentation ρ et on note encore ρ la représentation cuspidale de $GL(d_\rho, F)$ associée à ρ par la correspondance de Langlands.

Théorème 4.5.1. *Soit $\rho \otimes \sigma_a$ une sous-représentation de ψ telle que a_- soit défini. Alors, la restriction de $\epsilon_\mathcal{A}(\pi)$ à $Z_{\rho \otimes \sigma_a} \simeq \{\pm 1\}$ est le même caractère que la restriction de $\epsilon_\mathcal{A}(\pi)$ à $Z_{\rho \otimes \sigma_{a_-}}$ si et seulement si il existe une représentation π' du groupe $SO(2n - d_\rho(a+a_-)+1, F)$ et une inclusion*

$$\pi \hookrightarrow \rho||^{(a-1)/2} \times \cdots \times \rho||^{(a_-+1)/2} \times \pi'.$$

La conséquence de ce théorème est que l'on connaît très explicitement les modules de Jacquet des représentations dans $\Pi(\psi)$.

On considère la donnée endoscopique de G dont le groupe H est $SO(d_\rho(a + a_-) + 1, F) \times SO(2n - d_\rho(a + a_-) + 1, F)$. Et pour H on considère le morphisme $\psi_1 \times \psi_2$ de $W_F \times SL(2, \mathbb{C})$ à valeurs dans $Sp(d_\rho(a+a_-), \mathbb{C}) \times Sp(2n-d_\rho(a+a_-), \mathbb{C})$, où ψ_1 est la somme $\rho \otimes \sigma_a \oplus \rho \otimes \sigma_{a_-}$ et ψ_2 est la somme des autres sous-représentations incluses dans ψ.

On a besoin de connaître le paquet $\Pi(\psi_1)$; on sait par 4.4 qu'il a 2 éléments si $a_- \neq 0$ et 1 élément sinon. La situation est donc particulièrement simple. On sait que l'élément de $I^{st}_{cusp}(\tilde{G}L(d_\rho(a + a_-), F))$ qui définit ce paquet est l'image d'un prolongement à $\tilde{G}L(d_\rho(a + a_-), F)$ de la représentation tempérée $St(\rho, a) \times St(\rho, a_-)$. On écrit $St(\rho, a) \times St(\rho, a_-)$ comme l'unique sous-module irréductible de l'induite:

$$\rho|\,|^{(a-1)/2} \times \cdots \times \rho|\,|^{(a_-+1)/2} \times St(\rho, a_-)$$
$$\times St(\rho, a_-) \times \rho|\,|^{-(a_-+1)/2} \times \cdots \times \rho|\,|^{-(a-1)/2}.$$

Alors $\Pi(\psi_1)$ contient les sous-modules irréductibles de l'induite pour $SO(d_\rho(a + a_-) + 1, F)$:

$$\rho|\,|^{(a-1)/2} \times \cdots \times \rho|\,|^{(a_-+1)/2} \times St(\rho, a_-). \tag{1}$$

C'est un calcul de module de Jacquet expliqué en [13] 4.2 et analogue à ceux fait ici; on montre en loc.cit. qu'un module de Jacquet convenable de la distribution stable dans $\Pi(\psi_1)$ (relativement au facteur $GL(d_\rho, F)$) est non nul et a pour transfert stable $St(\rho, a_-) \times St(\rho, a_-)$; avec les notations de loc.cit. c'est $Jac_{(a-1)/2, \cdots, (a_-+1)/2}$. Ce module de Jacquet est donc l'induite à $SO(2a_- + 1, F)$ de la représentation $St(\rho, a_-)$. Si $a_- > 0$, cette induite est de longueur 2 car la parité de a_- est la bonne (cf. 2) et $\Pi(\psi_1)$ contient nécessairement les 2 sous-modules de (1). Si $a_- = 0$, (1) a un unique sous-module irréductible nécessairement dans $\Pi(\psi_1)$. On a donc décrit $\Pi(\psi_1)$ comme l'ensemble des sous-modules irréductibles de (1) et quand il y en a 2, la distribution stable est la somme de ces 2 sous-modules puisqu'un module de Jacquet d'une distribution stable est stable (cf. [13]4.2)

On ne peut évidemment pas donner une description aussi précise de $\Pi(\psi_2)$ et de la distribution stable qui est formée avec ses éléments; on fixe donc des nombres complexes c_{π_2} pour tout $\pi_2 \in \Pi(\psi_2)$ tel que $\sum_{\pi_2 \in \Pi(\psi_2)} c_{\pi_2} \pi_2$ soit stable. Ainsi la distribution stable correspondant à $\psi_1 \times \psi_2$ est:

$$\Big(\sum_{\pi_1 \in \Pi(\psi_1)} \pi_1 \Big) \otimes \Big(\sum_{\pi_2 \in \Pi(\psi_2)} c_{\pi_2} tr\, \pi_2 \Big). \tag{2}$$

Pour tout $\pi \in \Pi(\psi)$, on fixe $c_\pi \in \mathbb{C}$ tel que $\sum_{\pi \in \Pi(\psi)} c_\pi \pi$ soit stable. Et on pose $\epsilon_a(\pi) := \epsilon_{\mathcal{A}}(\pi)(z_{\rho \otimes \sigma_a})$, où $z_{\rho \otimes \sigma_a}$ est l'élément non trivial de $Z_{\rho \otimes \sigma_a}$;

on définit de même $\epsilon_{a_-}(\pi)$ si $a_- \neq 0$ sinon on pose $\epsilon_{a_-}(\pi) = 1$. D'après [2] (30.14), la distribution (2) se transfère en une distribution H-stable de G de la forme (à un scalaire près qui vient de l'imprécision dans le choix des c_π)

$$\sum_{\pi \in \Pi(\psi)} \epsilon_a(\pi)\epsilon_{a_-}(\pi)c_\pi tr\, \pi. \tag{3}$$

Ecrivons explicitement ce transfert sur les caractères vus comme fonctions localement L^1 sur les éléments semi-simples. Pour tout γ_G élément semi-simple de G, on l'égalité où les $\Delta(\gamma_G, \gamma_H)$ sont les facteurs de transfert et où γ_H parcourt un ensemble de représentant de classes de conjugaison dans H dont la classe stable se transfère en celle de γ_G:

$$\left(\sum_{\pi \in \Pi(\psi)} \epsilon_a(\pi)\epsilon_{a_-}(\pi)c_\pi tr\, \pi \right)(\gamma_G) =$$

$$\sum_{\gamma_H} \Delta(\gamma_G, \gamma_H)\left(\left(\sum_{\pi_1 \in \Pi(\psi_1)} tr\, \pi_1 \right)\left(\sum_{\pi_2 \in \Pi(\psi_2)} c_{\pi_2} tr\, \pi_2 \right) \right)(\gamma_H). \tag{4}$$

On va appliquer cette égalité en imposant à γ_G d'être dans le sous-groupe de Levi $M \simeq GL(d_\rho, F) \times SO(2(n - d_\rho) + 1, F)$ de G. On écrit un tel élément sous la forme $\gamma_G = m \times \gamma'$, où $m \in GL(d_\rho, F)$.

On note M_H^1 le sous-groupe de Levi de H isomorphe à

$$GL(d_\rho, F) \times SO(2(n_1 - d_\rho) + 1, F) \times SO(2n_2, F)$$

et M_H^2 celui qui est isomorphe à

$$SO(2n_1 + 1, F) \times GL(d_\rho, F) \times SO(2(n_2 - d_\rho) + 1, F).$$

Fixons $\gamma_G \in M$ comme ci-dessus mais on suppose que m est elliptique dans $GL(d_\rho, F)$. Dans (4) il suffit alors de sommer sur les éléments γ_H avec les propriétés imposées qui sont soit dans M_H^1 soit dans M_H^2. On note Z_M le centre de M et on applique cela non pas à γ_G mais à l'ensemble des $\gamma_G z$ où z parcourt Z_M. On prenant une limite convenable et on appliquant une formule de Casselman on peut remplacer dans (4), $tr\, \pi$ par $tr\, res_M(\pi)$ et les produits

$$\sum_{\gamma_H} \Delta(\gamma_G, \gamma_H)(tr\, \pi_1 tr\, \pi_2)(\gamma_H)$$

par

$$\sum_{\gamma_H^1} \Delta(\gamma_G, \gamma_H^1)tr\, (res_{M_H^1}\pi_1)tr\, \pi_2(\gamma_H^1)$$

$$+ \sum_{\gamma_H^2} \Delta(\gamma_G, \gamma_H^2)tr\, \pi_1 tr\, (res_{M_H^2}\pi_2)(\gamma_H^2),$$

où γ_H^1 est dans M_H^1 et γ_H^2 est dans M_H^2. En faisant varier m dans l'ensemble des éléments elliptiques de $GL(d_\rho, F)$ on projette sur le caractère de la représentation cuspidale $\rho||^{(a-1)/2}$; parce que ρ est cuspidale bien que l'on se limite aux éléments elliptiques, cela permet de faire disparaître toutes les autres représentations c'est-à-dire remplacer les restrictions par les restrictions suivies par cette projection sur $\rho||^{(a-1)/2}$. Il faut vérifier que cette opération appliquée à n'importe quel élément de $\Pi(\psi_2)$ donne 0. Cela résulte des propriétés standards des modules de Jacquet des paquets de séries discrètes que nous avons établies; pour que cette projection soit non nulle il faudrait que la sous-représentation $\rho \otimes \sigma_a$ de $W_F \times SL(2,\mathbb{C})$ soit incluse dans ψ_2, ce qui n'est pas le cas par choix de ψ_2. Comme nous allons en avoir besoin ci-dessous remarquons que cette propriété reste vraie en remplaçant $(a-1)/2$ par $(a'-1)/2$ pour tout $a' \in]a_-, a]$ car $\rho \otimes \sigma_{a'}$ pour un tel a' n'est pas sous représentation de ψ par minimalité de a_-. Si $a > a_- + 2$, on recommence ces opérations pour le Levi $M' = GL(d_\rho, F) \times SO(2(n-2d_\rho)+1, F)$ en projetant cette fois sur $\rho||^{(a-3)/2}$. On note

$$M_{a,a_-} \simeq GL(d_\rho, F) \times \cdots \times GL(d_\rho, F) \times SO(2n - d_\rho(a-a_-)+1, F)$$

le sous-groupe de Levi de G isomorphe où il y a $(a-a_-)/2$ copies de $GL(d_\rho, F)$. Et on note M_{a,a_-}^1 le sous-groupe de Levi de H:

$$Gl(d_\rho, F) \times \cdots \times GL(d_\rho, F) \times SO(2n_1 - d_\rho(a-a_-)+1, F)$$
$$\times SO(2n_2+1, F)$$

où il y a $(a-a_-)/2$ copies de $GL(d_\rho, F)$.

Il faut maintenant comparer les facteurs de transfert: on a $\gamma_G = \times_{i\in[1,(a-a_-)/2]}m_i \times \gamma'$ et $\gamma_H =_{i\in[1,(a-a_-)/2]} m_i \times \gamma_1' \times \gamma_2$ avec chaque m_i dans un groupe $GL(d_\rho, F)$, $\gamma' \in SO(2n - d_\rho(a-a_-)+1, F)$, $\gamma_1' \in SO(2n_1 - d_\rho(a-a_-)+1, F)$ et $\gamma_2 \in SO(2n_2+1, F)$. Et on est dans la situation où la classe stable de $\gamma_1' \times \gamma_2$ correspond à la classe stable de γ' dans l'endoscopie pour $SO(2n - d_\rho(a-a_-)+1, F)$. On vérifie que

$$\Delta(\gamma_G, \gamma_H) = \Delta(\gamma', \gamma_1'\gamma_2)$$

le premier facteur de transfert étant pour $SO(2n+1, F)$ et le second pour $SO(2n - d_\rho(a-a_-)+1, F)$; cela résulte du calcul des intégrales orbitales pour les éléments d'un Levi.

Finalement on établit que la distribution sur M_{a,a_-}

$$\sum_{\pi \in \Pi(\psi)} \epsilon_a(\pi)\epsilon_{a_-}(\pi)c_\pi proj_{\rho||^{(a_-+1)/2}} \cdots proj_{\rho||^{(a-1)/2}} res_{M_{a,a_-}} \pi$$

est un transfert de la distribution stable sur M^1_{a,a_-}:

$$\left(\sum_{\pi_1 \in \Pi(\psi_1)} proj_{\rho| \ |^{(a_-+1)/2}} \cdots proj_{\rho| \ |^{(a-1)/2}} res_{M^1_{a,a_-}} \pi_1 \right)$$

$$\times \left(\sum_{\pi_2 \in \Pi(\psi_2)} c_{\pi_2} tr\, \pi_2 \right).$$

Le résultat principal de [11] décrit exactement les termes

$$proj_{\rho| \ |^{(a_-+1)/2}} \cdots proj_{\rho| \ |^{(a-1)/2}} res_{M_{a,a_-}} \pi$$

quand π parcourt $\Pi(\psi)$. Exactement cette trace est 0 pour la moitié des représentations précisément celles pour lesquelles notre caractère ϵ_π associé à tout élément de $\Pi(\psi)$[10], [11] par vérifie $\epsilon_\pi(\rho, a) \neq \epsilon_\pi(\rho, a_-)$ et pour les autres est une représentation irréductible de la forme

$$\rho| \ |^{(a-1)/2} \otimes \cdots \otimes \rho| \ |^{(a_-+1)/2} \otimes \pi'$$

où π' est l'un des 2 sous-modules irréductibles de l'induite $St(\rho, a_-) \times \pi''$, avec π'' est un élément de $\Pi(\psi_2)$; si $a_- = 0$, il n'y a pas d'induite et on obtient directement π''. On note simplement $Jac_{(a-1)/2, \cdots, (a_-+1)/2}\pi$ cette représentation irréductible. On a décrit $\Pi(\psi_1)$ et avec cela on vérifie que:

$$\left(\sum_{\pi_1 \in \Pi(\psi_1)} proj_{\rho| \ |^{(a_-+1)/2}} \cdots proj_{\rho| \ |^{(a-1)/2}} res_{M^1_{a,a_-}} \pi_1 \right)$$

est $\otimes_{j \in [(a-1)/2, (a_-+1)/2]} \rho| \ |^j \otimes indSt(\rho, a_-)$ où l'induite est pour le groupe $SO(2a_- + 1, F)$ à partir du parabolique de Levi $GL(a_-, F)$. Ainsi la distribution pour $SO(2n - d_\rho(a - a_-) + 1, F)$

$$\sum_{\pi \in \Pi(\psi)} \epsilon_a(\pi)\epsilon_{a_-}(\pi)c_\pi Jac_{(a-1)/2, \cdots, (a_-+1)/2}\pi \tag{5}$$

est un transfert de la distribution (rappelons que $n_1 = (a + a_-)/2$):

$$\left(ind^{SO(2a_- +1, F)}_{GL(a - d_\rho, F)} St(\rho, a_-) \right) \otimes \left(\sum_{\pi_2 \in \Pi(\psi_2)} c_{\pi_2} \pi_2 \right). \tag{6}$$

Si $a_- = 0$, on peut tout de suite conclure: les groupes pour (5) et (6) sont les mêmes et le transfert est l'identité. Donc en particulier les coefficients c_{π_2} et c_π étant positifs on doit aussi avoir $\epsilon_a(\pi) = 1$ si $Jac_{(a-1)/2, \cdots, 1/2}\pi \neq 0$. De plus comme on l'a rappelé, la moitié des éléments de $\Pi(\psi)$ ont cette propriété de non nullité. On calcule le cardinal du groupe des caractères de $Cent_{Sp(2n, \mathbb{C})}(\psi)/Cent(Sp(2n, \mathbb{C}))$ triviaux sur

$Z_{\rho\otimes\sigma_a}$; ce groupe est facilement mis en bijection avec le groupe des caractères de $Cent_{Sp(2n-ad_\rho,\mathbb{C})}(\psi_2)/Cent(Sp(2n-ad_\rho,\mathbb{C}))$. Le cardinal de ce groupe est donc $2^{\ell_{\psi_2}-1}$ ce qui est la moitié du cardinal de $\Pi(\psi)$. Comme on a déjà montré que l'application $\pi \in \Pi(\psi)$ associe $\epsilon_{\mathcal{A}}$ est une bijection sur l'ensemble des caractères du centralisateur de ψ triviaux sur le centre de $Sp(2n,\mathbb{C})$ (cf. 4.4), on voit que la propriété $\epsilon_a(\pi) = 1$ caractérise les représentations π de $\Pi(\psi)$ telles que $Jac_{(a-1)/2,\cdots,1/2}\pi \neq 0$ ce qui est l'énoncé cherché dans ce cas.

Supposons maintenant que $a_- > 0$. Soient $\pi, \pi' \in \Pi(\psi)$ et $\pi_2 \in \Pi(\psi_2)$ tels que l'on ait l'égalité:

$$Jac_{(a-1)/2,\cdots,(a_-+1)/2}\pi \oplus Jac_{(a-1)/2,\cdots,(a_-+1)/2}\pi' = St(\rho,a_-) \times \pi_2.$$

On montre d'abord que $c_\pi = c_{\pi'} = c_{\pi_2}$; c'est un problème de transfert stable. D'une part on sait que $\sum_{\pi\in\Pi(\psi)} c_\pi \pi$ est le transfert d'un prolongement à $\tilde{G}L(2n,F)$ de la représentation tempérée $\pi(\psi)$ associée à ψ. D'autre part on sait que $\sum_{\pi_2\in\Pi(\psi_2)} c_{\pi_2}\pi_2$ est le transfert stable d'un prolongement à $\tilde{G}L(2n_2,F)$ de la représentation tempérée $\pi(\psi_2)$ de $GL(2n_2,F)$ associée à ψ_2. On sait que $\pi(\psi)$ est l'unique sous-module irréductible de l'induite

$$\rho||^{(a-1)/2} \times \cdots \times \rho||^{(a_-+1)/2} \times St(\rho,a_-) \times \pi(\psi_2)$$

$$\times St(\rho,a_-) \times \rho||^{-(a_-+1)/2} \times \cdots \times \rho||^{-(a-1)/2}.$$

Donc on passe de $\pi(\psi)$ à $St(\rho,a_-) \times \pi(\psi_2) \times St(\rho,a_-)$ en prenant des modules de Jacquet (cf. [13] 4.2) et on passe de $\pi(\psi_2)$ à la même représentation en induisant. Cela donne immédiatement l'égalité

$$\sum_{\pi\in\Pi(\psi)} c_\pi Jac_{(a-1)/2,\cdots,(a_--1)/2}\pi = \sum_{\pi_2\in\Pi(\psi_2)} c_{\pi_2}(St(\rho,a_-)\times\pi_2).$$

On en déduit l'assertion en décomposant chaque représentation comme déjà expliqué. Pour $\pi_2 \in \Pi(\psi_2)$, on note $\Pi(\pi_2)$ les 2 représentations, π', π'' de $\Pi(\psi)$ telles que

$$Jac_{(a-1)/2,\cdots,(a_-+1)/2}(\pi' \oplus \pi'') = St(\rho,a_-) \times \pi_2.$$

On a donc sur $SO(2n_2 + 2d_\rho a_- + 1, F)$ la distribution:

$$\sum_{\pi_2\in\Pi(\psi_2)} c_{\pi_2} \sum_{\pi\in\Pi(\pi_2)} \epsilon_a(\pi)\epsilon_{a_-}(\pi)Jac_{(a-1)/2,\cdots,(a_-+1)/2}\pi$$

qui est un transfert de la distribution sur $SO(2d_\rho a_- + 1, F) \times SO(2n_2 + 1, F)$:

$$\left(ind_{GL(d_\rho a_-, F)}^{SO(2d_\rho a_- + 1, F)} St(\rho, a_-) \right) \left(\sum_{\pi_2} c_{\pi_2} \pi_2 \right).$$

On écrit encore l'égalité qui en résulte pour tout élément

$$\gamma \in GL(d_\rho a_-) \times SO(2n_2 + 1, F)$$

dont la partie dans $GL(d_\rho a_-)$ est elliptique. Ensuite on projette sur le sous-espace vectoriel de $I_{cusp}(GL(d_\rho a_-))$ image de $St(\rho, a_-)$. On vérifie aisément que les éléments de $\Pi(\psi_2)$ n'ont pas de module de Jacquet qui peuvent contribuer à une telle projection. Le facteur de transfert $\Delta(\gamma, \gamma)$, où γ est d'abord vu comme un élément de $SO(2n_2 + 2d_\rho a_- + 1, F)$ puis du groupe endoscopique $SO(2d_\rho a_- + 1, F) \times SO(2n_2 + 1, F)$ vaut 1.

On obtient alors une égalité:

$$\sum_{\pi_2 \in \Pi(\psi_2)} c_{\pi_2} \left(\sum_{\pi \in \Pi(\pi_2)} (\epsilon_a(\pi) \epsilon_{a_-}(\pi)) \right) \pi_2 =$$

$$2 \times \left(\sum_{\pi_2 \in \Pi(\psi_2)} c_{\pi_2} \right) \pi_2 \Big).$$

D'où $\epsilon_a \epsilon_{a_-}(\pi) = 1$ pour tout π_2 et tout $\pi \in \Pi(\pi_2)$. Ensuite on conclut comme dans le cas où $a_- = 0$; la moitié des éléments, π, de $\Pi(\psi)$ vérifie à la fois $Jac_{(a-1)/2, \cdots, (a_- + 1)/2} \pi \neq 0$ et $\epsilon_a(\pi) \epsilon_{a_-}(\pi) = 1$. Mais chacune des 2 conditions est satisfaite par exactement la moitié des éléments de $\Pi(\psi)$, elles sont donc équivalentes. C'est l'assertion cherchée.

4.6. Paramètres des représentations cuspidales.

Théorème 4.6.1. *L'ensemble des représentations cuspidales de $SO(2n +1, F)$ est en bijection avec l'ensemble des couples (ψ, ϵ) où ψ est un morphisme sans trou de $W_F \times SL(2, \mathbb{C})$ dans $Sp(2n, \mathbb{C})$ et ϵ est un caractère alterné du centralisateur de ψ de restriction triviale au centre de $Sp(2n, \mathbb{C})$ (on renvoit à l'introduction pour une précision sur les notations).*

Soit π une représentation cuspidale et ψ le morphisme tel que $\pi \in \Pi(\psi)$; on a déjà vérifié que ψ est sans trou; on note $\epsilon_{\mathcal{A}}(\pi)$ le caractère associé par Arthur à π; il est alterné par 4.5. Réciproquement soit ψ un morphisme sans trou et $\pi \in \Pi(\psi)$. Soit ρ une représentation cuspidale irréductible et unitaire d'un $GL(d_\rho, F)$, $x \in \mathbb{R}$ et σ une représentation irréductible de $SO(2n - d_\rho + 1, F)$ tel que

$$\pi \hookrightarrow \rho| \ |^x \times \sigma. \tag{1}$$

Le fait que π soit cuspidal est exactement équivalent à ce qu'il ne soit pas possible de trouver de telles données satisfaisant (1). Supposons que π n'est pas cuspidal et montrons que $\epsilon_A \pi$ n'est pas alterné. Cela suffira. On fixe donc ρ, x satisfaisant (1); σ ne joue pas de rôle. On sait a priori que x est un demi-entier (cf. 1.5) et d'après [10] par exemple qu'en notant encore ρ la représentation irréductible de W_F correspondant à ρ la représentation définie par ψ de $W_F \times SL(2, \mathbb{C})$ contient comme sous-représentation la représentation $\rho \otimes \sigma_{2x+1}$; de plus comme π est une série discrète, on a sûrement $x > 0$. On écrit $a = 2x + 1$ et $a \geq 2$. Comme ψ est sans trou, a_- est défini et vaut $a - 2$. Il suffit d'appliquer 4.5 pour voir que $\epsilon_A(\pi)$ n'est pas alterné.

4.7. Au sujet de la combinaison linéaire stable dans un paquet. Soit encore ψ un morphisme θ-stable et θ discret de $W_F \times SL(2, \mathbb{C})$ dans $Sp(2n, \mathbb{C})$. On note $\sum_{\pi \in \Pi(\psi)} c_\pi \pi$ la distribution stable engendrée par les éléments du paquet $\Pi(\psi)$ avec c_π des réels positifs; pour fixer vraiment les c_π on demande à cette distribution d'être le transfert de la trace de la représentation tempérée $\pi(\psi)$ de $GL(2n, F)$ prolongé à $\tilde{GL}(2n, F)$. On remarque que ψ est uniquement déterminé par π et que c_π est donc aussi uniquement déterminé par π.

Il est naturel de conjecturer que les c_π sont tous égaux à 1. Ceci est démontré dans [20] si la restriction de ψ à W_F est triviale sur le groupe de ramification modérée. Par changement de base, on devrait pouvoir étendre le résultat de Waldspurger à tous les morphismes tels que la restriction de ψ à W_F se factorise par le groupe de Weil d'une extension résoluble de F. Mais dans l'état actuel, il semble difficile d'aller au delà et la remarque suivante est un substitut qui permettra d'écrire les distributions stables (et endoscopiques) pour tous les paquets pas seulement ceux qui sont tempérés. Pour cela, il faut rappeler ce qu'est le support cuspidal partiel d'une représentation irréductible, π d'un groupe classique, ici $SO(2n + 1, F)$: c'est l'unique représentation cuspidale, π_{cusp}, d'un groupe de la forme $SO(2n_{cusp} + 1, F)$ (ce qui définit l'entier n_{cusp}) tel que π soit un sous-quotient irréductible d'une induite de la forme $\sigma \times \pi_{cusp}$ où σ est une représentation irréductible de $GL(n - n_{cusp}, F)$. Comme une représentation cuspidale, comme π_{cusp}, est dans un unique paquet de série discrète, on a bien défini $c_{\pi_{cusp}}$ tout comme on a défini c_π ci-dessus.

Remarque 4.7.1. *Soit π une série discrète dont on note π_{cusp} le support cuspidal partiel, alors $c_\pi = c_{\pi_{cusp}}$.*

On le démontre par récurrence sur n et on réutilise la démonstration de 4.5. Dans la preuve de 4.5 on a déjà calculé c_π en fonction d'une série discrète d'un groupe $SO(2n' + 1, F)$ avec $n' < n$ sous l'hypothèse que la représentation définie par ψ contient une sous-représentation $\rho \otimes \sigma_a$ avec a_- défini et π vérifiant $\epsilon_a(\pi) = \epsilon_{a_-}(\pi)$ avec les notations de cette preuve. Il est immédiat de voir que π_{cusp} est aussi le support cuspidal partiel de la représentation π_2 de $SO(2n_2+1, F)$ pour laquelle on a montré que $c_\pi = c_{\pi_2}$. D'où le résultat par récurrence dans ce cas. On est donc ainsi ramené au cas où $\epsilon_{\mathcal{A}}(\pi)$ (le caractère associé par [2] 30.1 à π) est alterné. Faisons cette hypothèse sur $\epsilon_{\mathcal{A}}(\pi)$. D'après 4.4 et la description des paramétres des représentations cuspidales (4.6), π est cuspidal si et seulement si ψ est sans trou; il n'y a donc rien à démontrer dans le cas où ψ est sans trou.

Il faut donc voir encore le cas où ψ a des trous, ou encore, le cas où il existe $\rho \otimes \sigma_a$ une sous-représentation irréductible de ψ avec $a > 2$ telle que $\rho \otimes \sigma_{a-2}$ ne soit pas une sous-représentation de ψ. On fixe une telle représentation $\rho \otimes \sigma_a$ et on note d_ρ la dimension de la représentation ρ et on notera aussi ρ la représentation cuspidale de $GL(d_\rho, F)$ associée à ρ par la correspondance locale de Langlands ([6],[7]). On note ψ' le morphisme de $W_F \times SL(2, \mathbb{C})$ dans $Sp(2(n - d_\rho), \mathbb{C})$ qui est la somme des sous-représentations irréductibles incluses dans ψ sauf $\rho \otimes \sigma_a$ qui est remplacée par $\rho \otimes \sigma_{a-2}$. On a montré en [10] et [11] (cf. l'introduction de [11]) que l'application $Jac_{(a-1)/2}$ établit une bijection entre $\Pi(\psi)$ et $\Pi(\psi')$; rappelons ce qu'est cette application. Soit $\pi \in \Pi(\psi)$, alors il existe une unique représentation irréductible π' de $SO(2(n - d_\rho) + 1, F)$ telle que π soit un sous-module de l'induite $\rho| \, |^{(a-1)/2} \times \pi'$. Et on a montré que $\pi' \in \Pi(\psi')$ et que tous les éléments de $\Pi(\psi')$ sont obtenus ainsi exactement une fois.

Par la compatibilité du transfert stable à la prise de module de Jacquet ([13] 4.2), on obtient le fait que $\sum_{\pi \in \Pi(\psi)} c_\pi Jac_{(a-1)/2}\pi$ est un transfert d'un prolongement (bien déterminé) de la représentation tempérée $\pi(\psi')$ de $GL(2(n - d_\rho), F)$ à $\tilde{GL}(2(n - d_\rho), F)$. D'où l'égalité $c_\pi = c_{Jac_{(a-1)/2}\pi}$. Et on applique l'hypothèse de récurrence à $Jac_{(a-1)/2}\pi$. On obtient l'égalité $c_\pi = c_{\pi_{cusp}}$ puisqu'il est bien clair que le support cuspidal partiel de π est le même que celui de $Jac_{(a-1)/2}\pi$. Cela termine la preuve.

5. Le cas des Groupes Orthogonaux Impairs non Déployés

Ici $G(n)$ est un groupe orthogonal d'un espace de dimension $2n + 1$ d'une forme orthogonale ayant un noyau anisotrope de dimension 3. Les preuves ci-dessous n'ont rien d'originales et s'appliquent beaucoup plus généralement pour passer d'un groupe quasidéployé à une forme intérieure. Et on note $G_d(n)$ la forme déployée de $SO(2n + 1, F)$. On peut encore définir comme en 1.3.2, l'espace $I^{st}_{cusp}(G(n))$. Soit π une série discrète, l'argument donné en loc.cit. s'applique pour montrer que l'image du caractère de π dans $I^{st}_{cusp}(G(n))$ est non nul. Il existe donc une distribution stable de la forme

$$\sum_{\pi' \in \mathcal{P}} c_{\pi'} tr\, \pi', \tag{1}$$

où $\mathcal{P}$ est un ensemble de représentations elliptiques de $G(n)$ et $c_{\pi'}$ sont des éléments de $\mathbb{C}$ et tel que $\pi \in \mathcal{P}$.

D'après [3] 3.5 (un peu amelioré en [12] 4.5), le transfert induit un isomorphisme de $I^{st}_{cusp}(G(n))$ sur $I^{st}_{cusp}(G_d(n))$. Ainsi, il existe un ensemble $\mathcal{P}_d$ de représentations elliptiques de $G_d(n)$ et des nombres complexes, $c_{\pi'_d}$ pour $\pi'_d \in \mathcal{P}_d$ tels que

$$\sum_{\pi'_d \in \mathcal{P}_d} c_{\pi'_d} tr \pi'_d \tag{2}$$

soit un transfert stable de (1). Réciproquement, étant donné un paquet stable pour $G_d(n)$, c'est-à-dire une combinaison linéaire à coefficients complexes de caractères, stable, il existe une combinaison linéaire à coefficient complexes de caractères pour $G(n)$ qui est stable et qui se transfère en la combinaison de départ. Les caractères sont les caractères des représentations a priori elliptiques.

Ainsi pour π une série discrète de $G(n)$, il existe au moins un morphisme ψ de $W_F \times SL(2, \mathbb{C})$ dans $G^*(n)$ tel que π soit dans un paquet stable se tranférant en le paquet $\Pi(\psi)$ formé de séries discrètes de $G_d(n)$. Ceci suppose évidemment les hypothèses déjà faites au sujet des lemmes fondamentaux. On dira encore que $\pi \in \Pi(\psi)$, $\Pi(\psi)$ étant maintenant vu comme un ensemble de représentations elliptiques de $G(n)$.

Théorème 5.1. *Soit π_0 une représentation cuspidale de $G(n_0)$ et ρ une représentation cuspidale autoduale irréductible d'un $GL(d_\rho, F)$. Soit $x_0 \in \mathbb{R}$ tel que l'induite $\rho||^{x_0} \times \pi_0$ soit réductible. Alors $x_0 \in 1/2\mathbb{Z}$.*

On fixe π_0 et x_0 comme dans l'énoncé; si $x_0 = 0$, il n'y a rien à démontrer et on suppose donc que $x_0 > 0$. On note ici π l'unique sous-module irréductible de $\rho||^{x_0} \times \pi$. On reprend les notations, Jac_x de 1.5. On applique Jac_{x_0} à (1); soit $\pi' \in \mathcal{P}$ tel que $Jac_{x_0}\pi' \neq 0$ et contient π_0 dans sa décomposition en irréductible. Cela entraîne que π' est un sous-quotient irréductible de l'induite $\rho|\,|^{x_0} \times \pi_0$ et comme π' est elliptique cela nécessite que $\pi' = \pi$. Ainsi Jac_{x_0} appliqué à (1) est de la forme:

$$c_\pi \pi_0 \oplus \tau \qquad (3)$$

où τ dans (3) est une combinaison linéaire de représentations irréductibles dont aucune n'est équivalente à π_0. On vérifie que Jac_{x_0} appliqué à (2) est un transfert de (3). En particulier, Jac_{x_0} appliqué à (2) n'est pas nul. Ou encore, il existe une représentation elliptique de $G_d(n)$, π_d, tel que $Jac_{x_0}\pi_d \neq 0$. Mais comme on sait que le support cuspidal des représentations elliptiques de $G_d(n)$ est comme celui des séries discrètes, demi-entier, on en déduit que x_0 est demi-entier. D'où le théorème.

On démontre de la même façon les théorèmes ci-dessous:

Théorème 5.2. *Avec les hypothèses et notations du théorème précédent en particulier π_0 est une représentation cuspidale de $G(n)$. Soit ψ_0 un morphisme de $W_F \times SL(2,\mathbb{C})$ tel que $\pi_0 \in \Pi(\psi_0)$. Alors ψ_0 est θ-discret, sans trou et les points de réductibilité pour les induites de la forme $\rho|\,|^x \times \pi_0$ se calculent comme dans le cas déployé, c'est-à-dire que l'on a avec les notations de ce cas:*

$$x_{\rho,\pi_0} = (a_{\rho,\psi_0} + 1)/2.$$

Théorème 5.3. *Soient π une série discrète irréductible de $G(n)$ et ψ un morphisme de $W_F \times SL(2,\mathbb{C})$ dans $Sp(2n,\mathbb{C})$ tel que $\pi \in \Pi(\psi)$. Alors ψ est θ-discret. De plus ce morphisme ψ est uniquement déterminé par les blocs de Jordan de π comme dans le cas déployé et correspond à celui qui a été associé à π par [10] et [11]*

Si ψ n'est pas θ-discret, comme dans le cas déployé toute représentation de $\Pi(\psi)$ est sous-module d'une induite convenable. Ceci est exclu pour π qui est une série discrète, d'où le 3e théorème. La fin du théorème se démontre comme dans le cas déployé, nos références, [9], [10] et [11] ne font pas l'hypothèse que le groupe est déployé.

Pour terminer remarquons que l'on n'a pas ici calculé le nombre d'éléments d'un paquet $\Pi(\psi)$ et par voie de conséquence obtenu une paramétrisation des représentations cuspidales de $G(n)$; la méthode du cas déployé s'appuie sur les résultats de transfert endoscopique d'Arthur qu'il faudrait étendre. Il y a sans doute une méthode locale comme pour le transfert stable et cela vaudrait la peine de l'écrire en général.

References

[1] J. Arthur, Unipotent automorphic representations: conjectures in Orbites unipotentes et représentations II, *Astérisque 171-172* (1989), 13-72

[2] J. Arthur, An introduction to the trace formula *prépublication* (2005)

[3] J. Arthur, On local character relations, *Selecta Math. 2* (1996), 501-579

[4] I. N. Bernstein, A. V. Zelevinsky, Induced Representations of Reductive p-adic groups 1 *Ann de l'ENS* **10** (1977), 147-185

[5] L. Clozel, On limit multiplicities of discrete series representations in spaces of automorphic forms, *Inventiones math.* **83** (1986), 265-284

[6] M. Harris, R. Taylor, The geometry and cohomology of some simple Shimura varieties *Annals of Math Studies* **151**(2001), Princeton Univ. Press

[7] G. Henniart, Une preuve simple des conjectures de Langlands pour GL_n sur un corps p-adique, *Invent. Math.* **139** (2000), 439-455

[8] G. Lusztig, Classification of Unipotent Representations of Simple p-adic Groups, II,*arXiv:math. RT/0111248* (2001)

[9] C. Mœglin, Points de réductibilité pour les induites de cuspidales, *Journal of Algebra* **268** (2003)

[10] C. Mœglin, Classification des séries discrètes: paramètre de Langlands et exhaustivité,*JEMS* **4** (2002), 143-200

[11] C. Mœglin, M. Tadic, Construction of discrete series for classical p-adic groups, *journal de l'AMS* **15** (2002), 715-786

[12] C. Mœglin, J. -L. Waldspurger, Paquets stables de représentations tempérées et de rréduction unipotente pour $SO(2n+1)$, *Inventiones* **152** (2003), 461-623

[13] C. Mœglin, J. -L. Waldspurger, Sur le transfert des traces tordues d'un groupe linéaire à un groupe classique p-adique *prépublication,* http://www.math.jussieu.fr/~moeglin

[14] M. Schneider, U. Stuhler, Representation theory and sheaves on the Bruhat-Tits building *Publ. Math. IHES* **85** (1997), 97-191

[15] F. Shahidi, Local coefficients and intertwining operators for GL(n), *Compositio Math* **48** (1983) 271-295

[16] A. Silberger, Special representations of reductive p-adic groups are not integrable,*Ann. of Math.* **111** (1980), 571-587

[17] J. -L. Waldspurger, Transformation de Fourier et endoscopie, *J. Lie Theory* **10** (2000), 195-206

[18] J. -L. Waldspurger, La formule de Plancherel d'après Harish-Chandra *JIMJ* **2** (2003)

[19] J. -L. Waldspurger, Le groupe GL_N tordu, sur un corps p-adique, 1e partie, *prépublication* (2005)

[20] J. -L. Waldspurger, Le groupe GL_N tordu, sur un corps p-adique, 2e partie, *prépublication* (2005)

[21] A. V. Zelevinsky, Induced Representations of Reductive p-adic groups II *Ann de l'ENS* **13** (1980), 165-210

SOME ALGEBRAS OF ESSENTIALLY COMPACT DISTRIBUTIONS OF A REDUCTIVE p-ADIC GROUP

ALLEN MOY* and MARKO TADIĆ†

*Department of Mathematics
The Hong Kong University of Science and Technology
Clear Water Bay, Hong Kong
E-mail: amoy@ust.hk

†Department of Mathematics
University of Zagreb
Bijenička 30, 10000 Zagreb, Croatia
E-mail: tadic@math.hr

In this mainly expository paper, we review some convolution algebras for the category of smooth representations of G, and discuss their properties. Most important for us is the relation of these algebras with the Bernstein center algebra $\mathcal{Z}(G)$.

In honor of Roger Howe as a sexagenarian

1. Introduction

1.1. An indispensable tool in the representation theory of reductive Lie groups is to associate to an admissible representation π of a connected reductive group G a representation, also denoted as π, of the enveloping algebra $\mathfrak{U}(\mathrm{Lie}(G))$ of the Lie algebra $\mathrm{Lie}(G)$ of G. If the admissible representation π is irreducible, then Schur's lemma states the center $\mathfrak{U}(\mathrm{Lie}(G))$ acts as scalar operators. The center $\mathcal{Z}(\mathfrak{U}(\mathrm{Lie}(G)))$ of $\mathfrak{U}(\mathrm{Lie}(G))$ can be viewed as the differential operators on the manifold G which are left and right translation invariant, and this interpretation provides a concrete method to

1991 *Mathematics Subject Classification.* Primary 22E50, 22E35.

The first author is partly supported by Research Grants Council grants HKUST6112/02P, and CERG #602505. The second author is partly supported by Croatian Ministry of Science, Education and Sports grant # 037-0372794-2804.

realize elements of the center. Furthermore, a fundamental result of Harish-Chandra determines the algebraic structure of the center $\mathcal{Z}(\mathfrak{U}(\mathrm{Lie}(G)))$.

An analogue of the center of the enveloping algebra for the representation theory of reductive p-adic groups has taken much longer to emerge, and is due to Bernstein (see [BD]). Certain aspects of the Bernstein center, in particular, explicit construction of elements in the center are still in a stage of development. Suppose F is a non-archimedean local field of characteristic zero, i.e., a p-adic field, and $G = \mathsf{G}(F)$ the group of F-rational points of a connected reductive group G. Let $\mathcal{C}_c^\infty(G)$ denote the vector space of locally constant compactly supported (complex valued) functions on G. We follow standard terminology and refer to a linear functional $D : \mathcal{C}_c^\infty(G) \longrightarrow \mathbb{C}$ as a distribution (see section 2.1). Let $\mathcal{C}_c^\infty(G)^*$ denote the vector space of all distributions on G. Fix a choice of Haar measure on G. For $\theta \in C_c^\infty(G)$, let D_θ denote the distribution on G which is integration of a function in $C_c^\infty(G)$ against θ. If $f \in \mathcal{C}_c^\infty(G)$, then it is elementary (see section 2.3) the convolutions $f \star \theta$ and $\theta \star f$ can be expressed in terms of D_θ: Let $\check{f}$ denote the function $x \to f(x^{-1})$, and for $x \in G$, let λ_x, and ρ_x denote left and right translations by x^{-1}, and x respectively. Then

$$\theta \star f \;=\; x \to D_\theta(\lambda_x \check{f}) \quad \text{and} \quad f \star \theta \;=\; x \to D_\theta(\rho_{x^{-1}} \check{f}) \;.$$

These two formulae can then be extrapolated to provide a definition for the convolution of a distribution D with any $f \in \mathcal{C}_c^\infty(G)$, i.e., $D \star f := x \to D(\lambda_x \check{f})$ and $f \star D := x \to D(\rho_{x^{-1}} \check{f})$. In contrast to the case when the distribution arises from a $\theta \in \mathcal{C}_c^\infty(G)$, and the convolutions $\theta \star f$ and $f \star \theta$ belong to $\mathcal{C}_c^\infty(G)$, for an arbitrary distribution D, the functions $D \star f$ and $f \star D$ may not be compactly supported. A distribution D of $C_c^\infty(G)$ is said to be essentially compact if the convolutions $D \star f$ and $f \star D$ are compactly supported functions for all $f \in C_c^\infty(G)$. In [BD], and [B], Bernstein-Deligne and Bernstein consider the space of essentially compact distributions which are G-invariant. The space of such distributions forms a convolution algebra known as the Bernstein center $\mathcal{Z}(G)$ of G.

1.2. In this mainly expository paper, we review some convolution algebras for the category of smooth representations of G, and discuss their properties. Most important for us is the relation of these algebras with the Bernstein center algebra $\mathcal{Z}(G)$.

For example in Bernstein's notes [B], he considers the Hecke algebra $\mathcal{H}(G)$ of compactly supported locally constant distributions, as well as the algebra $\mathcal{U}_c(G)$ of compactly supported distributions, and the endomorphism algebra $\mathrm{End}_{\mathbb{C}}(\mathcal{C}_c^\infty(G))$ (see section 3.1). In [BD:§1.4], Bernstein, and Deligne consider the algebra

$$\mathcal{H}(G)\hat{} := \{\, D \in \mathcal{C}_c^\infty(G)^* \mid D \star f \in \mathcal{C}_c^\infty(G), \quad \forall f \in \mathcal{C}_c^\infty(G)\,\}\,.$$

As part of the authors' investigations into the Bernstein center, we recently introduced in [MT2] the algebra

$$\mathcal{U}(G):=\{D\in\mathcal{C}_c^\infty(G)^* \mid D\star f,\ \text{and}\ f\star D\in\mathcal{C}_c^\infty(G),\ \forall f\in\mathcal{C}_c^\infty(G)\}.$$

Obviously,

$$\mathcal{H}(G) \subset \mathcal{U}_c(G) \subset \mathcal{U}(G) \subset \mathcal{H}(G)\hat{}\,.$$

Furthermore, there is a natural monomorphism of $\mathcal{H}(G)\hat{}$ into the algebra $\mathrm{End}_{\mathbb{C}}(\mathcal{C}_c^\infty(G))$, but the map is not an isomorphism (see section 3.2). The algebra $\mathcal{U}(G)$ obviously has a more symmetrical definition than $\mathcal{H}(G)\hat{}$, but both $\mathcal{U}(G)$, and $\mathcal{H}(G)\hat{}$ share many properties. Obviously, the center $\mathcal{Z}(\mathcal{U}(G))$ of $\mathcal{U}(G)$, and the center $\mathcal{Z}(\mathcal{H}(G)\hat{})$ of $\mathcal{H}(G)\hat{}$, is the Bernstein center $\mathcal{Z}(G)$. We remark that $\mathcal{U}(G)$, like the enveloping algebra $\mathfrak{U}(\mathrm{Lie}(H))$ of a reductive Lie group H, has natural adjoint operation $*$. The algebra $\mathcal{H}(G)\hat{}$ does not have an adjoint operation.

We point out the similarity of the real and p-adic situations. The category of $\mathfrak{g}$-modules is equivalent to the category of $\mathfrak{U}(\mathfrak{g})$-modules. The center of this category, i.e., the algebra of all natural transformations of the identity functor, is isomorphic to the center of $\mathfrak{U}(\mathfrak{g})$. In particular, the center of $\mathfrak{g}$ is insufficient for describing the center of the category. A similar situation occurs in the p-adic case. The category $\mathrm{Alg}(G)$ of smooth representations of G is equivalent to the category of non-degenerate modules over the Hecke algebra $\mathcal{H}(G)$ of G. But, neither the center of G, nor the center of $\mathcal{H}(G)$, is sufficient to describe the center of the category $\mathrm{Alg}(G)$. However, $\mathrm{Alg}(G)$ is also equivalent to the category of non-degenerate $\mathcal{U}(G)$-modules, and its center $\mathcal{Z}(\mathrm{Alg}(G))$ is isomorphic to the center $\mathcal{Z}(\mathcal{U}(G)) = \mathcal{Z}(G)$ of $\mathcal{U}(G)$.

In [MT2], we mentioned some basic properties of $\mathcal{U}(G)$. Here, we provide proofs of those properties and establish additional properties of the convolution algebra $\mathcal{U}(G)$ and the closely related algebras mentioned above. We do this in section 3, after some preliminaries in section 2. Two highlights of section 3 are Theorem 3.4o and Theorem 3.5e. The former states, in particular, for any irreducible smooth representation (π, V) that $\pi(\mathcal{H}(G)\hat{})$ equals $\mathrm{End}_{\mathbb{C}}(V)$. The latter states every $D \in \mathcal{H}(G)\hat{}$ is tempered.

In section 4, we give some examples of explicit constructions of elements in the Bernstein center. It is rather hard but also rather important to describe explicitly distributions in the Bernstein center $\mathcal{Z}(G)$. These distributions are tempered and invariant. A big source of tempered invariant distributions are orbital integrals. These distributions are of principal interest in harmonic analysis on G, as well as in the modern theory of automorphic forms. Unfortunately, these distributions are rarely in the Bernstein center (see section 2.3). However, some natural linear combinations of the orbital integrals do belong the Bernstein center. In [MT2], the authors have constructed a large family of Bernstein center distributions in terms of orbital integrals. This is an interesting interplay between two types of very important distributions; namely, between orbital integral distributions, for which we have explicit formulas, but for which we do not have (in principle) explicit knowledge of their Fourier transforms, and Bernstein center distributions, for which (in principle) we know their Fourier transforms, but for which we have little explicit knowledge. We finish by formulating the main result of [MT2].

2. The Convolution Algebras $\mathcal{H}(G)\hat{\ }$ and $\mathcal{U}(G)$

2.1. Recall our already established notation from section 1: F is a non-archimedean local field of characteristic zero, i.e., a p-adic field, $G = \mathsf{G}(F)$ the group of F-rational points of a connected reductive group G, $\mathcal{C}_c^\infty(G)$ is the vector space of complex valued locally constant compactly supported functions on G, and $\mathcal{C}_c^\infty(G)^*$ is the space of complex linear functionals on $\mathcal{C}_c^\infty(G)$. The space $\mathcal{C}_c^\infty(G)$ can be viewed as having no topology. There is however a natural topology $\mathcal{T}$ on $\mathcal{C}_c^\infty(G)$, but all linear mappings are continuous with respect to $\mathcal{T}$. We briefly recall $\mathcal{T}$. Suppose X is a non-empty open compact subset of G and J is an open compact subgroup of G. Define

$$\mathcal{V}_{X,J} := \{ f \in \mathcal{C}_c^\infty(G) \mid \ (\text{i}) \quad \text{supp}(f) \subset X , \ (\text{ii}) \quad f \text{ is } J\text{-bi-invariant}\} . \tag{2.1a}$$

The sets $\mathcal{V}_{X,J}$ are finite dimensional vector spaces. They have a natural topology on them (given, for example, by the standard supreme norm $||f|| = \sup \{ |f(x)| \mid x \in X \}$). A sequence of functions f_n is said to converge to $f \in \mathcal{C}_c^\infty(G)$ precisely if there is a compact subset X of G and an open compact subgroup J of G so that all the f_n's and f are in $\mathcal{V}_{X,J}$, and we have convergence in that space. This defines the topology $\mathcal{T}$ on $\mathcal{C}_c^\infty(G)$.

It follows, in particular, that any linear functional $D : \mathcal{C}_c^\infty(G) \to \mathbb{C}$ is continuous with respect to $\mathcal{T}$. We can alternatively define the topology $\mathcal{T}$ as the inductive topology (in the category of locally convex vector spaces) determined by requiring all embeddings $\mathcal{V}_{X,J} \hookrightarrow \mathcal{C}_c^\infty(G)$ be continuous.

Following standard usage, we refer to a linear functional $D \in \mathcal{C}_c^\infty(G)^*$ as a distribution.

2.2. Define the left and right translation action of G on $\mathcal{C}_c^\infty(G)$ by

$$\lambda_g f := x \to f(g^{-1}x) \quad \text{and} \quad \rho_g f := x \to f(xg) \qquad (2.2a)$$

respectively. These two actions of G on $\mathcal{C}_c^\infty(G)$ obviously commute with one another. A distribution D is said to be G-invariant if $D(f) = D(\lambda_g \rho_g f)$ for all $g \in G$.

2.3. Suppose $\theta, f \in \mathcal{C}_c^\infty(G)$. Fix a choice of Haar measure on G. The convolution product $\theta \star f \in \mathcal{C}_c^\infty(G)$, which is a generalization of multiplication in the group algebra of a finite group, is defined as:

$$\theta \star f := x \longrightarrow \int_G \theta(g) f(g^{-1}x) \, dg . \qquad (2.3a)$$

The distribution

$$D_\theta(f) := \int_G \theta(g) f(g) \, dg \qquad (2.3b)$$

satisfies

$$D_\theta(f) = \int_G \theta(g) \check{f}(g^{-1}) \, dg , \quad \text{where } \check{f}(g) := f(g^{-1}) \qquad (2.3c)$$
$$= (\theta \star \check{f})(1) .$$

We deduce

$$(\theta \star f) = x \longrightarrow D_\theta(\lambda_x(\check{f})) . \qquad (2.3d)$$

With (2.3d) as a model, we define, for an arbitrary distribution D, and $f \in \mathcal{C}_c^\infty(G)$, the convolution $D \star f$ to be the function $G \to \mathbb{C}$ given by

$$D \star f := x \longrightarrow D(\lambda_x(\check{f})) \qquad (2.3e)$$
$$= x \longrightarrow D(\text{function } t \to f(t^{-1}x)) .$$

Similarly, we define

$$f \star D := x \longrightarrow D(\rho_{x^{-1}}(\check{f})) \qquad (2.3f)$$
$$= x \longrightarrow D(\text{function } t \to f(xt^{-1})) .$$

If D is G-invariant, then $D \star f = f \star D$. Both $D \star f$, and $f \star D$ are locally constant functions on G, but a-priori there is no reason they should be in

$\mathcal{C}_c^\infty(G)$. An illuminating example of this is an orbital integral. Suppose $y \in G$. Let $\mathcal{O} := \mathcal{O}(y)$ denote the conjugacy class of y. Then, $\mathcal{O}$ is a manifold isomorphic to the homogeneous space $G/C_G(y)$, where $C_G(y)$ is the centralizer of y in G, and there is a G-invariant measure $d\mu_\mathcal{O}$ on $\mathcal{O}$, which is unique up to scalar. Then,

$$\mu_\mathcal{O}(f) := \int_\mathcal{O} f(g)\, d\mu_\mathcal{O}(g) \tag{2.3g}$$

is a G-invariant distribution. If 1_J is the characteristic function of an open compact subgroup J, then $\lambda_g 1_J$ is the characteristic function of gJ, and

$$\int_\mathcal{O} 1_{gJ}\, d\mu_\mathcal{O} = \mu_\mathcal{O}(gJ \cap \mathcal{O})\,. \tag{2.3h}$$

In particular, the function $\mu_\mathcal{O} \star 1_J$ is compactly supported if and only if $\mathcal{O}$ is a compact orbit. An elementary argument then says for arbitrary $f \in \mathcal{C}_c^\infty(G)$, the convolution $\mu_\mathcal{O} \star f$ is compactly supported if and only if $\mathcal{O}$ is a compact orbit. An example of such compact orbits is the conjugacy class of a central element $z \in G$, for which the associated G-invariant distribution is the delta function δ_z.

2.4. The Hecke algebra $\mathcal{H}(G)$ is the subspace of distributions $D \in \mathcal{C}_c^\infty(G)^*$, satisfying

(i) supp(D) is compact, and

(ii) D is locally constant, i.e., there exists a compact open subgroup J_D of G so that $D(\lambda_g f) = D(f)$, and $D(\rho_g f) = D(f)\ \forall\ g \in J_D$. \hfill (2.4a)

A choice of Haar measure on G gives an identification of $\mathcal{H}(G)$ with $\mathcal{C}_c^\infty(G)$.

If D is a compactly supported distribution, and $f \in \mathcal{C}_c^\infty(G)$, it is elementary both $D\star f$ and $f\star D$ are compactly supported functions. Furthermore, the function $D \star f$ (resp. $f \star D$) is right (resp. left) J-invariant for a sufficiently small open compact subgroup J.

Definition 2.4b A distribution D is

(i) right essentially compact if $D \star f \in \mathcal{C}_c^\infty(G)$ for all $f \in \mathcal{C}_c^\infty(G)$,

(ii) left essentially compact if $f \star D \in \mathcal{C}_c^\infty(G)$ for all $f \in \mathcal{C}_c^\infty(G)$,

(iii) essentially compact if both $D \star f$ and $f \star D$ belong to $\mathcal{C}_c^\infty(G)$ for any $f \in \mathcal{C}_c^\infty(G)$.

We introduce three vector spaces of distributions. Following Bernstein and Deligne [BD:§1.4], set

$$\mathcal{H}(G)\hat{\ } := \{\, D \in \mathcal{C}_c^\infty(G)^* \mid D \star f \in \mathcal{C}_c^\infty(G) \quad \forall f \in \mathcal{C}_c^\infty(G) \,\} \,. \qquad (2.4c)$$

Suppose D_1, $D_2 \in \mathcal{H}(G)\hat{\ }$ and $f \in \mathcal{C}_c^\infty(G)$. To facilitate computations regarding compositions, let $C(f)$ denote the function $\check{f}$. Then, $D_2 \star (D_1 \star f) \in \mathcal{C}_c^\infty(G)$, and we have the formula:

$$\begin{aligned}
D_2 \star (D_1 \star f) &= x \to D_2(\lambda_x(C(D_1 \star f))) \\
&= x \to D_2(y \to (C(D_1 \star f))(x^{-1}y)) \\
&= x \to D_2(y \to ((D_1 \star f))(y^{-1}x)) \\
&= x \to D_2(y \to D_1(\lambda_{y^{-1}x}(C(f)))) \\
&= x \to D_2(y \to D_1(t \to C(f)((y^{-1}x)^{-1}t))) \\
&= x \to D_2(y \to D_1(t \to f(t^{-1}y^{-1}x))) \,.
\end{aligned} \qquad (2.4d)$$

If $D_1, D_2 \in \mathcal{H}(G)\hat{\ }$, define their convolution product $D_2 \star D_1$ as follows: For any $f \in \mathcal{C}_c^\infty(G)$,

$$\begin{aligned}
(D_2 \star D_1)(f) &:= (D_2 \star (D_1 \star \check{f}))(1) \\
&= D_2(y \to D_1(t \to f(yt))) \,.
\end{aligned} \qquad (2.4e)$$

In particular, the function $x \to (D_2 \star D_1)(\lambda_x(\check{f}))$ is precisely the function $D_2 \star (D_1 \star f)$. Thus, the convolution $(D_2 \star D_1)$ is again in $\mathcal{H}(G)\hat{\ }$. To see that the convolution product is associative, we compute:

$$\begin{aligned}
(D_3 \star (D_2 \star D_1))(f) &= D_3(x \to (D_2 \star D_1)(z \to f(xz))) \\
&= D_3(x \to D_2(y \to D_1(t \to f(xyt)))) \,,
\end{aligned}$$

and

$$\begin{aligned}
((D_3 \star D_2) \star D_1)(f) &= (D_3 \star D_2)(z \to D_1(t \to f(zt))) \\
&= D_3(x \to D_2(y \to D_1(t \to f(xyt)))).
\end{aligned} \qquad (2.4f)$$

The convolution product therefore makes $\mathcal{H}(G)\hat{\ }$ into an algebra. We note that for any $g \in G$, the delta distribution δ_g at g belongs to $\mathcal{H}(G)\hat{\ }$, and the delta function δ_{1_G} at the identity 1_G is the identity element of $\mathcal{H}(G)\hat{\ }$. The Hecke algebra $\mathcal{H}(G)$ is a left ideal of $\mathcal{H}(G)\hat{\ }$, i.e., invariant under left multiplication by $\mathcal{H}(G)\hat{\ }$.

The algebra $\mathcal{H}(G)\hat{}$ (see [BD:§1.4]) is a projective completion of the Hecke algebra $\mathcal{H}(G)$. As a (left-sided) analogue of the (right-sided) algebra $\mathcal{H}(G)\hat{}$, set

$$\hat{}\mathcal{H}(G) := \{\, D \in \mathcal{C}_c^\infty(G)^* \mid f \star D \in \mathcal{C}_c^\infty(G) \quad \forall f \in \mathcal{C}_c^\infty(G) \,\} \,. \qquad (2.4\text{g})$$

As an analogue of (2.4d), and (2.4e) we have

$$(f \star D_2) \star D_1 = x \to D_1(\, t \to D_2(\, y \to f(xt^{-1}y^{-1}) \,)\,)\,) \qquad (2.4\text{h})$$

and

$$
\begin{aligned}
(D_2 \star D_1)(f) &:= \big(\, (\, C(f) \star D_2 \,) \star D_1 \,\big)(1) \\
&= D_1(\, t \to (\, C(f) \star D_2 \,)(t^{-1}) \,) \\
&= D_1(\, t \to (\, D_2(y \to C(f)(t^{-1}y^{-1})) \,)\,) \\
&= D_1(\, t \to (\, D_2(y \to f(yt)) \,)\,)\,) \,.
\end{aligned}
\qquad (2.4\text{i})
$$

In particular, (2.4i) defines an associative convolution product on the space $\hat{}\mathcal{H}(G)$.

As a more symmetrical version of the two algebras $\mathcal{H}(G)\hat{}$, and $\hat{}\mathcal{H}(G)$, we set

$$
\begin{aligned}
\mathcal{U}(G) &:= \mathcal{H}(G)\hat{} \cap \hat{}\mathcal{H}(G) \\
&= \{\, D \in \mathcal{C}_c^\infty(G)^* \mid D \text{ essentially compact} \,\} \,.
\end{aligned}
\qquad (2.4\text{j})
$$

We remark that for $D_1, D_2 \in \mathcal{U}(G)$, formulae (2.4e) and (2.4i) provide two ways to define the convolution $D_2 \star D_1$; namely as

$$D_2 \star_r D_1(f) := (D_2 \star (D_1 \star C(f)))\,(1)\,, \qquad (2.4\text{k})$$

and

$$D_2 \star_l D_1(f) := ((C(f) \star D_2) \star D_1)\,(1) \,. \qquad (2.4\text{l})$$

We show these two are the same. We first recall the identity $f_1 \star f_2(1) = f_2 \star f_1(1)$ for any $f_1, f_2 \in \mathcal{C}_c^\infty(G)$. Now, given $f \in \mathcal{C}_c^\infty(G)$, choose an sufficiently small open compact subgroup J so that $e_J \star C(f) = C(f) = C(f) \star e_J$, $e_J \star (D_1 \star C(f)) = D_1 \star C(f) = (D_1 \star C(f)) \star e_J$, and $e_J \star (C(f) \star D_2) = C(f) \star D_2 = (C(f) \star D_2) \star e_J$. Then,

$$
\begin{aligned}
D_2 \star_r D_1(f) \;:=\; & (D_2 \star (D_1 \star C(f)))\,(1) \\
=\; & (D_2 \star (e_J \star (D_1 \star C(f))))\,(1) \\
=\; & ((D_2 \star e_J) \star (e_J \star (D_1 \star C(f))))\,(1) \\
=\; & ((D_2 \star e_J) \star (e_J \star (D_1 \star (e_J \star C(f)))))\,(1) \\
=\; & ((D_2 \star e_J) \star (e_J \star D_1)) \star (e_J \star C(f))\,(1) \\
=\; & (e_J \star C(f)) \star ((D_2 \star e_J) \star (e_J \star D_1))\,(1) \qquad (2.4\mathrm{m}) \\
=\; & C(f) \star ((D_2 \star e_J) \star (e_J \star D_1))\,(1) \\
=\; & (C(f) \star (D_2 \star e_J)) \star (e_J \star D_1)\,(1) \\
=\; & ((C(f) \star D_2) \star e_J) \star (e_J \star D_1)\,(1) \\
=\; & ((C(f) \star D_2) \star D_1)\,(1) \\
=\; & D_2 \star_l D_1(f)\,.
\end{aligned}
$$

The Hecke algebra $\mathcal{H}(G)$ is a right ideal of $\widehat{}\mathcal{H}(G)$ and a two-sided ideal of $\mathcal{U}(G)$.

The center $\mathcal{Z}(\mathcal{U}(G))$ of $\mathcal{U}(G)$ is the subspace:

$$
\begin{aligned}
\mathcal{Z}(\mathcal{U}(G)) \;=\; & G\text{-invariant essentially compact distributions on } G \\
\;=\; & \mathcal{Z}(G)\,, \quad \text{the Bernstein center}\,.
\end{aligned}
\qquad (2.4\mathrm{n})
$$

This is also the center of $\mathcal{H}(G)\widehat{}$ and $\widehat{}\mathcal{H}(G)$.

2.5. Let $\mathcal{A}$ denote either $\mathcal{H}(G)\widehat{}$, or $\mathcal{U}(G)$. For any $g \in G$, the delta function δ_g at g belongs to $\mathcal{A}$. From this, we deduce that any (left) $\mathcal{A}$-module V is a representation of the group G. Recall that if J is an open compact subgroup of G, then the function

$$
e_J = \frac{1}{\mathrm{meas} J} 1_J \in \mathcal{C}_c^\infty(G) \qquad (2.5\mathrm{a})
$$

is an idempotent of $\mathcal{C}_c^\infty(G)$, i.e., $e_J \star e_J = e_J$. An $\mathcal{A}$-module V is said to be non-degenerate if for any $v \in V$ there exists an open compact subgroup J_v so that $e_{J_v} v = v$. Since $\delta_g \star e_{J_v} = e_{J_v}$ for all $g \in J_v$, it follows $\delta_g v = v$ for all $g \in J_v$. Thus, a non-degenerate representation of $\mathcal{A}$ is a smooth representation of G. Note that an $\mathcal{A}$-module V is non-degenerate if and only if $V = \pi((\mathcal{H}(G))(V)$. The only if part is obvious. To see the if part, suppose $V = \pi((\mathcal{H}(G))(V)$, and $v \in V$. Write v as $v = f\,w$, and take L to be an open compact subgroup so that f is L-left-invariant. Then, $\delta_g v = \delta_g (f\,w) = (\delta_g \star f)\,w = f\,w = v$.

Conversely, we now explain how a smooth representation (π, V) leads to a non-degenerate representation of $\mathcal{A}$. Suppose (π, V) is smooth, $v \in V$

and $D \in \mathcal{A}$. Choose a compact open subgroup J so that $\pi(J)v = v$. The convolution product, $D \star e_J$, lies in the subspace $\mathcal{C}_c(G/J) \subset \mathcal{C}_c^\infty(G)$ of right J-invariant functions. Define

$$\pi(D) v := \pi(D \star e_J) v = \int_G (D \star e_J)(g) \, \pi(g) \, (v) \, dg . \tag{2.5b}$$

To see that $\pi(D)$ is well-defined is an elementary calculation. Suppose L is an open compact subgroup of J. Then

$$e_J = \frac{1}{\mathrm{meas}(J)} \sum_{Lg \in L \backslash J} 1_L \star \delta_g, \text{ so}$$

$$D \star e_J = \frac{1}{\mathrm{meas}(J)} \sum_{Lg \in L \backslash J} D \star 1_L \star \delta_g. \tag{2.5c}$$

Thus,

$$\pi(D \star e_J) \, (v) = \frac{1}{\mathrm{meas}(J)} \sum_{Lg \in L \backslash J} \pi(D \star 1_L) \, \pi(g) \, (v)$$

$$= \frac{1}{[J : L] \cdot \mathrm{meas}(L)} \sum_{Lg \in L \backslash J} \pi(D \star 1_L) \, (v) \tag{2.5d}$$

$$= \pi(D \star e_L) \, (v) .$$

It follows $v \mapsto \pi(D) \, v$ is a well-defined action of the $\mathcal{A}$ on V. It is then elementary to show $\pi(D_1)(\pi(D_2)v) = \pi(D_1 \star D_2)v$, i.e., $\pi : \mathcal{A} \longrightarrow \mathrm{End}_{\mathbb{C}}(V)$ is a representation. Thus, a smooth representation of G is precisely the an $\mathcal{A}$-module V which is non-degenerate.

If (π_1, V_1) and (π_2, V_2) are two smooth representations, and $T : V_1 \longrightarrow V_2$ is a G-map, then

$$T \, \pi_1(D) = \pi_2(D) \, T, \text{ for any } D \in \mathcal{A}. \tag{2.5e}$$

To see this, suppose $v \in V_1$. Consider v, and $T(v)$. Choose an open compact subgroup J which fixes both v and $T(v)$, and consider $D \star e_J$. We have $\pi_1(D)(v) = \pi_1(D \star e_J)(v)$; hence, $T(\pi_1(D))(v) = (T \circ \pi_1)(D \star e_J)(v) = \pi_2(D \star e_J)(T(v)) = \pi_2(D)(T(v))$. When $D \in \mathcal{Z}(G)$, the operator $\pi(D)$ commutes with the action of π, i.e., $\pi(D) \in \mathrm{End}_G(V)$ so $\pi(D)$ is itself a G-morphism. In this way, to each G-invariant essentially compact distribution, there is a naturally attached endomorphism of each object in the category of smooth reprsentations, which commutes with the morphisms of the category.

2.6. The algebra $\mathcal{U}(G)$ is easily made into a $\star$-algebra as follows: For $f \in \mathcal{C}_c^\infty(G)$, define the adjoint $f^\star \in \mathcal{C}_c^\infty(G)$ to be $f^\star(g) := \overline{f(g^{-1})}$, and for $D \in \mathcal{U}(G)$, define the adjoint $D^\star$ to be the distribution $D^\star(f) := \overline{D(f^\star)}$. In particular, the adjoint of the delta distribution δ_g is the delta distribution $\delta_{g^{-1}}$. It is not hard to see the $\star$-involution swaps $\mathcal{H}(G)\hat{}$, and $\hat{}\mathcal{H}(G)$.

3. Some Properties of the Convolution Algebras $\mathcal{H}(G)\hat{}$ and $\mathcal{U}(G)$

3.1. In this section we compare the algebras $\mathcal{H}(G)\hat{}$, $\hat{}\mathcal{H}(G)$, and $\mathcal{U}(G)$ to several related algebras. These other algebras are as follows.

ALGEBRA OF DISTRIBUTIONS WITH COMPACT SUPPORT. This algebra of distributions is defined as

$$\mathcal{U}_c(G) := \{ D \in \mathcal{U}(G) \mid \text{supp}(D) \text{ is compact } \} . \qquad (3.1a)$$

Clearly, $\mathcal{H}(G) \subset \mathcal{U}_c(G) \subset \mathcal{U}(G)$. We note that $\delta_{1_G} \in \mathcal{U}_c(G) \setminus \mathcal{H}(G)$.

ALGEBRA OF LINEAR ENDOMORPHISM OF $\mathcal{C}_c^\infty(G)$. Set

$$\text{End}_{\mathbb{C}}(\mathcal{C}_c^\infty(G)) := \{ \text{ linear endomorphism of } \mathcal{C}_c^\infty(G) \} . \qquad (3.1b)$$

Formula (2.2a) defines two commuting G-actions on $\text{End}_{\mathbb{C}}(\mathcal{C}_c^\infty(G))$; in particular, we can view $\text{End}_{\mathbb{C}}(\mathcal{C}_c^\infty(G))$ as a $G \times G$-module.

$$(g,h)f := x \to (\lambda_g \circ \rho_h)(f)(x) = f(g^{-1}xh) \qquad (3.1c)$$

where $(g,h) \in G \times G$, $f \in \mathcal{C}_c^\infty(G)$. Set

$$\text{End}_{G \times G}(\mathcal{C}_c^\infty(G)) := \{ T \in \text{End}_{\mathbb{C}}(\mathcal{C}_c^\infty(G)) \mid T \circ \lambda_g = \lambda_g \circ T,$$
$$\text{and } T \circ \rho_g = \rho_g \circ T, \forall g \in G\}. \qquad (3.1d)$$

3.2. Let $\mathcal{A}$ be either $\mathcal{H}(G)\hat{}$ or $\mathcal{U}(G)$. Given $D \in \mathcal{A}$, we obtain an element $T_D \in \text{End}_{\mathbb{C}}(\mathcal{C}_c^\infty(G))$ as follows:

$$T_D(f) := D \star f \quad \text{which, by definition, is } x \to D(\lambda_x(\check{f})) . \qquad (3.2a)$$

We have $T_{D_1}(T_{D_2}(f)) = D_1 \star (D_2 \star f) = (D_1 \star D_2) \star f$, so the map

$$D \to T_D \qquad (3.2b)$$

is an algebra homomorphism of $\mathcal{A}$ into $\text{End}_{\mathbb{C}}(\mathcal{C}_c^\infty(G))$. Since we can recover the essentially compact linear functional D from T_D by the formula

$$D(f) = T_D(\check{f})(1) , \qquad (3.2c)$$

the algebra homomorphism is an injection.

For arbitrary $T \in \mathrm{End}_{\mathbb{C}}(\mathcal{C}_c^\infty(G))$, an extrapolation of formula (3.2c) defines a linear functional D_T on $\mathcal{C}_c^\infty(G)$ as

$$D_T(f) := T(\check{f})(1) . \tag{3.2d}$$

We apply formula (3.2a) to D_T:

$$T_{D_T}(f) = x \to D_T(\lambda_x(\check{f})) = T((\lambda_x(\check{f}))\check{\ })(1) . \tag{3.2e}$$

Since $\lambda_x(\check{f}) = g \to \check{f}(x^{-1}g) = f(g^{-1}x)$, and so $(\lambda_x(\check{f}))\check{\ } = g \to f(gx)$, i.e., $(\lambda_x(\check{f}))\check{\ } = \rho_x(f)$. So,

$$T_{D_T}(f) = x \to T(\rho_x(f))(1) . \tag{3.2f}$$

It can be seen from this that, in general, the linear functional D_T is not essentially compact. In particular, the algebra monomorphism (3.2b), considered on $\mathcal{H}(G)\check{\ }$, is not onto. If T satisfies $T \circ \rho_y = \rho_y \circ T$ for all $y \in G$, we conclude T_{D_T} has the property that $T_{D_T} \star f$ equals $T(f)$ and so belongs to $\mathcal{C}_c^\infty(G)$ for all $f \in \mathcal{C}_c^\infty(G)$. This is one half the definition for the linear functional D_T to be essentially compact. Similarly, if $T \circ \lambda_y = \lambda_y \circ T$ for all $y \in G$, then $f \star T_{D_T}$ equals $T(f)$. In particular, if $T \circ \rho_y = \rho_y \circ T$ and $T \circ \lambda_y = \lambda_y \circ T$ for all $y \in G$, then $f \star T_{D_T} = T(f) = T_{D_T} \star f$; so, the linear functional D_T is both essentially compact and G-invariant, i.e., in the center of $\mathcal{A}$. Thus, the map (3.2b) is an isomorphism of $\mathcal{Z}(\mathcal{A})$ with $\mathrm{End}_{G \times G}(\mathcal{C}_c^\infty(G))$, see [B].

At this point it is natural to recall the following theorem of Bernstein (see [BD:§1.9.1] as well as [B:§4.2]).

Proposition 3.2g. *The center of the category* $\mathrm{Alg}(G)$ *of smooth representations of* G *is isomorphic to* $\mathcal{Z}(G)$.

Proof. An element z of the center $\mathcal{Z}(\mathrm{Alg}(G))$, also called an endomorphism of the catgeory, is an assignment to each object, i.e., smooth representation (π, V), a morphism $z(\pi) : V \to V$ so that if (π_1, V_1) and (π_2, V_2) are two smooth representations and $\phi : V_1 \to V_2$ is a morphism, then the following diagram commutes.

$$
\begin{array}{ccc}
V_1 & \xrightarrow{\;\phi\;} & V_2 \\
{\scriptstyle z(\pi_1)}\big\downarrow & & \big\downarrow{\scriptstyle z(\pi_2)} \\
V_1 & \xrightarrow[\;\phi\;]{} & V_2
\end{array}
\tag{3.2h}
$$

Suppose $D \in \mathcal{Z}(G)$. If (π_1, V_1) and (π_2, V_2) are smooth representations, and $\phi : V_1 \longrightarrow V_2$ is a G-map, by (2.5e) we have $\pi_2(D) \circ \phi = \phi \circ \pi_1(D)$. Therefore, $\Gamma(D) := \pi \mapsto \pi(D)$ is an endomorphism of the category $\mathrm{Alg}(G)$. The map $D \to \Gamma(D)$, from $\mathcal{Z}(G)$ to $\mathcal{Z}(\mathrm{Alg}(G))$, is clearly a homomorphism of rings. We prove it is an isomorphism. We view $\mathcal{C}_c^\infty(G)$ as a smooth representation of G via left translations λ.

CLAIM. The map $D \to \lambda(D)$ from $\mathcal{Z}(G)$ to $\{ z(\lambda) \mid z \in \mathcal{Z}(\mathrm{Alg}(G)) \}$ is an isomorphism.

For $D \in \mathcal{Z}(G)$, we have $\lambda(D)(f) = D \star f$; therefore, $D \to \lambda(D)$ is an injection. Conversely, suppose $T \in \mathrm{End}_\lambda(\mathcal{C}_c^\infty(G))$ satisfies $T \circ \phi = \phi \circ T$ for any G-endomorphism of $\mathcal{C}_c^\infty(G)$. Any right translation ρ_g is a G-endomorphism of $\mathcal{C}_c^\infty(G)$; therefore, $T \circ \rho_g = \rho_g \circ T$. Hence, we deduce $T \in \mathrm{End}_{G \times G}(\mathcal{C}_c^\infty(G))$, and so there exists $D \in \mathcal{Z}(G)$ so that $T = T_D$. This proves the claim.

In particular, it follows the map Γ is an injection. To prove Γ is an isomorphism, it suffices to show any $z \in \mathcal{Z}(\mathrm{Alg}(G))$ is completely determined by $z(\lambda)$ (see also the remark in [BDK:§2.2]). To do this, choose $D \in \mathcal{Z}(G)$ so that $z(\lambda) = T_D$. Suppose (π, V) is a smooth representation, $v \in V$, and v is fixed by the open compact subgroup J. The map $\phi_v : \mathcal{C}_c^\infty(G) \to V$, defined as $\phi_v(f) := \pi(f)v$ is a G-map, and $\phi_v(e_J) = v$. Hence, $z(V) \circ \phi_v = \phi_v \circ z(\lambda)$; so, $z(V)(v) = z(V)(\phi_v(e_J)) = \phi_v(z(\lambda)(e_J)) = \phi_v(D \star e_J) = \pi(D)(\phi_v(e_J)) = \pi(D)(v)$. We conclude $z(V) = \pi(D)$, and thus Γ is an isomorphism as required. $\qquad\square$

3.3. Partition of the delta distribution δ_{1_G}. Recall a sequence $\mathcal{J} = \{J_i\}$ of decreasing open compact subgroups of G, i.e.,

$$\mathcal{J} = \{J_i\} \quad \text{with} \quad J_1 \supset J_2 \supset \cdots \supset J_i \supset \cdots \qquad (3.3a)$$

is cofinal among the neighborhoods of the identity, if given a neighborhood $\mathcal{V}$ of the identity, there exists a J_r so that $\mathcal{V} \supset J_r$. For such a cofinal sequence $\mathcal{J}$, set

$$
\begin{aligned}
e_i \quad &= e_{\{\mathcal{J}, i\}} \quad := \frac{1}{\mathrm{meas}(J_i)} \, 1_{J_i}, \\[2mm]
\Delta_i \quad &= \Delta_{\{\mathcal{J}, i\}} \quad := \begin{cases} e_1 & i = 1 \\ e_i - e_{i-1} & i > 1, \end{cases} \qquad (3.3b) \\[2mm]
D_{\Delta_i} \quad &= D_{\Delta_{\{\mathcal{J}, i\}}} \quad := \text{distribution associated} \\
&\qquad\qquad\qquad\qquad \text{to } \Delta_i \text{ as in (2.3b).}
\end{aligned}
$$

Note that

$$\Delta_i \star \Delta_j = \delta_{i,j}\Delta_i, \tag{3.3c}$$

$$e_i \star \Delta_j = \Delta_j \star e_i = \begin{cases} \Delta_j & j \le i, \\ 0 & j > i. \end{cases} \tag{3.3d}$$

Furthermore, if (π, V) is a smooth representation of G, and $\mathrm{Im}(\pi(\Delta_i))$ denotes the image subspace of the operator $\pi(\Delta_i)$, then V decomposes as a direct sum

$$V = \bigoplus_{i=1}^{\infty} \mathrm{Im}(\pi(\Delta_i)), \tag{3.3e}$$

and we have

$$\pi(\Delta_j)v = \delta_{i,j}v \quad \text{for} \quad v \in \mathrm{Im}(\pi(\Delta_i)). \tag{3.3f}$$

Proposition 3.3g. *Suppose $\mathcal{J} = \{J_i\}$ is a decreasing sequence of compact open subgroups of G which is cofinal among the neighborhoods of the identity, and define e_i, Δ_i, and $D_{\Delta_i} = D_{\Delta_{\{\mathcal{J},i\}}}$ as in (3.3b). Then, for any $f \in \mathcal{C}_c^{\infty}(G)$, we have:*

(i) *For i sufficiently large $\Delta_i \star f = 0 = f \star \Delta_i$. Equivalently, $D_{\Delta_i}(f) = 0$ for i sufficiently large.*

(ii) *$\sum_{i=1}^{\infty} \Delta_i \star f = f = \sum_{i=1}^{\infty} f \star \Delta_i$, and $\sum_{i=1}^{\infty} D_{\Delta_i}(f) = f(1) = \delta_{1_G}(f)$. In particular, we have a decomposition of the delta distribution δ_{1_G} as*

$$\delta_{1_G} = \sum_{i=1}^{\infty} D_{\Delta_i}. \tag{3.3h}$$

Proof. Choose N so that the function f is J_N-bi-invariant. If $i \ge N$, then $J_i \subset J_N$, so $e_i \star f = f = f \star e_i$. This immediately implies (i) holds for $i > N$. The series of part (ii), when evaluated at $f \in \mathcal{C}_c^{\infty}(G)$, has only a finite number of non-zero terms; therefore, the equality is obvious. $\square$

3.4. The motivation for the next result is to take $D \in \mathcal{U}(G)$ and two partitions of the delta distribution at the identity, $\delta_{1_G} = \sum D_{\Delta_{\{\mathcal{J},i\}}}$, and $\delta_{1_G} = \sum D_{\Delta_{\{\mathcal{K},j\}}}$, and then justify the identity $D = \delta_{1_G} \star D \star \delta_{1_G} = \sum_{i,j} D_{\Delta_{\{\mathcal{J},i\}}} \star D \star D_{\Delta_{\{\mathcal{K},j\}}}$.

Definition 3.4a. Let C be an Abelian group, and $g_{i,j} \in C$ a two parameter family of elements of C. We say this family is locally finite if when we fix i_0, then the cardinality of $\{ j \mid g_{i_0,j} \ne 0 \}$ is finite, and when we fix j_0, then the cardinality of $\{ i \mid g_{i,j_0} \ne 0 \}$ is finite.

Proposition 3.4b. *Suppose* $\mathcal{J} = \{J_i\}$ *and* $\mathcal{K} = \{K_j\}$ *are two decreasing sequences of compact open subgroups of* G, *with each sequence cofinal among the neighborhoods of the identity. Define* $e_{\{\mathcal{J},i\}}$, $\Delta_{\{\mathcal{J},i\}}$, $D_{\Delta_{\{\mathcal{J},i\}}}$ *and* $e_{\{\mathcal{K},j\}}$, $\Delta_{\{\mathcal{K},j\}}$, $D_{\Delta_{\{\mathcal{K},j\}}}$ *as in (3.3b). For any* $D \in \mathcal{U}(G)$, *set*

$$\Delta_{\{D,(\mathcal{J},\mathcal{K}),(i,j)\}} := \Delta_{\{\mathcal{J},i\}} \star D \star \Delta_{\{\mathcal{K},j\}} \in \mathcal{C}_c^\infty(G) , \qquad (3.4c)$$

and to $\Delta_{\{D,(\mathcal{J},\mathcal{K}),(i,j)\}}$, *let* $D_{\{(\mathcal{J},\mathcal{K}),(i,j)\}}$ *be the associated distribution as in (2.3b). Then,*

(i) *Suppose* $f \in \mathcal{C}_c^\infty(G)$. *For* $i + j$ *sufficiently large, the two convolutions*

$$\Delta_{\{D,(\mathcal{J},\mathcal{K}),(i,j)\}} \star f, \text{ and } f \star \Delta_{\{D,(\mathcal{J},\mathcal{K}),(i,j)\}} \qquad (3.4d)$$

equal the zero function.

(ii) *We have a decomposition of* D *as*

$$D = \sum_{i,j} D_{\{(\mathcal{J},\mathcal{K}),(i,j)\}} . \qquad (3.4e)$$

Moreover, both of the two-parameter families $\Delta_{\{D,(\mathcal{J},\mathcal{K}),(i,j)\}}$, *as well as* $D_{\{(\mathcal{J},\mathcal{K}),(i,j)\}}$, *are locally finite.*

(iii) *Suppose* $g_{i,j} \in \mathcal{C}_c^\infty(G)$ *for* $i, j \geq 1$ *is a locally finite collection of smooth functions. Set*

$$\Delta_{\{(\mathcal{J},\mathcal{K}),g_{i,j}\}} := \Delta_{\{\mathcal{J},i\}} \star g_{i,j} \star \Delta_{\{\mathcal{K},j\}}, \qquad (3.4f)$$

and let $D_{\Delta_{\{(\mathcal{J},\mathcal{K}),g_{i,j}\}}}$ *be the associated distribution as in (2.3b). Then, for any* $f \in \mathcal{C}_c^\infty(G)$, *for* $i + j$ *sufficiently large, we have* $f \star \Delta_{\{(\mathcal{J},\mathcal{K}),g_{i,j}\}} = 0 = \Delta_{\{(\mathcal{J},\mathcal{K}),g_{i,j}\}} \star f$. *In particular, holds* $D_{\Delta_{\{(\mathcal{J},\mathcal{K}),g_{i,j}\}}}(f) = 0$, *and*

$$D := \sum_{i,j} D_{\Delta_{\{(\mathcal{J},\mathcal{K}),g_{i,j}\}}} \qquad (3.4g)$$

defines an essentially compact distribution.

(iv) *Every essentially compact distribution is realizable in the form (3.4g).*

(v) *Suppose* $g_j \in \mathcal{C}_c^\infty(G)$ *for* $j \geq 1$ *is a collection of smooth functions. Set*

$$\Delta_{\{(\mathcal{K},\mathcal{J}),g_j\}} := \Delta_{\{\mathcal{J},j\}} \star g_j \star \Delta_{\{\mathcal{K},j\}}, \qquad (3.4h)$$

and let $D_{\Delta_{\{(\mathcal{K},\mathcal{J}),g_j\}}}$ be the associated distribution as in (2.3b). Then, for any $f \in \mathcal{C}_c^\infty(G)$, for j sufficiently large we have $f \star \Delta_{\{(\mathcal{K},\mathcal{J}),g_j\}} = 0 = \Delta_{\{(\mathcal{K},\mathcal{J}),g_j\}} \star f$. In particular, we have $D_{\Delta_{\{(\mathcal{K},\mathcal{J}),g_j\}}}(f) = 0$, and

$$D := \sum_j D_{\Delta_{\{(\mathcal{K},\mathcal{J}),g_j\}}} \tag{3.4i}$$

is an essentially compact distribution.

Proof. (i) To prove (i), we have $\Delta_{\{D,(\mathcal{J},\mathcal{K}),(i,j)\}} \star f = \Delta_{\{\mathcal{J},i\}} \star D \star \Delta_{\{\mathcal{K},j\}} \star f$. Choose j_0 so that $\Delta_{\{\mathcal{K},j\}} \star f = 0$ for $j \geq j_0$. For each j in the range $1 \leq j < j_0$, choose N_j so that if $i \geq N_j$, then $\Delta_{\{\mathcal{J},i\}} \star D \star \Delta_{\{\mathcal{K},j\}} = 0$. Then, for $i + j \geq N_r := \max\{N_j \mid 1 \leq j < j_0\} + j_0$, we have either $\Delta_{\{\mathcal{K},j\}} \star f = 0$ or $\Delta_{\{\mathcal{J},i\}} \star D \star \Delta_{\{\mathcal{K},j\}} = 0$, hence $\Delta_{\{D,(\mathcal{J},\mathcal{K}),(i,j)\}} \star f = 0$. Similarly, there is a N_l so that for $i + j \geq N_l$, we have $f \star \Delta_{\{D,(\mathcal{J},\mathcal{K}),(i,j)\}} = 0$. Thus, if $i+j \geq \max(N_r, N_l)$ we have both $f \star \Delta_{\{D,(\mathcal{J},\mathcal{K}),(i,j)\}} = 0 = \Delta_{\{(\mathcal{J},\mathcal{K}),(i,j)\}} \star f$, i.e., the assertion (i).

(ii) Formula (3.4d) is an immediate consequence of (i). Fix i_0. Then $\Delta_{\{\mathcal{J},i_0\}} \star D \in C_c^\infty(G)$. Chose j_0 such that $\Delta_{\{\mathcal{J},i_0\}} \star D$ is constant on left K_{j_0}-classes, i.e., on each gK_{j_0}, $g \in G$. Then, for $j > j_0$ we have $\Delta_{\{D,(\mathcal{J},\mathcal{K}),(i_0,j)\}} = (\Delta_{\{\mathcal{J},i_0\}} \star D) \star \Delta_{\{\mathcal{K},j\}} = (\Delta_{\{\mathcal{J},i_0\}} \star D) \star e_{K_{j_0}} \star \Delta_{\{\mathcal{K},j\}} = 0$ by (3.3d). In the same way one proves the second property for local finiteness. This implies the family $\Delta_{\{D,(\mathcal{J},\mathcal{K}),(i,j)\}}$ is locally finite, and so the family $D_{\{(\mathcal{J},\mathcal{K}),(i_0,j)\}}$ is locally finite.

(iii) Choose i_0 so that f is constant on left J_{i_0}-classes. Then, for $i > i_0$, by (3.3d), we have $f \star \Delta_{\{\mathcal{J},i\}} = 0$, and so $f \star \Delta_{\{\mathcal{J},i\}} \star g_{i,j} \star \Delta_{\{\mathcal{K},j\}} = 0$. Since $g_{i,j}$ is a locally finite family, we can find j_0 such that if $j > j_0$, then $g_{i,j} = 0$ for all $i \leq i_0$. This imples that $f \star \Delta_{\{\mathcal{J},i\}} \star g_{i,j} \star \Delta_{\{\mathcal{K},j\}} = 0$ for $i+j > i_0+j_0$. Similarly, we prove $\Delta_{\{\mathcal{J},i\}} \star g_{i,j} \star \Delta_{\{\mathcal{K},j\}} \star f = 0$ for sufficienly large $i + j$.

Consider now $f \star D = f \star \left(\sum_{i,j} D_{\Delta_{\{(\mathcal{J},\mathcal{K}),g_{i,j}\}}} \right) = \sum_{i,j} f \star D_{\Delta_{\{(\mathcal{J},\mathcal{K}),g_{i,j}\}}}$. By the previous paragraph, this is a finite sum. Furthermore, $f \star D_{\Delta_{\{(\mathcal{J},\mathcal{K}),g_{i,j}\}}}$ are compactly supported smooth functions; therefore, $f \star D \in \mathcal{C}_c^\infty(G)$. Similarly, $D \star f \in \mathcal{C}_c^\infty(G)$. This proves $D \in U(G)$.

(iv) For $D \in \mathcal{U}(G)$ take $g_{i,j} = \Delta_{\{\mathcal{J},i\}} \star D \star \Delta_{\{\mathcal{K},j\}}$, use that $\Delta_{\{\mathcal{J},i\}}$ and $\Delta_{\{\mathcal{K},j\}}$ are idempotents, and apply (ii).

(v) This assertion is a special case of (iii). $\qquad\square$

We now give a description of the algebra $\mathcal{H}(G)\hat{}$ analogous to Proposition 3.4b for $\mathcal{U}(G)$,

Proposition 3.4j. *Let* $\mathcal{K} = \{K_j\}$ *be a decreasing sequence of compact open subgroups of* G, *cofinal among the neighborhoods of the identity. Define* $e_{\{\mathcal{K},j\}}$, $\Delta_{\{\mathcal{K},j\}}$, $D_{\Delta_{\{\mathcal{K},j\}}}$ *as in (3.3b). For* $D \in \mathcal{H}(G)\hat{}$, *set*

$$\Delta_{\{D,\mathcal{K},j\}} := D \star \Delta_{\{\mathcal{K},j\}} \in \mathcal{C}_c^\infty(G), \tag{3.4k}$$

and to $\Delta_{\{D,\mathcal{K},j\}}$, *let* $D_{\{D,\mathcal{K},j\}}$ *be the associated distribution as in (2.3b). Then:*

(i) *Suppose* $f \in \mathcal{C}_c^\infty(G)$. *For* j *sufficiently large, the convolution* $\Delta_{\{D,\mathcal{K},j\}} \star f$ *is the zero function.*

(ii) *We have a decomposition of* D *as*

$$D = \sum_j D_{\{D,\mathcal{K},j\}} . \tag{3.4l}$$

(iii) *Suppose* $\{\, g_j \in \mathcal{C}_c^\infty(G) \,|\, j \geq 1 \,\}$ *is a sequence of smooth functions. Let* $D_{g_j \star \Delta_{\{\mathcal{K},j\}}}$ *be in (2.3b). Then, for any* $f \in \mathcal{C}_c^\infty(G)$, *for sufficiently large* j, $(g_j \star \Delta_{\{\mathcal{K},j\}}) \star f = 0$, *and so* $D_{g_j \star \Delta_{\{\mathcal{K},j\}}}(f) = 0$, *and*

$$D := \sum_j D_{g_j \star \Delta_{\{\mathcal{K},j\}}} \tag{3.4m}$$

is in $\mathcal{H}(G)\hat{}$.

(iv) *Every distribution in* $\mathcal{H}(G)\hat{}$ *is realizable in the form (3.4m).*

Proof. Observe that (i) follows from (i) of Proposition 3.3g. Further, (ii) follows from (ii) of the same proposition. The third claim follows from (i) of the same proposition. The last claim follows from the first two claims. $\square$

Set $S_k := \sum_{j=1}^k D_{g_j \star \Delta_{\{\mathcal{K},j\}}}$. We observe that

$$S_{k+1} * e_{\{\mathcal{K},k\}} = S_k. \tag{3.4n}$$

Therefore, in the above proposition we are working implicitly with elements of the projective limit.

Recall that if V is an infinite dimensional vector space and W a nontrivial vector space, then the space $\mathrm{Hom}_{\mathbb{C}}(V, W)$ has dimension at least the continuum.

Theorem 3.4o. *Suppose (π, V) is a smooth representation of a (connected) reductive p-adic group G. Write π also for the associated non-degenerate representation of $\mathcal{H}(G)\hat{}$ on V. Then, we have:*

(i)

$$\{\pi(D) \mid D \in \mathcal{U}(G) \text{ and } \dim_{\mathbb{C}}(\pi(D)V) < \infty\}$$
$$\subset \{\pi(D) \mid D \in \mathcal{H}(G)\}, \tag{3.4p}$$

Equality holds if π is admissible.

(ii)

$$\pi(\mathcal{H}(G)) \subset \pi(\mathcal{U}_c(G)) \subset \pi(\mathcal{U}(G)) \subset \pi(\mathcal{H}(G)\hat{}). \tag{3.4q}$$

(iii) *If π is irreducible, then*

$$\mathrm{End}_{\mathbb{C}}(V) = \pi(\mathcal{H}(G)\hat{}). \tag{3.4r}$$

(iv) *If π is irreducible infinite dimensional, then all inclusions in (3.4q) are strict.*

(v) *If π is an irreducible finite dimensional representation, then all inclusions in (3.4q) are actually equalities.*

Proof. (i) Suppose D belongs to the left hand side of (3.4p). The finite dimensionality hypothesis means we can choose an open compact subgroup J so that $\pi(D)(V)$ is contained in the J-invariants V^J. Obviously, $\pi(D) = \pi(e_J)\pi(D) = \pi(e_J \star D)$, and $e_J \star D$ is in Hecke algebra, since $D \in \mathcal{U}(G)$. This proves (i).

(ii) This assertion is obvious, since we know these inclusions for the corresponding algebras.

(iii) Suppose J is an open compact subgroup of G. Write V^J for the finite dimensional subspace of V fixed by J. For convenience, we fix a Haar measure on G, and therefore an identification of the subspace $\mathcal{C}_c(G//J) \subset \mathcal{C}_c^\infty(G)$ of J-bi-invariant functions with the Hecke algebra $\mathcal{H}(G//J)$ of compactly supported locally constant distributions which are J-bi-invariant. Let π^J denote the representation of $\mathcal{H}(G//J)$ on V^J. The irreducibility hypothesis on V means $\mathrm{End}_{\mathbb{C}}(V^J) = \pi^J(\mathcal{H}(G//J))$. In particular, if V is finite dimensional, then $V = V^J$ for some J, and (3.4r) follows.

Suppose V is infinite dimensional. Recall that V must have countable dimension. Indeed, for any choice of a sequence $\mathcal{K} = \{K_\ell\}$ as in (3.3a), each of the subspaces $\mathrm{Im}(\pi(\Delta_\ell))$ is finite dimensional (since an irreducible smooth representation is admissible), and so V has countable dimension. Set $s_0 := 0$. For $\ell > 0$, set

$$s_\ell := \dim(\operatorname{Im}(\pi(\Delta_1 + \cdots + \Delta_\ell))) , \qquad (3.4s)$$

and select a basis $v_{s_{\ell-1}+1}, \ldots, v_{s_\ell}$ for $\operatorname{Im}(\pi(\Delta_\ell))$. That (3.3e) holds means the sequence $\{\, v_k \,\}$ is a basis for V. To show (3.4r), it is enough to show for an arbitrary sequence of V-vectors $\{\, w_k \,\}$, the existence of a $D \in \mathcal{H}(G)^{\widehat{}}$ so that $w_k = \pi(D)(v_k)$ for all k. Take a sequence $\mathcal{J} = \{\, J_\ell \,\}$ as in (3.3a) so that $w_1, \ldots, w_{s_\ell}$ are fixed by J_ℓ. We can then find $Q_\ell \in \mathcal{H}(J_\ell \backslash G / K_\ell)$, the subspace of compactly supported J_ℓ-left-invariant and K_ℓ-right-invariant distributions, so that $\pi(Q_\ell)(v_i) = w_i$ for $1 \le i \le s_\ell$. Let

$$D = \sum_{j=1}^{\infty} D_{Q_j \star \Delta_{\{\mathcal{K},j\}}}. \qquad (3.4t)$$

Then, $D \in \mathcal{H}(G)^{\widehat{}}$ by (iii) of Proposition 3.4j.

Fix $k \ge 1$. Take $j \ge 1$ so that $s_{j-1} + 1 \le k \le s_j$. Then

$$\begin{aligned}
\pi(D)v_k &= \sum_{i=1}^{\infty} \pi(Q_i)\pi(\Delta_{\{\mathcal{K},i\}})v_k \\
&= \pi(Q_j)\pi(\Delta_{\{\mathcal{K},j\}})v_k = \pi(Q_j)v_k = w_k.
\end{aligned} \qquad (3.4u)$$

This proves (3.4r) when V is infinite dimensional.

(iv) Suppose π is irreducible and infinite dimensional. We observe that $\pi(\mathcal{H}(G))$ consists of finite rank operators, while $\pi(\mathcal{U}_c(G))$ contains some operators with infinite dimensional rank, and therefore, $\pi(\mathcal{H}(G)) \subsetneq \pi(\mathcal{U}_c(G))$.

We now prove $\pi(\mathcal{U}(G)) \subsetneq \pi(\mathcal{H}(G)^{\widehat{}})$. Recall that the space of finite rank operators in $\operatorname{End}_{\mathbb{C}}(V)$ has dimension the continuum. Therefore, since $\mathcal{H}(G)$ is countable dimensional, we can find a finite rank operator A on V which is not in $\pi(\mathcal{H}(G))$. By (i), A is not in $\pi(\mathcal{U}(G))$, while (iii) imples $A \in \pi(\mathcal{H}(G)^{\widehat{}})$. This proves $\pi(\mathcal{U}(G)) \subsetneq \pi(\mathcal{H}(G)^{\widehat{}})$.

Since $\pi(\mathcal{U}_c(G)) \subset \pi(\mathcal{U}(G))$, and $\operatorname{End}_{\mathbb{C}}(V) = \pi(\mathcal{H}(G)^{\widehat{}})$, the above strict inclusion $\pi(\mathcal{U}(G)) \subsetneq \pi(\mathcal{H}(G)^{\widehat{}})$ means

$$\pi(\mathcal{U}_c(G)) \subsetneq \operatorname{End}_{\mathbb{C}}(V) . \qquad (3.4v)$$

We now give a direct proof of this statement, which we will then modify to show $\pi(\mathcal{U}_c(G)) \subsetneq \pi(\mathcal{U}(G))$).

Suppose $D \in \mathcal{U}_c(G)$ is a compactly supported distribution. Take $X \subset G$ to be a compact subset containing $\operatorname{supp}(D)$. Suppose $v \in V$. Choose an open compact subgroup J which fixes v. Write the product set XJ as a disjoint union

$$XJ = \bigsqcup_{i=1}^{M} g_i J \ . \tag{3.4w}$$

Clearly, the distribution $D \star e_J$ has support contained in XJ, and is J-right-invariant. It follows

$$\pi(D)(v) = \pi(D \star e_J)(v) \in V_{\{X,v\}}$$
$$:= \operatorname{span}\{\, \pi(g_1)(v), \dots, \pi(g_M)(v) \,\} \ . \tag{3.4x}$$

Therefore, we have proved for a fixed compact subset $X \subset G$, and $v \in V$, there exists a finite dimensional subspace $V_{\{X,v\}} \subset V$ so that if $D \in (C_c^\infty(G))^*$ has support in X, then $\pi(D)v \in V_{\{X,v\}}$.

We now apply Cantor's diagonal argument. Take a basis $\{v_i\}$ of V, and write $G = \cup_{i=1}^\infty X_i$ as a union of increasing compact subsets X_i. For each X_i, and v_i choose a finite dimensional space $V_{\{X_i,v_i\}}$ so that if $D \in \mathcal{U}_c(G)$ with $\operatorname{supp}(D) \subset X_i$, then $\pi(D)(v_i) \in V_{\{X_i,v_i\}}$. Choose $w_i \in V$ so that $w_i \notin V_{\{X_i,v_i\}}$. There exists a linear transformation T of V so that $T(v_i) = w_i$.

CLAIM. If D is a compactly supported distribution, then $\pi(D) \neq T$.

We prove the claim by contradiction. Suppose $D \in \mathcal{U}_c(G)$ is such that $\pi(D) = T$. Take i, so that $\operatorname{supp}(D) \subset X_i$. Then, $\pi(D)(v_i) \in V_{\{X_i,v_i\}}$, but $w_i = T(v_i) = \pi(D)(v_i) \notin V_{\{X_i,v_i\}}$. This is a contradiction. So, the claim is proved, and (3.4v) follows immediately.

We now refine the above proof of (3.4v), to show that $\pi(\mathcal{U}_c(G)) \subsetneq \pi(\mathcal{U}(G))$. Define a strictly increasing sequence of indexes $t_1 < t_2 < \dots$ as follows: Let $s_\ell := \dim(\operatorname{Im}(\pi(\Delta_1 + \dots + \Delta_\ell)))$ be as in (3.4s). Choose $t_1 \geq 1$ so that $v_{s_{t_1}} \notin V_{\{X_{s_1},v_{s_1}\}}$. Then, recursively choose $t_{i+1} > t_i$ so $v_{s_{t_{i+1}}} \notin V_{\{X_{s_{i+1}},v_{s_{i+1}}\}}$. For each $i \geq 1$ choose $g_i \in \mathcal{H}(G)$ such that $\pi(g_i)v_{s_i} = v_{s_{t_i}}$. Form the distribution

$$D := \sum_{i=1}^{\infty} D_{\Delta_{\{\mathcal{K},t_i\}} \star g_i \star \Delta_{\{\mathcal{K},i\}}} \ . \tag{3.4y}$$

By (v) of Proposition 3.4b, this is an essentially compact distribution, i.e., $D \in \mathcal{U}(G)$. For all $i = 1, 2, \dots$, we have

$$\begin{aligned}
\pi(D)v_{s_i} &= \sum_{k=1}^{\infty} \pi(\Delta_{\{\mathcal{K},t_k\}} \star g_k \star \Delta_{\{\mathcal{K},k\}})v_{s_i} \\
&= \pi(\Delta_{\{\mathcal{K},t_i\}} \star g_i \star \Delta_{\{\mathcal{K},i\}})v_{s_i} \\
&= \pi(\Delta_{\{\mathcal{K},t_i\}} \star g_i)v_{s_i} = \pi(\Delta_{\{\mathcal{K},t_i\}})v_{s_{t_i}} = v_{s_{t_i}} \ .
\end{aligned} \tag{3.4z}$$

Suppose $\pi(D) = \pi(D_c)$ for some $D_c \in \mathcal{U}_c(G)$. Choose $i \geq 1$ such that $\mathrm{supp}(D_c) \subset X_{s_i}$. Then $\pi(D_c)v_{s_i} \in V_{\{X_{s_i}, v_{s_i}\}}$ by the choice of $V_{\{X_{s_i}, v_{s_i}\}}$. But $\pi(D_c)v_{s_i} = \pi(D)v_{s_i} = v_{s_{t_i}} \notin V_{\{X_{s_i}, v_{s_i}\}}$ by the choice of t_i. This is a contradiction, and therefore $\pi(\mathcal{U}_c(G)) \subsetneqq \pi(\mathcal{U}(G))$. The proof of (iv) is now complete.

(v) Since π is irreducible and finite dimensional, we have $\pi(\mathcal{H}(G)) = \mathrm{End}_\mathbb{C}(V)$. $\qquad\square$

3.5. In this section we show any essentially compact distribution is tempered, i.e., extends to a continuous linear functional of the Schwartz space $\mathscr{C}(G)$ of G. We begin by briefly recalling its definition. More details and proofs can be found in [W].

Let $A_\emptyset$ be denote a maximal split F-torus in G, and $M_\emptyset$ its F-centralizer. Denote the maximal compact subgroup of $M_\emptyset$ by $^\circ M_\emptyset$. Fix a minimal F-parabolic subgroup P of G containing $A_\emptyset$. Let K be a special good maximal compact subgroup of G. The selection of P determines the set of simple roots (with respect to $A_\emptyset$), which further defines a cone $M_\emptyset^+$ in $A_\emptyset$. Then we have Cartan decomposition

$$G = \bigsqcup_{m \in M_\emptyset^+ / ^\circ M_\emptyset} KmK \quad \text{(disjoint decomposition)}. \qquad (3.5a)$$

Thus, we have a bijection $K\backslash G/K$ onto $M_\emptyset^+/^\circ M_\emptyset \subset M_\emptyset/^\circ M_\emptyset$. This bijection we denote by σ. The quotient $M_\emptyset/^\circ M_\emptyset$ is a lattice, and we fix a norm $\|\ \|$ on this lattice, which is invariant for the action of the Weyl group of $A_\emptyset$. Denote by δ_P the modular character of P. Extend δ_P to a K-invariant function on G via the Iwasawa decomposition, i.e., by the formula $\delta_P(pk) = \delta_P(p)$ for $p \in P$ and $k \in K$. Set Ξ to be the K-spherical function

$$\Xi(g) = \int_K \delta_P(kg)^{1/2} dk . \qquad (3.5b)$$

We recall that Ξ is the matrix coefficient of the K-spherical vector in the unitary principal series induced representation from the trivial character of $M_\emptyset$.

Denote by $\mathcal{C}^\infty(G)$ the space of complex locally constant functions on G. For r a positive integer, and $f \in \mathcal{C}^\infty(G)$, set

$$v_r(f) := \sup\left\{ \frac{|f(g)|\,(1 + \|\sigma(g)\|)^r}{\Xi(g)} \ \Big|\ g \in G \right\} . \qquad (3.5c)$$

For a fixed open compact subgroup J of K, set

$$\mathscr{C}(G, J) := \{ f \in \mathcal{C}^\infty(G) \mid \text{(i)} \ f \text{ is } J\text{-bi-invariant,} \\ \text{(ii)} \ \text{for every } r, \quad v_r(f) < \infty \}. \tag{3.5d}$$

The functions v_r define semi-norms on $\mathscr{C}(G, J)$, and the collection of these semi-norms yields a topology on $\mathscr{C}(G, J)$ so that it is a Fréchet space. Furthermore, functions in $\mathscr{C}(G, J)$ are square integrable, and thus the convolution of two such functions can be defined by the usual formula. The convolution again belongs to $\mathscr{C}(G, J)$, and multiplication is continuous. In this way, $\mathscr{C}(G, J)$ is a Fréchet algebra.

The system $\mathscr{C}(G, J)$, as J runs over the open subgroups of K, is an inductive system in the category of locally convex topological vector spaces, and the Schwartz space $\mathscr{C}(G)$ is the inductive limit of this family. The Schwartz space is a complete locally convex space. Since the spaces $\mathscr{C}(G, J)$ are Fréchet algebras, the mapping $(f_1, f_2) \mapsto f_1 \star f_2$ is a continuous linear mapping $\mathscr{C}(G) \to \mathscr{C}(G)$ whenever we fix either f_1 or f_2.

Clearly, $\mathcal{C}_c^\infty(G) \subset \mathscr{C}(G)$. A distribution D on G is said to be tempered, if it extends to a continuous linear functional on $\mathscr{C}(G)$. Each compactly supported distribution is tempered. We shall see that this is a special case of a more general fact:

Theorem 3.5e. *Any distribution in $\mathcal{H}(G)\hat{\ }$ is tempered. In particular, any essentially compact distribution, and therefore, any $D \in \mathcal{Z}(\mathcal{U}(G))$, is tempered.*

Proof. Let $D \in \mathcal{H}(G)\hat{\ }$. Suppose f is in the Schwartz space $\mathscr{C}(G)$. Then, there exists an open compact subgroup J so that f is J-bi-invariant. Since $D \in \mathcal{H}(G)\hat{\ }$, we have $D \star e_J \in \mathcal{C}_c^\infty(G)$, and so the convolution $(D \star e_J) \star \check{f}$ is defined. Set

$$D^\#(f) := \left((D \star e_J) \star \check{f} \right)(1). \tag{3.5f}$$

We observe that if f has compact support, then $D^\#(f) = D(f)$. Furthermore, if L is an open compact subgroup of J, by associativity of convolution, and the hypothesis f, hence $\check{f}$ is J-bi-invariant, we have

$$\begin{aligned}
(D \star e_L) \star \check{f} &= (D \star e_L) \star (e_J \star \check{f}) = ((D \star e_L) \star e_J) \star \check{f} \\
&= (D \star (e_L \star e_J)) \star \check{f} \\
&= (D \star e_J) \star \check{f},
\end{aligned} \tag{3.5g}$$

and so

$$\left((D \star e_L) \star \check{f} \right)(1) = \left((D \star e_J) \star \check{f} \right)(1). \tag{3.5h}$$

In particular, we conclude $D^\#$ is a well-defined extension of the linear functional D to elements $f \in \mathscr{C}(G)$. To prove $D^\#$ defines a continuous extension, it is enough to prove its restriction to the subspace $\mathscr{C}(G, J)$ of J-bi-invariant functions is continuous. The map $D^\# : \mathscr{C}(G, J) \longrightarrow \mathbb{C}$ is the composition of three continuous maps

$$f \mapsto \check{f} \mapsto (D \star e_J) \star \check{f} \mapsto \big((D \star e_J) \star \check{f} \big)(1) \tag{3.5i}$$

and therefore continuous. $\square$

We remark that by slight modification, this proof also applies to the algebra $\widehat{\mathcal{H}(G)}$ too.

4. Some Explicit G-invariant Essentially Compact Distributions

4.1. The results of the sections 2 and 3 establish the algebras $\widehat{\mathcal{H}(G)}$ and $\mathcal{U}(G)$ as suitable p-adic analogues of the enveloping algebra of the Lie algebra of a connected reductive Lie group. The center of each is precisely the Bernstein center of G-invariant essentially compact distributions. In the notes [B], Bernstein raised the problem of explicit construction of G-invariant essentially compact distributions. In this section we give examples of such distributions, ending with recent results of the authors [MT2].

4.2. We begin with an example of Bernstein's from his notes [B].

4.2a. Bernstein's example. *Suppose* $G = \mathrm{SL(n)}(F)$, $\psi : F \to \mathbb{C}$ *a nontrivial additive character, and* θ *is the continuous G-invariant function* $\theta(g) := \psi(\mathrm{trace}(g))$. *Then, the G-invariant distribution*

$$D_\theta(f) := \int_G \theta(g) f(g) \, dg \ , \ f \in \mathcal{C}_c^\infty(G) \tag{4.2b}$$

is essentially compact.

Proof. We observe that it is enough to show $\theta \star 1_J$ is compactly supported for any open compact subgroup J. This is because given $f \in \mathcal{C}_c^\infty(G)$, there exists an open compact J such that $e_J \star f = f = f \star e_J$. So, if $D_\theta \star e_J \in \mathcal{C}_c^\infty(G)$, then $D_\theta \star f = D_\theta \star (e_J \star f) = (D_\theta \star e_J) \star f \in \mathcal{C}_c^\infty(G)$. As a second observation, we note that it is enough to restrict J to be congruence subgroups K_m of the maximal compact $K = \mathrm{SL(n)}(\mathcal{R}_F)$. Here, $\mathcal{R}_F$ is the ring of integers in F. So, suppose $J = K_m$. To show

$$D_\theta \star 1_J := g \mapsto \int_J \theta(gx) \, dx \tag{4.2c}$$

is compactly supported, we use the Cartan decomposition $G = KA^+K$ to write g as $g = k_1 d k_2$, where $k_1, k_2 \in K$, and d is a diagonal matrix with ascending powers of the uniformizing element ϖ on the diagonal

$$d = \operatorname{diag}(\varpi^{-a_1}, \varpi^{-a_2}, \ldots, \varpi^{-a_n}) \quad , \quad a_1 \geq a_2 \geq \cdots \geq a_n . \qquad (4.2\mathrm{d})$$

Then,

$$
\begin{aligned}
\int_J \psi(\operatorname{tr}(gx))\, dx &= \int_J \psi(\operatorname{tr}(k_1 d k_2 x))\, dx \\
&= \int_J \psi(\operatorname{tr}(d k_2 x k_1))\, dx \\
&= \int_J \psi(\operatorname{tr}(d k x))\, dx \quad , \quad k := k_2 k_1 .
\end{aligned}
\qquad (4.2\mathrm{e})
$$

In the last line, we have used the fact that K normalizes the subgroup $J = K_m$. To see why the integral vanishes for g, i.e., d outside a bounded set, we consider the case of SL(2). This case illustrates the basic idea. Let $\wp$ denote the prime ideal in $\mathcal{R}_F$. We have:

$$d k x = \begin{bmatrix} \varpi^{-a} & 0 \\ 0 & \varpi^{a} \end{bmatrix} \begin{bmatrix} k_{1,1} & k_{1,2} \\ k_{2,1} & k_{2,2} \end{bmatrix} \begin{bmatrix} 1 + x_{1,1} & x_{1,2} \\ x_{2,1} & 1 + x_{2,2} \end{bmatrix}, x_{i,j} \in \wp^m. \qquad (4.2\mathrm{f})$$

So,

$$
\begin{aligned}
\operatorname{tr}(dkx) = \varpi^{-a}&(k_{1,1}(1 + x_{1,1}) + k_{1,2}x_{2,1}) + \\
&\varpi^{a}(k_{2,1}(x_{1,2}) + k_{2,2}(1 + x_{2,2})).
\end{aligned}
\qquad (4.2\mathrm{g})
$$

We have

$$
\begin{aligned}
\psi(\operatorname{tr}(dkx)) &= \psi(\varpi^{-a}(k_{1,1}(1 + x_{1,1}) + k_{1,2}x_{2,1})) \cdot \\
&\quad \psi(\varpi^{a}(k_{2,1}(x_{1,2}) + k_{2,2}(1 + x_{2,2}))) \\
&= \psi(\varpi^{-a}k_{1,1}) \cdot \psi(\varpi^{-a}(k_{1,1}x_{1,1} + k_{1,2}x_{2,1})) \cdot \\
&\quad \psi(\varpi^{a}(k_{2,1}(x_{1,2} + k_{2,2}x_{2,2}))) \cdot \psi(\varpi^{a}k_{2,2}).
\end{aligned}
\qquad (4.2\mathrm{h})
$$

If g is sufficiently large, i.e., the integer a is large positive, then will $\psi(\varpi^{a}(k_{2,1}(x_{1,2} + k_{2,2}x_{2,2})))$ and $\psi(\varpi^{a}k_{2,2})$ be identically 1 for all elements $x_{1,2}, x_{2,2} \in \wp^m$. Thus, for a sufficiently large positive, we have

$$\psi(\operatorname{tr}(dkx)) = \psi(\varpi^{-a}k_{1,1}) \cdot \psi(\varpi^{-a}(k_{1,1}x_{1,1} + k_{1,2}x_{2,1})) . \qquad (4.2\mathrm{i})$$

The important term is the 2nd term. We coordinatize the group J by elements $x_{1,1}, x_{1,2}, x_{2,1} \in \wp^m$. Then

$$\int_J \psi(\operatorname{tr}(dkx))\, dx =$$

$$\int_{\wp^m \times \wp^m \times \wp^m} \psi(\varpi^{-a} k_{1,1}) \cdot \psi(\varpi^{-a}(k_{1,1}x_{1,1} + k_{1,2}x_{2,1}))dx_{1,1}dx_{2,1}dx_{1,2} =$$

$$\int_{\wp^m} \psi(\varpi^{-a}k_{1,1})\Big(\int_{\wp^m \times \wp^m} \psi(\varpi^{-a}(k_{1,1}x_{1,1} + k_{1,2}x_{2,1}))dx_{1,1}dx_{2,1}\Big)dx_{1,2}.$$

$$(4.2j)$$

For a sufficiently large, since $k \in \operatorname{SL}(2)(\mathcal{R}_F)$, the inner integral over $\wp^m \times \wp^m$ is clearly zero. Therefore, the distribution D_θ is essentially compact. □

4.3. It is very tempting to try to generalize the distribution $g \mapsto \psi(\operatorname{trace}(g))$ as follows:

(1) For $x \in G = \operatorname{SL}(n)(F)$, let $c_1(x)$ denote the trace of x, and more generally $c_k(x)$ the coefficient of the t^{n-k} in the characteristic polynomial $p_x(t)$ of x. Consider the class functions and distributions

$$\theta_k(x) := \psi(c_k(x))$$
$$D_k(f) = D_{\theta_k}(f) := \int_G \theta_k(x)f(x)\, dx.$$

$$(4.3a)$$

Which D_k belong to the Bernstein center? The class function $g \to c_k(x)$ is in fact the character of an irreducible finite dimensional F-representation of $\operatorname{SL}(n)$. Take $V = F^n$ to be the standard defining representation of $G = \operatorname{SL}(n)(F)$. For $0 \le k \le n$, consider the exterior power $\Lambda^k V$ representation of G. Then,

  (i) It is an irreducible miniscule representation.
  (ii) The trace of $g \in G$ on $\Lambda^k V$ is $c_k(g)$.

If a distribution D is essentially compact, then, it is obvious, the distribution $\check{D} : f \mapsto D(\check{f})$ is also essentially compact. For $g \in \operatorname{SL}(n)(F)$, we have $c_k(g^{-1}) = c_{n-k}(g)$. It follows $D_k \in \mathcal{Z}(G)$ if and only if $D_{n-k} \in \mathcal{Z}(G)$. In particular, since $D_1 \in \mathcal{Z}(G)$, we have $D_{n-1} \in \mathcal{Z}(G)$.

(2) More generally, suppose $\rho : G \to \operatorname{GL}(m)(F)$ is an irreducible representation of G. Does the class function

$$\theta_\rho(g) := \psi(\operatorname{trace}(\rho(g))),$$

$$(4.3b)$$

define a distribution in the Bernstein center?

The next example shows these two generalizations are false.

4.3c. SL(4)(F) and the coefficient c_2. *The distribution Θ associated to the class function $g \mapsto \psi(c_2(g))$ is not essentially compact.*

Proof. Fix a positive integer m so that $\wp^m$ lies in the kernel of ψ. Take $J = K_m$, the conguence subgroup of level m. For

$$dk = \begin{bmatrix} \varpi^{-t} & 0 & 0 & 0 \\ 0 & 1 & 0 & 0 \\ 0 & 0 & 1 & 0 \\ 0 & 0 & 0 & \varpi^t \end{bmatrix} \begin{bmatrix} 0 & 0 & 0 & 1 \\ 0 & 0 & 1 & 0 \\ 0 & 1 & 0 & 0 \\ 1 & 0 & 0 & 0 \end{bmatrix}, \; t > 0 \; , \qquad (4.3d)$$

we show $\Theta \star 1_J(dk) \neq 0$ for arbitrarily large t. For

$$g = \begin{bmatrix} a & b & c & d \\ e & f & g & h \\ i & j & k & l \\ m & n & o & p \end{bmatrix} , \qquad (4.3e)$$

$c_2(g)$ is

$$c_2(g) = (be + ci + dm + gj + hn + lo) \\ - (af + ak + ap + fk + fp + kp) \; . \qquad (4.3f)$$

Now, we have

$$dkx = \begin{bmatrix} \varpi^{-t}x_{4,1} & \varpi^{-t}x_{4,2} & \varpi^{-t}x_{4,3} & \varpi^{-t}(1+x_{4,4}) \\ x_{3,1} & x_{3,2} & 1+x_{3,3} & x_{3,4} \\ x_{2,1} & 1+x_{2,2} & x_{2,3} & x_{2,4} \\ \varpi^t(1+x_{1,1}) & \varpi^t x_{1,2} & \varpi^t x_{1,3} & \varpi^t x_{1,4} \end{bmatrix} . \qquad (4.3g)$$

So,

$$\begin{aligned} c_2(dkx) &= \varpi^{-t}\big(x_{4,2}x_{3,1} + x_{4,3}x_{2,1} - x_{4,1}x_{3,2} - x_{4,1}x_{2,3}\big) \\ &+ \big((1+x_{4,4})(1+x_{1,1}) \\ &+ (1+x_{3,3})(1+x_{2,2}) - x_{3,2}x_{2,3} - x_{4,1}x_{1,4}\big) \\ &+ \varpi^t\big(x_{3,4}x_{1,2} + x_{2,4}x_{1,3} - x_{3,2}x_{1,4} - x_{2,3}x_{1,4}\big) \; . \end{aligned} \qquad (4.3h)$$

The assumption $\wp^m \subset \text{Ker}(\psi)$ means

$$\psi(c_2(dkx)) = \\ \psi(\varpi^{-t}\big(x_{4,2}x_{3,1} + x_{4,3}x_{2,1} - x_{4,1}x_{3,2} - x_{4,1}x_{2,3}\big)) \cdot \psi(1)^2 \; . \qquad (4.3i)$$

The variables $x_{4,4}$, $x_{4,3}$, $x_{4,2}$, $x_{4,1}$, $x_{3,2}$, $x_{3,1}$, $x_{2,3}$, $x_{2,1}$ run freely over $\wp^m$. The resulting integral is a Kloosterman sum, and it is non-zero for $t >> 0$. Hence, $\Theta \star 1_J(dk) \neq 0$ for $t >> 0$, so the distribution Θ is not essentially compact. □

Remark 4.3j. The above proof and counterexamples can be adapted to the following situations.

(1) Suppose $G = \mathrm{Sp}(2m)$, and $\rho : G \longrightarrow \mathrm{GL}(2m)(F)$ the natural defining representation. Then, the G-invariant distribution associated to the class function $g \mapsto \psi(\mathrm{trace}(\rho(g)))$ is essentially compact.

(2) Suppose E/F is a quadratic extension of F and $G = \mathrm{SU}(2,1)$, and $\rho : G \longrightarrow \mathrm{GL}(2m)(E)$ the natural defining representation. Then, the G-invariant distribution associated (using Haar measure) to the class function $g \mapsto \psi(\mathrm{trace}_{E/F}(\mathrm{trace}(\rho(g))))$ is not essentially compact.

4.4. One plentiful, but mysterious source of elements in the Bernstein center is the set of irreducible supercuspidal representations.

4.4a. Supercuspidal characters. *Suppose $G = \mathsf{G}(F)$ is a semisimple group. If (π, V) is an irreducible supercuspidal representation of G, then the character θ_π of π is an element of the Bernstein center.*

Proof. We may assume π is infinite dimensional. The hypothesis G is semisimple means π is unitary. Let $\langle, \rangle$ be a G-invariant hermitian form on the space V of π, and let $\{\, v_i \; i \in \mathbb{N} \,\}$ be an orthonormal basis. We have

$$\theta_\pi(g) \;=\; \sum_i \langle v_i, \pi(g)v_i \rangle \;. \tag{4.4b}$$

Suppose J is an open compact subgroup of G. We have

$$
\begin{aligned}
\theta_\pi \star e_J(h) \;&=\; \frac{1}{\mathrm{meas}(J)} \int_J \theta_\pi(hx)\, dx \\
&=\; \frac{1}{\mathrm{meas}(J)} \int_J \sum_i \langle v_i, \pi(hx)v_i \rangle\, dx \\
&=\; \frac{1}{\mathrm{meas}(J)} \int_J \sum_i \langle \pi(h^{-1})v_i, \pi(x)v_i \rangle\, dx \;.
\end{aligned}
\tag{4.4c}
$$

So,

$$\theta_\pi \star e_J(h) \;=\; \sum_i \langle \pi(h^{-1})v_i, \pi(e_J)v_i \rangle \;. \tag{4.4d}$$

The operator $\pi(e_J)$ projects V_π to the finite dimensional space of J-fixed vectors. We may choose the orthogonal basis so the span$\{v_1, \ldots, v_r\}$ is V_π^J. Then

$$\theta_\pi \star e_J(h) \;=\; \sum_{i=1}^r \langle v_i, \pi(h^{-1})v_i \rangle \;. \tag{4.4e}$$

The assumption that π is supercuspidal means each of the matrix coefficients

$$h \mapsto \langle \pi(h^{-1})v_i, v_i \rangle \qquad (4.4\mathrm{f})$$

is supported on a compact set. In particular, their finite sum, i.e., $\theta_\pi \star e_J$ has compact support. □

4.5. As mentioned in section 2, if $\mathcal{O}$ is a conjugacy class in a connected reductive p-adic group, the orbital integral distribution (2.3g) is essentially compact if and only if $\mathcal{O}$ is compact. The authors have discovered for non-compact classes in $\mathrm{SL}(2)(F)$ that certain linear combination of orbital integral are essentially compact (see [MT1]). These combinations can be predicted by the asymptotical behavior of the orbits at infinity. Furthermore, the authors have obtained a generalization of the $\mathrm{SL}(2)(F)$ results to hyperbolic conjugacy classes in quasi-split groups. We finish by formulating the main result of [MT2].

We assume G is the group of F-rational points of a connected reductive quasi-split F-group G. Let $A_\emptyset$ be a maximal split F-torus, $M_\emptyset = C_G(A_\emptyset)$, and $B = P_\emptyset$ a Borel F-subgroup containing $M_\emptyset$. Let $D : M_\emptyset \longrightarrow \mathbb{R}$ denote the Weyl denominator.

For $t \in M_\emptyset$, define the normalized orbital integral of the conjugacy class $\mathrm{Ad}(G)(t)$ in the usual way, i.e.,

$$F_f^{M_\emptyset}(t) = D(t)^{1/2} \int_{G/M_\emptyset} f(hth^{-1}) \, dh . \qquad (4.5\mathrm{a})$$

Then, the main result of [MT2] is the following:

4.5d. Linear combination of orbital integrals. *Let γ_0, $\gamma \in M_\emptyset$. Suppose that $\gamma_0 (w \cdot \gamma)$ is regular for every $w \in W_G(A_\emptyset)$. It means that if $w' \in W$, and $w'(\gamma_0 \, w(\gamma)) = \gamma_0 \, w(\gamma)$, then $w' = 1$. Then, the distribution*

$$f \mapsto \sum_{w \in W_G(A_\emptyset)} \mathrm{sgn}(w) \; F_f^{M_\emptyset}(\gamma_0 \, w(\gamma)), \quad \forall \, f \in C_c^\infty(G) \qquad (4.5\mathrm{b})$$

belongs to the Bernstein center.

Acknowledgment

The authors thank Dan Barbasch, and Gordan Savin for helpful discussions in regards to the examples of section 4. The second author completed the last stages of this work while visiting the E. Schrödinger Institute in Vienna. He thanks the institute, and its director J. Schwermer for their hospitality.

References

[B] Bernstein, J. (written by K. Rumelhart), *Draft of: Representations of p-adic groups*, preprint.

[BD] Bernstein, J.,rédigé par Deligne, P., *Le "centre" de Bernstein*, in book "Représentations des Groupes Réductifs sur un Corps Local" written by J.-N. Bernstein, P. Deligne, D. Kazhdan and M.-F. Vignéras, Hermann, Paris, 1984.

[BDK] Bernstein, J., Deligne, P. and Kazhdan, D., *Trace Paley-Wiener theorem for reductive p-adic groups*, J. Analyse Math 47 (1986), 180-192.

[MT1] Moy, A., Tadić, M., *Conjugacy class asymptotics, orbital integrals, and the Bernstein center: the case of SL(2)*, Represent. Theory 9 (2005), 327-353.

[MT2] Moy, A., Tadić, M., *A construction of elements in the Bernstein center for quasi-split groups*, preprint (August 2006).

[W] Waldspurger, J.-L., *La formule de Plancherel pour les groupes p-adiques*, J. Inst. Math. Jussieu 2 (2003), 235-333.

ANNIHILATORS OF GENERALIZED VERMA MODULES OF THE SCALAR TYPE FOR CLASSICAL LIE ALGEBRAS

TOSHIO OSHIMA

Graduate School of Mathematical Sciences
University of Tokyo, Tokyo 153-8914, Japan
E-mail: oshima@ms.u-tokyo.ac.jp

Dedicated to Roger Howe on the occasion of his 60th birthday.

We construct a generator system of the annihilator of a generalized Verma module of a classical reductive Lie algebra induced from a character of a parabolic subalgebra as an analogue of the minimal polynomial of a matrix. In a classical limit it gives a generator system of the defining ideal of any semisimple co-adjoint orbit of the Lie algebra. We also give some applications to integral geometry.

Keywords and phrases. Verma module, primitive ideal, universal enveloping algebra.

1. Introduction

In [O3] generalized Capelli operators are defined in the universal enveloping algebra of $GL(n, \mathbb{C})$ and it is shown that they characterize the differential equations satisfied by the functions in degenerate principal series representations of $GL(n, \mathbb{R})$. The operators are useful to formulate boundary value problems for various boundaries of the symmetric space $GL(n, \mathbb{R})/O(n)$ and to construct generalized hypergeometric equations related to Radon transformations on Grassmannian manifolds. In [O4] using these operators we construct a generator system of the annihilator of the generalized Verma module for $\mathfrak{gl}(n, \mathbb{C})$ induced from any character of any parabolic subalgebra.

In this paper the generator system is constructed for the classical Lie algebra $\mathfrak{g}$. Here $\mathfrak{g}$ is $\mathfrak{gl}(n, \mathbb{C})$, $\mathfrak{o}(2n, \mathbb{C})$, $\mathfrak{o}(2n + 1, \mathbb{C})$ or $\mathfrak{sp}(n, \mathbb{C})$. In the case

2000 Mathematics Subject Classification. Primary 16S34; Secondary 17B35, 17B20

The author was partially supported by Grant-in-Aid for Scientific Researches (B No. 16340034) Japan Society of Promotion of Science.

of $\mathfrak{gl}(n, \mathbb{C})$ the generator system in [O4] is an analogue of minors and elementary divisors. The generator system here is an analogue of the minimal polynomial of a matrix and different from the one constructed in [O4]. For the generator system of the center of the universal enveloping algebra the former corresponds to Capelli identity in [C1] and [C2] and the latter to the trace of the power of a matrix with components in the Lie algebra which is presented by [Ge].

In §2 we introduce a square matrix F with components in $\mathfrak{g}$ or the universal enveloping algebra $U(\mathfrak{g})$ associated to a finite dimensional representation of a Lie algebra $\mathfrak{g}$ and define a *minimal polynomial* of F with respect to a $\mathfrak{g}$-module (cf. Definition 2.4).

In this paper we examine the minimal polynomial of F associated to the natural representation of the classical Lie algebra $\mathfrak{g}$.

In §3 we calculate the Harish-Chandra homomorphism of certain polynomials of F. It is a little complicated but elementary. A complete answer is given in Theorem 4.19 when $\mathfrak{g}$ is $\mathfrak{gl}(n, \mathbb{C})$. For the calculation, in §4 we introduce some polynomials of F and study their action on the generalized Verma module.

Then we construct a two-sided ideal of $U(\mathfrak{g})$ generated by the components $q(F)_{ij}$ for the minimal polynomial $q(x)$ of F and prove Theorem 4.4, which is the main result in this paper. It says (cf. Remark 4.5 ii)) that the ideal describes the *gap* between the generalized Verma module and the usual Verma module (cf. (5.1) and (5.7)) if at least the infinitesimal character of the Verma module is regular (resp. strongly regular) in the case when $\mathfrak{g}$ is $\mathfrak{gl}(n, \mathbb{C})$, $\mathfrak{o}(2n + 1, \mathbb{C})$ or $\mathfrak{sp}(n, \mathbb{C})$ (resp. $\mathfrak{o}(2n, \mathbb{C})$). The main motivation to write this paper is to construct a two-sided ideal with this property originated in the problem in [O1].

It follows from this theorem that the ideal coincides with the annihilator of the generalized Verma module of the scalar type for the classical Lie algebra if at least the infinitesimal character is (strongly) regular (cf. Corollary 4.6).

We will use the homogenized universal enveloping algebra $U^\epsilon(\mathfrak{g})$ introduced in [O4] so that we can compare the generator system with that of the annihilator of a co-adjoint orbit in the dual of $\mathfrak{g}$. As a classical limit we get the generator system of any semisimple co-adjoint orbit for a classical Lie algebra, which is described in Theorem 4.11 (cf. Remark 4.12).

In §5 we show some applications of our two-sided ideals to integral transformations of sections of a line bundle over a generalized flag manifold. For example, Theorem 5.1 is a typical application, which shows that the system of differential equations defined by our two-sided ideal characterizes the image of the Poisson transform of the functions on any boundary of the Riemannian symmetric space of the non-compact type.

In §6 we discuss the infinitesimal character which is excluded in the results in §4 and present some problems.

In the subsequent paper [OO] which has since appeared, we work in the same context and give a simple explicit formula of minimal polynomials of generalized Verma modules of the scalar type for any reductive Lie algebra and study the same problem as in this paper.

In order to explain our idea, suppose $G = GL(2n, \mathbb{C})$ and put $A = \begin{pmatrix} \lambda I_n & 0 \\ B & \mu I_n \end{pmatrix} \in \mathfrak{g} = \mathrm{Lie}(G)$. Here λ, $\mu \in \mathbb{C}$ and B is a generic element of $M(n, \mathbb{C})$. Note that A is conjugate to $\lambda I_n \oplus \mu I_n$ if $\lambda \neq \mu$ and to $\begin{pmatrix} \lambda & 0 \\ 1 & \lambda \end{pmatrix} \oplus \cdots \oplus \begin{pmatrix} \lambda & 0 \\ 1 & \lambda \end{pmatrix}$ otherwise. We will identify $\mathfrak{g}$ and its dual $\mathfrak{g}^*$ by the symmetric bilinear form $\langle X, Y \rangle = \mathrm{Trace}\, XY$. Let $I_\Theta (\subset S(\mathfrak{g}))$ be the defining ideal of the closure $\bar{V}_\Theta$ of the conjugacy class $V_\Theta = \bigcup_{g \in G} \mathrm{Ad}(g)A$ with $\mathrm{Ad}(g)X = gXg^{-1}$.

Note that $I_\Theta = I_\Theta^0$ by denoting

$$
\begin{cases}
I_\Theta^\epsilon = \bigcap_{g \in G} \mathrm{Ad}(g) J_\Theta^\epsilon, \\
J_\Theta^\epsilon = \sum_{X_1, X_2, X_3 \in M(n, \mathbb{C})} U^\epsilon(\mathfrak{g}) \Big(\begin{pmatrix} X_1 & 0 \\ X_3 & X_2 \end{pmatrix} - \lambda \, \mathrm{Trace}\, X_1 - \mu \, \mathrm{Trace}\, X_2 \Big).
\end{cases}
$$

Here $U^\epsilon(\mathfrak{g})$ is the quotient of the tensor algebra of $\mathfrak{g}$ by the two-sided ideal generated by elements of the form $X \otimes Y - Y \otimes X - \epsilon[X, Y]$. Then $U^0(\mathfrak{g})$ is the symmetric algebra $S(\mathfrak{g})$ of $\mathfrak{g}$ and $U^1(\mathfrak{g})$ is the universal enveloping algebra of $\mathfrak{g}$. We call a generalization of I_Θ^ϵ a quantization of I_Θ^0 and the quantization I_Θ^1 is nothing but the annihilator of the generalized Verma module $U(\mathfrak{g})/J_\Theta^1$.

Since $\mathrm{rank}(X - \lambda I_{2n}) \leq n$ and $\mathrm{rank}(X - \mu I_{2n}) \leq n$ for $X \in \bar{V}_\Theta$, the $(n+1)$-minors $(\in S(\mathfrak{g}))$ of $\big((E_{ij}) - \lambda I_{2n}\big)$ and $\big((E_{ij}) - \mu I_{2n}\big)$ are in I_Θ. On the contrary, they generate I_Θ if $\lambda \neq \mu$. The quantizations of the minors are generalized Capelli operators studied by [O3].

If $\lambda = \mu$, the derivatives of $(n + 1)$-minors of $\big((E_{ij}) - xI_{2n}\big)$ at $x = \lambda$ are also in I_Θ and in general the generators are described by using the elementary divisors. In [O4], we define their quantizations, namely, we explicitly construct the corresponding generators for any generalized Verma module of the scalar type for $\mathfrak{gl}(n, \mathbb{C})$ using generalized Capelli operators and quantized elementary divisors. Moreover in [O4] we determine the condition that the annihilator describes the gap between the generalized Verma module and the usual Verma module. In the example here, the equality

$$
J_\Theta^\epsilon \;=\; I_\Theta^\epsilon + \sum_{i>j} U^\epsilon(\mathfrak{g}) E_{ij} + \sum_{i=1}^{n} U^\epsilon(\mathfrak{g})(E_{ii} - \lambda) + \sum_{i=n+1}^{2n} U^\epsilon(\mathfrak{g})(E_{ii} - \mu)
$$

$$(1.1)$$

holds if and only if $\lambda - \mu \notin \{\epsilon, 2\epsilon, \ldots, (n-1)\epsilon\}$. When $\epsilon = 1$, this condition is satisfied if $U(\mathfrak{g})/J_\Theta^\epsilon$ has a regular infinitesimal character, which is equivalent to the condition that $\lambda - \mu \notin \{1, 2, \ldots . 2n - 1\}$. If (1.1) holds, the quantized generators are considered to be the differential equations which characterize the representations of the group G related to the generalized Verma module. Hence they are important and the motivation of our study in this note is this fact.

Now since $(x - \lambda)(x - \mu)$ is the minimal polynomial of A, all the components of $\big((E_{ij}) - \lambda I_{2n}\big)\big((E_{ij}) - \mu I_{2n}\big)$ are in J_Θ^0. They generate I_Θ^0 together with $\sum_{i=1}^{2n} E_{ii} - n\lambda - n\mu$ if $\lambda \neq \mu$. We can quantize this minimal polynomial and the quantized minimal polynomial in this example is $q^\epsilon(x) = (x - \lambda)(x - \mu - n\epsilon)$. We can show that the $4n^2$ components of the matrix $q^\epsilon\big((E_{ij})\big) \in M\big(2n, U^\epsilon(\mathfrak{g})\big)$ and the element $\sum_{i=1}^{2n} E_{ii} - n\lambda - n\mu$ generate I_Θ^ϵ if $\lambda - \mu \notin \{\epsilon, 2\epsilon, \ldots, (n-1)\epsilon\}$. When we identify $U(\mathfrak{g})$ with the ring of left invariant holomorphic differential operators on $GL(n, \mathbb{C})$, we have $q^1\big((E_{ij})\big) = ({}^t X\partial - \lambda)({}^t X\partial - \mu - n)$ with the matrices $X = (x_{ij}) \in GL(n, \mathbb{C})$ and $\partial = (\frac{\partial}{\partial x_{ij}})$.

The main topic in this paper is to construct the elements in $U(\mathfrak{g})$ which kills the generalized Verma module of the scalar type for the classical Lie algebra by using the quantized minimal polynomial.

The author expresses his sincere gratitude to Mittag-Leffler Institute. The result in this paper for $\mathfrak{g} = \mathfrak{gl}(n, \mathbb{C})$ was obtained when the author was invited there from September until November in 1995 and it is reported in [O2].

2. Minimal Polynomials

For a module $\mathfrak{A}$ and positive integers N and N', we denote by $M(N, N', \mathfrak{A})$ the set of matrices of size $N \times N'$ with components in $\mathfrak{A}$. If $N = N'$, we simply denote it by $M(N, \mathfrak{A})$ and then $M(N, \mathfrak{A})$ is naturally an associative algebra if so is $\mathfrak{A}$.

We use the standard notation $\mathfrak{gl}_n$, $\mathfrak{o}_n$ and $\mathfrak{sp}_n$ for classical Lie algebras over $\mathbb{C}$. For a Lie algebra $\mathfrak{g}$ we denote by $U(\mathfrak{g})$ and $S(\mathfrak{g})$ the universal enveloping algebra and the symmetric algebra of $\mathfrak{g}$, respectively. For a non-negative integer k let $S(\mathfrak{g})^{(k)}$ be the subspace of $S(\mathfrak{g})$ formed by elements of degree at most k. If we fix a Poincare-Birkhoff-Witt base of $U(\mathfrak{g})$, we can identify $U(\mathfrak{g})$ and $S(\mathfrak{g})$ as vector spaces and we denote by $U(\mathfrak{g})^{(k)}$ the subspace of $U(\mathfrak{g})$ corresponding to $S(\mathfrak{g})^{(k)}$.

The Lie algebra $\mathfrak{gl}_N$ is identified with $M(N, \mathbb{C}) \simeq \mathrm{End}(\mathbb{C}^N)$ by $[X, Y] = XY - YX$. Let $E_{ij} = \left(\delta_{\mu i}\delta_{\nu j}\right)_{\substack{1 \leq \mu \leq N \\ 1 \leq \nu \leq N}} \in M(N, \mathbb{C})$ be the standard matrix units. Note that the symmetric bilinear form

$$\langle X, Y \rangle = \mathrm{Trace}\, XY \quad \text{for} \quad X, Y \in \mathfrak{gl}_N \tag{2.1}$$

on $\mathfrak{gl}_N$ is non-degenerate and satisfies

$$\langle E_{ij}, E_{\mu\nu} \rangle = \delta_{i\nu}\delta_{j\mu},$$

$$X = \sum_{i,j} \langle X, E_{ji} \rangle E_{ij},$$

$$\langle \mathrm{Ad}(g)X, \mathrm{Ad}(g)Y \rangle = \langle X, Y \rangle \quad \text{for } X, Y \in \mathfrak{gl}_N \text{ and } g \in GL(N, \mathbb{C}).$$

Lemma 2.1. *Let $\mathfrak{g}$ be a Lie algebra over $\mathbb{C}$ and let $(\pi, \mathbb{C}^N)$ be a representation of $\mathfrak{g}$. We denote by $U(\pi(\mathfrak{g}))$ the subalgebra of the universal enveloping algebra $U(\mathfrak{gl}_N)$ of $\mathfrak{gl}_N$ generated by $\pi(\mathfrak{g})$. Let p be a linear map of $\mathfrak{gl}_N$ to $U(\pi(\mathfrak{g}))$ satisfying*

$$p([X, Y]) = [X, p(Y)] \quad \text{for } X \in \pi(\mathfrak{g}) \text{ and } Y \in \mathfrak{gl}_N, \tag{2.2}$$

that is, $p \in \mathrm{Hom}_{\pi(\mathfrak{g})}(\mathfrak{gl}_N, U(\pi(\mathfrak{g})))$.
Fix $f(x) \in \mathbb{C}[x]$ and put

$$\begin{cases} F = \left(p(E_{ij})\right)_{\substack{1 \leq i \leq N \\ 1 \leq j \leq N}} \in M(N, U(\pi(\mathfrak{g}))), \\[2mm] \left(Q_{ij}\right)_{\substack{1 \leq i \leq N \\ 1 \leq j \leq N}} = f(F) \in M(N, U(\pi(\mathfrak{g}))). \end{cases} \tag{2.3}$$

Then

$$\left(p\big(\mathrm{Ad}(g)E_{ij}\big)\right)_{\substack{1\le i\le N \\ 1\le j\le N}} = {}^tg\,F\,{}^tg^{-1} \quad for \ \ g\in GL(n,\mathbb{C}) \qquad (2.4)$$

and

$$[X,Q_{ij}] = \sum_{\nu=1}^{N} X_{\nu i}Q_{\nu j} - \sum_{\nu=1}^{N} X_{j\nu}Q_{i\nu}$$

$$= \sum_{\nu=1}^{N}\langle X,E_{i\nu}\rangle Q_{\nu j} - \sum_{\nu=1}^{N} Q_{i\nu}\langle X,E_{\nu j}\rangle \quad for \ X = \big(X_{ij}\big)_{\substack{1\le i\le N \\ 1\le j\le N}}\in\pi(\mathfrak{g})$$

$$(2.5)$$

with $X_{ij}\in\mathbb{C}$.

Proof. Put $g = \big(g_{ij}\big)$ and $g^{-1} = \big(g'_{ij}\big)$. Then

$$\left(p\big(\mathrm{Ad}(g)E_{ij}\big)\right)_{\substack{1\le i\le N \\ 1\le j\le N}} = \left(p(\sum_{\mu,\nu} g_{\mu i}g'_{j\nu}E_{\mu\nu})\right)_{\substack{1\le i\le N \\ 1\le j\le N}} = {}^tg\,F\,{}^tg^{-1}.$$

Fix $X\in\pi(\mathfrak{g})$. Since

$$[X,E_{ij}] = [\sum_{\mu,\nu} X_{\mu\nu}E_{\mu\nu}, E_{ij}] = \sum_{\mu=1}^{N} X_{\mu i}E_{\mu j} - \sum_{\nu=1}^{N} X_{j\nu}E_{i\nu},$$

we have (2.5) for $f(x)=x$ by (2.2).

Suppose $\big(Q^1_{ij}\big)$ and $\big(Q^2_{ij}\big)\in M\big(N,U(\pi(\mathfrak{g}))\big)$ satisfy (2.5). Put $Q^3_{ij} = \sum_{k=1}^{N} Q^1_{ik}Q^2_{kj}$ in $U(\pi(\mathfrak{g}))$. Then

$$[X,Q^3_{ij}] = \sum_{k=1}^{N}[X,Q^1_{ik}]Q^2_{kj} + \sum_{k=1}^{N} Q^1_{ik}[X,Q^2_{kj}]$$

$$= \sum_{k=1}^{N}\left(\sum_{\mu=1}^{N} X_{\mu i}Q^1_{\mu k}Q^2_{kj} - \sum_{\nu=1}^{N} X_{k\nu}Q^1_{i\nu}Q^2_{kj}\right)$$

$$+ \sum_{k=1}^{N}\left(\sum_{\mu=1}^{N} Q^1_{ik}X_{\mu k}Q^2_{\mu j} - \sum_{\nu=1}^{N} Q^1_{ik}X_{j\nu}Q^2_{k\nu}\right)$$

$$= \sum_{\mu=1}^{N} X_{\mu i}Q^3_{\mu j} - \sum_{\nu=1}^{N} X_{j\nu}Q^3_{i\nu}$$

and therefore the elements $\big(Q_{ij}\big)$ of $M\big(N,U(\pi(\mathfrak{g}))\big)$ satisfying (2.5) form a subalgebra of $M\big(N,U(\pi(\mathfrak{g}))\big)$. $\qquad\square$

Definition 2.2. If the symmetric bilinear form (2.1) is non-degenerate on $\pi(\mathfrak{g})$, the orthogonal projection of $\mathfrak{gl}_N$ onto $\pi(\mathfrak{g})$ satisfies the assumption for p in Lemma 2.1, which we call the *canonical projection* of $\mathfrak{gl}_N$ to $\pi(\mathfrak{g})$.

Remark 2.3. Suppose that $\mathfrak{g}$ is reductive. Let G be a connected and simply connected Lie group with the Lie algebra $\mathfrak{g}$ and let G_U be a maximal compact subgroup of G. Assume that the finite dimensional representation (π, V) in Lemma 2.1 can be lifted to the representation of G_U. Let $\mathfrak{g} = \bar{\mathfrak{n}} \oplus \mathfrak{a} \oplus \mathfrak{n}$ be a triangular decomposition of $\mathfrak{g}$ such that $\exp \mathfrak{a} \cap G_U$ is a maximal torus of G_U. Let $\Sigma(\mathfrak{a})$ and $\Sigma(\mathfrak{a})^+$ be the sets of the roots for the pair $(\mathfrak{g}, \mathfrak{a})$ and $(\mathfrak{n}, \mathfrak{a})$, respectively, and let $\Psi(\mathfrak{a})$ denote the fundamental system of $\Sigma(\mathfrak{a})^+$. We fix a Hermitian inner product on V so that π is a unitary representation of G_U. Moreover let $\{v_1, \ldots, v_N\}$ be an orthonormal basis of V such that v_j is a weight vector of a weight ϖ_j with respect to the Cartan subalgebra $\mathfrak{a}$. We may assume that $\varpi_i - \varpi_j \in \Sigma(\mathfrak{a})^+$ means $i > j$. Hence ϖ_1 is the lowest weight and ϖ_N is the highest weight of the representation π. Under this basis we identify $\pi(X) = \big(\pi(X)_{ij}\big) \in M(N, \mathbb{C}) \simeq \mathrm{End}(\mathbb{C}^N) \simeq \mathfrak{gl}_N$ for $X \in \mathfrak{g}$ by $\pi(X)v_j = \sum_{i=1}^N \pi(X)_{ij} v_i$. Note that $\pi(\mathfrak{a}) \subset \mathfrak{a}_N$, $\pi(\mathfrak{n}) \subset \mathfrak{n}_N$ and $\pi(\bar{\mathfrak{n}}) \subset \bar{\mathfrak{n}}_N$ by denoting

$$\mathfrak{a}_N = \sum_{j=1}^N \mathbb{C}E_{ii}, \quad \mathfrak{n}_N = \sum_{1 \le j < i \le N} \mathbb{C}E_{ij} \quad \text{and} \quad \bar{\mathfrak{n}}_N = \sum_{1 \le i < j \le N} \mathbb{C}E_{ij}. \quad (2.6)$$

Since $\pi(X)$ is skew Hermitian for the element X in the Lie algebra $\mathfrak{g}_U$ of G_U and $\mathbb{C}\pi(\mathfrak{g}_U) = \pi(\mathfrak{g})$, we have ${}^t\overline{\pi(\mathfrak{g})} = \pi(\mathfrak{g})$. Hence the symmetric bilinear form (2.1) is non-degenerate on $\pi(\mathfrak{g})$ and there exists the canonical projection of $\mathfrak{gl}_N$ to $\pi(\mathfrak{g})$.

Definition 2.4 (Characteristic polynomials and minimal polynomials). Given a Lie algebra $\mathfrak{g}$ and a faithful finite dimensional representation $(\pi, \mathbb{C}^N)$ of $\mathfrak{g}$, we identify $\mathfrak{g}$ as a subalgebra of $\mathfrak{gl}_N$ through π. Let $U(\mathfrak{g})$ and $Z(\mathfrak{g})$ be the universal enveloping algebra of $\mathfrak{g}$ and the center of $U(\mathfrak{g})$, respectively. Suppose a $\mathfrak{g}$-homomorphism p of $\mathrm{End}(\mathbb{C}^N) \simeq \mathfrak{gl}_N$ to $U(\mathfrak{g})$ is given. For $F = \big(p(E_{ij})\big) \in M(N, U(\mathfrak{g}))$, we say $q_F(x) \in Z(\mathfrak{g})[x]$ is the *characteristic polynomial* of F if it is a non-zero polynomial of x satisfying

$$q_F(F) = 0 \quad (2.7)$$

with the minimal degree.

Suppose moreover a $\mathfrak{g}$-module M is given. Then we call $q_{F,M}(x) \in \mathbb{C}[x]$ the *minimal polynomial* of F with respect to M if it is the monic polynomial with the minimal degree which satisfies

$$q_{F,M}(F)M = 0. \tag{2.8}$$

If p is the canonical projection in Definition 2.2, we sometimes denote F_π, q_π and $q_{\pi,M}$ in place of F, q_F and $q_{F,M}$, respectively.

Remark 2.5. i) After the results in this paper was obtained, the author was informed that [Go2] studied the characteristic polynomial of F_π for the irreducible representation π of the reductive Lie algebra.

ii) If $\mathfrak{g}$ is reductive, the characteristic polynomial is uniquely determined by (π, p) up to a constant multiple of the element of $Z(\mathfrak{g})$ since $Z(\mathfrak{g})$ is an integral domain.

iii) If $\mathfrak{g}$ is reductive and M has an infinitesimal character χ, that is, χ is an algebra homomorphism of $Z(\mathfrak{g})$ to $\mathbb{C}$ with $(D - \chi(D))M = 0$ for $D \in Z(\mathfrak{g})$, then $\chi(q_F(x)) \in \mathbb{C}[x]q_{F,M}(x)$.

iv) The characteristic polynomial and minimal polynomial of a matrix in the linear algebra can be regarded as a classical limit of our definition. See the proof of Proposition 4.16.

Theorem 2.6. *Let $\mathfrak{g}$ be a reductive Lie algebra and let F be a matrix of $U(\mathfrak{g})$ defined from a representation of π under* Definition 2.4.

i) *There exists the characteristic polynomial $q_F(x)$ whose degree is not larger than $\sum_\varpi m_\pi(\varpi)^2$. Here ϖ runs through the weights of π and $m_\pi(\varpi)$ denotes the multiplicity of the generalized weight ϖ in π.*

ii) *The minimal polynomial $q_{F,M}(x)$ exists if a $\mathfrak{g}$-module M has a finite length or an infinitesimal character. Its degree is not larger than that of the characteristic polynomial $q_F(x)$ if M has an infinitesimal character.*

Proof. Let $\hat{Z}(\mathfrak{g})$ denote the quotient field of $Z(\mathfrak{g})$ and put $\hat{U}(\mathfrak{g}) = \hat{Z}(\mathfrak{g}) \otimes_{Z(\mathfrak{g})} U(\mathfrak{g})$. Owing to [Ko] it is known that $U(\mathfrak{g}) = \Lambda(H(\mathfrak{g})) \otimes Z(\mathfrak{g})$, where $H(\mathfrak{g})$ is the space of $\mathfrak{g}$-harmonic polynomials of $S(\mathfrak{g})$ and Λ is the map of the symmetrization of $S(\mathfrak{g})$ onto $U(\mathfrak{g})$. It is also known that $H(\mathfrak{g}) \simeq \sum_{\tau \in \hat{\mathfrak{g}}_f} m_\tau(0)\tau$ as a representation space of $\mathfrak{g}$ by denoting $\hat{\mathfrak{g}}_f$ the equivalence classes of the finite dimensional irreducible representations of $\mathfrak{g}$.

Hence the dimension of the $\mathfrak{g}$-homomorphisms of $\pi \otimes \pi^*$ to $\hat{U}(\mathfrak{g})$ over the field $\hat{Z}(\mathfrak{g})$ is not larger than $\sum_{\tau \in \hat{\mathfrak{g}}_f}[\pi \otimes \pi^*, \tau]m_\tau(0)$. Here $[\pi \otimes \pi^*, \tau]$ is the multiplicity of τ appearing in $[\pi \otimes \pi^*]$ in the sense of the Grothendieck group. Moreover it is clear that $\sum_{\tau \in \hat{\mathfrak{g}}_f}[\pi \otimes \pi^*, \tau]m_\tau(0) = m_{\pi \otimes \pi^*}(0) = \sum_\varpi m_\pi(\varpi)^2$. Here we note that Lemma 2.1 says that the space $V_k = \sum_{i,j} \mathbb{C}F_{ij}^k$ is naturally a subrepresentation of the representation of $\mathfrak{g}$ which is realized in $M(N, \mathbb{C})$ and belongs to $\pi \otimes \pi^*$ and that the map $T_k : E_{ij} \mapsto$

F_{ij}^k defines a $\mathfrak{g}$-homomorphism of $M(N,\mathbb{C})$ to $U(\mathfrak{g})$. Hence $T_1,\ldots,T_m$ are linearly dependent over $\hat{Z}(\mathfrak{g})$ if $m > \sum_{\varpi} m_{\pi}(\varpi)^2$. Thus we have proved the existence of the characteristic polynomial with the required degree.

For the existence of the minimal polynomial it is sufficient to prove the existence of a non-zero polynomial $f(x)$ with $f(F)M = 0$. Considering the irreducible subquotients of M in Definition 2.4, we may assume M has an infinitesimal character λ. Let $q_F(x)$ be the characteristic polynomial. We can choose $\mu \in \mathfrak{a}^*$ so that $\bar{\omega}(q_F(x))(\lambda + \mu t) \in \mathbb{C}[x,t]$ is not zero. Here $\bar{\omega}$ is the Harish-Chandra homomorphism defined by (4.10). Put $I_\lambda = \sum_{Z \in Z(\mathfrak{g})} U(\mathfrak{g})(Z - \bar{\omega}(Z)(\lambda))$. We can find a non-negative integer k such that $f(x,t) = t^{-k}\bar{\omega}(q_F(x)) \in \mathbb{C}[x,t]$ and $f(x,0)$ is not zero. We define $h(t) \in M(N, H(\mathfrak{g}) \otimes \mathbb{C}[t])$ so that $f(F,t) - \Lambda(h(t)) \in M(N, I_{\lambda+\mu t})$. Since $q_F(F)(\lambda + \mu t) \in M(N, I_{\lambda+\mu t})$, $h(t) = 0$ for $t \neq 0$ and hence $h = 0$ and therefore $f(F,0)(U(\mathfrak{g})/I_\lambda) = 0$. Hence $f(F,0)M = 0$ because $\mathrm{Ann}(M) \supset I_\lambda$. $\qquad\square$

Hereafter in this note we assume

$$\begin{cases} \pi \text{ is injective,} \\ p(\mathfrak{gl}_N) \subset \mathfrak{g}, \\ p(X) = CX \quad \text{for} \quad X \in \mathfrak{g} \end{cases} \qquad (2.9)$$

in Lemma 2.1 with a suitable non-zero constant C. Then we have the following.

Remark 2.7. i) Since π is faithful, $\mathfrak{g}$ is identified with the Lie subalgebra $\pi(\mathfrak{g})$ of $\mathfrak{gl}_N$ and $U(\pi(\mathfrak{g}))$ is identified with the universal enveloping algebra $U(\mathfrak{g})$ of $\mathfrak{g}$. We note that the existence of p with (2.9) is equivalent to the existence of a $\mathfrak{g}$-invariant subspace of $\mathfrak{gl}_N$ complementary to $\mathfrak{g}$.

ii) Fix $g \in GL(N,\mathbb{C})$. If we replace $(\pi, \mathbb{C}^N)$ by $(\pi^g, \mathbb{C}^N)$ with $\pi^g(X) = \mathrm{Ad}(g)\pi(X)$ for $X \in \mathfrak{g}$ in Lemma 2.1, $(F_{ij}) \in M(N, \mathfrak{g})$ naturally changes into ${}^t g^{-1}(F_{ij}) {}^t g$ and therefore the corresponding characteristic polynomial and minimal polynomial does not depend on the realization of the representation π. In fact, the map p^g of $U(\mathfrak{g})$ to $\pi(\mathfrak{g})$ is naturally defined by $\pi^g(X) = \mathrm{Ad}(g)(p(\mathrm{Ad}(g)^{-1}X))$ and hence $p^g(E) = \mathrm{Ad}(g)(p(\mathrm{Ad}(g^{-1})F) = \mathrm{Ad}(g)(p({}^t g^{-1} F {}^t g)) = \pi^g({}^t g^{-1}(F_{ij}){}^t g)$.

iii) Suppose $\mathfrak{g}$ is semisimple. Then the existence of p is clear because any finite dimensional representation of $\mathfrak{g}$ satisfies the assumption in Remark 2.3.

iv) Let σ be an involutive automorphism of $\mathfrak{gl}_N$. Put

$$\mathfrak{g} = \{X \in \mathfrak{gl}_N; \ \sigma(X) = X\}.$$

Let π be the inclusion map of $\mathfrak{g} \subset \mathfrak{gl}_N$. Since $\mathfrak{q} = \{X \in \mathfrak{gl}_N; \ \sigma(X) = -X\}$ is $\mathfrak{g}$-stable, we may put $p(X) = \frac{X + \sigma(X)}{2}$ in Lemma 2.1, which is the canonical projection with respect to the bilinear form of $\mathfrak{gl}_N$.

v) For a positive integer k and complex numbers $\lambda_1, \ldots, \lambda_k$, the vector space spanned by the N^2 components of the matrix $(p(E) - \lambda_1 I_N) \cdots (p(E) - \lambda_k I_N)$ is ad($\mathfrak{g}$)-invariant. Moreover the trace of the matrix is a central element of $U(\mathfrak{g})$, which is clear from Lemma 2.1 and studied by [Ge] and [Go1] etc.

3. Projection to the Cartan Subalgebra

Now we consider the natural realization of classical simple Lie algebras. Denoting

$$\tilde{I}_n = \left(\delta_{i,n+1-j}\right)_{\substack{1 \le i \le n \\ 1 \le j \le n}} = \begin{pmatrix} & & 1 \\ & \cdot^{\displaystyle\cdot^{\displaystyle\cdot}} & \\ 1 & & \end{pmatrix} \quad \text{and} \quad \tilde{J}_n = \begin{pmatrix} & \tilde{I}_n \\ -\tilde{I}_n & \end{pmatrix},$$

we naturally identify

$$\begin{aligned} \mathfrak{o}_n &= \{X \in \mathfrak{gl}_n; \sigma_{\mathfrak{o}_n}(X) = X\} \quad \text{with } \sigma_{\mathfrak{o}_n}(X) = -\tilde{I}_n {}^t X \tilde{I}_n, \\ \mathfrak{sp}_n &= \{X \in \mathfrak{gl}_{2n}; \sigma_{\mathfrak{sp}_n}(X) = X\} \quad \text{with } \sigma_{\mathfrak{sp}_n}(X) = -\tilde{J}_n {}^t X \tilde{J}_n. \end{aligned} \tag{3.1}$$

Definition 3.1. Let $\mathfrak{g} = \mathfrak{gl}_n$ or $\mathfrak{o}_{2n}$ or $\mathfrak{o}_{2n+1}$ or $\mathfrak{sp}_n$ and put $N = n$ or $2n$ or $2n + 1$ or $2n$, respectively, so that $\mathfrak{g}$ is a subalgebra of $\mathfrak{gl}_N$. Put

$$\bar{i} = N + 1 - i \tag{3.2}$$

for any integer i and define

$$\epsilon_i = \begin{cases} 0 & \text{if} \quad \mathfrak{g} = \mathfrak{gl}_n, \\ 1 & \text{if} \quad \mathfrak{g} = \mathfrak{o}_N, \\ 1 & \text{if} \quad \mathfrak{g} = \mathfrak{sp}_n \quad \text{and} \quad i \le n, \\ -1 & \text{if} \quad \mathfrak{g} = \mathfrak{sp}_n \quad \text{and} \quad i > n. \end{cases} \tag{3.3}$$

Then the involutions $\sigma_{\mathfrak{g}}$ of $\mathfrak{gl}_N$ defining $\mathfrak{g}$ with $\mathfrak{g} = \mathfrak{o}_N$ and $\mathfrak{sp}_n$ satisfy

$$\sigma_{\mathfrak{g}}(E_{ij}) = -\epsilon_i \epsilon_j E_{\bar{j}\bar{i}}.$$

We moreover define

$$F = \left(F_{ij}\right)_{\substack{1 \le i \le N \\ 1 \le j \le N}} = \left(E_{ij} - \epsilon_i \epsilon_j E_{\bar{j}\bar{i}}\right)_{\substack{1 \le i \le N \\ 1 \le j \le N}}. \qquad (3.4)$$

This definition of F means $C = 2$ in (2.9) if $\mathfrak{g} = \mathfrak{o}_N$ or $\mathfrak{sp}_n$. We will denote F_i in place of F_{ii} for simplicity. Then $\mathfrak{g} = \sum_{i,j} \mathbb{C}F_{ij}$ and

$$[X, F_{ij}] = \sum_{\nu=1}^{N} \left(X_{\nu i}F_{\nu j} - X_{j\nu}F_{i\nu}\right) \quad \text{for} \quad X = \left(X_{ij}\right) \in \mathfrak{g} \subset M(N, \mathbb{C}) \quad (3.5)$$

by Lemma 2.1.

Use the notation (2.6) and define $\mathfrak{a} = \mathfrak{g} \cap \mathfrak{a}_N$, $\mathfrak{n} = \mathfrak{g} \cap \mathfrak{n}_N$ and $\bar{\mathfrak{n}} = \mathfrak{g} \cap \bar{\mathfrak{n}}_N$. Then

$$\mathfrak{g} = \bar{\mathfrak{n}} \oplus \mathfrak{a} \oplus \mathfrak{n} \qquad (3.6)$$

is a triangular decomposition of $\mathfrak{g}$.

Definition 3.2. For a positive integer k and complex numbers $\lambda_1, \ldots, \lambda_k$ put

$$F^k(\lambda_1, \ldots, \lambda_k) = (F - \lambda_1 I_N) \cdots (F - \lambda_k I_N)$$

and define an element $\bar{F}^k(\lambda_1, \ldots, \lambda_k)$ in $M(N, U(\mathfrak{a}))$ by

$$F^k(\lambda_1, \ldots, \lambda_k) \equiv \bar{F}^k(\lambda_1, \ldots, \lambda_k) \mod M(N, \bar{\mathfrak{n}}U(\mathfrak{g}) + U(\mathfrak{g})\mathfrak{n}) \qquad (3.7)$$

In this section we will study the image $\bar{F}^k(\lambda_1, \ldots, \lambda_k)$ of $F^k(\lambda_1, \ldots, \lambda_k)$ under the Harish-Chandra homomorphism with respect to (3.6). First we note that if

$$F_{ij} \in \begin{cases} \bar{\mathfrak{n}} & \text{if } i < j, \\ \mathfrak{a} & \text{if } i = j, \\ \mathfrak{n} & \text{if } i > j, \end{cases} \qquad (3.8)$$

we have

$$
F_{ij}^k(\lambda_1,\dots,\lambda_k) \equiv \sum_{\mu=1}^{j} F_{i\mu}^{k-1}(\lambda_1,\dots,\lambda_{k-1})(F_{\mu j} - \lambda_k \delta_{\mu j}) \quad \mathrm{mod}\ U(\mathfrak{g})\mathfrak{n}
$$

$$
= F_{ij}^{k-1}(\lambda_1,\dots,\lambda_{k-1})(F_j - \lambda_k)
$$

$$
+ \sum_{\mu=1}^{j-1}\Big(F_{\mu j} F_{i\mu}^{k-1}(\lambda_1,\dots,\lambda_{k-1}) - [F_{\mu j}, F_{i\mu}^{k-1}(\lambda_1,\dots,\lambda_{k-1})]\Big)
$$

$$
= F_{ij}^{k-1}(\lambda_1,\dots,\lambda_{k-1})(F_j - \lambda_k) + \sum_{\mu=1}^{j-1}\Big(F_{\mu j} F_{i\mu}^{k-1}(\lambda_1,\dots,\lambda_{k-1})
$$

$$
- \sum_{\nu=1}^{i-1}\langle F_{\mu j}, E_{i\nu}\rangle F_{\nu\mu}^{k-1}(\lambda_1,\dots,\lambda_{k-1})
$$

$$
+ \sum_{\nu=\mu+1}^{N}\langle F_{\mu j}, E_{\nu\mu}\rangle F_{i\nu}^{k-1}(\lambda_1,\dots,\lambda_{k-1})\Big)
$$

$$\tag{3.9}$$

by Lemma 2.1.

The following is clear by the induction on k.

Remark 3.3. i) The highest homogeneous part of $\bar{F}^k(\lambda_1,\dots,\lambda_k)$ with the degree k is given by

$$
\bar{F}^k(\lambda_1,\dots,\lambda_k) \equiv \Big(\delta_{ij} F_{ii}^k\Big)_{\substack{1\le i\le N \\ 1\le j\le N}} \quad \mathrm{mod}\ M(N, U(\mathfrak{a})^{(k-1)}).
$$

ii) If $\mathfrak{g} = \mathfrak{gl}_n$ or $\mathfrak{o}_{2n+1}$ or $\mathfrak{sp}_n$ and π is the natural representation of $\mathfrak{g}$, it is clear that Trace F_π^k for $k = 1, 2,\dots, n$ or $k = 2, 4,\dots, 2n$ or $2, 4,\dots, 2n$, respectively, generate $Z(\mathfrak{g})$ as an algebra. In particular for any $D \in Z(\mathfrak{g})$ there uniquely exists a polynomial $f(x)$ with Trace $f(F) = D$. In the case when $\mathfrak{g} = \mathfrak{o}_{2n}$ we use both the natural representation π and the half-spin representation π' of $\mathfrak{g}$ and then Trace F_π^k for $k = 2, 4,\dots, 2(n-1)$ and Trace $F_{\pi'}^n$ generate $Z(\mathfrak{g})$.

iii) The Killing form of $\mathfrak{g}$ is a positive constant multiple of the restriction of the bilinear form (2.1) to $\mathfrak{g}$ if $\mathfrak{g}$ is simple.

Hereafter suppose that $\mathfrak{g} = \mathfrak{gl}_n$ or $\mathfrak{o}_{2n}$ or $\mathfrak{o}_{2n+1}$ or $\mathfrak{sp}_n$ and that F is given by (3.4). Then (3.4) means

$$
\langle F_{\mu j}, E_{i\nu}\rangle = \delta_{ij}\delta_{\mu\nu} - \epsilon_\mu \epsilon_j \delta_{\bar\mu i}\delta_{\bar j\nu} \quad \text{and} \quad \langle F_{\mu j}, E_{\nu\mu}\rangle = \delta_{j\nu} - \epsilon_\mu \epsilon_j \delta_{\bar\mu\nu}\delta_{\bar j\mu}
$$

and therefore it follows from (3.9) that

$$F_{ij}^k(\lambda_1, \ldots, \lambda_k) \equiv F_{ij}^{k-1}(\lambda_1, \ldots, \lambda_{k-1})(F_j - \lambda_k + j - 1)$$

$$+ \sum_{\mu=1}^{j-1} \Big(F_{\mu j} F_{i\mu}^{k-1}(\lambda_1, \ldots, \lambda_{k-1})$$

$$- \delta_{ij} F_{\mu\mu}^{k-1}(\lambda_1, \ldots, \lambda_{k-1}) + \epsilon_\mu \epsilon_j \delta_{\mu \bar{i}} F_{\bar{j}\bar{i}}^{k-1}(\lambda_1, \ldots, \lambda_{k-1})$$

$$- \epsilon_\mu \epsilon_j \delta_{\mu \bar{j}} F_{ij}^{k-1}(\lambda_1, \ldots, \lambda_{k-1}) \Big) \mod U(\mathfrak{g})\mathfrak{n}.$$

$$(3.10)$$

Since $[U(\mathfrak{g})\mathfrak{n}, U(\mathfrak{a})] \subset U(\mathfrak{g})\mathfrak{n}$ and since $F_{ij} \in \mathfrak{n}$ and $F_{\bar{j}\bar{i}} \in \mathfrak{n}$ for $i > j$, the equation (3.10) shows $F_{ij}^k(\lambda_1, \ldots, \lambda_k) \equiv 0 \mod U(\mathfrak{g})\mathfrak{n}$ for $i > j$ by the induction on k. Similarly we have $F_{ij}^k(\lambda_1, \ldots, \lambda_k) \in \bar{\mathfrak{n}} U(\mathfrak{g})$ if $i < j$. Hence by denoting

$$\omega_i = \begin{cases} 0 & \text{if } i \leq n, \\ \epsilon_i & \text{if } i > n, \end{cases}$$

$$\omega_j' = \begin{cases} 0 & \text{if } j \leq n \quad \text{or} \quad \bar{j} \geq j, \\ \epsilon_j & \text{if } j > n \quad \text{and} \quad \bar{j} < j, \end{cases}$$

$$(3.11)$$

we have

$$F_{ii}^k(\lambda_1, \ldots, \lambda_k) \qquad\qquad (3.12)$$

$$\equiv F_{ii}^{k-1}(\lambda_1, \ldots, \lambda_{k-1})(F_i - \lambda_k + i - 1 - \omega_i) + \omega_i F_{\bar{i}\bar{i}}^{k-1}(\lambda_1, \ldots, \lambda_{k-1})$$

$$- \sum_{\mu=1}^{i-1} F_{\mu\mu}^{k-1}(\lambda_1, \ldots, \lambda_{k-1}) \mod U(\mathfrak{g})\mathfrak{n},$$

$$F_{ii+1}^k(\lambda_1, \ldots, \lambda_k) \qquad\qquad (3.13)$$

$$\equiv F_{ii+1}^{k-1}(\lambda_1, \ldots, \lambda_{k-1})(F_{i+1} - \lambda_k + i - \omega_{i+1}') + \omega_i F_{\overline{i-1}\,\bar{i}}^{k-1}(\lambda_1, \ldots, \lambda_{k-1})$$

$$+ F_{ii+1} F_{ii}^{k-1}(\lambda_1, \ldots, \lambda_{k-1}) \mod U(\mathfrak{g})\mathfrak{n}.$$

Now we give the main result in this section:

Proposition 3.4. *Suppose that* $\mathfrak{g} = \mathfrak{gl}_n$ *or* $\mathfrak{o}_{2n}$ *or* $\mathfrak{o}_{2n+1}$ *or* $\mathfrak{sp}_n$ *and that* F *is given by* (3.4). *Let* $\Theta = \{n_1 < n_2 < \cdots < n_L = n\}$ *be a sequence of positive integers. Put* $n_\nu' = n_\nu - n_{\nu-1}$ *for* $\nu = 1, \ldots, L$ *and fix a positive number* k *and a sequence of complex numbers* $\mu_1, \mu_2, \ldots$. *Put* $n_0 = 0$ *and* $n_\nu = n$ *for* $\nu > L$ *and define*

$$\iota_\Theta(\nu) = p \quad \text{if} \quad n_{p-1} < \nu \leq n_p,$$

$$\tilde{J}(\mu)_i = U(\mathfrak{g})\mathfrak{n} + \sum_{\nu=1}^{i} U(\mathfrak{g})(F_\nu - \mu_{\iota\Theta(\nu)} + n_{\iota\Theta(\nu)-1}).$$

If $\mathfrak{g} = \mathfrak{gl}_n$, *we put* $H(\Theta, \mu_1, \ldots, \mu_L) = F^L(\mu_1, \ldots, \mu_L)$.
If $\mathfrak{g} = \mathfrak{sp}_n$ *or* $\mathfrak{o}_{2n}$, *we put*

$$H(\Theta, \mu_1, \ldots, \mu_L) = F^{2L}(\mu_1, \ldots, \mu_L,$$
$$-\mu_1 - n_1' + 2n + \delta, \ldots, -\mu_L - n_L' + 2n + \delta).$$

If $\mathfrak{g} = \mathfrak{o}_{2n+1}$, *we put*

$$H(\Theta, \mu_1, \ldots, \mu_L) = F^{2L+1}(\mu_1, \ldots, \mu_L, n,$$
$$-\mu_1 - n_1' + 2n, \ldots, -\mu_L - n_{L-1}' + 2n).$$

Moreover we define

$$\tilde{H}(\Theta, \mu_1, \ldots, \mu_{L-1}) = F^{2L-1}(\mu_1, \ldots, \mu_{L-1}, n_{L-1},$$
$$-\mu_1 - n_1' + 2n + \delta, \ldots, -\mu_{L-1} - n_{L-1}' + 2n + \delta).$$

Here

$$\delta = \begin{cases} 1 & \text{if } \mathfrak{g} = \mathfrak{sp}_n, \\ 0 & \text{if } \mathfrak{g} = \mathfrak{o}_{2n+1} \text{ or } \mathfrak{gl}_n, \\ -1 & \text{if } \mathfrak{g} = \mathfrak{o}_{2n}. \end{cases} \tag{3.14}$$

i) *The off-diagonal elements of* $F^k(\mu_1, \ldots, \mu_k)$ *satisfy*

$$F_{ij}^k(\mu_1, \ldots, \mu_k) \equiv 0 \mod U(\mathfrak{g})\mathfrak{n} \quad \text{if } i > j,$$
$$F_{ij}^k(\mu_1, \ldots, \mu_k) \equiv 0 \mod \bar{\mathfrak{n}}U(\mathfrak{g}) \quad \text{if } i < j.$$

ii) *If* $i \leq n$, *then*

$$F_{ii}^k(\mu_1, \ldots, \mu_k)$$
$$\equiv \begin{cases} 0 & \mod \tilde{J}(\mu)_i \quad \text{if } i \leq n_k, \\ \prod_{\nu=1}^{k}(\mu_{k+1} - \mu_\nu - n_\nu') & \mod \tilde{J}(\mu)_i \quad \text{if } n_k < i \leq n_{k+1}. \end{cases}$$

iii) *If* $i < n$, *then*

$$F_{ii+1}^k(\mu_1, \ldots, \mu_k)$$
$$\equiv \Big(\prod_{\nu=1}^{\ell-1}(\mu_\ell - \mu_\nu - n_\nu' - n_{\ell-1} + i) \prod_{\nu=\ell+1}^{k}(\mu_\ell - \mu_\nu - n_{\ell-1} + i) \Big) F_{ii+1}$$
$$\mod \tilde{J}(\mu)_i \quad \text{if } n_{\ell-1} < i < n_\ell \text{ and } k \geq \ell.$$

iv) *We have*

$$H_{ii}(\Theta, \mu_1, \ldots, \mu_L) \equiv 0 \mod \tilde{J}(\mu)_n \quad for \quad i = 1, \ldots, N.$$

In particular, if $\mu_L = n_{L-1}$ and $\mathfrak{g} = \mathfrak{o}_{2n}$ or $\mathfrak{o}_{2n+1}$ or $\mathfrak{sp}_n$, then

$$\tilde{H}_{ii}(\Theta, \mu_1, \ldots, \mu_{L-1}) \equiv 0 \mod \tilde{J}(\mu)_n \quad for \quad i = 1, \ldots, N$$

and

$$\tilde{H}_{nn+1}(\Theta, \mu_1, \ldots, \mu_{L-1})$$
$$\equiv (-1)^{L-1} \Big(\prod_{\nu=1}^{L-1} (\mu_\nu + n'_\nu - n)(\mu_\nu + n'_\nu - n - \delta) \Big) F_{nn+1}$$
$$\mod U(\mathfrak{g}) \tilde{J}(\mu)_n.$$

Proof. Put $F_{ij}^k(\mu) = F_{ij}^k(\mu_1, \ldots, \mu_k)$ for simplicity. If $i < n$, it follows from (3.12) that

$$F_{i+1i+1}^k(\mu) - F_{ii}^k(\mu) \equiv F_{i+1i+1}^{k-1}(\mu)(F_{i+1} - \mu_k + i)$$
$$- F_{ii}^{k-1}(\mu)(F_i - \mu_k + i) \mod U(\mathfrak{g})\mathfrak{n}$$

and therefore by the induction on k we have

$$F_{ii}^k(\mu) \equiv F_{i+1i+1}^k(\mu) \mod U(\mathfrak{g})\mathfrak{n} + U(\mathfrak{g})(F_{i+1} - F_i). \tag{3.15}$$

Here we note that $F_{\nu+1} - F_\nu \in \tilde{J}(\mu)_{n_\ell}$ if $n_{\ell-1} < \nu < n_\ell$. Hence we have

$$F_{ii}^k(\mu) + \tilde{J}(\mu)_{n_\ell} = F_{n_\ell n_\ell}^k(\mu) + \tilde{J}(\mu)_{n_\ell} \quad for \quad n_{\ell-1} < i \le n_\ell \quad and \quad 1 \le \ell \le L.$$

Put $s_\nu = n_\nu - n_{\nu-1}$ and introduce polynomials $f(k, \ell)$ of $(\mu_1, \ldots, \mu_L, s_1, \ldots, s_L)$ with $\ell \le L$ so that

$$F_{n_\ell n_\ell}^k(\mu_1, \ldots, \mu_k) \equiv f(k, \ell) \mod \tilde{J}(\mu)_{n_\ell}. \tag{3.16}$$

Similarly for i with $n_{\ell-1} < i < n_\ell$, we put $t = \mu_\ell - n_{\ell-1} + i$ and define polynomials $g(k, \ell)$ of $(t, \mu_1, \ldots, \mu_L, s_1, \ldots, s_L)$ so that

$$F_{ii+1}^k(\mu_1, \ldots, \mu_k) \equiv g(k, \ell) E_{ii+1} \mod \tilde{J}(\mu)_i. \tag{3.17}$$

Then we have

$$f(k, \ell) = \begin{cases} 1 & \text{if } k = 0, \\ f(k-1, \ell)(\mu_\ell - \mu_k) - \sum_{\nu=1}^{\ell-1} s_\nu f(k-1, \nu) & \text{if } k \ge 1, \end{cases} \tag{3.18}$$

$$g(k, \ell) = \begin{cases} 1 & \text{if } k = 1, \\ g(k-1, \ell)(t - \mu_k) + f(k-1, \ell) & \text{if } k > 1. \end{cases}$$

We will first prove $f(k, \ell) = 0$ if $k \geq \ell$ by the induction on ℓ. Putting $\ell = 1$ in (3.18), we have $f(k, 1) = f(k - 1, 1)(\mu_1 - \mu_k)$ and $f(1, 1) = 0$ and therefore $f(k, 1) = 0$ for $k \geq 1$. Then if $k \geq \ell + 1$, we have $f(k, \ell + 1) = f(k-1, \ell+1)(\mu_{\ell+1} - \mu_k) - \sum_{\nu=1}^{\ell} s_\nu f(k-1, \nu) = f(k-1, \ell+1)(\mu_{\ell+1} - \mu_k)$ by the hypothesis of the induction. Hence we have $f(k, \ell+1) = 0$ for $k \geq \ell+1$ by the induction on k.

Putting $\mu_\ell = \mu_{\ell-1} + s_{\ell-1}$ in (3.18), we have $f(k, \ell) - f(k, \ell - 1) = f(k-1, \ell) - f(k-1, \ell) = \cdots = 0$ and therefore $f(\ell-1, \ell)|_{\mu_\ell = \mu_{\ell-1} + s_{\ell-1}} = 0$. Hence there exist polynomials $h(\ell)$ with $f(\ell-1, \ell) = h(\ell)(\mu_\ell - \mu_{\ell-1} - s_{\ell-1})$. Then (3.18) shows

$$h(\ell)(\mu_\ell - \mu_{\ell-1} - s_{\ell-1}) = f(\ell - 2, \ell)(\mu_\ell - \mu_{\ell-1}) - s_{\ell-1}f(\ell - 2, \ell - 1).$$

It follows from (3.18) that $f(k, \ell)$ is a polynomial of degree at most 1 with respect to $s_{\ell-1}$ because $f(k, \nu)$ does not contain $s_{\ell-1}$ for $\nu < \ell$. Hence $h(\ell) = f(\ell-2, \ell)|_{s_{\ell-1}=0}$. Moreover by putting $s_{\ell-1} = 0$ in (3.18), it is clear that $f(\ell - 2, \ell)|_{s_{\ell-1}=0}$ does not contain $\mu_{\ell-1}$. Hence $h(\ell) = f(\ell - 2, \ell - 1)|_{\mu_{\ell-1} \mapsto \mu_\ell}$ and we get

$$f(\ell - 1, \ell) = \prod_{\nu=1}^{\ell-1} (\mu_\ell - \mu_\nu - s_\nu) \tag{3.19}$$

by the induction on ℓ. Thus we have ii).

Now we put

$$f(\ell - 1, \ell) = \sum_{\nu=0}^{\ell-1} c(\nu, \ell)(\mu_\ell - \mu_{\nu+1})(\mu_\ell - \mu_{\nu+2}) \cdots (\mu_\ell - \mu_{\ell-1}) \tag{3.20}$$

with homogeneous polynomials $c(\nu, \ell)$ of $(\mu_1, \ldots, \mu_{\ell-1}, s_1, \ldots, s_{\ell-1})$ with degree ν. Here $c(\nu, \ell)$ does not contain μ_ℓ. Then by the induction on $k = \ell - 1, \ell - 2, \ldots, 0$, (3.18) shows

$$f(k, \ell) = \sum_{\nu=0}^{k} c(\nu, \ell)(\mu_\ell - \mu_{\nu+1})(\mu_\ell - \mu_{\nu+2}) \cdots (\mu_\ell - \mu_k),$$

$$-\sum_{\nu=1}^{\ell-1} s_\nu f(k - 1, \nu) = c(k, \ell) \tag{3.21}$$

because $\sum_{\nu=1}^{\ell-1} s_\nu f(k - 1, \nu)$ does not contain μ_ℓ. We will show

$$g(\ell, \ell) = \sum_{k=0}^{\ell-1} (t - \mu_\ell)(t - \mu_{\ell-1}) \cdots (t - \mu_{k+2}) f(k, \ell) \tag{3.22}$$

$$= \prod_{\nu=1}^{\ell-1}(t - \mu_\nu - s_\nu). \tag{3.23}$$

Note that (3.22) is a direct consequence of (3.18). Denoting

$$g_k(\ell) = \sum_{\nu=0}^{\ell-1} c(\nu, \ell)(\mu_\ell - \mu_{\nu+1}) \cdots (\mu_\ell - \mu_{k-1})(\mu_\ell - \mu_k)(t - \mu_{k+1}) \cdots (t - \mu_{\ell-1})$$

for $k = 0, \dots, \ell - 1$, we have

$$g_{k-1}(\ell) - g_k(\ell)$$

$$= \sum_{\nu=0}^{k-1} c(\nu, \ell)(\mu_\ell - \mu_{\nu+1}) \cdots (\mu_\ell - \mu_{k-1})(t - \mu_\ell)(t - \mu_{k+1}) \cdots (t - \mu_{\ell-1})$$

$$= (t - \mu_\ell)(t - \mu_{\ell-1}) \cdots (t - \mu_{k+1}) f(k - 1, \ell)$$

from (3.21) and therefore (3.22) shows

$$g(\ell, \ell) = g_{\ell-1}(\ell) + \sum_{k=1}^{\ell-1}(g_{k-1}(\ell) - g_k(\ell)) = g_0(\ell) = f(\ell - 1, \ell)|_{\mu_\ell \mapsto t},$$

which implies (3.23). Since $f(k, \ell) = 0$ for $k \geq \ell$, (3.18) shows

$$g(k, \ell) = \prod_{\nu=1}^{\ell-1}(t - \mu_\nu - s_\nu) \prod_{\nu=\ell+1}^{k}(t - \mu_\nu) \quad \text{if } k \geq \ell, \tag{3.24}$$

from which iii) follows.

In general we have proved the following lemma.

Lemma 3.5. *The functions $f(k, \ell)$ and $g(k, \ell)$ of $\mu_1, \mu_2, \dots, s_1, s_2, \dots$ and t which are recursively defined by (3.18) satisfy (3.19), (3.24) and $f(k, \ell) = 0$ for $k \geq \ell \geq 1$.*

Now suppose $\mathfrak{g} = \mathfrak{sp}_n$ or $\mathfrak{o}_{2n}$. Then

$$F_{n+1n+1}^k(\mu) \equiv F_{n+1n+1}^{k-1}(\mu)(F_{n+1} - \mu_k) + \sum_{\nu=1}^{n}(F_{n+1n+1}^{k-1}(\mu) - F_{\nu\nu}^{n-1}(\mu))$$

$$+ \delta(F_{n+1n+1}^{k-1}(\mu) - F_{nn}^{k-1}(\mu)) \equiv 0 \mod U(\mathfrak{g})\mathfrak{n}. \tag{3.25}$$

Hence

$$F_{n+1n+1}^k(\mu) - F_{nn}^k(\mu)$$

$$\equiv F_{n+1n+1}^{k-1}(\mu)(F_{n+1} - \mu_k + n + \delta) - F_{nn}^{k-1}(\mu)(F_n - \mu_k + n + \delta)$$

$$\equiv 0 \mod U(\mathfrak{g})\mathfrak{n} + U(\mathfrak{g})(F_{n+1} - F_n)$$

by the induction on k and

$$
F_{n+1n+1}^k(\mu) \equiv 0 \quad \mathrm{mod} \sum_{\nu=1}^n U(\mathfrak{g})F_{\nu\nu}^{k-1}(\mu) + U(\mathfrak{g})\mathfrak{n} \tag{3.26}
$$
$$
+ U(\mathfrak{g})(F_{n+1} - \mu_k + n + \delta).
$$

Since $F_{n+1} = -F_n$, we have from (3.25)

$$
F_{n+1n+1}^L(\mu_1, \ldots, \mu_{L-1}, n_{L-1}) \equiv 0 \quad \mathrm{mod}\ \tilde{J}(\mu)_{n_{L-1}} + \sum_{\nu=n_{L-1}+1}^n U(\mathfrak{g})F_\nu
$$

in the case $\mu_L = n_{L-1}$ and from (3.26) with $-(\mu_L - n_{L-1}) - \mu_{L+1} + n + \delta = 0$

$$
F_{n+1n+1}^{L+1}(\mu_1, \ldots, \mu_L, -\mu_L + n_{L-1} + n + \delta) \equiv 0 \quad \mathrm{mod}\ \tilde{J}(\mu)_n.
$$

Suppose $i < n$. Then

$$
F_{ii}^k(\mu) \equiv F_{ii}^{k-1}(\mu)(F_{\bar{i}} - \mu_k) + \sum_{\nu=1}^{\bar{i}-1}\big(F_{\bar{i}\bar{i}}^{k-1}(\mu) - F_{\nu\nu}^{k-1}(\mu)\big)
$$
$$
+ \delta(F_{\bar{i}\bar{i}}^{k-1}(\mu) - F_{ii}^{k-1}(\mu)) \quad \mathrm{mod}\ U(\mathfrak{g})\mathfrak{n}
$$

and therefore

$$
\begin{aligned}
F_{i+1\overline{i+1}}^k(\mu) - F_{ii}^k(\mu) &\equiv F_{i+1\overline{i+1}}^{k-1}(\mu)(F_{\bar{i}+1} - \mu_k + \bar{i} + \delta) \\
&\quad - F_{ii}^{k-1}(\mu)(F_{\bar{i}} - \mu_k + \bar{i} + \delta) \\
&\quad + \delta(F_{i-1\overline{i-1}}^{k-1}(\mu) - F_{ii}^{k-1}(\mu)) \quad \mathrm{mod}\ U(\mathfrak{g}) \\
&\equiv 0 \quad \mathrm{mod}\ U(\mathfrak{g})\mathfrak{n} + U(\mathfrak{g})(F_i - F_{i-1}) \\
&\quad + U(\mathfrak{g})(F_{\overline{i-1}} - F_{\bar{i}}),
\end{aligned} \tag{3.27}
$$
$$
F_{\bar{n}_p\bar{n}_p}^k(\mu) \equiv 0 \quad \mathrm{mod} \sum_{\nu=1}^{\bar{n}_p-1} U(\mathfrak{g})F_{\nu\nu}^{k-1}(\mu) + U(\mathfrak{g})\mathfrak{n}
$$
$$
+ U(\mathfrak{g})(F_{\bar{n}_p} - \mu_k + \bar{n}_p - 1 + \delta).
$$

Note that $F_{\bar{n}_p} - \mu_k + \bar{n}_p - 1 + \delta = (F_{\bar{n}_p} + \mu_p - n_{p-1}) - \mu_k - \mu_p + n_{p-1} - n_p + 2n + \delta$. Since $F_{\bar{\nu}} = -F_\nu$, we have

$$
F_{i+1\overline{i+1}}^k(\mu) \equiv F_{ii}^k(\mu) \quad \mathrm{mod}\ \tilde{J}(\mu)_n \quad \text{for}\quad \bar{n}_p \le \bar{i} < \bar{n}_{p-1} \tag{3.28}
$$

and hence by the induction on $p = L, L-1, \ldots, 1$, we have

$$
F_{\bar{n}_p\bar{n}_p}^{2L+1-p}(\mu_1, \ldots, \mu_L, -\mu_L + n_{L-1} - n_L + 2n + \delta,
$$
$$
\ldots, -\mu_p + n_{p-1} - n_p + 2n + \delta) \equiv 0 \quad \mathrm{mod}\ \tilde{J}(\mu)_n \tag{3.29}
$$

and if $\mu_L = n_{L-1}$, then

$$F_{\tilde{n}_p \tilde{n}_p}^{2L-p}(\mu_1, \ldots, \mu_{L-1}, n_{L-1}, \mu_{L-1} + n_{L-2} - n_{L-1} + 2n + \delta,$$
$$\ldots, -\mu_p + n_{p-1} - n_p + 2n + \delta) \equiv 0 \mod \tilde{J}(\mu)_n. \quad (3.30)$$

Suppose $\mu_L = n_{L-1}$ and $\mathfrak{g} = \mathfrak{sp}_n$. Then from (3.13) we have

$$F_{nn+1}^k(\mu) \equiv F_{nn+1}^{k-1}(\mu)(F_{n+1n+1} - \mu_k + n + \delta)$$
$$+ F_{nn+1}F_{nn}^{k-1}(\mu) \mod U(\mathfrak{g})\mathfrak{n}$$

$$\equiv F_{nn+1}^L(\mu) \prod_{\nu=L+1}^{k} (-\mu_\nu + n + \delta) \mod \tilde{J}(\mu)_n \quad \text{if} \quad k \geq L.$$
$$(3.31)$$

It follows from Lemma 3.5 with $t = n + 1$ that

$$\bar{H}_{nn+1}(\Theta, \mu) \equiv F_{nn+1} \prod_{\nu=1}^{L-1} (-\mu_\nu + n_{\nu-1} - n_\nu + n + \delta) \prod_{\nu=1}^{L-1} (\mu_\nu - n_{\nu-1} + n_\nu - n)$$
$$\mod \tilde{J}(\mu)_n.$$

Thus we have proved iv).

Lastly suppose $\mathfrak{g} = \mathfrak{o}_{2n+1}$. Note that $F_{n+1} = 0$ and $F_{n+2} = -F_n$. Then

$$F_{n+1n+1}^k(\mu) \equiv F_{n+1n+1}^{k-1}(F_{n+1} - \mu_k)$$
$$+ \sum_{\nu=1}^{n} (F_{n+1n+1}^{k-1}(\mu) - F_{\nu\nu}^{k-1}(\mu)) \mod U(\mathfrak{g})\mathfrak{n}$$

$$\equiv 0 \mod \sum_{\nu=1}^{n} U(\mathfrak{g})F_{\nu\nu}^{k-1}(\mu) + U(\mathfrak{g})(-\mu_k + n),$$

$$F_{n+2n+2}^k(\mu) \equiv F_{n+2n+2}^{k-1}(F_{n+2} - \mu_k) + \sum_{\nu=1}^{n+1} (F_{n+2n+2}^{k-1}(\mu) - F_{\nu\nu}^{k-1}(\mu))$$
$$- (F_{n+2n+2}^{k-1}(\mu) - F_{nn}^{k-1}(\mu)) \mod U(\mathfrak{g})\mathfrak{n}$$

$$\equiv 0 \mod \sum_{\nu=1}^{n+1} U(\mathfrak{g})F_{\nu\nu}^{k-1}(\mu) + U(\mathfrak{g})(-F_n - \mu_k + n)$$

and

$$F_{n+1n+1}^{L+1}(\mu_1, \ldots, \mu_L, n) \equiv 0 \mod \tilde{J}(\mu)_n,$$
$$F_{n+2n+2}^{L+2}(\mu_1, \ldots, \mu_L, n, -\mu_L + n_{L-1} + n) \equiv 0 \mod \tilde{J}(\mu)_n.$$

Since

$$F^k_{n+1n+1}(\mu) - F^k_{nn}(\mu) \equiv F^{k-1}_{n+1n+1}(F_{n+1} - \mu_k + n) - F^{k-1}_{nn}(F_n - \mu_k + n)$$
$$\mathrm{mod}\ U(\mathfrak{g})\mathfrak{n}$$

and

$$F^k_{n+2n+2}(\mu) - F^k_{n+1n+1}(\mu) \equiv F^{k-1}_{n+2n+2}(F_{n+2} - \mu_k + n)$$
$$- F^{k-1}_{n+1n+1}(F_{n+1} - \mu_k + n)$$
$$- (F^{k-1}_{n+1n+1}(\mu) - F^{k-1}_{nn}(\mu)) \quad \mathrm{mod}\ U(\mathfrak{g})\mathfrak{n},$$

we have

$$F^k_{n+2n+2}(\mu) \equiv F^k_{n+1n+1}(\mu) \equiv F^k_{nn}(\mu) \quad \mathrm{mod}\ U(\mathfrak{g})\mathfrak{n} + U(\mathfrak{g})F_n$$

and

$$F^L_{n+1n+1}(\mu) \equiv F^L_{n+2n+2}(\mu) \equiv 0 \quad \mathrm{mod}\ \tilde{J}(\mu)_n \quad \text{if}\quad \mu_L = n_{L-1}.$$

Note that (3.27) is valid if $\bar{i} < n$. But since $F_{\bar{n}_p} - \mu_k + \bar{n}_p - 1 + \delta = -\mu_k - (F_{n_p} - \mu_p + n_{p-1}) - \mu_p + n_{p-1} - n_p + 2n$, we have

$$F^{2L+2-p}_{\bar{n}_p \bar{n}_p}(\mu_1, \ldots, \mu_L, n, -\mu_L + n_{L-1} - n_L + 2n, \ldots, -\mu_p + n_{p-1} - n_p + 2n) \equiv 0$$
$$\mathrm{mod}\ \tilde{J}(\mu)_n$$

for $p = L, L-1, \ldots, 1$. Similarly we have (3.30) with $\delta = 0$ if $\mu_L = n_{L-1}$. Moreover (3.31) is valid with $\delta = 0$ and we have iv) as in the case of $\mathfrak{g} = \mathfrak{sp}_n$. $\qquad\square$

4. Generalized Verma Modules

In this section we define by (4.14) a two-sided ideal of $U^\epsilon(\mathfrak{g})$ associated to every generalized Verma module of the scalar type for the classical Lie algebra $\mathfrak{g}$. We have Theorem 4.4 which shows that the ideal describes the gap between the generalized Verma module and the usual Verma module and then Corollary 4.6 says that the ideal coincides with the annihilator of the generalized Verma module. In the classical limit, an explicit generator system of every semisimple co-adjoint orbit of $\mathfrak{g}$ is given in Theorem 4.11.

The ideal is constructed from a polynomial of a matrix with elements in $\mathfrak{g}$ and the polynomial is proved to be a minimal polynomial in a certain sense (cf. Proposition 4.16 and 4.18). Lastly in this section Theorem 4.19 gives a description of the image of the Harish-Chandra homomorphism of any polynomial function of the matrix $(E_{ij}) \in M(n, \mathfrak{g})$.

Retain the notation in the previous section. Let $\Theta = \{(0 <)n_1 < n_2 < \cdots < n_L(= n)\}$ be the sequence of strictly increasing positive integers ending at n. Put

$$H_\Theta = \sum_{k=1}^{L} \sum_{i=1}^{n_k} F_i \quad \text{and} \quad H_{\bar\Theta} = \sum_{k=1}^{L-1} \sum_{i=1}^{n_k} F_i.$$

Recall that $F_i = F_{ii}$, $F = (F_{ij}) \in M(N, \mathfrak{g})$, $\mathfrak{n} = \sum_{i>j} \mathbb{C}F_{ij}$, $\mathfrak{a} = \sum_i \mathbb{C}F_i$, $\bar{\mathfrak{n}} = \sum_{i<j} \mathbb{C}F_{ij}$ and $\mathfrak{g} = \mathfrak{n} \oplus \mathfrak{a} \oplus \bar{\mathfrak{n}}$. Note that $F_{ij} = E_{ij}$ in the case $\mathfrak{g} = \mathfrak{gl}_n$ and $F_{ij} = E_{ij} + \sigma_\mathfrak{g}(E_{ij})$ in the case $\mathfrak{g} = \mathfrak{o}_{2n+1}$, $\mathfrak{sp}_n$ or $\mathfrak{o}_{2n}$. Here $\sigma_\mathfrak{g}$ is the involution of $\mathfrak{gl}_N$ to define $\mathfrak{g}$ in (3.1) so that $\mathfrak{g}$ is the subalgebra of $\mathfrak{gl}_N$ fixed by $\sigma_\mathfrak{g}$. Let G be the analytic subgroup of $GL(N, \mathbb{C})$ with the Lie algebra $\mathfrak{g}$. Namely $G = GL(n, \mathbb{C})$, $SO(2n + 1, \mathbb{C})$, $Sp(n, \mathbb{C})$ or $SO(2n, \mathbb{C})$.

Define

$$\begin{cases} \mathfrak{m}_\Theta = \{X \in \mathfrak{g};\ \mathrm{ad}(H_\Theta)X = 0\}, \\ \mathfrak{n}_\Theta = \{X \in \mathfrak{n};\ \langle X, \mathfrak{m}_\Theta \rangle = 0\},\ \bar{\mathfrak{n}}_\Theta = \{X \in \bar{\mathfrak{n}};\ \langle X, \mathfrak{m}_\Theta \rangle = 0\}, \qquad (4.1) \\ \mathfrak{p}_\Theta = \mathfrak{m}_\Theta + \mathfrak{n}_\Theta. \end{cases}$$

We similarly define $m_{\bar\Theta}$, $\mathfrak{n}_{\bar\Theta}$, $\bar{\mathfrak{n}}_{\bar\Theta}$ and $\mathfrak{p}_{\bar\Theta}$ replacing Θ by $\bar\Theta$ in the above definition. Then $\mathfrak{n} = \mathfrak{n}_{\{1,2,\dots,n\}}$, $\bar{\mathfrak{n}} = \bar{\mathfrak{n}}_{\{1,2,\dots,n\}}$, $\mathfrak{a} = \mathfrak{a}_{\{1,2,\dots,n\}}$ and $\mathfrak{p}_\Theta$ and $\mathfrak{p}_{\bar\Theta}$ are parabolic subalgebras of $\mathfrak{g}$ containing the Borel subalgebra $\mathfrak{b} = \mathfrak{n} + \mathfrak{a}$.

Let $\{e_1, \dots, e_n\}$ be the dual bases of $\{F_1, \dots, F_n\}$. Then the fundamental system $\Psi(\mathfrak{a})$ for the pair $(\mathfrak{n}, \mathfrak{a})$ is

$$\Psi(\mathfrak{a}) = \begin{cases} \{e_2 - e_1, e_3 - e_2, \dots, e_n - e_{n-1}\} & \text{if } \mathfrak{g} = \mathfrak{gl}_n, \\ \{e_2 - e_1, e_3 - e_2, \dots, e_n - e_{n-1}, -e_n\} & \text{if } \mathfrak{g} = \mathfrak{o}_{2n+1}, \\ \{e_2 - e_1, e_3 - e_2, \dots, e_n - e_{n-1}, -2e_n\} & \text{if } \mathfrak{g} = \mathfrak{sp}_n, \\ \{e_2 - e_1, e_3 - e_2, \dots, e_n - e_{n-1}, -e_n - e_{n-1}\} & \text{if } \mathfrak{g} = \mathfrak{o}_{2n}. \end{cases}$$
$$(4.2)$$

We put $\alpha_j = e_{j+1} - e_j$ for $j = 1, \dots, n-1$ and $\alpha_n = -e_n$ or $-2e_n$ or $-e_n - e_{n-1}$ if $\mathfrak{g} = \mathfrak{o}_{2n+1}$ or $\mathfrak{sp}_n$ or $\mathfrak{o}_{2n}$, respectively. Then the fundamental system for $(\mathfrak{m}_\Theta \cap \mathfrak{n}, \mathfrak{a})$ is $\Psi(\mathfrak{a}) \setminus \{\alpha_{n_1}, \dots, \alpha_{n_{L-1}}\}$ and that for $(\mathfrak{m}_{\bar\Theta} \cap \mathfrak{n}, \mathfrak{a})$ is

$$\begin{cases} \Psi(\mathfrak{a}) \setminus \{\alpha_{n_1}, \dots, \alpha_{n_{L-1}}, \alpha_n\} & \text{if } \mathfrak{g} = \mathfrak{o}_{2n+1} \text{ or } \mathfrak{sp}_n, \\ \Psi(\mathfrak{a}) \setminus \{\alpha_{n_1}, \dots, \alpha_{n_{L-1}}\} & \text{if } \mathfrak{g} = \mathfrak{o}_{2n} \text{ and } n_{L-1} \neq n - 1, \\ \Psi(\mathfrak{a}) \setminus \{\alpha_{n_1}, \dots, \alpha_{n_{L-1}}, \alpha_n\} & \text{if } \mathfrak{g} = \mathfrak{o}_{2n} \text{ and } n_{L-1} = n - 1. \end{cases}$$

The Dynkin diagram of the root system of $\mathfrak{g}$ is as follows:

$$\mathfrak{gl}_n \quad \underset{\alpha_1}{\circ}\!-\!\underset{\alpha_2}{\circ}\!-\cdots-\!\underset{\alpha_{n-2}}{\circ}\!-\!\underset{\alpha_{n-1}}{\circ} \qquad\qquad \mathfrak{o}_{2n+1} \quad \underset{\alpha_1}{\circ}\!-\!\underset{\alpha_2}{\circ}\!-\cdots-\!\underset{\alpha_{n-1}}{\circ}\!\Rightarrow\!\underset{\alpha_n}{\circ}$$

$$\mathfrak{sp}_n \quad \underset{\alpha_1}{\circ}\!-\!\underset{\alpha_2}{\circ}\!-\cdots-\!\underset{\alpha_{n-1}}{\circ}\!\Leftarrow\!\underset{\alpha_n}{\circ} \qquad\qquad \mathfrak{o}_{2n}$$

$$(4.3)$$

Fix $\lambda = (\lambda_1, \ldots, \lambda_L) \in \mathbb{C}^L$ and define a character λ_Θ of $\mathfrak{p}_\Theta$

$$\lambda_\Theta\left(X + \sum_{i=1}^{n} C_i F_i\right) = \sum_{i=1}^{n} C_i \lambda_{\iota_\Theta(i)} \quad \text{for } X \in \mathfrak{n}_\Theta + [\mathfrak{m}_\Theta, \mathfrak{m}_\Theta]. \qquad (4.4)$$

We similarly define a character $\lambda_{\bar\Theta}$ of $\mathfrak{p}_{\bar\Theta}$ if $\lambda_L = 0$.

We introduce the *homogenized universal enveloping algebra*

$$U^\epsilon(\mathfrak{g}) = \left(\sum_{k=0}^{\infty} \otimes^k \mathfrak{g}\right) / \langle X \otimes Y - Y \otimes X - \epsilon[X, Y]; \ X, \ Y \in \mathfrak{g}\rangle \qquad (4.5)$$

of $\mathfrak{g}$ as in [O4]. Here ϵ is a parameter of a complex number or a central element of $U^\epsilon(\mathfrak{g})$. Let $U^\epsilon(\mathfrak{g})^{(m)}$ be the image of $\sum_{k=0}^{m} \otimes^k \mathfrak{g}$ in $U^\epsilon(\mathfrak{g})$ and let $Z^\epsilon(\mathfrak{g})$ be the subalgebra of G-invariants of $U^\epsilon(\mathfrak{g})$. Fix generators $\Delta_1, \ldots, \Delta_n$ of $Z^\epsilon(\mathfrak{g})$ so that

$$\begin{cases} \Delta_j \in U^\epsilon(\mathfrak{g})^{(j)} \quad (1 \le j \le n) & \text{if } \mathfrak{g} = \mathfrak{gl}_n, \\ \Delta_j \in U^\epsilon(\mathfrak{g})^{(2j)} \quad (1 \le j \le n) & \text{if } \mathfrak{g} = \mathfrak{o}_{2n+1} \text{ or } \mathfrak{sp}_n, \\ \Delta_j \in U^\epsilon(\mathfrak{g})^{(2j)} \quad (1 \le j < n), \ \ \Delta_n \in U^\epsilon(\mathfrak{g})^{(n)} & \text{if } \mathfrak{g} = \mathfrak{o}_{2n}. \end{cases}$$

$$(4.6)$$

If $\mathfrak{g} = \mathfrak{o}_{2n}$, we assume that Δ_n changes into $-\Delta_n$ by the outer automorphism of $\mathfrak{o}_{2n}$ which maps $(F_1, \ldots, F_{n-1}, F_n)$ to $(F_1, \ldots, F_{n-1}, -F_n)$. Moreover put

$$\begin{cases} J_\Theta^\epsilon(\lambda) = \sum_{X \in \mathfrak{p}_\Theta} U^\epsilon(\mathfrak{g})(X - \lambda_\Theta(X)), & M_\Theta^\epsilon(\lambda) = U^\epsilon(\mathfrak{g})/J_\Theta^\epsilon(\lambda), \\ J_{\bar\Theta}^\epsilon(\lambda) = \sum_{X \in \mathfrak{p}_{\bar\Theta}} U^\epsilon(\mathfrak{g})(X - \lambda_{\bar\Theta}(X)), & M_{\bar\Theta}^\epsilon(\lambda) = U^\epsilon(\mathfrak{g})/J_{\bar\Theta}^\epsilon(\lambda), \\ J^\epsilon(\lambda_\Theta) = \sum_{X \in \mathfrak{b}} U^\epsilon(\mathfrak{g})(X - \lambda_\Theta(X)), & M^\epsilon(\lambda_\Theta) = U^\epsilon(\mathfrak{g})/J^\epsilon(\lambda_\Theta). \end{cases}$$

$$(4.7)$$

For a $U^\epsilon(\mathfrak{g})$-module M the annihilator of M is denoted by $\mathrm{Ann}(M)$ and put $\mathrm{Ann}_G(M) = \bigcap_{g \in G} \mathrm{Ad}(g)\,\mathrm{Ann}(M)$. Note that $\mathrm{Ann}_G(M) = \mathrm{Ann}(M)$ if $\epsilon \ne 0$. When $\epsilon = 1$, $U^\epsilon(\mathfrak{g})$ is the universal enveloping algebra $U(\mathfrak{g})$ of $\mathfrak{g}$

and we will sometimes omit the superfix ϵ for $J_\Theta^\epsilon(\lambda)$ and $M_\Theta^\epsilon(\lambda)$ etc. Then $M_\Theta(\lambda)$ and $M_{\bar\Theta}(\lambda)$ are generalized Verma modules which are quotients of the Verma module $M(\lambda_\Theta)$.

Remark 4.1. i) Suppose $\mathfrak{g} = \mathfrak{o}_{2n}$. Then we have not considered the parabolic subalgebra $\mathfrak{p}$ such that the fundamental system for $(\mathfrak{m}_\mathfrak{p}, \mathfrak{a})$ contains α_{n-1} and does not contain α_n. But this is reduced to the case when it contains α_n and does not contain α_{n-1} by the outer automorphism of $\mathfrak{o}_{2n}$.

ii) Considering the above remark, the parabolic subalgebra $\mathfrak{p}$ containing the Borel subalgebra $\mathfrak{b}$ corresponds to $\mathfrak{p}_\Theta$ or $\mathfrak{p}_{\bar\Theta}$ and therefore we will sometimes use the notation $\mathfrak{m}_\mathfrak{p}$, $\mathfrak{n}_\mathfrak{p}$, $\bar{\mathfrak{n}}_\mathfrak{p}$, $\lambda_\mathfrak{p}$, $J_\mathfrak{p}^\epsilon(\lambda)$, $M_\mathfrak{p}^\epsilon(\lambda)$ and $M^\epsilon(\lambda_\mathfrak{p})$ for $\mathfrak{m}_{\Theta'}$, $\mathfrak{n}_{\Theta'}$, $\bar{\mathfrak{n}}_{\Theta'}$, λ_Θ, $J_{\Theta'}^\epsilon(\lambda)$, $M_{\Theta'}^\epsilon(\lambda)$ and $M^\epsilon(\lambda_\Theta)$, respectively, by this correspondence. Note that $\Theta' = \bar\Theta$ means $\lambda_n = 0$.

Let $\rho \in \mathfrak{a}^*$ with $\rho(H) = \frac{1}{2}\,\mathrm{Trace}(\mathrm{ad}(H))|_\mathfrak{n}$ for $H \in \mathfrak{a}$. Then with δ in (3.14)

$$
\rho = \begin{cases}
\displaystyle\sum_{\nu=1}^n \left(\nu - \tfrac{n+1}{2}\right)e_\nu & \text{if } \mathfrak{g} = \mathfrak{gl}_n, \\[2mm]
\displaystyle\sum_{\nu=1}^n \left(\nu - n - \tfrac{1}{2}\right)e_\nu = \sum_{\nu=1}^n \left(\nu - n - \tfrac{\delta+1}{2}\right)e_\nu & \text{if } \mathfrak{g} = \mathfrak{o}_{2n+1}, \\[2mm]
\displaystyle\sum_{\nu=1}^n \left(\nu - n - 1\right)e_\nu = \sum_{\nu=1}^n \left(\nu - n - \tfrac{\delta+1}{2}\right)e_\nu & \text{if } \mathfrak{g} = \mathfrak{sp}_n, \\[2mm]
\displaystyle\sum_{\nu=1}^n \left(\nu - n\right)e_\nu = \sum_{\nu=1}^n \left(\nu - n - \tfrac{\delta+1}{2}\right)e_\nu & \text{if } \mathfrak{g} = \mathfrak{o}_{2n}.
\end{cases}
\tag{4.8}
$$

We define $\bar\lambda = (\bar\lambda_1, \ldots, \bar\lambda_n) \in \mathbb{C}^n$ by

$$
\lambda_\Theta|_\mathfrak{a} + \epsilon\rho = \bar\lambda_1 e_1 + \bar\lambda_2 e_2 + \cdots + \bar\lambda_n e_n.
\tag{4.9}
$$

For $P \in U^\epsilon(\mathfrak{g})$ let $\omega(P)$ and $\bar\omega(P)$ denote the elements of $S(\mathfrak{a}) \simeq U^\epsilon(\mathfrak{a})$ with

$$
\begin{aligned}
P - \omega(P) &\in \bar{\mathfrak{n}}U^\epsilon(\mathfrak{g}) + U^\epsilon(\mathfrak{g})\mathfrak{n}, \\
\bar\omega(P)(\mu + \epsilon\rho) &= \omega(P)(\mu) \quad \text{for } \forall\mu \in \mathfrak{a}^*.
\end{aligned}
\tag{4.10}
$$

Then $\bar\omega$ induces the Harish-Chandra isomorphism

$$
\bar\omega : Z^\epsilon(\mathfrak{g}) \ \xrightarrow{\sim}\ S(\mathfrak{a})^W.
\tag{4.11}
$$

Here W is the Weyl group for the pair $(\mathfrak{g}, \mathfrak{a})$ and $S(\mathfrak{a})^W$ denotes the totality of W-invariants in $S(\mathfrak{a})$.

.

Definition 4.2. Retain the above notation and define polynomials

$$\begin{cases} q_\Theta^\epsilon(\mathfrak{gl}_n; x, \lambda) = \prod_{j=1}^{L} (x - \lambda_j - n_{j-1}\epsilon), \\ q_\Theta^\epsilon(\mathfrak{o}_{2n+1}; x, \lambda) = (x - n\epsilon) \prod_{j=1}^{L} (x - \lambda_j - n_{j-1}\epsilon)(x + \lambda_j + (n_j - 2n)\epsilon), \\ q_\Theta^\epsilon(\mathfrak{sp}_n; x, \lambda) = \prod_{j=1}^{L} (x - \lambda_j - n_{j-1}\epsilon)(x + \lambda_j + (n_j - 2n - 1)\epsilon), \\ q_\Theta^\epsilon(\mathfrak{o}_{2n}; x, \lambda) = \prod_{j=1}^{L} (x - \lambda_j - n_{j-1}\epsilon)(x + \lambda_j + (n_j - 2n + 1)\epsilon) \end{cases}$$

$$(4.12)$$

and if $\mathfrak{g} = \mathfrak{sp}_n$ or $\mathfrak{o}_{2n+1}$ or $\mathfrak{o}_{2n}$,

$$q_\Theta^\epsilon(\mathfrak{g}; x, \lambda) = (x - n_{L-1}\epsilon) \prod_{j=1}^{L-1} (x - \lambda_j - n_{j-1}\epsilon)(x + \lambda_j + (n_j - 2n - \delta)\epsilon) \quad (4.13)$$

with the δ given by (3.3). Furthermore define two-sided ideals of $U^\epsilon(\mathfrak{g})$

$$\begin{cases} I_\Theta^\epsilon(\lambda) = \sum_{i=1}^{N} \sum_{j=1}^{N} U^\epsilon(\mathfrak{g}) q_\Theta^\epsilon(\mathfrak{g}; F, \lambda)_{ij} + \sum_{j \in J} U^\epsilon(\mathfrak{g})\Big(\Delta_j - \omega(\Delta_j)(\lambda_\Theta)\Big), \\ I_{\bar\Theta}^\epsilon(\lambda) = \sum_{i=1}^{N} \sum_{j=1}^{N} U^\epsilon(\mathfrak{g}) q_{\bar\Theta}^\epsilon(\mathfrak{g}; F, \lambda)_{ij} + \sum_{j \in \bar J} U^\epsilon(\mathfrak{g})\Big(\Delta_j - \omega(\Delta_j)(\lambda_\Theta)\Big) \end{cases}$$

$$(4.14)$$

with (3.4) and

$$\begin{cases} J = \{1, 2, \ldots, L - 1\} & \text{if } \mathfrak{g} = \mathfrak{gl}_n, \\ J = \{1, 2, \ldots, L\}, \ \bar J = \{1, 2, \ldots, L - 1\} & \text{if } \mathfrak{g} = \mathfrak{o}_{2n+1}, \\ J = \bar J = \{1, 2, \ldots, L - 1\} & \text{if } \mathfrak{g} = \mathfrak{sp}_n, \\ J = \bar J = \{1, 2, \ldots, L - 1\} \cup \{n\} & \text{if } \mathfrak{g} = \mathfrak{o}_{2n}. \end{cases}$$

$$(4.15)$$

Remark 4.3. i) Let $p(x)$ and $q(x)$ be monic polynomials with $q(x) \in \mathbb{C}[x]p(x)$. Then

$$\begin{cases} \sum_{i=1}^{N} p(F)_{ii} \in Z^\epsilon(\mathfrak{g}), \\ \sum_{i=1}^{N} p(F)_{ii} - \sum_{i=1}^{N} F_i^{\deg p} \in \bar{\mathfrak{n}} U^\epsilon(\mathfrak{g}) + U^\epsilon(\mathfrak{g})\mathfrak{n} + U^\epsilon(\mathfrak{g})^{(\deg p - 1)}, \\ q(F)_{ij} \in \sum_{\substack{1 \le \mu \le N \\ 1 \le \nu \le N}} U^\epsilon(\mathfrak{g}) p(F)_{\mu\nu}. \end{cases}$$

Hence it is clear

$$I^\epsilon_{\Theta'}(\lambda) \supset \sum_{D \in Z^\epsilon(\mathfrak{g})} U^\epsilon(\mathfrak{g})\big(D - \omega(D)(\lambda_\Theta)\big) \quad \text{for } \Theta' = \Theta \text{ and } \bar\Theta. \qquad (4.16)$$

Note that it is known that the right hand side of the above coincides with $\mathrm{Ann}_G(M^\epsilon(\lambda_\Theta))$.

ii) $I^\epsilon_\Theta(\lambda)$ and $I^\epsilon_{\bar\Theta}(\lambda)$ are homogeneous ideals with respect to $(\mathfrak{g}, \lambda, \epsilon)$.

Now we give the main theorem in this paper:

Theorem 4.4. i) *Let* $\mathfrak{g} = \mathfrak{gl}_n$, $\mathfrak{o}_{2n+1}$, $\mathfrak{sp}_n$ *or* $\mathfrak{o}_{2n}$. *Then*

$$\begin{cases} I^\epsilon_{\bar\Theta}(\lambda) \subset \mathrm{Ann}\big(M^\epsilon_{\bar\Theta}(\lambda)\big), \\ q^\epsilon_{\bar\Theta}(\mathfrak{g}; F, \lambda)_{ii+1} \equiv r^\epsilon_i(\mathfrak{g}; \Theta, \lambda) F_{ii+1} \mod J^\epsilon(\lambda_\Theta) & \text{if } n_{k-1} < i < n_k, \\ J^\epsilon_{\bar\Theta}(\lambda) = I^\epsilon_{\bar\Theta}(\lambda) + J^\epsilon(\lambda_\Theta) & \text{if } r^\epsilon(\mathfrak{g}; \Theta, \lambda) \neq 0. \end{cases} \qquad (4.17)$$

Here

$$r^\epsilon_i(\mathfrak{gl}_n; \Theta, \lambda) = \prod_{\nu=1}^{k-1}\big(\lambda_k - \lambda_\nu - (n_\nu - i)\epsilon\big) \prod_{\nu=k+1}^{L}\big(\lambda_k - \lambda_\nu - (n_{\nu-1} - i)\epsilon\big)$$

$$= \prod_{\nu=1}^{k-1}(\bar\lambda_i - \bar\lambda_{n_\nu}) \prod_{\nu=k+1}^{L}(\bar\lambda_{i+1} - \bar\lambda_{n_{\nu-1}+1}),$$

$$r^\epsilon_i(\mathfrak{sp}_n; \Theta, \lambda) = r^\epsilon_i(\mathfrak{o}_{2n}; \Theta, \lambda)$$

$$= r^\epsilon_i(\mathfrak{gl}_n; \Theta, \lambda) \prod_{\nu=1}^{L}\big(\lambda_k + \lambda_\nu + (n_\nu - 2n - \delta + i)\epsilon\big)$$

$$= \prod_{\nu=1}^{k-1}(\bar\lambda_i - \bar\lambda_{n_\nu}) \prod_{\nu=k+1}^{L}(\bar\lambda_{i+1} - \bar\lambda_{n_{\nu-1}+1}) \prod_{\nu=1}^{L}(\bar\lambda_{i+1} + \bar\lambda_{n_\nu}),$$

$$r^\epsilon_i(\mathfrak{o}_{2n+1}; \Theta, \lambda) = r^\epsilon_i(\mathfrak{gl}_n; \Theta, \lambda)\big(\lambda_k - (n - i)\epsilon\big)\prod_{\nu=1}^{L}\big(\lambda_k + \lambda_\nu + (n_\nu - 2n + i)\epsilon\big)$$

$$= \frac{1}{2}(\bar\lambda_i + \bar\lambda_{i+1}) \prod_{\nu=1}^{k-1}(\bar\lambda_i - \bar\lambda_{n_\nu}) \prod_{\nu=k+1}^{L}(\bar\lambda_{i+1} - \bar\lambda_{n_{\nu-1}+1}) \prod_{\nu=1}^{L}(\bar\lambda_{i+1} + \bar\lambda_{n_\nu})$$

if $n_{k-1} < i < n_k$ *and then*

$$r^\epsilon(\mathfrak{g}; \Theta, \lambda) = \prod_{k=1}^{L} \prod_{n_{k-1} < i < n_k} r^\epsilon_i(\mathfrak{g}; \Theta, \lambda). \qquad (4.18)$$

ii) Suppose $\lambda_L = 0$ and $\mathfrak{g} = \mathfrak{sp}_n$ or $\mathfrak{o}_{2n+1}$ or $\mathfrak{o}_{2n}$. Then

$$
\begin{cases}
I_{\bar{\Theta}}^\epsilon(\lambda) \subset \mathrm{Ann}\big(M_{\bar{\Theta}}^\epsilon(\lambda)\big), & \\
q_{\bar{\Theta}}^\epsilon(\mathfrak{g};F,\lambda)_{ii+1} \equiv r_i^\epsilon(\mathfrak{g};\bar{\Theta},\lambda)F_{ii+1} \mod J^\epsilon(\lambda_\Theta) & \text{if } \iota_\Theta(i) = \iota_\Theta(i+1), \\
q_{\bar{\Theta}}^\epsilon(\mathfrak{g};F,\lambda)_{nn+1} \equiv \bar{r}^\epsilon(\mathfrak{g};\bar{\Theta},\lambda)F_{nn+1} \mod J^\epsilon(\lambda_\Theta) & \text{if } \mathfrak{g} \neq \mathfrak{o}_{2n}, \\
J_{\bar{\Theta}}^\epsilon(\lambda) = I_{\bar{\Theta}}^\epsilon(\lambda) + J^\epsilon(\lambda_\Theta) & \text{if } r^\epsilon(\mathfrak{g};\bar{\Theta},\lambda) \neq 0
\end{cases}
\tag{4.19}
$$

with denoting

$$
\begin{aligned}
r_i^\epsilon(\mathfrak{g};\bar{\Theta},\lambda) &= \prod_{\nu=1}^{k-1}\big(\lambda_k - \lambda_\nu - (n_\nu - i)\epsilon\big)\prod_{\nu=k+1}^{L}\big(\lambda_k - \lambda_\nu - (n_{\nu-1} - i)\epsilon\big) \\
&\quad \cdot \prod_{\nu=1}^{L-1}\big(\lambda_k + \lambda_\nu + (n_\nu - 2n - \delta + i)\epsilon\big) \\
&= \prod_{\nu=1}^{k-1}(\bar{\lambda}_i - \bar{\lambda}_{n_\nu})\prod_{\nu=k+1}^{L}(\bar{\lambda}_{i+1} - \bar{\lambda}_{n_{\nu-1}+1})\prod_{\nu=1}^{L-1}(\bar{\lambda}_{i+1} + \bar{\lambda}_{n_\nu}) \\
&\qquad \text{if } n_{k-1} < i < n_k,
\end{aligned}
$$

$$
\begin{aligned}
\bar{r}_o^\epsilon(\mathfrak{g};\bar{\Theta},\lambda) &= (-1)^{L-1}\prod_{\nu=1}^{L-1}\big(\lambda_\nu + (n_\nu - n)\epsilon\big)\big(\lambda_\nu + (n_\nu - n - \delta)\epsilon\big) \\
&= \begin{cases}
(-1)^{L-1}\prod_{\nu=1}^{L-1}\bar{\lambda}_{n_\nu}(\bar{\lambda}_{n_\nu} - \bar{\lambda}_n) & \text{if } \mathfrak{g} = \mathfrak{sp}_n, \\
(-1)^{L-1}\prod_{\nu=1}^{L-1}(\bar{\lambda}_{n_\nu} - \bar{\lambda}_n)^2 & \text{if } \mathfrak{g} = \mathfrak{o}_{2n+1},
\end{cases}
\end{aligned}
$$

and

$$
\begin{cases}
r^\epsilon(\mathfrak{o}_{2n};\bar{\Theta},\lambda) = \prod_{k=1}^{L}\prod_{n_{k-1}<i<n_k} r_i^\epsilon(\mathfrak{g};\bar{\Theta},\lambda), & \\
r^\epsilon(\mathfrak{g};\bar{\Theta},\lambda) = r^\epsilon(\mathfrak{o}_{2n};\bar{\Theta},\lambda)\bar{r}_o^\epsilon(\mathfrak{g};\bar{\Theta},\lambda) & \text{if } \mathfrak{g} = \mathfrak{sp}_n \text{ or } \mathfrak{o}_{2n+1}.
\end{cases}
\tag{4.20}
$$

Proof. Define the parameters μ_ν in Proposition 3.4 by

$$\mu_\nu = \lambda_\nu + n_{\nu-1} \quad \text{for } 1 \leq \nu \leq L,$$
$$\mu_{\nu+L} = -\lambda_\nu - n_\nu + 2n + \delta = -\mu_\nu - n_\nu' + 2n + \delta \quad \text{for } 1 \leq \nu \leq L,$$
$$\mu_{2L+1} = -n.$$

Then in the proposition $H(\Theta, \mu_1, \ldots, \mu_L) = \prod_{\nu=1}^{L'}(F - \mu_\nu)$ with $L' = L$, $2L$, $2L$ or $2L + 1$ if $\mathfrak{g} = \mathfrak{gl}_n$, $\mathfrak{sp}_n$, $\mathfrak{o}_{2n}$ or $\mathfrak{o}_{2n+1}$, respectively. Moreover $\tilde{H}(\Theta, \mu_1, \ldots, \mu_{L-1}) = \prod_{\nu=1}^{2L-1}(F - \mu_\nu)$ with $\lambda_L = 0$. Note that if $\ell \le L$,

$$\mu_\ell - \mu_\nu - n'_\nu - n_{\ell-1} + i = \lambda_\ell + i - \lambda_\nu - n_\nu,$$

$$\mu_\ell - \mu_\nu - n_{\ell-1} + i = \lambda_\ell + i - \lambda_\nu - n_{\nu-1},$$

$$\mu_\ell - \mu_{\nu+L} - n_{\ell-1} + i = \lambda_\ell + i + \lambda_\nu + n'_\nu - 2n - \delta,$$

$$\mu_\ell - \mu_{2L+1} - n_{\ell-1} + i = \lambda_\ell + i + n.$$

For $\Theta' = \Theta$ or $\bar{\Theta}$, Proposition 3.4 and the isomorphism $U(\mathfrak{g}) \simeq U^\epsilon(\mathfrak{g})$ show that

$$q_{\Theta'}^\epsilon(\mathfrak{g}; F, \lambda)_{ij} \equiv 0 \mod \bar{\mathfrak{n}} U^\epsilon(\mathfrak{g}) + J^\epsilon(\lambda_{\Theta'}),$$

which assures $I_{\Theta'}^\epsilon(\lambda) \subset \mathrm{Ann}\big(M_{\Theta'}^\epsilon(\lambda)\big)$ (cf. [O4, Lemma 2.11 and Remark 2.12]) because $M_{\Theta'}^\epsilon(\lambda)$ is irreducible $\mathfrak{g}$-module for generic (λ, ϵ) and $\sum \mathbb{C} q_{\Theta'}^\epsilon(\mathfrak{g}; F, \lambda)_{ij}$ is $\mathfrak{g}$-invariant.

Suppose $\mathfrak{g} = \mathfrak{o}_{2n}$. Since $I_{\bar{\Theta}}^\epsilon + J^\epsilon(\lambda_\Theta)$ is stable under the outer automorphism which maps α_{n-1} to α_n, $I_{\bar{\Theta}}^\epsilon + J^\epsilon(\lambda_\Theta) \ni F_{ii+1}$ if and only if $I_{\bar{\Theta}}^\epsilon + J^\epsilon(\lambda_\Theta) \ni F_{ii+2}$.

Hence the claims of the theorem are direct consequences of Proposition 3.4. Note that the functions in the theorem are homogeneous with respect to $(x, \mathfrak{g}, \lambda, \epsilon)$. $\qquad\square$

Any zero of $r^\epsilon(\mathfrak{g}; \Theta', \lambda)$ in the above theorem corresponds to the hypersurface defined by a root in $\Sigma(\mathfrak{g})$ except for the term $\bar{\lambda}_{i+1} + \bar{\lambda}_{n_\nu}$ with $i = n_\nu - 1$ and $\nu = k$ in the case when $\mathfrak{g} = \mathfrak{o}_{2n}$. Hence we have the following remark.

Remark 4.5. i) Suppose $\mathfrak{g} = \mathfrak{gl}_n$. Considering the weights with respect to $\mathfrak{a}$, we have

$$\Big(I_\Theta^\epsilon(\lambda) + J^\epsilon(\lambda_\Theta)\Big) \cap \mathbb{C} E_{ii+1} = \Big(q_\Theta^\epsilon(\mathfrak{g}; F, \lambda)_{ii+1} + J^\epsilon(\lambda_\Theta)\Big) \cap \mathbb{C} E_{ii+1}$$

as in the argument in [O4, §3]. Hence $J_\Theta^\epsilon(\lambda) = I_\Theta^\epsilon(\lambda) + J^\epsilon(\lambda_\Theta)$ if and only if $r^\epsilon(\mathfrak{g}; \Theta, \lambda) \ne 0$.

ii) If the infinitesimal character of $M^\epsilon(\lambda_\Theta)$ is regular (resp. strongly regular) in the case when $\mathfrak{g} = \mathfrak{gl}_n$, $\mathfrak{o}_{2n+1}$ or $\mathfrak{sp}_n$ (resp. $\mathfrak{o}_{2n}$), then $r^\epsilon(\mathfrak{g}; \Theta, \lambda) \ne 0$ and $r^\epsilon(\mathfrak{g}; \bar{\Theta}, \lambda) \ne 0$.

Here we defined that the infinitesimal character μ is *strongly regular* if it is not fixed by any non-trivial element of the group of automorphisms of the root system. Note that the group is generated by the reflections with respect to the simple roots and the automorphisms of the Dynkin diagram.

iii) Suppose $\mathfrak{g} = \mathfrak{o}_{2n}$. Then $r^\epsilon(\mathfrak{g}; \bar{\Theta}, \lambda) \neq 0$ if

$$\langle \lambda_\Theta|_\mathfrak{a} + \epsilon\rho, \alpha \rangle \neq 0 \quad \text{for } \alpha \in \Sigma(\mathfrak{a}),$$

$$\langle \lambda_\Theta|_\mathfrak{a} + \epsilon\rho, 2\alpha_j + \cdots + 2\alpha_{n-2} + \alpha_{n-1} + \alpha_n \rangle \neq 0 \quad \text{for } j = 2, \ldots, n-1$$

$$(4.21)$$

under the notation in (4.3).

It is proved by [BG] and [Jo] that for $\mu \in \mathfrak{a}^*$ the map

$$\{I; I \text{ are the two-sided ideals of } U(\mathfrak{g}) \text{ with } I \supset \text{Ann}\big(M(\mu)\big)\}$$

$$\ni I \mapsto I + J(\mu) \in \{J; J \text{ are the left ideals of } U(\mathfrak{g}) \text{ with } J \supset J(\mu)\}$$

$$(4.22)$$

is injective if $\mu + \rho$ is dominant

$$2\frac{\langle \mu + \rho, \alpha \rangle}{\langle \alpha, \alpha \rangle} \notin \{-1, -2, \ldots\} \quad \text{for any root } \alpha \text{ for the pair } (\mathfrak{n}, \mathfrak{a}). \quad (4.23)$$

Under the notation (4.9) with $\mu = \lambda_\Theta|_\mathfrak{a}$ and $\epsilon = 1$ the condition (4.23) is equivalent to

$$\begin{cases} \bar{\lambda}_i - \bar{\lambda}_j \notin \{1, 2, \ldots\} \quad (1 \le i < j \le n) & \text{if } \mathfrak{g} = \mathfrak{gl}_n, \\ \bar{\lambda}_i \pm \bar{\lambda}_j \notin \{1, 2, \ldots\} \quad (1 \le i < j \le n) & \text{if } \mathfrak{g} = \mathfrak{o}_{2n}, \\ \bar{\lambda}_i \pm \bar{\lambda}_j, \ \bar{\lambda}_k \notin \{1, 2, \ldots\} \quad (1 \le i < j \le n, \ 1 \le k \le n) & \text{if } \mathfrak{g} = \mathfrak{sp}_n, \\ \bar{\lambda}_i \pm \bar{\lambda}_j, \ 2\bar{\lambda}_k \notin \{1, 2, \ldots\} \quad (1 \le i < j \le n, \ 1 \le k \le n) & \text{if } \mathfrak{g} = \mathfrak{o}_{2n+1}. \end{cases}$$

$$(4.24)$$

Note that

$$\bar{\lambda}_i - \bar{\lambda}_{i+1} = -1 \quad \text{if } n_{k-1} < i < n_k \text{ and } 1 \le k \le L. \quad (4.25)$$

Corollary 4.6. i) *If $\lambda_\Theta|_\mathfrak{a} + \rho$ is dominant and $r^1(\mathfrak{g}; \Theta', \lambda) \neq 0$, then*

$$\text{Ann}\big(M_{\Theta'}(\lambda)\big) = I^1_{\Theta'}(\lambda) \quad (4.26)$$

for $\Theta' = \Theta$ or $\bar{\Theta}$.

ii) *Suppose the infinitesimal character $M_{\Theta'}(\lambda)$ is regular (resp. strongly regular) in the case when $\mathfrak{g} = \mathfrak{o}_{2n}$ with $\Theta' = \bar{\Theta}$ or $\mathfrak{g} = \mathfrak{gl}_n$ or $\mathfrak{g} = \mathfrak{o}_{2n+1}$ or $\mathfrak{g} = \mathfrak{sp}_n$ (resp. $\mathfrak{g} = \mathfrak{o}_{2n}$ with $\Theta' = \Theta$). Then (4.26) holds.*

Proof. i) Since $J_{\Theta'}(\lambda) = I_{\Theta'}(\lambda) + J(\lambda_{\Theta'}) \subset \text{Ann}\big(M_{\Theta'}(\lambda)\big) + J(\lambda_{\Theta'}) \subset J_{\Theta'}(\lambda)$ by Theorem 4.4, we have (4.26) by the injectivity of (4.22).

ii) Let $\Psi(\mathfrak{m}_\Theta, \mathfrak{a})$ denote the fundamental system for $(\mathfrak{m}_\Theta, \mathfrak{a})$. Here $\Psi(\mathfrak{m}_\Theta, \mathfrak{a}) \subset \Psi(\mathfrak{a})$. Fix $w \in W$ satisfying $w\Psi(\mathfrak{m}_\Theta, \mathfrak{a}) \subset \Psi(\mathfrak{a})$. Define $w.\Theta'$ and $w.\lambda$ so that $\mathfrak{m}_{w.\Theta'} = \text{Ad}(w)\mathfrak{m}_{\Theta'}$ and $\overline{w.\lambda} = w\bar{\lambda} \in \mathbb{C}^n$. Then [Ja, Corollary 15.27] says

$$\text{Ann}\big(M_{\Theta'}(\lambda)\big) = \text{Ann}\big(M_{w.\Theta'}(w.\lambda)\big). \tag{4.27}$$

For example, $\text{Ann}\big(M_{\{k,n\}}(\lambda_1, \lambda_2)\big) = \text{Ann}\big(M_{\{n-k,n\}}(\lambda_2 + k, \lambda_1 - n + k)\big)$. Here we note that $q_{\Theta'}(\mathfrak{g}; x, \lambda)$ does not change under this transformation.

Case $\mathfrak{g} = \mathfrak{gl}_n$. By a permutation of the L blocks $\big\{\{n_{k-1}+1, \ldots, n_k\}; k = 1, \ldots, L\big\}$, we may assume $\Re\bar{\lambda}_{n_0+1} \leq \Re\bar{\lambda}_{n_1+1} \leq \cdots \leq \Re\bar{\lambda}_{n_{L-1}+1}$. Here $\Re c$ denotes the real part of c for $c \in \mathbb{C}$. Since $M_\Theta(\lambda)$ has a regular infinitesimal character, (4.24) and (4.25) assure that $\lambda_\Theta|_\mathfrak{a} + \rho$ is dominant and $r^1(\mathfrak{g}; \Theta, \lambda) \neq 0$. Hence we have ii) from i).

Case $\mathfrak{g} = \mathfrak{sp}_n$ or $\mathfrak{o}_{2n+1}$. First suppose $\Theta' = \Theta$. We may assume $\Re\bar{\lambda}_{n_0+1} \leq \Re\bar{\lambda}_{n_1+1} \leq \cdots \leq \Re\bar{\lambda}_{n_{L-1}+1} \leq 0$. Since $M_\Theta(\lambda)$ has a regular infinitesimal character, $\lambda_\Theta|_\mathfrak{a} + \rho$ is dominant and we have ii). Here we note that if $\bar{\lambda}_{n_k+1} < 0$ and $\bar{\lambda}_{n_{k+1}} > 0$, then $\bar{\lambda}_{n_{k+1}} \notin 2\mathbb{Z}$ because $\bar{\lambda}_i + \bar{\lambda}_{i+1} \neq 0$ if $n_k < i < n_{k+1}$.

Next suppose $\Theta' = \bar{\Theta}$. Then we may assume $\Re\bar{\lambda}_{n_0+1} \leq \Re\bar{\lambda}_{n_1+1} \leq \cdots \leq \Re\bar{\lambda}_{n_{L-2}+1} \leq 0$ and $\bar{\lambda}_{n_L} = -\frac{\delta+1}{2}$ (cf. (3.14)). Hence we similarly have ii).

Case $\mathfrak{g} = \mathfrak{o}_{2n}$. The map $e_n \mapsto -e_n$ corresponds to an outer automorphism of $\mathfrak{o}_{2n}$ which does not change $\mathfrak{b}$. Combining the corresponding automorphism of the universal enveloping algebra with the above argument, we may have the same assumption on $\bar{\lambda}$ as in the previous case. Thus we similarly have ii). $\qquad\square$

Remark 4.7. Suppose $\mathfrak{g} = \mathfrak{gl}_n$.

i) In [O4] another generator system of $\text{Ann}_G\big(M_\Theta^\epsilon(\lambda)\big)$ is given for every $(\Theta, \epsilon, \lambda)$. It is interesting to express them by the generators constructed in this note, which is done by [Sa] when $\mathfrak{p}_\Theta$ is a maximal parabolic subalgebra. In the case of the minimal parabolic subalgebra, that is, in the case of the central elements of $U(\mathfrak{g})$, it is studied by [I1], [I2] and [Um]. In general, it may be considered as a generalization of Newton's formula for symmetric polynomials.

ii) Put $\Theta = \{k, n\}$ with $1 \leq k < n$ and fix $\lambda = (\lambda_1, \lambda_2) \in \mathbb{C}^2$. Then

$$\lambda \text{ is regular} \Leftrightarrow \lambda_1 - \lambda_2 \notin \big\{1, 2, \ldots, n-1\big\},$$

$$\lambda \text{ is dominant } \Leftrightarrow \lambda_1 - \lambda_2 \notin \{2,3,4,\dots\},$$

$$J_\Theta(\lambda) = I_\Theta(\lambda) + J(\lambda_\Theta) \Leftrightarrow \lambda_1 - \lambda_2 \notin \{1,2,\dots,\max\{k,n-k\}-1\},$$

$$J_\Theta(\lambda) = \mathrm{Ann}\big(M_\Theta(\lambda)\big) + J(\lambda_\Theta) \Leftrightarrow \lambda_1 - \lambda_2 \notin \{1,2,\dots,\min\{k,n-k,\tfrac{n}{2}-1\}\}.$$

Here the last equivalence follows from [O4, Theorem 3.1] and the one before last is clear from Remark 4.5 i).

Remark 4.8. Considering the m-th exterior product of the natural representation of $\mathfrak{gl}_n$, we may put $p(E) = \big(E_{IJ}\big)_{\#I=\#J=m} \in M\big(\binom{n}{m}, U(\mathfrak{g})\big)$ in Lemma 2.1, where $I = \{i_1,\dots,i_m\}$, $J = \{j_1,\dots,j_m\}$ with $1 \le i_1 < \cdots < i_m \le n$ and $1 \le j_1 < \cdots < j_m \le n$ and $E_{IJ} = \det\big(E_{i_\mu j_\nu} + (\mu - m)\epsilon\delta_{i_\mu j_\nu}\big)_{\substack{1 \le \mu \le m \\ 1 \le \nu \le m}}$. Here $\det\big(A_{ij}\big) = \sum_{\sigma \in \mathfrak{S}_n} A_{\sigma(1)1} \cdots A_{\sigma(n)n}$. The study of $f(p(E))$ for polynomials $f(x)$ may be interesting because it may be a *quantization* of the ideals of the *rank varieties* (cf. [ES]) defined by the condition rank $f(A) = m$ for $A \in M(n,\mathbb{C})$.

Remark 4.9. For $\mathfrak{g} = \mathfrak{o}_n$ or $\mathfrak{sp}_n$ we may expect an explicit generator system for $\mathrm{Ann}_G\big(M_\mathfrak{p}^\epsilon(\lambda)\big)$ which are of the same type given by [O4] for $\mathfrak{gl}_n$. It should be a *quantization* of determinants and Pfaffians (and elementary divisors for the singular case). The quantization of determinants and Pfaffians for $\mathfrak{o}_n$ is studied by [HU], [I2], [IU] and [Od] etc. It is shown by [Od] that it gives $\mathrm{Ann}\big(M_\mathfrak{p}(\lambda)\big)$ for the expected $\mathfrak{p}$.

Remark 4.10. We have considered $\sum_{i,j} \mathbb{C}f(p(E))_{ij}$ for the construction of a two-sided ideal of $U(\mathfrak{g})$ with a required property. We may pick up a $\mathfrak{g}$-invariant subspace V of $\sum_{i,j} \mathbb{C}f(p(E))_{ij}$ to get a refined result. Moreover for a certain problem (cf. [O1]) related to a symmetric pair $(\mathfrak{g}, \mathfrak{k})$ it is useful to study $\mathfrak{k}$-invariant subspaces of $\sum_{i,j} \mathbb{C}f(p(E))_{ij}$ which should have required zeros under the map of Harish-Chandra homomorphism for the pair. This will be discussed in another paper [OSh].

In the case when $\epsilon = 0$ we have the following.

Theorem 4.11. *Let $\lambda \in \mathfrak{a}$ and suppose that the centralizer of λ in $\mathfrak{g}$ is $\mathfrak{m}_{\Theta'}$ with $\Theta' = \Theta$ or $\bar{\Theta}$. If $\mathfrak{g} = \mathfrak{o}_{2n}$, we moreover assume $\lambda_i \neq 0$ for $i = 1,\dots,n$. Then*

$$I_{\Theta'}^0(\lambda) = \{f \in S(\mathfrak{g}); \, f|_{\mathrm{Ad}(G)\lambda} = 0\}.$$

Proof. It is clear from Theorem 4.4 that the element of $I_{\Theta'}^0(\lambda)$ vanishes on λ and therefore $I_{\Theta'}^0(\lambda)$ vanishes on $\mathrm{Ad}(G)\lambda$ because $I_{\Theta'}^0(\lambda)$ is G-stable.

We will prove that the dimension of the space $\sum_{i=1}^{N} \sum_{j=1}^{N} \mathbb{C} dq_{\Theta'}^0(\mathfrak{g}; F, \lambda)_{ij}|_\lambda$ is not smaller than $\dim \mathfrak{m}_{\Theta'}$. This is shown by the direct calculation and it is almost the same in any case and therefore we give it in the case when $\mathfrak{g} = \mathfrak{sp}_n$ and $\Theta' = \bar{\Theta}$.

Put $\Theta = \{n_1, \ldots, n_L\}$ and $\lambda = (\lambda_1, \ldots, \lambda_L)$. Note that $\lambda_L = 0$ and $q_{\bar{\Theta}}^0(\mathfrak{sp}_n; x, \lambda) = x \prod_{1 \leq \nu < L} (x - \lambda_\nu)(x + \lambda_\nu)$. If $n_{k-1} \leq i < n_k$ and $n_{k-1} \leq j < n_k$ and $k < L$, we have

$$dq_{\bar{\Theta}}^0(\mathfrak{sp}_n; F, \lambda)_{ij}|_{\lambda_\Theta} = 2\lambda_k^2 \prod_{1 \leq \nu < L, \ \nu \neq k} (\lambda_k - \lambda_\nu)(\lambda_k + \lambda_\nu) dF_{ij}.$$

If $n_{L-1} \leq i < 2n - n_{L-1}$ and $n_{L-1} \leq 2n - n_{L-1}$, then

$$dq_{\bar{\Theta}}^0(\mathfrak{sp}_n; F, \lambda)_{ij}|_{\lambda_\Theta} = \prod_{1 \leq \nu < L} (-\lambda_\nu)(\lambda_\nu) dF_{ij}.$$

The assumption of the proposition implies $\lambda_k \neq 0$, $\lambda_\nu \neq 0$ and $\lambda_k^2 \neq \lambda_\nu^2$ in the above and therefore we get the required result.

Put $V = \{X \in \mathfrak{g}; f(X) = 0 \ (\forall f \in I_{\Theta'}(\lambda))\}$. Since $[\lambda, \mathfrak{g}] = \mathfrak{n}_\Theta + \bar{\mathfrak{n}}_\Theta$, the tangent space of $\mathrm{Ad}(G)\lambda$ at λ is isomorphic to $\mathfrak{n}_\Theta + \bar{\mathfrak{n}}_\Theta$. Since $\mathrm{Ad}(G)\lambda \subset V$, it follows from the above calculation of the dimension that $\mathrm{Ad}(G)\lambda$ and V are equal in a neighborhood of λ. In particular, V is non-singular at λ.

Let $X \in \mathfrak{g}$ with $f(X) = 0$ for all $f \in I_{\Theta'}^0(\lambda)$. We will show $X \in \mathrm{Ad}(G)\lambda$, which completes the proof of the theorem. Let $X = X_s + X_n$ be the Jordan decomposition of X. Here X_s is semisimple and X_n is nilpotent. By the action of the element of $\mathrm{Ad}(G)$, we may assume $X_s \in \mathfrak{a}$ and $X_n \in \mathfrak{n}$. Then it is clear that $f(X_s + tX_n) = 0$ for all $f \in I_{\Theta'}^0(\lambda)$ and $t \in \mathbb{C}$. Moreover it is also clear that X_s is a transformation of λ_Θ under a suitable element of the Weyl group of the root system for the pair $(\mathfrak{g}, \mathfrak{a})$ and therefore we may assume $X_s = \lambda$. Since the tangent space of V at λ is isomorphic to $\mathfrak{n}_\Theta + \bar{\mathfrak{n}}_\Theta$, we have $X_n \in \mathfrak{n}_\Theta$. Hence $X_n = 0$ because $[X_s, X_n] = 0$. Therefore $X \in \mathrm{Ad}(G)\lambda$. $\qquad\square$

Remark 4.12. Theorem 4.11 shows that we have constructed a generator system of the defining ideal of the adjoint orbit of any semisimple element of any classical Lie algebra. In fact, for any $\lambda \in \mathfrak{a}$ in the orbit the centralizer of λ in $\mathfrak{g}$ is $\mathfrak{m}_\Theta$ or $\mathfrak{m}_{\bar{\Theta}}$ or $\mathfrak{g}$ with a suitable Θ.

In [O4] we constructed a generator system of the ideal corresponding to the closure of an arbitrary conjugacy class of $\mathfrak{gl}_n$, which is of a different type from the one given here.

We will generalize the Cayley-Hamilton theorem in the linear algebra. Put

$$
\bar{d}_{\mathfrak{g}}(x) = \begin{cases}
\displaystyle\prod_{i=1}^{n}\left(x - F_i - \tfrac{n-1}{2}\epsilon\right) & \text{if } \mathfrak{g} = \mathfrak{gl}_n, \\[2ex]
\displaystyle\prod_{i=1}^{n}\left(x - F_i - n\epsilon\right)\left(x + F_i - n\epsilon\right) & \text{if } \mathfrak{g} = \mathfrak{sp}_n, \\[2ex]
\displaystyle\prod_{i=1}^{n}\left(x - F_i - (n-1)\epsilon\right)\left(x + F_i - (n-1)\epsilon\right) & \text{if } \mathfrak{g} = \mathfrak{o}_{2n}, \\[2ex]
(x - n\epsilon)\displaystyle\prod_{i=1}^{n}\left(x - F_i - (n-\tfrac{1}{2})\epsilon\right)\left(x + F_i - (n-\tfrac{1}{2})\epsilon\right) & \text{if } \mathfrak{g} = \mathfrak{o}_{2n+1}.
\end{cases}
$$

Here we note that if $\Theta = \{1, 2, \ldots, n\}$, then $n_j = j$, $\lambda_i + n_{i-1}\epsilon = \bar{\lambda}_i + (n + \tfrac{\delta-1}{2})\epsilon$ and $\lambda_i + (n_i - 2n - \delta)\epsilon = \bar{\lambda}_i - (n + \tfrac{\delta-1}{2})\epsilon$. Since $\bar{d}_{\mathfrak{g}}(x) \in S(\mathfrak{a})^W[x]$, there exists $d_{\mathfrak{g}}(x) \in Z^\epsilon(\mathfrak{g})[x]$ with

$$\bar{\omega}(d_{\mathfrak{g}}(x)) = \bar{d}_{\mathfrak{g}}(x), \qquad (4.28)$$

which is equivalent to $d_{\mathfrak{g}}(x) \equiv \bar{d}_{\mathfrak{g}}(x)(\mu) \mod J^\epsilon(\mu - \epsilon\rho)$. Then Theorem 4.4 assures $d_{\mathfrak{g}}(F) \equiv \bar{d}_{\mathfrak{g}}(F)(\mu) \equiv 0 \mod J^\epsilon(\mu - \epsilon\rho)$. Hence $\bar{\omega}(d_{\mathfrak{g}}(F))(\mu) = 0$ for any $\mu \in \mathfrak{a}^*$ and therefore $\bar{\omega}(d_{\mathfrak{g}}(F)) = 0$, which assures $d_{\mathfrak{g}}(F) = 0$ because $\sum_{i,j}\mathbb{C}d_{\mathfrak{g}}(F)_{ij}$ is $\mathfrak{g}$-invariant (cf. [O4, Lemma 2.12]). Thus we have the following corollary.

Corollary 4.13 (The Cayley-Hamilton theorem for the natural representation of the classical Lie algebra $\mathfrak{g}$).

$$d_{\mathfrak{g}}(F) = 0. \qquad (4.29)$$

Remark 4.14. This result for $\mathfrak{gl}_n$ and $\mathfrak{o}_n$ is given by [Um] and [I2], respectively. A more general result is given by [Go2] (cf. [OO]).

Remark 4.15. Suppose $\mathfrak{g} = \mathfrak{gl}_n$. Then it follows from [O3] that

$$d_{\mathfrak{g}}(x) = \det\left(x - F_{ij} - (i - n)\epsilon\delta_{ij}\right)_{\substack{1 \leq i \leq n \\ 1 \leq j \leq n}}. \qquad (4.30)$$

In [O4] we define another generator system of $\mathrm{Ann}_G(M_\Theta^\epsilon(\lambda))$ for any $(\Theta, \lambda, \epsilon)$ by using "elementary divisors" in place of the "minimal polynomial" $q_\Theta^\epsilon(\mathfrak{gl}_n; x, \lambda)$.

Proposition 4.16. *Suppose $\mathfrak{g} = \mathfrak{gl}_n$ and let π be its natural representation. Then the characteristic polynomial of $F = (E_{ij})$ in $U^\epsilon(\mathfrak{g})$ is $d_{\mathfrak{g}}^\epsilon(x)$ and the minimal polynomial of $(F, M_\Theta^\epsilon(\lambda))$ is $q_\Theta^\epsilon(\mathfrak{g}; x, \lambda)$.*

Proof. Suppose $\epsilon = 0$ and identify the dual space $\mathfrak{g}^*$ of $\mathfrak{g}$ with $\mathfrak{g}$ by the bilinear form (2.1). Put $J_\Theta^0(\lambda)^\perp = \{X \in \mathfrak{g}^*; \langle X, Y \rangle = 0 \ (\forall Y \in J_\Theta^0(\lambda))\}$. Then the condition $q(F)M_\Theta^0(\lambda) = 0$ for a polynomial $q(x)$ is equivalent to

$q(F)(J^0_\Theta(\lambda)^\perp) = 0$, which also equivalent to $q(A_\Theta(\lambda)) = 0$ for a generic element $A_{\Theta,\lambda}$ of $J^0_\Theta(\lambda)^\perp$ because the closure of $\bigcup_{g\in GL(n,\mathbb{C})} gA_{\Theta,\lambda}g^{-1}$ is $\bigcup_{g\in GL(n,\mathbb{C})} g(J^0_\Theta(\lambda)^\perp)g^{-1}$ (cf. [O4, §2]). In fact

$$A_{\Theta,\lambda} = \begin{pmatrix} \lambda_1 I_{n'_1} & & & & 0 \\ A_{21} & \lambda_2 I_{n'_2} & & & \\ A_{31} & A_{32} & \lambda_3 I_{n'_3} & & \\ \vdots & \vdots & \vdots & \ddots & \\ A_{L1} & A_{L2} & A_{L3} & \cdots & \lambda_L I_{n'_L} \end{pmatrix}$$

with generic $A_{ij} \in M(n'_i, n'_j, \mathbb{C})$. Hence our minimal polynomial is the same as that of $A_{\Theta,\lambda}$ in the linear algebra and the claim in the lemma for the minimal polynomial is clear.

We may assume $\epsilon = 1$. Let $p(x)$ be the minimal polynomial of $(F, M_\Theta(\lambda))$ with a fixed λ. Define a homogeneous and monic polynomial $p(x, \epsilon)$ of (x, ϵ) with $p(x) = p(x, 1)$. Then $p(x, \epsilon)M^\epsilon_\Theta(\epsilon\lambda) = 0$ for $\epsilon \in \mathbb{C}$. If follows from the result in the case $\epsilon = 0$ that the degree of $p(x)$ should not be smaller than that of $q_\Theta(\mathfrak{g}; x, \lambda)$. Hence $q_\Theta(\mathfrak{g}; x, \lambda)$ is the minimal polynomial for $M^\epsilon_\Theta(\lambda)$.

Since the degree of the minimal polynomial $q_\Theta(\mathfrak{g}; x, \lambda)$ for $\Theta = \{1, 2, \ldots, n\}$ is n, the degree of the characteristic polynomial is not smaller than n. Hence $d_\mathfrak{g}(x)$ is the characteristic polynomial. □

Definition 4.17. The non-zero element $q(x, \lambda, \epsilon) \in \mathbb{C}[x, \lambda, \epsilon] \simeq S(\mathfrak{a}_{\Theta'})[x, \epsilon]$ is called the *global minimal polynomial* of $(F, M^\epsilon_\mathfrak{p}(\lambda))$ if $q(x, \lambda, \epsilon)$ satisfies $q(F, \lambda, \epsilon)M^\epsilon_\mathfrak{p}(\lambda) = 0$ for any (λ, ϵ) in the parameter space $\mathfrak{a}^*_{\Theta'} \times \mathbb{C}$ and any other non-zero polynomial whose degree with respect to x is smaller than that of $q(x, \lambda, \epsilon)$ does not satisfies this.

Proposition 4.18. *The polynomials $q^\epsilon_{\Theta'}(\mathfrak{g}; x, \lambda)$ in Definition 4.2 are the global minimal polynomials of $(F, M^\epsilon_{\Theta'}(\lambda))$ for $\Theta' = \Theta$ and $\bar{\Theta}$.*

Proof. Let $q(x, \lambda, \epsilon)$ be a global minimal polynomial of $(F, M^\epsilon_{\Theta'}(\lambda))$. We may assume $q(x, \lambda, 0)$ is not zero by dividing by ϵ^k with an integer k if necessary. Put $\epsilon = 0$ and consider the generic λ. Since $q(\langle F, \lambda\rangle, \lambda, 0) = 0$, the minimality of the degree is clear. Here $\langle F, \lambda\rangle = \big(\lambda(F_i)\delta_{ij}\big)_{\substack{1\le i\le N \\ 1\le j\le N}}$ is a diagonal matrix in $M(N, \mathbb{C})$. □

Let $f(x)$ be any polynomial in $\mathbb{C}[x]$. We will give a characterization of the image $\bar{\omega}(f(F))$ under the Harish-Chandra homomorphism $\bar{\omega}$ defined by (4.10) in the case of the natural representation of $\mathfrak{g} = \mathfrak{gl}_n$.

Theorem 4.19. *Put $f(x) = \prod_{i=1}^m (x - \lambda_i)$. For sets A and B we denote by $\mathrm{Map}(A, B)$ the totality of the maps of A to B. For $\tau \in$*

$\mathrm{Map}\big(\{1,\ldots,n\},\{1,\ldots,m\}\big)$ *we put*

$$
\begin{cases}
I_\tau(\lambda) = \displaystyle\sum_{j=1}^{n} \mathbb{C}[\mathfrak{a},\lambda,\epsilon]\Big(E_{jj} - \lambda_{\tau(j)} - \big(m_\tau(j) - \frac{n-1}{2}\big)\epsilon\Big), \\[2mm]
m_\tau(j) = \#\{\nu \in \{1,\ldots,j-1\}; \, \tau(\nu) = \tau(j)\} \quad \text{for } j = 1,\ldots,n.
\end{cases}
$$

Then for $F = \Big(E_{ij}\Big)_{\substack{1\leq i\leq n \\ 1\leq j\leq n}} \in M(n,\mathfrak{gl}_n)$ *we have*

$$
\sum_{j=1}^{n} \mathbb{C}[\mathfrak{a},\lambda,\epsilon]\bar\omega(f(F))_{jj} = \bigcap_{\tau\in\mathrm{Map}(\{1,\ldots,n\},\{1,\ldots,m\})} I_\tau(\lambda). \tag{4.31}
$$

Moreover $\bar\omega(f(F)_{ij}) = 0$ *if* $i \neq j$ *and the polynomial* $\bar\omega(f(F))_{jj})$ *is the unique homogeneous element in* $\mathbb{C}[\mathfrak{a},\lambda,\epsilon]$ *belonging to the right hand side of* (4.31) *such that* $\bar\omega(f(F)_{jj})|_{\lambda=\epsilon=0} = E_{jj}^m$.

Proof. First note that $\bar\omega(f(F)_{ij})|_{\epsilon=0} = \delta_{ij}\prod_{\nu=1}^{m}(E_{jj} - \lambda_\nu)$ and $\bar\omega(f(F)_{jj})$ are of homogeneous of degree m with respect to $(\mathfrak{a},\lambda,\epsilon)$.

We have already proved that $\bar\omega(f(F)_{ij}) = 0$ if $i \neq j$ and $\bar\omega(f(F)_{jj}) \equiv F_{jj}^m$ mod $U^\epsilon(\mathfrak{a})^{(m-1)}[\lambda]$. Hence for any fixed λ and ϵ, the system of the equations

$$
\bar\omega(f(F)_{11}) = \cdots = \bar\omega(f(F)_{nn}) = 0 \text{ for } (E_{11},\ldots,E_{nn}) \in \mathbb{C}^n \tag{4.32}
$$

is in the complete intersection and has m^n roots counting their multiplicities. Hence to prove (4.31) it is sufficient to show that

$$
\Big(\lambda_{\tau(1)} + \big(m_\tau(1) - \frac{n-1}{2}\big)\epsilon, \ldots, \lambda_{\tau(n)} + \big(m_\tau(n) - \frac{n-1}{2}\big)\epsilon\Big) \tag{4.33}
$$

is a root for any $\tau \in \mathrm{Map}\big(\{1,\ldots,n\},\{1,\ldots,m\}\big)$ and a generic $(\lambda,\epsilon) \in \mathbb{C}^{m+1}$.

Fix any $\sigma \in \mathrm{Map}\big(\{1,\ldots,n\},\{1,\ldots,m\}\big)$ which satisfies $\sigma(i) \leq \sigma(j)$ if $1 \leq i < j \leq n$. Define $\Theta = \{n_1,\ldots,n_L\}$ so that $0 = n_0 < n_1 < \cdots < n_L = n$ and that the condition $\sigma(i) = \sigma(n_k)$ is equivalent to $n_{k-1} < i \leq n_k$ for $k = 1,\ldots,L$. Put $\mu = (\lambda_{\sigma(n_1)} + n_0\epsilon,\ldots,\lambda_{\sigma(n_L)} + n_{L-1}\epsilon)$. Then Theorem 4.4 shows that $f(F)_{ij}$ is in $\mathrm{Ann}(M/J_\Theta^\epsilon(\mu))$. Here we note that $\mu_\Theta + \epsilon\rho = \sum_{j=1}^{n}\big(\lambda_{\sigma(j)} + (m_\sigma(j) + \frac{n-1}{2})\epsilon\big)e_j$. Recall that [O4, Theorem 2.12] determines the common zeros of $\bar\omega\big(\mathrm{Ann}(M/J_\Theta^\epsilon(\mu))\big)$ for $\epsilon \neq 0$. Namely they are (4.33) with τ in

$$
\{\tau \in \mathrm{Map}\big(\{1,\ldots,n\},\{1,\ldots,m\}\big);
$$
$$
\#\{\nu; \tau(\nu) = \sigma(j)\} = n_j - n_{j-1} \text{ for } j = 1,\ldots,L\}.
$$

For any τ we have the above σ so that τ is in this set and we have the theorem. □

5. Integral Transforms on Generalized Flag Manifolds

Let $\mathfrak{g}$ be a complex reductive Lie algebra and $\mathfrak{p}$ be a parabolic subalgebra containing a Borel subalgebra $\mathfrak{b}$. For a holomorphic character λ of $\mathfrak{p}$ we define left ideals

$$\begin{cases} J_{\mathfrak{p}}(\lambda) = \sum_{X \in \mathfrak{p}} (X - \lambda(X)), \\ J_{\mathfrak{b}}(\lambda) = \sum_{X \in \mathfrak{b}} (X - \lambda(X)) \end{cases} \tag{5.1}$$

of the universal enveloping algebra $U(\mathfrak{g})$ of $\mathfrak{g}$. Let $I_{\mathfrak{p}}(\lambda)$ be the two-sided ideal of $U(\mathfrak{g})$ which satisfies

$$I_{\mathfrak{p}}(\lambda) \subset J_{\mathfrak{p}}(\lambda). \tag{5.2}$$

Let G be a connected real semisimple Lie group and let P be a parabolic subgroup of G such that the complexifications of $\mathrm{Lie}(G)$ and $\mathrm{Lie}(P)$ are $\mathfrak{g}$ and $\mathfrak{p}$, respectively. Let L_λ be a line bundle over G/P such that the local section of L_λ is killed by $J_{\mathfrak{p}}(\lambda)$. Then the image of any $\mathfrak{g}$-equivalent map of the space of sections of L_λ over an open subset of G/P is killed by $I_{\mathfrak{p}}(\lambda)$. Here the element of $I_{\mathfrak{p}}(\lambda)$ is identified with a left invariant differential operator but it may be identified with a right invariant differential operator through the anti-automorphism of $U(\mathfrak{g})$ ($X \mapsto -X$, $XY \mapsto (-Y)(-X)$ for $X, Y \in \mathfrak{g}$) because $I_{\mathfrak{p}}(\lambda)$ is a two-sided ideal. If the $\mathfrak{g}$-equivariant map is an integral transform to the space of functions on a homogeneous space X of G or sections of a vector bundle over X, it is a natural question how the system of differential equations corresponding to $I_{\mathfrak{p}}(\lambda)$ characterizes the image.

The same problem may be considered when L_λ is the holomorphic line bundle over the complexification of G/P.

5.1. Penrose transformations.
Let $G_{\mathbb{C}}$ be a reductive complex Lie group with the Lie algebra $\mathfrak{g}$. Let G be a real form of $G_{\mathbb{C}}$ and let $P_{\mathbb{C}}$ be a parabolic subalgebra of $G_{\mathbb{C}}$ with the Lie algebra $\mathfrak{p}$ and let V be a G-orbit in $G_{\mathbb{C}}/P_{\mathbb{C}}$. Suppose $\mathcal{O}_\lambda$ is a holomorphic line bundle over $G_{\mathbb{C}}/P_{\mathbb{C}}$ which is killed by $J_{\mathfrak{p}}(\lambda)$. Here the element of $J_{\mathfrak{p}}(\lambda)$ is identified with a right invariant holomorphic differential operator on $G_{\mathbb{C}}$. Then the image of any G-equivariant map

$$\mathcal{T} : H_V^*(G_{\mathbb{C}}/P_{\mathbb{C}}, \mathcal{O}_\lambda) \to E \tag{5.3}$$

is killed by $I_{\mathfrak{p}}(\lambda)$. Here E is usually a space of sections of a certain line (or vector) bundle over a homogeneous space of G. In this case $I_{\mathfrak{p}}(\lambda)$ is identified with a system of holomorphic differential equations and we may identify the element of $I_{\mathfrak{p}}(\lambda)$ as a left invariant differential operator on G through the anti-automorphism of the universal enveloping algebra or a right invariant differential operator on G. An interesting example is discussed in [Se].

5.2. Poisson transformations. Let G be a connected semisimple Lie group with finite center, let K be a maximal compact subgroup of G and let P be a parabolic subalgebra of G with the Langlands decomposition $P = MAN$ and let P_o be a minimal parabolic subgroup with the Langlands decomposition $P_o = M_o A_o N_o$ satisfying $M_o \subset M$, $A_o \supset A$, $N_o \supset N$ and $P_o \subset P$. Let λ be an element of the complexification $\mathfrak{a}^*$ of the dual of the Lie algebra of A and put

$$\mathcal{B}(G/P, L_\lambda) = \{f \in \mathcal{B}(G); f(xman) = a^\lambda f(x) \ (\forall m \in M, \ \forall a \in A, \ \forall n \in N)\}$$

which is the space of hyperfunction sections of spherical degenerate principal series. Let $\mathfrak{p}$ be a complexification of the Lie algebra of P. The Poisson transformation of the space $\mathcal{B}(G/P, L_\lambda)$ is defined by

$$\mathcal{P}^\lambda : \mathcal{B}(G/P, L_\lambda) \to \mathcal{B}(G/K), \ f \mapsto (\mathcal{P}^\lambda f)(x) = \int_K f(xk)dk \qquad (5.4)$$

with the normalized Haar measure dk on K. Let $\mathbb{D}(G/K)$ be the ring of invariant differential operator of G and let χ_λ be the algebra homomorphism of $\mathbb{D}(G/K)$ to $\mathbb{C}$ so that the image of $\mathcal{P}_\lambda$ is in the solution space $\mathcal{A}(G/K, \mathcal{M}_\lambda)$ of the system

$$\mathcal{M}_\lambda : Du = \chi_\lambda(D)u \quad (\forall D \in \mathbb{D}(G/K)) \qquad (5.5)$$

for $u \in \mathcal{A}(G/K)$. Here $\mathcal{A}(G/K)$ denotes the space of real analytic functions on G/K.

Note that $\mathcal{B}(G/P, L_\lambda)$ is the subspace of the space of hyperfunction sections of spherical principal series

$$\mathcal{B}(G/P_o, L_\lambda) = \{f \in \mathcal{B}(G); f(xman) = a^\lambda f(x)$$
$$(\forall m \in M_o, \ \forall a \in A_o, \ \forall n \in N_o)\}.$$

Here λ is extended to the complexification $\mathfrak{a}_o^*$ of the dual of the Lie algebra $\mathfrak{a}_0$ of A_o so that it takes the value 0 on $\mathrm{Lie}(M) \cap \mathrm{Lie}(A_o)$.

Theorem 5.1. *Suppose that the Poisson transform*

$$\mathcal{P}_o^\lambda : \mathcal{B}(G/P_o, L_\lambda) \to \mathcal{A}(G/K, \mathcal{M}_\lambda), \ f \mapsto (\mathcal{P}_o^\lambda f)(x) = \int_K f(xk)dk \quad (5.6)$$

for the boundary G/P_o of G/K is bijective. Assume the condition

$$J_\mathfrak{p}(\lambda) = I_\mathfrak{p}(\lambda) + J_\mathfrak{b}(\lambda) \quad (5.7)$$

for a two-sided ideal $I_\mathfrak{p}(\lambda)$ of $U(\mathfrak{g})$. Then the Poisson transform $\mathcal{P}^\lambda$ for the boundary G/P is a G-isomorphism onto the simultaneous solution space of the system $\mathcal{M}_\lambda$ and the system defined by $I_\mathfrak{p}(\lambda)$.

Proof. Since $\mathcal{B}(G/P, L_\lambda)$ is a subspace of $\mathcal{B}(G/P_o, L_\lambda)$ and $\mathcal{P}^\lambda$ is a G-equivariant map, the image of $\mathcal{P}_o^\lambda$ satisfies the systems $\mathcal{M}_\lambda$ and $I_\mathfrak{p}(\lambda)$.

Suppose the function $u \in \mathcal{A}(G/K, \mathcal{M}_\lambda)$ satisfies $I_\mathfrak{p}(\lambda)$. Since the function $(\mathcal{P}_o^\lambda)^{-1}u \in \mathcal{B}(G/P_o, L_\lambda)$ also satisfies $I_\mathfrak{p}(\lambda)$, the condition (5.7) assures $(\mathcal{P}_o^\lambda)^{-1}u \in \mathcal{B}(G/P, L_\lambda)$ because we may assume $\mathbb{C} \otimes_\mathbb{R} \text{Lie}(P_o) \supset \mathfrak{b}$. $\square$

Remark 5.2. i) The above theorem with its proof is based on the idea given by [O3] which explains it in the case when $G = GL(n, \mathbb{R})$.

ii) The bijectivity of $\mathcal{P}_o^\lambda$ is equivalent to the condition $e(\lambda + \rho) \neq 0$ by [K–]. This condition is introduced by [He] for the injectivity of $\mathcal{P}_o^\lambda$. Here

$$e(\lambda) = \prod_{\alpha \in \Sigma_o^+} \left\{ \Gamma\left(\frac{\langle \lambda, \alpha \rangle}{2\langle \alpha, \alpha \rangle} + \frac{m_\alpha}{4} + \frac{1}{2} \right) \Gamma\left(\frac{\langle \lambda, \alpha \rangle}{2\langle \alpha, \alpha \rangle} + \frac{m_\alpha}{4} + \frac{m_{2\alpha}}{2} \right) \right\}, \quad (5.8)$$

Σ^+ is the set of the positive roots for the pair $(\mathfrak{g}, \mathfrak{a}_0)$ so that $\text{Lie}(N)$ corresponds to Σ^+. Moreover $\Sigma_o^+ = \{\alpha \in \Sigma^+; \frac{1}{2}\alpha \notin \Sigma^+\}$, m_α is the multiplicity of the root $\alpha \in \Sigma^+$ and $\rho = \frac{1}{2}\sum_{\alpha \in \Sigma^+} m_\alpha \alpha$.

iii) Suppose G is simple and of the classical type and suppose the condition $e(\lambda + \rho) \neq 0$. Let $I_\mathfrak{p}(\lambda)$ be the system given by (4.14). Then if moreover the infinitesimal character of $\mathcal{B}(G/P, L_\lambda)$ is regular (or strongly regular in the case when $\text{Lie}(G)$ is $\mathfrak{o}_{2n}$ or its real form), $\mathcal{P}^\lambda$ is G-isomorphic to the solution space of the system of differential equations $I_\mathfrak{p}(\lambda)$ on G/K since Theorem 4.4 assures (5.7). This is because the natural map of $Z(\mathfrak{g})$ to $\mathbb{D}(G/K)$ is surjective and therefore it follows from Remark 4.3 i) that $\mathcal{M}_\lambda$ is contained in $I_\mathfrak{p}(\lambda)$. Here the function on G/K is identified with the right K-invariant function on G. Note that all the assumptions are valid when $\lambda = 0$ and $\text{Lie}(G)$ is not $\mathfrak{o}_{2n}$ or its real form.

iv) Let $\mathfrak{p}_o$ be the complexification of the Lie algebra of P_o and put $J_{\mathfrak{p}_o}(\lambda) = \sum_{X \in \mathfrak{p}_o}(X - \lambda(X))$. Then we may replace the assumption (5.7) by

$$J_\mathfrak{p}(\lambda) = I_\mathfrak{p}(\lambda) + J_{\mathfrak{p}_o}(\lambda) \quad (5.9)$$

in Theorem 5.1, which is clear from its proof. Suppose that $I_{\mathfrak{p}}(\lambda)$ is the system given by (4.14). Then under the notation in Theorem 4.4, (5.9) is valid if $r_i^\epsilon(\mathfrak{g}; \Theta', \lambda) \neq 0$ for $i \notin \Theta$ and moreover if $\bar{r}^\epsilon(\mathfrak{g}; \Theta', \lambda) \neq 0$ for $\mathfrak{g} = \mathfrak{o}_{2n+1}$ or $\mathfrak{sp}_n$ in the case $\Theta' = \bar{\Theta}$.

v) Owing to [K–] the abstract existence of the system of differential equations characterizing the image of $\mathcal{P}^\lambda$ is clear (cf. [OSh]) but a certain existence theorem of the system in the case $\lambda = 0$ is given by [Jn]. A more precise study for this problem including the relation to the *Hua equations* will be discussed in [OSh].

vi) See [BOS] for other functions spaces.

5.3. Intertwining operators. Retain the notation in §5.2. Let $\tilde{\lambda}$ and $\tilde{\mu}$ be characters of P and P_0, respectively. Put

$$\mathcal{B}(G/Q, L_\tau) = \{f \in \mathcal{B}(G); f(xq) = \tau(q)^{-1}f(x) \ (\forall q \in Q)\} \qquad (5.10)$$

for $(Q, \tau) = (P, \tilde{\lambda})$, $(P_0, \tilde{\lambda})$ and $(P_0, \tilde{\mu})$. Let $\mathfrak{j}$ be a Cartan subalgebra of Lie(G) containing $\mathfrak{a}$ and let λ be an element of the complexification of the dual of $\mathfrak{j}$. Assume (5.7) and $\tilde{\lambda}(e^H) = e^{-\lambda(H)}$ for $H \in \mathfrak{j}$. If there exists a G-equivariant bijective map

$$\mathcal{T}_{\tilde{\mu}}^{\tilde{\lambda}} : \mathcal{B}(G/P_0, L_{\tilde{\lambda}}) \xrightarrow{\sim} \mathcal{B}(G/P_0, L_{\tilde{\mu}}),$$

$\mathcal{T}_{\tilde{\mu}}^{\tilde{\lambda}}\big(\mathcal{B}(G/P, L_{\tilde{\lambda}})\big)$ is identical with the space of solutions of the system of differential equations on $\mathcal{B}(G/P_0, L_{\tilde{\mu}})$ defined by $I_{\mathfrak{p}}(\lambda)$.

5.4. Radon transformations. Let G be a connected real semisimple Lie group and let P_1 and P_2 be maximal parabolic subgroups of G. If there exists a G-equivariant map $\mathcal{R} : \mathcal{B}(G/P_1, L_{\lambda_1}) \to \mathcal{B}(G/P_2, L_{\lambda_2})$ for certain characters λ_j of P_j under the notation (5.10), the image of $\mathcal{R}$ satisfies the system $I_{\mathfrak{p}}(\lambda)$. Here $\mathfrak{p}$ and λ correspond to P_1 and λ_1, respectively. An example is given in [O3].

5.5. Hypergeometric functions. Some special cases of Radon or Penrose transformations and their relations to Aomoto-Gelfand hypergeometric functions are discussed in [O3], [Se] and [Ta].

5.6. Whittaker models. A Whittaker model is a realization of a representation of a semisimple Lie group G in a functions space induced from a non-trivial character of a nilpotent subgroup of G. The Whittaker model of a degenerate principal series is given by a certain integral transform of the series and therefore our two sided ideal $I_{\mathfrak{p}}(\lambda)$ is useful to study it (cf. [O5]).

6. Closure of Ideals

Now we will consider the non-regular λ which are excluded in Theorem 4.4. We begin with a general consideration.

Definition 6.1. Let M be a C^∞-manifold and let U be a connected open subset of $\mathbb{C}^\ell$. We denote by $\mathcal{D}'(M)$ the space of distributions on M. Suppose that meromorphic functions $f_1(\lambda), \ldots, f_n(\lambda)$ of U valued in $\mathcal{D}'(M)$ are given. Moreover suppose there exists a non-zero holomorphic function $r(\lambda)$ on U such that $f_1, \ldots, f_n$ are holomorphic on $U_r = \{\lambda \in U;\ r(\lambda) \neq 0\}$ and $\dim V_\lambda = m$ for any $\lambda \in U_r$ by putting $V_\lambda = \sum_{j=1}^n f_j(\lambda)$. For $\lambda \in U$ we define

$$\bar{V}_\mu = \{f(0);\ f \text{ is a holomorphic function on } \{t \in \mathbb{C};\ |t| < 1\} \text{ valued in}$$
$$\mathcal{D}'(M) \text{ and there exists a holomorphic curve } c : \{t \in \mathbb{C};\ |t| < 1\} \to U$$
$$\text{such that } c(t) \in U_r \text{ and } f(t) \in V_{c(t)} \text{ for } 0 < |t| \ll 1 \text{ and } c(0) = \mu\}.$$

We call $\bar{V}_\mu$ the *closure* of the holomorphic family of the spaces V_λ ($\lambda \in U_r$) at μ. It follows from [OS, Proposition 2.21] that $\dim \bar{V}_\mu \geq m$. We define that a point $\mu \in U \setminus U_r$ is a *removable* (resp. *un-removable*) *singular point* if $\dim V_\mu = m$ (resp. $\dim V_\mu > m$). Note that $\bar{V}_\lambda = V_\lambda$ if $\lambda \in U_r$, which follows from the last statement in Lemma 6.3 by replacing μ and U_r by λ ($\in U_r$) and $U_r \setminus \{\lambda\}$.

Example 6.2. The origin $\lambda = (\lambda_1, \lambda_2) = 0$ is a removable singular point of $V_\lambda = \mathbb{C}(x + \lambda_1) + \mathbb{C}(\lambda_2 x + \lambda_1 y + \lambda_1^2)$ and an un-removable singular point of $V_\lambda = \mathbb{C}(\lambda_1 x + \lambda_2 y)$.

Lemma 6.3. i) *If μ is a removable singular point of the spaces V_λ, then there exist a neighborhood U_μ of μ and holomorphic functions $h_1(\lambda), \ldots, h_m(\lambda)$ on U_μ valued in $\mathcal{D}'(M)$ such that they are linearly independent for any $\lambda \in U_\mu$ and they span V_λ for any $\lambda \in U_\mu \cap U_r$. Conversely the existence of $h_j(\lambda)$ ($j = 1, \ldots, m$) with these properties implies that μ is a removable singular point.*

ii) *If U is convex and there is no un-removable singular point in U, we may choose $U_\mu = U$ in* i).

Proof. i) Suppose $\dim \bar{V}_\mu = m$. We may assume $f_1(\lambda), \ldots, f_m(\lambda)$ are linearly independent for a generic point λ in U_r. Fix a curve c to U with $c(0) = \mu$ and $c(t) \in U_r$ for $0 < |t| \ll 1$. Then [OS, Proposition 2.21] assures the existence of holomorphic functions $v_i(t)$ ($1 \leq i \leq m$) on $\{t \in \mathbb{C};\ |t| < 1\}$ valued in $\mathcal{D}'(M)$ and a holomorphic curve $c : \{t \in \mathbb{C};\ |t| < 1\} \to U$ such that $c(0) = \mu$, $c(t) \in U_r$ and $v_i(t) \in V_{c(t)}$ for $0 < |t| \ll 1$ and $v_1(t), \ldots, v_m(t)$ are linearly independent for any t. Then the set $\{v_1(0), \ldots, v_m(0)\}$ is a basis of V_μ. Fix test functions $\phi_1, \ldots, \phi_m$ so that $\langle v_i(0), \phi_j \rangle = \delta_{ij}$ and put $c_{ij}(\lambda) = \langle f_i(\lambda), \phi_j \rangle$. If $0 < |t| \ll 1$, then $v_1(t), \ldots, v_m(t)$ span $V_{c(t)} = \sum_{i=1}^m \mathbb{C} f_i(c(t))$ and therefore $f_i(c(t)) = \sum_{j=1}^m c_{ij}(c(t)) v_j(t)$, which means $\det\big(c_{ij}(\lambda)\big)$ is not identically zero. Let $\big(d_{ij}(\lambda)\big)$ be the inverse of $\big(c_{ij}(\lambda)\big)$ and define $h_i(\lambda) = \sum_{j=1}^m d_{ij}(\lambda) f_j(\lambda)$ so that $\langle h_i(c(t)), \phi_j \rangle = \delta_{ij}$ for $0 < |t| \ll 1$.

Suppose $h_k(\lambda)$ has a pole at $\lambda = \mu$. Then there exists a test function ϕ such that $\langle h_k(\lambda), \phi \rangle$ has a pole at μ. Then it follows from Weierstrass' preparation theorem that there exists a curve $c(t)$ as above and moreover $\langle h_k(c(t)), \phi \rangle$ has a pole at the origin. Choose a positive integer ℓ so that the function $\tilde{h}(t) = t^\ell h_k(c(t))$ is holomorphically extended to $t = 0$ and $\tilde{h}(0) \neq 0$. Since $\langle \tilde{h}(t), \phi_j \rangle = t^\ell \delta_{kj}$, $\langle \tilde{h}(0), \phi_j \rangle = 0$ for $j = 1, \ldots, m$, which contradicts to the facts $\langle v_i(0), \phi_j \rangle = \delta_{ij}$ because $0 \neq \tilde{h}(0) \in V_\mu = \sum_{i=1}^m \mathbb{C} v_i(0)$ by definition.

Thus we have proved that $h_i(\lambda)$ are holomorphic functions on λ in a neighborhood of U_μ of μ. Since $\langle h_i(\lambda), \phi_j \rangle = \delta_{ij}$, they are the required functions. In fact, $f_i(\lambda) = \sum_{j=1}^m \langle f_i(\lambda), \phi_j \rangle h_j(\lambda)$ for generic λ and therefore $V_\lambda \subset \sum_{j=1}^m \mathbb{C} h_j(\lambda)$ for $\lambda \in U_\mu \cap U_r$.

Now suppose the existence of $h_1, \ldots, h_m$ and consider the function f to define $\bar{V}_\mu$ in Definition 6.1. Then under the above notation, $f(t) = \sum_{j=1}^m \langle f(t), \phi_j \rangle h_j(c(t))$ for $0 < |t| \ll 1$ and therefore $f(0) = \sum_{j=1}^m \langle f(0), \phi_j \rangle h_j(c(0))$, which means $\dim V_\mu = m$.

ii) The claim in i) reduces the global existence of h_i to the second problem of Cousin and it is solved for the convex open domain by Oka's principle. □

Remark 6.4. i) Replacing "meromorphic" and "holomorphic" by "rational" and "regular", respectively, we have also Lemma 6.3 in the algebraic sense.

ii) When M is a finite set in Lemma 6.3, $\mathcal{D}'(M)$ is a finite dimensional vector space V over $\mathbb{C}$ and $f_i(\lambda)$ are the elements of V with a meromorphic parameter λ.

Definition 6.5. Fix a base $\{X_1, \ldots, X_m\}$ of $\mathfrak{g}$. Let

$$q_\nu(\lambda, \epsilon) = \sum_{\alpha_1 \geq 0, \ldots, \alpha_m \geq 0} q_{\nu, \alpha}(\lambda, \epsilon) X_1^{\alpha_1} \cdots X_m^{\alpha_m} \qquad (6.1)$$

be elements of $U^\epsilon(\mathfrak{g})$ for $(\lambda, \epsilon) \in \mathbb{C}^{r+1}$ and $\nu = 1, \ldots, k$. Here $q_{\nu, \alpha}$ are polynomial functions of (λ, ϵ) and $q_{\nu, \alpha} = 0$ if $\alpha_1 + \cdots + \alpha_m$ is sufficiently large. Let $I(\lambda, \epsilon)$ is the left ideal of $U^\epsilon(\mathfrak{g})$ generated by q_ν for $\nu = 1, \ldots, k$. Put $d_j = \max_{\lambda, \epsilon} \dim I(\lambda, \epsilon) \cap U^\epsilon(\mathfrak{g})^{(j)}$ for $j = 1, 2, \ldots$. Then we can find

$$p_{j,\mu}(\lambda, \epsilon) = \sum_{\alpha_1 \geq 0, \ldots, \alpha_m \geq 0} p_{j,\mu,\alpha}(\lambda, \epsilon) X_1^{\alpha_1} \cdots X_m^{\alpha_m}$$

such that $p_{j,\mu}(\lambda, \epsilon) \in I(\lambda, \epsilon) \cap U^\epsilon(\mathfrak{g})^{(j)}$ for any (λ, ϵ), $p_{j,\mu,\alpha}(\lambda, \epsilon)$ are poly-nomial functions and $p_{j,1}(\lambda, \epsilon), \ldots, p_{j,d_j}(\lambda, \epsilon)$ are linearly independent for generic (λ, ϵ). Then we denote by $\bar{I}(\lambda, \epsilon)^{(j)}$ the closure of the holomor-phic family $\sum_{\mu=1}^{d_j} \mathbb{C}p_{j,\mu}$ at (λ, ϵ) and put $\bar{I}(\lambda, \epsilon) = \bigcup_{j=1}^{\infty} \bar{I}(\lambda, \epsilon)^{(j)}$. We call $\bar{I}(\lambda, \epsilon)$ the closure of the ideal $I(\lambda, \epsilon)$ with respect to the parameter (λ, ϵ). We call a point $(\lambda, \epsilon) \in \mathbb{C}^{r+1}$ an un-removable singular point if (λ, ϵ) is an un-removable singular point of $\sum_{\mu=1}^{d_j} \mathbb{C}p_{j,\mu}$ for a certain j. Note that $\bar{I}(\lambda, \epsilon)$ does not depend on the choice of $\{X_1, \ldots, X_m\}$ or $p_{j,\mu}$.

Let $\bar{I}_{\Theta'}^\epsilon(\lambda)$ be the closure of the two-sided ideal $I_{\Theta'}^\epsilon(\lambda)$ given by (4.14) for $\Theta' = \Theta$ or $\bar{\Theta}$. Then we give the following two problems for a given $\mathfrak{g}$ and Θ'.

Problem 1. Does there exist no un-removable singular point in the pa-rameter (λ, ϵ) of the holomorphic family $I_{\Theta'}^\epsilon(\lambda)$?

Problem 2. Does the equality $\bar{I}_{\Theta'}(\lambda) = \mathrm{Ann}\big(M_{\Theta'}(\lambda)\big)$ hold?

Remark 6.6. i) It is clear that $\bar{I}_{\Theta'}^1(\lambda) \subset \mathrm{Ann}\big(M_{\Theta'}(\lambda)\big)$.

ii) The non-existence of the un-removable singularity in Problem 1 is equivalent to the following conditions. We can choose $\{q_\nu(\lambda, \epsilon); \nu = 1, \ldots, k\}$ of the form (6.1) such that $\bar{I}_{\Theta'}^\epsilon(\lambda) = \sum_{\nu=1}^{k} U^\epsilon(\mathfrak{g})q_\nu(\lambda, \epsilon)$ for any fixed (λ, ϵ). It is also equivalent to the fact that the graded ring

$$\mathrm{gr}\big(\bar{I}_{\Theta'}^\epsilon(\lambda)\big) = \bigoplus_{j=1}^{\infty} \Big(\bar{I}_{\Theta'}^\epsilon(\lambda) \cap U^\epsilon(\mathfrak{g})^{(j)} \big/ \bar{I}_{\Theta'}^\epsilon(\lambda) \cap U^\epsilon(\mathfrak{g})^{(j-1)}\Big)$$

does not depend on (λ, ϵ) because the space is spanned by homogeneous elements with respect to $(\mathfrak{g}, \lambda, \epsilon)$ and the conditions are also equivalent to the fact that the dimension of the vector space $\bar{I}_{\Theta'}^\epsilon(\lambda) \cap U^\epsilon(\mathfrak{g})^{(j)}$ does not depend on (λ, ϵ).

iii) Problem 1 and Problem 2 are affirmative if $\mathfrak{g} = \mathfrak{gl}_n$ because there exist $q_\nu(\lambda, \epsilon)$ $(\nu = 1, \ldots, k)$ of the form (6.1) such that $\mathrm{Ann}_G\big(M_\Theta^\epsilon(\lambda)\big)$ is generated by $q_\nu(\lambda, \epsilon)$ $(\nu = 1, \ldots, k)$ for any (λ, ϵ) (cf. [O4]). In this case $\mathrm{gr}\big(I_\Theta^\epsilon(\lambda)\big)$ is a prime ideal of $S(\mathfrak{g})$ but this is not true in general.

iv) If $\mathrm{gr}\big(I_{\Theta'}^0(\lambda)\big)$ is a prime ideal for generic λ, then Problem 1 and Problem 2 are affirmative, which is proved by the same argument as in [O4]. Note that $I_{\Theta'}^0(\lambda)$ is the defining ideal of $\mathrm{Ad}(G)\lambda$ for generic $\lambda \in \mathfrak{a}_{\Theta'}$ by Theorem 4.11.

v) Problem 2 is affirmative if the infinitesimal character is (strongly) regular.

References

[BOS] S. Ben Saïd, T. Oshima and N. Shimeno, *Fatou's theorems and Hardy-type spaces for eigenfunctions of the invariant differential operators on symmetric spaces*, Intern. Math. Research Notices **196**(2005), 915–931.

[BG] J. N. Bernstein and S. I. Gelfand, *Tensor products of finite and infinite dimensional representations of semisimple Lie algebras*, Comp. Math. **41**(1980), 245–285.

[C1] A. Capelli, *Über die Zurückführung der Cayley'schen Operation Ω auf gewöhnliche Polar Operationen*, Math. Ann. **29** (1887), 331–338.

[C2] ———, *Sur les opérations dans la théorie des formes algébriques*, Math. Ann. **37**(1890), 1–37.

[ES] D. Eisenbud and D. Saltman, *Rank variety of matrices*, Commutative algebra, Math. Sci. Res. Inst. Publ. **15**, 173–212, Springer-Verlag, 1989.

[Ge] I. M. Gelfand, *Center of the infinitesimal groups*, Mat. Sb. Nov. Ser. **26**(68)(1950), 103–112; English transl. in "Collected Papers", Vol. II, pp.22–30.

[Go1] M. D. Gould, *A trace formula for semi-simple Lie algerbras*, Ann. Inst. Henri Poincaré, Sect. A **32**(1980), 203–219.

[Go2] ———, *Characteristic identities for semi-simple Lie algebras*, J. Austral Math. Ser. B **26**(1985), 257–283.

[He] S. Helgason, *A duality for symmetric spaces with applications to group representations II*, Advances in Math. **22**(1976), 187–219.

[HU] R. Howe and T. Umeda, *Capelli identity, the double commutant theorem, and multiplicity-free actions*, Math. Ann. **290**(1991), 565–619.

[I1] M. Itoh, *Explicit Newton's formula for $\mathfrak{gl}_n$*, J. Alg. **208**(1998), 687–697.

[I2] ———, *Capelli elements for the orthogonal Lie algebras*, J. Lie Theory **10**(2000), 463–489.

[IU] M. Itoh and T. Umeda, *On the central elements in the universal enveloping algebra of the orthogonal Lie algebras*, Compositio Math. **127**(2001), 333–359.

[K–] M. Kashiwara, A. Kowata, K. Minemura, K. Okamoto, T. Oshima and M. Tanaka, *Eigenfunctions of invariant differential operators on a symmetric space*, Ann. of Math. **107**(1978), 1–39.

[Ko] B. Kostant, *Lie group representations on polynomial rings*, Amer. J. Math. **85**(1963), 327–404.

[Ja] J. C. Jantzen, *Einhüllende Algebren halbeinfacher Lie-Algebren*, Springer-Verlag, 1983, pp. 298.

[Jn] K. D. Johnson, *Generalized Hua operators and parabolic subgroups*, Ann. of Math. **120**(1984), 477–495.

[Jo] A. Joseph, *Dixmier's problem for Verma and principal series submodules*, J. London Math. Soc. **20**(1979), 193–204.

[KW] K. Kinoshita and M. Wakayama, *Explicit Capelli identities for skew symmetric matrices*, Proc. Edinb. Math. Soc. (2)**45**(2002), 449–465.

[Od] H. Oda, *Annihilator operators of the degenerate principal series for simple Lie groups of type (B) and (D)*, Doctor thesis presented to the University of Tokyo, 2000.

[OO] H. Oda and T. Oshima, *Minimal polynomials and annihilators of generalized Verma modules of the scalar type*, J. Lie Theory **16**(2006), 155–219.

[O1] T. Oshima, *Boundary value problems for various boundaries of symmetric spaces*, RIMS Kôkyûroku, Kyoto Univ. **281**(1976), 211–226, in Japanese.

[O2] _____, *Capelli identities, degenerate series and hypergeometric functions*, Proceedings of a symposium on Representation Theory at Okinawa, 1995, 1–19.

[O3] _____, *Generalized Capelli identities and boundary value problems for GL(n)*, Structure of Solutions of Differential Equations, World Scientific, 1996, 307–335.

[O4] _____, *A quantization of conjugacy classes of matrices*, Adv. in Math. **196**(2005), 124–146.

[O5] _____, *Whittaker models of degenerate principal sereis*, RIMS Kôkyûroku, Kyoto Univ. **1476**(2006), 71–78, in Japanese.

[OS] T. Oshima and J. Sekiguchi, *Eigenspaces of invariant differential operators on an affine symmetric spaces*, Invent. Math. **57**(1980), 1–81.

[OSh] T. Oshima and N. Shimeno, *Boundary value problems on Riemannian symmetric spaces of the noncompact type*, in preparation.

[Sa] H. Sakaguchi, *U(𝔤)-modules associated to Grassmannian manifolds*, Master thesis presented to the University of Tokyo, 1999.

[Se] H. Sekiguchi, *The Penrose transform for certain non-compact homogeneous manifolds of U(n,n)*, J. Math. Sci. Univ. Tokyo **3**(1996), 655-697.

[Ta] T. Tanisaki, *Hypergeometric systems and Radon transforms for Hermitian symmetric spaces*, Adv. Studies in Pure Math. **26**(2000), 235–263.

[Um] T. Umeda, *Newton's formula for $\mathfrak{gl}_n$*, Proc. Amer. Math. Soc. **126** (1998), 3169–3175.

BRANCHING TO A MAXIMAL COMPACT SUBGROUP

DAVID A. VOGAN, JR.

Department of Mathematics, MIT
Cambridge, MA 02139, USA
E-mail: dav@math.mit.edu

Suppose $G = \mathbf{G}(\mathbb{R})$ is the group of real points of a complex connected reductive algebraic group, and K is a maximal compact subgroup of G. The classical branching problem in this setting is to determine the restriction to K of a standard representation of G. Implicit in the branching problem are the problems of parametrizing both irreducible representations of K and standard representations of G. We address these three problems, looking for answers amenable to computer implementation.

Contents

Author's work partially supported by the National Science Foundation grant number DMS-0554278.

1. General Introduction

Suppose G is a real reductive Lie group and K is a maximal compact subgroup. (We will make our hypotheses more precise and explicit beginning in Definition 2.3 below.) Harish-Chandra proved that any (nice) irreducible representation of G decomposes on restriction to K into irreducible components having finite multiplicity. Our goal is to describe an algorithm for computing those multiplicities: that is, for solving the "branching problem" from G to K.

The traditional wisdom is that this branching problem was solved with the proof of the Blattner Conjecture by Hecht and Schmid more than thirty years ago. What the Blattner formula actually accomplishes is a reduction of the branching problem from G to K to various branching problems from K to subgroups $S \subset K$. (The subgroups S that arise are the intersections of K with parabolic subgroups of G.) For these compact group branching problems, one can hope to use the Weyl character formula in one form or another. Often this works very well; there are many widely available software packages for doing such calculations. There are two fundamental difficulties, however. First, if K is disconnected, then applying the theory of highest weights and the Weyl character formula to K requires some care. Second, if S is disconnected, these problems are multiplied.

A good example to keep in mind is $G = GL_n(\mathbb{R})$, the group of invertible $n \times n$ matrices with real entries. The maximal compact subgroup is the orthogonal group $K = O_n$. The diagonal matrices in K form a subgroup $S = O_1^n$ isomorphic to $\{\pm 1\}^n$. A principal series representation of G is determined by (among other things) a character δ of S. If μ is an irreducible representation of K, then the multiplicity of μ in this principal series is equal to the multiplicity of δ in $\mu|_S$. That is, to solve the branching problem for GL_n (even for principal series) we must solve the branching problem from O_n to O_1^n. This latter problem is "elementary" but painful, because the (abelian) group S is not contained in a maximal torus of K. In this special case there are more or less classical solutions available, but finding generalizations sufficient to deal cleanly with general branching from G to K looks very difficult.

We will approach the problem indirectly. Recall first of all (from [8], for example) that each irreducible representation π of G has a (very small) finite set

$$A(\pi) \subset \text{irreducible representations of } K \tag{1.1}$$

of *lowest K-types*. These are the representations that actually appear in the restriction of π to K, and which have minimal norm with respect to that requirement. (The precise definition of the norm of an irreducible representation of K is a little complicated, but it is approximately the length of the highest weight.) It turns out that the cardinality of $A(\pi)$ is always a power of two; the power is bounded by the rank of K.

For our purposes, the most convenient way to parametrize irreducible representations of K is by listing certain irreducible representations of G in which they appear as lowest K-types. Here is a quick-and-dirty version of the result we need (taken from [10]); a more complete and careful statement is in Theorem 11.9 below.

Theorem 1.2. *Suppose G is a real reductive group with maximal compact subgroup K and Cartan involution θ. Then there are natural bijections among the following three sets.*

(1) *Tempered irreducible representations π of G, having real infinitesimal character.*

(2) *Irreducible representations τ of K.*

(3) *Discrete final limit parameters Φ (Definition 11.2) attached to θ-stable Cartan subgroups H of G, modulo conjugation by K.*

The bijection from (1) to (2) sends π to the unique lowest K-type $\tau(\pi)$ of π. That from (3) to (1) is the Knapp-Zuckerman parametrization of irreducible tempered representations: if Φ is a discrete final limit parameter, we write $\pi(\Phi)$ for the corresponding tempered representation, and $\tau(\Phi)$ for its unique lowest K-type.

From a theoretical point of view, what is most interesting about this theorem is the bijection between (1) and (2). For our algorithmic purposes, what is most important is the bijection $\Phi \mapsto \tau(\Phi)$ from (3) (a set which is explicitly computable) to the set of representations of K we wish to study.

To a first approximation, a discrete final limit parameter attached to the θ-stable Cartan subgroup H is simply a character of the compact group $H \cap K$. We are therefore parametrizing irreducible representations of K by characters of the compact parts of Cartan subgroups of G (up to conjugation). This is in the spirit of the Cartan-Weyl parametrization of irreducible representations of a compact connected Lie group K by characters of a maximal torus T (up to conjugation); indeed we will see in Example 1.5 that the Cartan-Weyl result may be regarded as a special case of Theorem 1.2.

Example 1.3. Suppose $G = SL_2(\mathbb{R})$. Recall that $K = SO_2$, and that the irreducible representations of K are the one-dimensional characters

$$\tau_m \begin{pmatrix} \cos\theta & \sin\theta \\ -\sin\theta & \cos\theta \end{pmatrix} = e^{im\theta}. \tag{1.4}$$

Here is the list of the tempered irreducible representations of G having real infinitesimal character.

(1) For each strictly positive integer $n > 0$ there is a holomorphic discrete series representation $\pi(n)$ with Harish-Chandra parameter n. Its lowest K-type is τ_{n+1}, and its full restriction to K is

$$\pi(n)|_K = \tau_{n+1} + \tau_{n+3} + \tau_{n+5} + \cdots$$

(2) For each strictly positive integer $n > 0$ there is an antiholomorphic discrete series representation $\pi(-n)$ with Harish-Chandra parameter $-n$. Its lowest K-type is τ_{-n-1}, and its full restriction to K is

$$\pi(-n)|_K = \tau_{-n-1} + \tau_{-n-3} + \tau_{-n-5} + \cdots$$

(3) There is a holomorphic limit of discrete series representation $\pi(0^+)$. Its lowest K-type is τ_1, and its full restriction to K is

$$\pi(0^+)|_K = \tau_1 + \tau_3 + \tau_5 + \cdots$$

(4) There is an antiholomorphic limit of discrete series representation $\pi(0^-)$. Its lowest K-type is τ_1, and its full restriction to K is

$$\pi(0^-)|_K = \tau_{-1} + \tau_{-3} + \tau_{-5} + \cdots$$

(5) There is the spherical principal series representation π_{sph} with continuous parameter zero. Its lowest K-type is τ_0, and its full restriction to K is

$$\pi_{sph}|_K = \tau_0 + \tau_2 + \tau_{-2} + \tau_4 + \tau_{-4} + \cdots$$

The bijection of Theorem 1.2 identifies each non-trivial character τ_m as the lowest K-type of a (limit of) discrete series representation, holomorphic if $m > 0$ and antiholomorphic if $m < 0$. The trivial character τ_0 is the lowest K-type of the spherical principal series representation.

In terms of part (3) of Theorem 1.2, the discrete series and limits of discrete series correspond to the compact Cartan subgroup SO_2, and the spherical principal series to the split Cartan subgroup of diagonal matrices.

Example 1.5. Suppose that G is connected complex reductive algebraic group. A maximal compact subgroup K of G is the same thing as a compact form of G; K can be any compact connected Lie group. If T is a maximal torus in K, then the centralizer H of T in G is a θ-stable Cartan subgroup of G, and $H \cap K = T$. The inclusion of K in G provides an identification

$$W(K,T) = N_K(T)/T \simeq N_G(H)/H = W(G,H). \qquad (1.6a)$$

There is a unique θ-stable complement A for T in H, so that $H \simeq T \times A$. The group A is isomorphic via the exponential map to its Lie algebra.

Fix a Borel subgroup $B = HN$ of G. The Iwasawa decomposition is $G = KAN$. Any character ξ of H extends uniquely to B, and so gives rise to a principal series representation

$$\pi_B(\xi) = \operatorname{Ind}_B^G(\xi). \qquad (1.6b)$$

As a consequence of the Iwasawa decomposition,

$$\pi_B(\xi)|_K = \operatorname{Ind}_T^K(\xi|_T) = \sum_{\tau \text{ irr of } K} (\text{mult of } \xi|_T \text{ in } \tau)\tau. \qquad (1.6c)$$

If ξ is unitary, then $\pi_B(\xi)$ is tempered, irreducible, and (up to equivalence) independent of the choice of B. It follows that

$$\pi_B(w \cdot \xi) \simeq \pi_B(\xi) \qquad \xi \text{ unitary, } w \in W(G,H). \qquad (1.6d)$$

In this way the tempered irreducible representations of G are identified with Weyl group orbits of unitary characters of H. We may drop the subscript B since that choice does not matter.

The tempered representation $\pi(\xi)$ has real infinitesimal character if and only if $\xi|A$ is trivial. Characters of H trivial on A are the same as characters of the compact torus T. We have therefore described a bijection

$$\{\text{temp irr of } G, \text{ real infl char}\} \leftrightarrow \{\text{chars of } T \text{ mod } W(K,T)\}; \qquad (1.6e)$$

this is the bijection between (1) and (3) in Theorem 1.2. It follows from (1.6c) that the lowest K-type of $\pi(\xi)$ is the smallest representation of K containing the weight $\xi|T$. This is the representation $\tau(\xi|_T)$ of extremal weight ξ_T. The bijection between (2) and (3) in Theorem 1.2 is therefore precisely the Cartan-Weyl parametrization by Weyl group orbits of extremal weights.

We can now give notation for the branching problem we wish to solve. Given an irreducible representation τ of K and a tempered irreducible representation π of G with real infinitesimal character, write

$$m(\tau, \pi) = \text{multiplicity of } \tau \text{ in } \pi|_K \qquad (1.7)$$

$$= \dim \text{Hom}_K(\tau, \pi), \qquad (1.8)$$

a non-negative integer; this is the kind of multiplicity we wish to compute. (Originally we were interested in replacing π by an arbitrary standard representation π'. But it is easy explicitly to write $\pi'|_K$ as a (small) finite sum of $\pi_j|_K$, with each π_j tempered with real infinitesimal character. (One needs a single term π_j for each lowest K-type of π'.) The special case considered in (1.7) is therefore enough to solve the general problem of branching to K.)

If G is a complex group, then the discussion in Example 1.5 shows that $m(\tau, \pi(\xi))$ is the multiplicity of the T-weight $\xi|_T$ in the K-representation τ. The branching law we are studying here therefore generalizes the classical problem of finding weight multiplicities for representations of a compact group.

We want to compute the matrix $m(\tau, \pi)$. What is interesting now is that Theorem 1.2 identifies the index set for the rows of the matrix m with the index set for the columns: we may regard m as a square matrix, indexed by any of the three sets in Theorem 1.2. The definition of lowest K type implies that m is upper triangular (with respect to the ordering of representations of K used to define lowest). The fact that lowest K-types have multiplicity one implies that m has ones on the diagonal; and of course the entries of m are integers.

As a formal consequence of these observations, we find that the matrix m is necessarily invertible. A little more explicitly,

Proposition 1.9. *The multiplicity matrix m of 1.7 has a two-sided inverse M. That is, for every irreducible representation τ of K and every tempered irreducible representation π of G with real infinitesimal character, there is an integer $M(\pi, \tau)$ having the following properties:*

(1) $M(\pi, \tau) = 0$ *unless τ is less than or equal to the lowest K-type of π. If equality holds, then $M(\pi, \tau) = 1$.*

(2) *If τ and τ' are two irreducible representations of K, then*

$$\sum_{\pi \text{ temp real infl char}} m(\tau, \pi) M(\pi, \tau') = \delta_{\tau, \tau'}.$$

*Here the summands can be nonzero only for tempered representa-
tions π whose lowest K-type is less than or equal to τ, a finite set.*

(3) *If π and π' are two irreducible tempered representations of real infinitesimal character, then*

$$\sum_{\tau \text{ irr of } K} M(\pi, \tau) m(\tau, \pi') = \delta_{\pi, \pi'}.$$

*Here the summands can be nonzero only for τ less than or equal to
the lowest K-type of π', a finite set.*

*The matrix M expresses each irreducible representation of K as an integer
combination of tempered representations of G:*

$$\tau = \sum_{\pi \text{ temp real infl char}} M(\pi, \tau) \pi.$$

*This sum extends over tempered representations π whose lowest K-type is
greater than or equal to τ, and so may in principle be infinite.*

The final assertion of the proposition is just a reformulation of (2).

Inverting an upper triangular matrix with diagonal entries equal to 1 is
easy; so either of the matrices m and M is readily calculated from the other.
Since both matrices are infinite, this statement requires a little care to make
into an algorithm. In practice we can consider only some finite upper left
corner of the matrix, corresponding to the finitely many representations of
K less than or equal to some bound, and the tempered representations with
lowest K-type in that finite set.

Example 1.10. We return to the example of $SL_2(\mathbb{R})$. In the preceding
example we wrote down the multiplicity matrix m: for example,

$$m(\tau_n, \pi_{sph}) = \begin{cases} 1 & n \text{ even} \\ 0 & n \text{ odd}. \end{cases}$$

Using the interpretation of the inverse matrix M given at the end of Proposition 1.9, it is a simple matter to compute the inverse matrix M:

$$\tau_0 = \pi_{sph} - \pi(1) - \pi(-1)$$
$$\tau_1 = \pi(0^+) - \pi(2)$$
$$\tau_{-1} = \pi(0^-) - \pi(-2)$$
$$\tau_n = \pi(n-1) - \pi(n+1) \qquad (n \geq 1)$$
$$\tau_{-n} = \pi(-n+1) - \pi(-n-1) \qquad (n \geq 1).$$

Explicitly, this means that for example

$$M(\pi(j), \tau_n) = \begin{cases} 1 & j = n-1 \\ -1 & j = n+1 \\ 0 & \text{otherwise} \end{cases}$$

for any $n \geq 2$.

The central observation we wish to make about Examples 1.3 and 1.10 is that the formulas in the second example (for the inverse matrix M) are much simpler. What we will do in this paper (essentially in Theorem 16.6) is give a closed formula for M in the general case. The formula is due to Zuckerman in the case of the trivial representation of K; the generalization here is straightforward. In order to compute an explicit branching law, one can then invert (some finite upper left corner of) the upper triangular matrix M.

Implementing this branching law within Fokko du Cloux's `atlas` computer program is a current project of the NSF Focused Research Group "Atlas of Lie groups and representations." I have therefore included in this paper comments about how the algorithms described are related to things that the `atlas` software already computes.

Here is what the formula for M looks like in the case of a complex group.

Example 1.11. Suppose that G is connected complex reductive algebraic group; use the notation of Example 1.5 above. Fix a set $\Delta^+(\mathfrak{k}, T)$ of positive roots for T in $\mathfrak{k}$; these are non-trivial characters of T. Suppose τ is an irreducible representation of K, of highest weight ξ (a character of T). We want to write τ as an integer combination of restrictions to K of tempered representations of G with real infinitesimal character; equivalently, as an integer combination of representations of K induced from characters of T. This is accomplished by the Weyl character formula for τ. We can give the answer in two forms; the equivalence of the two is the Weyl denominator formula. First, we have

$$\tau = \sum_{w \in W(K,T)} \text{sgn}(w) \pi(\xi + (\rho - w\rho)).$$

Here ρ is half the sum of the positive roots of T in $\mathfrak{k}$. This formula appears to have $|W|$ terms, but some of the weights $\xi + (\rho - w\rho)$ may be conjugate by W (so there may be fewer terms).

The second formula is

$$\tau = \sum_{S \subset \Delta^+(\mathfrak{k},T)} (-1)^{|S|} \pi(\xi + 2\rho(S)).$$

Here $2\rho(S)$ is the sum of the positive roots in S. This formula appears to have $2^{|\Delta^+|}$ terms, but there is cancellation (even before the question of W conjugacy is considered). After cancellation is taken into account, the second formula reduces precisely to the first.

2. Technical Introduction

For a Lie group H we will write H_0 for the identity component, $\mathfrak{h}_0$ for the real Lie algebra, and $\mathfrak{h} = \mathfrak{h}_0 \otimes_\mathbb{R} \mathbb{C}$ for its complexification. If $\mathbf{H}$ is a complex algebraic group, we sometimes write simply $\mathbf{H}$ instead of $\mathbf{H}(\mathbb{C})$. When $\mathbf{H}$ is defined over $\mathbb{R}$, the group of real points will be written

$$H = \mathbf{H}(\mathbb{R}). \tag{2.1}$$

The notation allows for various kinds of ambiguity, since for example the same real Lie group H may appear as the group of real points of several distinct algebraic groups $\mathbf{H}$.

We write a subscript e for the identity component functor on complex algebraic groups. The group of complex points of the identity component is the classical identity component of the group of complex points:

$$\mathbf{H}_e(\mathbb{C}) = [\mathbf{H}(\mathbb{C})]_0. \tag{2.2a}$$

The group of real points of a connected algebraic group may however be disconnected, so we have in general only an inclusion

$$\mathbf{H}_e(\mathbb{R}) \supset [\mathbf{H}(\mathbb{R})]_0; \tag{2.2b}$$

Briefly (but perhaps less illuminatingly)

$$H_e \supset H_0. \tag{2.2c}$$

If $\mathbf{H}(\mathbb{R})$ is compact, then equality holds:

$$\mathbf{H}_e(\mathbb{R}) = [\mathbf{H}(\mathbb{R})]_0 \qquad (\mathbf{H}(\mathbb{R}) \text{ compact}). \tag{2.2d}$$

That is,

$$H_e = H_0 \qquad (H \text{ compact}). \tag{2.2e}$$

Definition 2.3. We fix once and for all a complex connected reductive algebraic group $\mathbf{G}$, together with a *real form*; that is, a complex conjugate-linear involutive automorphism

$$\sigma \colon \mathbf{G}(\mathbb{C}) \to \mathbf{G}(\mathbb{C}).$$

(When this is all translated into the `atlas` setting with several real forms, it will be convenient to define a *strong real form* to be a particular choice of representative for σ in an appropriate extended group containing $\mathbf{G}(\mathbb{C})$

as a subgroup of index two. I am avoiding that by speaking always of a single real form.) The corresponding *group of real points* is

$$\mathbf{G}(\mathbb{R}, \sigma) = \mathbf{G}(\mathbb{R}) = \mathbf{G}(\sigma) = G$$
$$= \mathbf{G}(\mathbb{C})^{\sigma};$$

here the second line defines the notation introduced in the first.

Definition 2.4. A *Cartan involution* for the real group G is an algebraic involutive automorphism θ of $\mathbf{G}$ subject to

(1) The automorphisms σ and θ commute: $\sigma\theta = \theta\sigma$.
(2) The composite (conjugate-linear) involutive automorphism $\sigma\theta$ has as fixed points a compact real form U of $\mathbf{G}$.

The second requirement is equivalent to

(2′) The group of fixed points of θ on G is a maximal compact subgroup K of G:

$$G^\theta = K.$$

Under these conditions, the complexification of the compact Lie group K is equal to $\mathbf{G}^\theta$:

$$\mathbf{K} = \mathbf{G}^\theta.$$

The action of θ on the complex Lie algebra $\mathfrak{g}$ defines an eigenspace decomposition

$$\mathfrak{g} = \mathfrak{k} + \mathfrak{s}, \tag{2.5}$$

with $\mathfrak{s}$ the -1 eigenspace of θ. This decomposition is inherited by any θ-stable real or complex subspace of $\mathfrak{g}$. The letter $\mathfrak{s}$ is chosen to suggest "symmetric," since when G is $GL_n(\mathbb{R})$ we can take θ to be negative transpose on the Lie algebra.

Almost all the real Lie groups here are going to be groups of real points of complex algebraic groups. In general an identification of a real Lie group with such a group of real points need not be unique even if it exists. But there is one very important special case when the algebraic group *is* unique. Any compact (possibly disconnected) Lie group S is a compact real form of a canonically defined complex reductive algebraic group $\mathbf{S}$. The complex algebra of regular functions on $\mathbf{S}$ is the complex algebra of matrix coefficients of finite-dimensional representations of S (and so $\mathbf{S}$ is by definition the spectrum of this ring).

This explicit description of $\mathbf{S}$ shows that if S is a closed subgroup of another compact Lie group K, then $\mathbf{S}$ is canonically an algebraic subgroup of $\mathbf{K}$ defined over $\mathbb{R}$.

The complexification of the unitary group U_n may be naturally identified with the complex general linear group $\mathbf{GL}_n(\mathbb{C})$. (Another dangerous bend here: the corresponding real form of $\mathbf{GL}_n(\mathbb{C})$ is given by inverse conjugate transpose. This is not the standard real form, given by complex conjugation, which is (one of the reasons) why U_n is not the same as $GL_n(\mathbb{R})$.) Pursuing this a bit shows that a continuous morphism from a compact group S to U_n is the restriction of a unique algebraic morphism of $\mathbf{S}$ into $\mathbf{GL}_n$. That is, continuous representations of the compact Lie group S may be identified with algebraic representations of $\mathbf{S}$.

The involutive automorphism θ is unique up to conjugation by G. Again in the setting of the atlas software we will keep a *strong Cartan involution*, a choice of representative for θ in an extended group for $\mathbf{G}$. (This extended group is not the same as the one containing σ, since that one contains conjugate-linear automorphisms).

Here is some general notation for representations.

Definition 2.6. Suppose $\mathbf{H}$ is a complex algebraic group. The set of equivalence classes of irreducible (necessarily finite-dimensional) algebraic representations of $\mathbf{H}$ is written

$$\Pi^*_{alg}(\mathbf{H}) = \Pi^*(\mathbf{H}),$$

the *algebraic dual* of $\mathbf{H}$. If $\mathbf{H}$ is a complex torus, then these representations are all one-dimensional, providing a canonical identification

$$\Pi^*(\mathbf{H}) \simeq X^*(\mathbf{H})$$

(the term on the right being the lattice of characters of $\mathbf{H}$).

Suppose $\mathbf{H}$ is reductive and defined over $\mathbb{R}$, and that H is compact. According to the remarks after Definition 2.4, restriction to H provides a canonical identification

$$\Pi^*(\mathbf{H}) \simeq \widehat{H},$$

with $\widehat{H}$ denoting the set of equivalence classes of continuous irreducible representations of H. Because of this identification, we will use the more suggestive notation

$$\Pi^*_{adm}(H) = \Pi^*(H)$$

for $\widehat{H}$, calling it the *admissible dual* of H. Then the canonical identification looks like

$$\Pi^*_{alg}(\mathbf{H}) \simeq \Pi^*_{adm}(H) \qquad (H \text{ compact}).$$

The term "continuous dual" makes more sense than "admissible dual" for H compact, but we have chosen the latter to fit with the following generalization.

Suppose that $\mathbf{H}$ is reductive and defined over $\mathbb{R}$, and that L is a maximal compact subgroup of H. The *admissible dual* of H is

$$\Pi^*_{adm}(H) = \Pi^*(H) = \text{equivalence classes of irreducible } (\mathfrak{h}, L)\text{-modules};$$

this set is often written $\widehat{H}$.

If L' is another maximal compact subgroup of H, then we know that L' is conjugate to L by $\mathrm{Ad}(h)$, for some element $h \in H$. Twisting by h carries (irreducible) $(\mathfrak{h}, L)$-modules to (irreducible) $(\mathfrak{h}, L')$-modules, and the correspondence of irreducibles is independent of the choice of h carrying L to L'. For this reason we can omit the dependence of $\Pi^*(H)$ on L.

Our goal is to parametrize $\Pi^*(G)$ and $\Pi^*(K)$ (that is, $\widehat{G}$ and $\widehat{K}$) in compatible and computer-friendly ways. Such parametrizations involve a variety of different "ρ-shifts." The most fundamental of these is that Harish-Chandra's parameter for a discrete series representation is not the differential of a character of the compact Cartan subgroup, but rather such a differential shifted by a half-sum of positive roots. There is a convenient formalism for keeping track of these shifts, that of "ρ-double covers" of maximal tori. But we have avoided it in these notes.

3. Highest Weights for K

The phrase "maximal torus" can refer to a product of circles in a compact Lie group; or to a product of copies of $\mathbb{C}^\times$ in a complex reductive algebraic group; or to the group of real points of a complex maximal torus (which is isomorphic to a product of circles, $\mathbb{R}^\times$, and $\mathbb{C}^\times$). I will use all three meanings, sometimes with an adjective "compact" or "complex" or "real" to clarify.

Definition 3.1. Choose a compact maximal torus

$$T_{f,0} \subset K_0;$$

that is, a maximal connected abelian subgroup. This torus is unique up to conjugation by K_0. Then $T_{f,0}$ is a product of circles, and its complexification may be identified canonically with a complex maximal torus

$$\mathbf{T}_{f,0} \subset \mathbf{K}_e$$

The centralizer

$$\mathbf{H}_f = Z_{\mathbf{G}}(\mathbf{T}_{f,0})$$

is a θ-stable maximal torus in $\mathbf{G}$ (this requires proof, but is not difficult) that is defined over $\mathbb{R}$. The group $H_f = \mathbf{H}_f(\mathbb{R})$ is a real maximal torus of G, called a *fundamental Cartan subgroup*; it is the centralizer in G of $T_{f,0}$, and is unique up to conjugation by K_0.

Define

$$\mathbf{T}_f = \mathbf{H}_f^\theta = \mathbf{H}_f \cap \mathbf{K} = Z_\mathbf{K}(\mathbf{T}_{f,0}).$$

The group of real points is

$$T_f = H_f^\theta = H_f \cap K = Z_K(T_{f,0}),$$

a compact abelian subgroup of $K(\mathbb{R})$ containing the maximal torus $T_{f,0}$. In fact it is easy to see that $T_f(\mathbb{R})$ is a maximal abelian subgroup of $K(\mathbb{R})$ (and similarly over $\mathbb{C}$); it might reasonably be called a *maximally toroidal maximal abelian subgroup*, but I will follow [9] and call it a *small Cartan subgroup*.

The definition of T_f as $Z_K(T_{f,0})$ is internal to K, and so makes sense for any compact Lie group. In that generality T_f is a compact Lie group with identity component the compact torus $T_{f,0}$; but T_f need not be abelian. If we define

$$K^\sharp = \{k \in K \mid \mathrm{Ad}(k) \text{ is an inner automorphism of } \mathfrak{k}\},$$

then

$$K^\sharp = T_f K_0, \qquad K^\sharp/T_f \simeq K_0/T_{f,0};$$

all of these equalities remain true for complex points.

Clearly $K^\sharp$ is a normal subgroup of K, of finite index. We may therefore define the *R-group of K* as

$$R(K) = K/K^\sharp \simeq \mathbf{K}/\mathbf{K}^\sharp.$$

The R-group is a finite group, equal to the image of K in the outer automorphism group of the root datum of K_0. It will play an important part in our description of representations of K. For the special groups K appearing in our setup, we will see that the R-group is a product of copies of $\mathbb{Z}/2\mathbb{Z}$; the number of copies is bounded by the cardinality of a set of orthogonal simple roots in the Dynkin diagram of $\mathbf{G}$.

Definition 3.2. Recall from Definition 2.6 that the *group of characters of T_f* is

$$\Pi^*(T_f) = \text{continuous characters } \xi\colon T_f \to U_1$$

$$\simeq \Pi^*(\mathbf{T}_f) = \text{algebraic characters } \xi\colon \mathbf{T}_f \to \mathbb{C}^\times;$$

identification of the second definition with the first is by restriction to T_f. (Here U_1 is the circle group, the compact real form of $\mathbb{C}^\times$.) The group $\Pi^*(T_f)$ is a finitely generated abelian group. More precisely, restriction to $T_{f,0}$ defines a short exact sequence of finitely generated abelian groups

$$0 \to \Pi^*(T_f/T_{f,0}) \to \Pi^*(T_f) \to \Pi^*(T_{f,0}) \to 0;$$

the image is a lattice (free abelian group) of rank equal to the dimension of the torus, and the kernel is the (finite) group of characters of the component group $T_f/T_{f,0}$.

We will occasionally refer to the *lattice of cocharacters of* T_f

$$\Pi_*(T_f) = \text{continuous homomorphisms } \xi \colon U_1 \to T_f$$

$$= \text{algebraic homomorphisms } \xi \colon \mathbb{C}^\times \to \mathbf{T}_f \,.$$

Clearly such homomorphisms automatically have image in the identity component, so $\Pi_*(T_f) = \Pi_*(T_{f,0})$. There is a natural bilinear pairing

$$\langle , \rangle \colon \Pi^*(T_f) \times \Pi_*(T_f) \to \mathbb{Z};$$

this pairing is trivial on the torsion subgroup $\Pi^*(T_f/T_{f,0})$, and descends to the standard identification of $\Pi_*(T_{f,0})$ as the dual lattice of $\Pi^*(T_{f,0})$.

Whenever V is a complex representation of T_f, we write

$$\Delta(V, T_f) = \text{set of weights of } T_f \text{ on } V$$

regarded as a multiset of elements of $\Pi^*(T_f)$.

The most important example of such a set is the *root system of* T_f *in* K

$$\Delta(K, T_f) = \Delta\left(\mathfrak{k}/\mathfrak{t}_f\right).$$

In one-to-one correspondence with the root system is the system of *coroots of* T_f *in* K

$$\Delta^\vee(K, T_f) \subset \Pi_*(T_f);$$

we write $\alpha^\vee$ for the coroot corresponding to the root α.

The difference from the classical definition of root system is that we are keeping track of the action of the disconnected group T_f. Because the roots of $T_{f,0}$ in K have multiplicity one, the restriction map from $\Pi^*(T_f)$ to $\Pi^*(T_{f,0})$ is necessarily one-to-one on this root system. That is, each root of the connected torus $T_{f,0}$ in the connected group K_0 extends to a unique root of T_f. The point of this discussion is that $\Delta(K, T_f)$ is in canonical one-to-one correspondence with a root system in the classical sense (that of $T_{f,0}$ in K_0). We may therefore speak of classical concepts like "system of positive roots" without fear of reproach.

Definition 3.3. While we are feeling bold, let us therefore choose a system

$$\Delta^+(K, T_f) \subset \Delta(K, T_f)$$

of positive roots. Because so much of what we will do depends on this choice, it will appear constantly in the notation. Simply for notational convenience, we will therefore use the shorthand

$$\Delta_c^+ = \Delta^+(K, T_f);$$

the subscript c stands for "compact." A representation (μ, E_μ) of T_f is called *dominant* (or Δ_c^+-*dominant*) if for every positive root $\alpha \in \Delta_c^+$, and every character $\xi \in \Pi^*(T_f)$ occurring in E_μ, we have $\langle \xi, \alpha^\vee \rangle \geq 0$. We write

$$\Pi^*_{\Delta_c^+-dom}(T_f) = \Pi^*_{dom}(T_f)$$

for the set of dominant weights.

Choice of Δ_c^+ is equivalent to the choice of a Borel subalgebra

$$\mathfrak{b}_c = \mathfrak{t}_f + \mathfrak{n}_c.$$

We want to use $\mathfrak{b}_c$ to construct representations of K, and for that purpose it will be convenient to take the roots of the Borel subalgebra to be the *negative* roots:

$$\Delta(\mathfrak{n}_c, T_f) = -\Delta_c^+.$$

Define

$$\mathbf{B}_c^\sharp = \mathbf{T}_f \mathbf{N}_c \subset \mathbf{K}^\sharp,$$

which we call a *small Borel subgroup of* $\mathbf{K}$. There are several things to notice here. First, $\mathbf{B}_c^\sharp$ and its unipotent radical $\mathbf{N}_c$ are not preserved by the complex structure, so they are not defined over $\mathbb{R}$. Second, the group $\mathbf{B}_c^\sharp$ is disconnected if $\mathbf{T}_f$ is; we have

$$\mathbf{B}_{c,0} = \mathbf{T}_{f,0} \mathbf{N}_c \subset \mathbf{K}_0,$$

a Borel subgroup in the usual sense.

From time to time we will have occasion to refer to the Borel subalgebra whose roots are the positive roots; the nil radical is $\sigma \mathfrak{n}_c$ (with σ the complex conjugation). The corresponding small Borel subgroup is $\sigma(\mathbf{B}_{c,0})$.

Our first goal is to parametrize representations of K in terms of elements of $\Pi^*(T_f)$, together with a little more information. The point of this is that $\Pi^*(T_f)$ is accessible to the computer. I will recall why (in a slightly more general context) in Section 6.

Before embarking on a precise discussion of representations of K, we need a few more definitions.

Definition 3.4. In the setting of Definition 3.3, the *large Cartan subgroup of K* is

$$T_{fl} = \text{normalizer in } K \text{ of } \mathfrak{b}_c$$
$$= \text{normalizer in } K \text{ of } \mathfrak{n}_c.$$

A representation μ_l of T_{fl} is called *dominant* if its restriction to T_f is dominant (Definition 3.3). If μ_l is irreducible, this is equivalent to requiring that just one weight of T_f in μ_l be dominant. We write

$$\Pi^*_{\Delta^+_c - dom}(T_{fl}) = \Pi^*_{dom}(T_{fl})$$

for the set of dominant irreducible representations of T_{fl}.

Because $\mathfrak{b}_c \cap \sigma \mathfrak{b}_c = \mathfrak{t}_f$, the group T_{fl} normalizes $T_{f,0}$. Its complexification is

$$\mathbf{T}_{fl} = \text{normalizer of } \mathbf{T}_{f,0} \text{ and } \mathbf{B}_{c,0} \text{ in } \mathbf{K}$$
$$= \text{normalizer of } \mathbf{T}_{f,0} \text{ and } \mathfrak{n}_c \text{ in } \mathbf{K}.$$

Define

$$\mathbf{B}_{cl} = \text{normalizer of } \mathbf{B}_{c,0} \text{ in } \mathbf{K}$$
$$= \text{normalizer of } \mathfrak{n}_c \text{ in } \mathbf{K}$$
$$= \mathbf{T}_{fl}\mathbf{N}_c.$$

The equality of the three definitions is an easy exercise; the last description is a semidirect product with the unipotent factor normal.

Next we introduce the flag varieties that control the representation theory of K.

Definition 3.5. One of the most fundamental facts about connected compact Lie groups is the identification (defined by the obvious map from left to right)

$$K_0/T_{f,0} \simeq \mathbf{K}_0/\mathbf{B}_{c,0}. \tag{3.6a}$$

This homogeneous space is called the *flag variety of K_0* (or of $\mathbf{K}_0$); it may be identified with the projective algebraic variety of Borel subalgebras of $\mathfrak{k}$. Especially when we have this interpretation in mind, we may write the space as $X(K_0)$ or $X(\mathbf{K}_0)$.

For disconnected K we will be interested in at least two versions of this space. The first is the *large flag variety for K* (or $\mathbf{K}$)

$$X_{large}(K) = K/T_f \simeq \mathbf{K}/\mathbf{B}_c^\sharp. \tag{3.6b}$$

The right side is a complex projective algebraic variety. This variety has (many) K-equivariant embeddings in the flag variety for $\mathbf{G}$, corresponding to the finitely many extensions of $\mathfrak{b}_c$ to a Borel subalgebra of $\mathfrak{g}$. But we will also need to consider the *small flag variety for K* (or $\mathbf{K}$)

$$X_{small}(K) = K/T_{fl} \simeq \mathbf{K}/\mathbf{B}_{cl}. \tag{3.6c}$$

Since $\mathbf{B}_{cl}$ is by definition the normalizer of $\mathfrak{b}_c$, the small flag variety may be identified with the variety of Borel subalgebras of $\mathfrak{k}$. It follows that the inclusion of $\mathbf{K}_0$ in $\mathbf{K}$ defines a $\mathbf{K}_0$-equivariant identification

$$K_0/T_{f,0} \simeq \mathbf{K}_0/\mathbf{B}_{c,0} \simeq \mathbf{K}/\mathbf{B}_{cl} \simeq K/T_{fl}. \tag{3.6d}$$

Because T_f is a subgroup of T_{fl}, there is a natural K-equivariant surjection

$$K/T_f = X_{large}(K) \rightarrow X_{small}(K) = K/T_{fl}. \tag{3.6e}$$

The equivariant geometry here is slightly confusing. Because $T_f \cap K_0 = T_{f,0}$, K/T_f is (as an algebraic variety) a union of card$(K/K^\sharp)$ copies of $K_0/T_{f,0}$. (Here we write card S for the cardinality of a set S.) The reason this is the number of copies is that $K_0 T_f = K^\sharp$. The projection of (3.6e) sends each copy isomorphically onto $K/T_{fl} \simeq K_0/T_{f,0}$.

Proposition 3.7. *Suppose we are in the setting of Definition 3.4. Then the group T_{fl} meets every component of K. We have*

$$T_{fl} \cap K_0 = T_{f,0}, \qquad T_{fl} \cap K^\sharp = T_f.$$

Consequently the inclusion of T_{fl} in K defines natural isomorphisms

$$T_{fl}/T_{f,0} \simeq K/K_0, \qquad T_{fl}/T_f \simeq K/K^\sharp = R(K)$$

(Definition 3.1).

We can now begin to talk about representations of K.

Definition 3.8. Recall from Definition 2.6 the identification

$$\Pi^*(K) = \text{equiv. classes of continuous irr. representations of } K$$

$$\simeq \text{equiv. classes of algebraic irr. representations of}$$

$$\mathbf{K} = \Pi^*(\mathbf{K}).$$

The identification of the second definition with the first is by restriction to K.

Typically I will write (τ, V_τ) for a representation of K (not necessarily irreducible). The *highest weight space of τ* is

$$V_\tau^h = V_\tau/\mathfrak{n}_c V_\tau,$$

the subspace of coinvariants for the Lie algebra $\mathfrak{n}_c$ in V_τ. Recall that $\mathfrak{n}_c$ corresponds to negative roots; for that reason, V_τ^h may be naturally identified with the subspace of vectors annihilated by the positive root vectors. Explicitly,

$$V_\tau^{\sigma\mathfrak{n}_c} \simeq V_\tau^h$$

by composing the inclusion of left side in V_τ with the quotient map. Of course it is this second definition of highest weight space that is more commonly stated, but the first is going to be more natural for us. It is clear from Definition 3.4 that the large Cartan subgroup T_{fl} acts on V_τ^h; write

$$\mu_l(\tau)\colon T_{fl} \to \mathrm{End}(V_\tau^h)$$

for the corresponding representation. Of course we may also regard $\mu_l(\tau)$ as a representation of T_f, $T_{f,0}$, $\mathbf{T}_{fl}$, $\mathbf{B}_{cl}$, and so on.

Proposition 3.9. *Suppose we are in the setting of Definitions 2.4 and 3.4.*

(1) *Passage to highest weight vectors (Definition 3.8) defines a bijection*

$$\Pi^*(K) \to \Pi^*_{dom}(T_{fl}), \qquad \tau \mapsto \mu_l(\tau)$$

from irreducible representations of K to dominant irreducible representations (Definition 3.4) of the large Cartan subgroup T_{fl}.

(2) *Suppose (μ_l, E_{μ_l}) is a dominant representation of T_{fl}. Extend μ_l to an algebraic representation of $\mathbf{B}_{cl}$ (on the same space E_{μ_l}) by making $\mathbf{N}_c$ act trivially. Write $\mathcal{O}(\mu_l)$ for the corresponding $\mathbf{K}$-equivariant sheaf on the small flag variety $X_{small}(K)$. Then the space*

$$V(\mu_l) = H^0(X_{small}, \mathcal{O}(\mu_l))$$

of global sections is a finite-dimensional representation of K, irreducible if and only if μ_l is. Evaluation of sections at the base point of $X_{small}(K)$ defines a natural isomorphism

$$V(\mu_l)^h \simeq E_{\mu_l};$$

This construction therefore inverts the bijection of (1).

(3) *Suppose $\mu \in \Pi^*_{dom}(T_f)$ is a dominant character of T_f. Write $\mathcal{O}(\mu)$ for the corresponding $K(\mathbb{C})$-equivariant sheaf on the large flag variety $X_{large}(K)$, and define*

$$V(\mu) = H^0(X_{large}, \mathcal{O}(\mu)),$$

a finite-dimensional representation of K. Finally define

$$\mu_l = \mathrm{Ind}_{T_f}^{T_{fl}} \mu,$$

a representation of dimension equal to the cardinality of $T_{fl}/T_f \simeq$ $R(K)$ (Proposition 3.7). Then there is a natural isomorphism

$$H^0(X_{large}, \mathcal{O}(\mu)) \simeq H^0(X_{small}, \mathcal{O}(\mu_l));$$

that is, $V(\mu) \simeq V(\mu_l)$.

Our goal was to relate irreducible representations of K to dominant characters of T_f. What remains is to relate dominant irreducible representations of T_{fl} to dominant characters of T_f. We will see that "most" dominant irreducible representations of T_{fl} are induced from dominant irreducible characters of T_f. Part (3) of the Proposition shows how to construct irreducible representations of K from such dominant characters of T_f. We will see what happens in the remaining cases in Theorem 4.10.

4. The R-Group of K and Irreducible Representations of the Large Cartan

In this section we elucidate the structure of $R(K)$, and find a (computable) parametrization of $\widehat{K}$ in terms of T_{fl}.

Definition 4.1. In the setting of Definition 3.1, the *Weyl group of H_f in G* is

$$W(G, H_f) = N_G(H_f)/H_f.$$

This acts on $\mathbf{H}_f$ respecting the real structure and θ; it is a subgroup of the Weyl group of the root system of $\mathbf{H}_f$ in $\mathbf{G}$. Each coset in this Weyl group has a representative in K. Furthermore any element of K normalizing $T_{f,0}$ actually normalizes all of H_f; this is immediate from the definition of H_f.

The *Weyl group of T_f in K* is

$$W(K, T_f) = N_K(T_f)/T_f.$$

According to the remarks in the preceding paragraph, inclusion defines an isomorphism

$$W(K, T_f) \simeq W(G, H_f).$$

Because T_f is a normal subgroup of T_{fl} (Definition 3.4), Proposition 3.7 provides a natural inclusion

$$T_{fl}/T_f \simeq R(K) \hookrightarrow W(K, T_f).$$

This inclusion depends on our fixed choice of Δ_c^+. If necessary to avoid confusion, we may write the image as $R(K)_{dom}$ or $R(K)_{\Delta_c^+}$. A little more explicitly, it is clear from the definition of T_{fl} that

$$R(K)_{dom} = \{w \in W(K, T_f) \mid w(\Delta_c^+) = \Delta_c^+\}.$$

Finally, since $T_f \cap K_0 = T_{f,0}$, there is a natural inclusion

$$W(K_0, T_{f,0}) \hookrightarrow W(K, T_f).$$

Proposition 4.2. *In the setting of Definition 4.1, there is a semidirect product decomposition*

$$W(K, T_f) = R(K)_{dom} \ltimes W(K_0, T_{f,0}).$$

This is an immediate consequence of the fact that $W(K_0, T_{f,0})$ acts in a simply transitive way on choices of positive roots for $T_{f,0}$ in K.

We turn next to an explicit description of $R(K)_{dom}$. Define the *modular character* $2\rho_c$ of T_{fl} to be the determinant of the adjoint action of T_{fl} on the holomorphic tangent space

$$2\rho_c(t) = \det\left(\mathrm{Ad}(t) \text{ acting on } \mathfrak{k}/\mathfrak{b}_c\right) \tag{4.3a}$$

The notation is a little misleading, since there may not be a character ρ_c with square equal to $2\rho_c$. But it is traditional. I will use the same notation $2\rho_c$ for the restriction to T_f or to $T_{f,0}$. We could avoid this abuse of notation by writing $2\rho_{cl}$, $2\rho_c$, and $2\rho_{c,0}$ for the characters of T_{fl}, T_f, and $T_{f,0}$ respectively. But the ambiguity will not lead to problems.

As a character of T_f, we have

$$2\rho_c = \sum_{\alpha \in \Delta_c^+} \alpha. \tag{4.3b}$$

But T_{fl} may act non-trivially on $\Delta(K, T_f)$, so the summands need not extend to characters of T_{fl}.

For the moment the most important property of the modular character for us is

$$\Delta_c^+ = \left\{\alpha \in \Delta(K, T_f) \mid \langle 2\rho_c, \alpha^\vee \rangle > 0\right\}. \tag{4.3c}$$

More precisely,

$$\text{simple roots of } T_f \text{ in } \Delta_c^+ = \left\{\alpha \in \Delta(K, T_f) \mid \langle 2\rho_c, \alpha^\vee \rangle = 2\right\}. \tag{4.3d}$$

Proposition 4.4. *In the setting of (4.3), regard $2\rho_c$ as an element of $\mathfrak{t}_f^*$; extend it to a θ-fixed element of $\mathfrak{h}_f^*$, as is possible uniquely. Define*

$$\Delta_c^\perp = \{\beta \in \Delta(\mathbf{G}, \mathbf{H}_f) \mid \langle 2\rho_c, \beta^\vee \rangle = 0\}.$$

(1) *The set of roots $\Delta_c^\perp$ is a root system. If we fix any positive system for $\Delta(\mathbf{G}, \mathbf{H}_f)$ making $2\rho_c$ dominant, then $\Delta_c^\perp$ is spanned by simple roots. We may therefore define*

$$W_c^\perp = W(\Delta_c^\perp),$$

a Levi subgroup of $W(\mathbf{G}, \mathbf{H}_f)$.

(2) *The stabilizer of $2\rho_c$ in $W(\mathbf{G}, \mathbf{H}_f)$ is equal to $W_c^\perp$.*

(3) *As a subgroup of $W(K, T_f)$ (Definition 4.1), $R(K)_{dom}$ is the intersection with $W_c^\perp$:*

$$R(K)_{dom} = W(K, T_f) \cap W_c^\perp.$$

(4) *The root system $\Delta_c^\perp$ is of type A_1^r. That is,*

$$\Delta_c^\perp = \{\pm\beta_1, \dots, \pm\beta_r\},$$

with $P = \{\beta_i\}$ a collection of strongly orthogonal noncompact imaginary roots. Consequently $W_c^\perp \simeq (\mathbb{Z}/2\mathbb{Z})^r$. Elements of $W_c^\perp$ are in one-to-one correspondence with subsets $A \subset P$: to the subset A corresponds the Weyl group element

$$w_A = \prod_{\beta_i \in A} s_{\beta_i}.$$

(5) *The group $R(K)_{dom}$ acts on the collection of weights (Definition 3.3) of T_f. Each $R(K)_{dom}$ orbit of weights has a unique element μ satisfying*

$$\sum_{\beta_i \in A} \langle \mu, \beta_i^\vee \rangle \geq 0 \qquad \text{(all } w_A \in R(K)_{dom}\text{)}.$$

(6) *Suppose $\mu \in \Pi_{dom}^*(T_f)$. Define*

$$P(\mu) = \{\beta_i \in P \mid \langle \mu, \beta_i^\vee \rangle = 0\}$$

$$R(K, \mu)_{dom} = \{w_A \mid w_A \in R(K)_{dom}, \ A \subset P(\mu)\}.$$

Then $R(K, \mu)_{dom}$ is the stabilizer of μ in $R(K)_{dom}$.

Proof. An appropriate analogue of (1) holds with $2\rho_c$ replaced by any weight in $\Pi^*(\mathbf{H}_f)$; and (still in that generality) (2) is Chevalley's theorem on the stabilizer of a weight in the Weyl group. Part (3) is now clear from Definition 4.1. For (4), it is clear that $\Delta_c^\perp$ is preserved by θ but has no compact roots (since compact coroots cannot vanish on $2\rho_c$). Now the assertion that $\Delta_c^\perp$ is of type A_1^r, with all roots noncompact imaginary, can be proved either by a short direct argument, or by inspection of the classification of real reductive groups. Parts (5) and (6) are elementary. $\square$

The $R(K)_{dom}$ orbit representatives defined in (5) are called *P-positive*, or *$\{\beta_i\}$-positive*. Of course the notion depends on the choice of $\{\beta_i\}$; that is, on the choice of one root from each pair $\pm\beta_i$.

Not all of the roots in P contribute to $R(K)_{dom}$, and occasionally it will be helpful to single out the ones (called *essential*) that do. In the notation of the preceding Proposition, set

$$P_{ess} = \bigcup_{w_A \in R(K)_{dom}} A \subset P = \{\beta_1, \ldots, \beta_r\}. \qquad (4.5)$$

We may write this set as $P_{ess}(K) \subset P(K)$ if necessary. Similarly we define $P_{ess}(\mu)$. Clearly the notion of P-positive depends only on P_{ess}, so we may call it P_{ess}-*positive*.

Corollary 4.6. *In the setting of Definition 4.1, each orbit of $W(K, T_f)$ on $\Pi^*(T_f)$ has a unique representative which is dominant for Δ_c^+ (Definition 3.3 and P-positive (Proposition 4.4).*

Definition 4.7. Suppose $\mu \in \Pi^*_{dom}(T_f)$. In terms of the isomorphism $R(K)_{dom} \simeq T_{fl}/T_f$ of Proposition 3.7, define $T_{fl}(\mu)$ to be the inverse image of $R(K, \mu)_{dom}$ (Proposition 4.4). With respect to the natural action of T_{fl} on characters of its normal subgroup T_f, we have

$$T_{fl}(\mu) = \{x \in T_{fl} \mid x \cdot \mu = \mu\}.$$

We are interested in representations of $T_{fl}(\mu)$ extending the character μ of T_f. We will show in a moment (Lemma 4.8) that the quotient group $T_{fl}(\mu)/\ker(\mu)$ is abelian, so irreducible representations of this form are one-dimensional. We can therefore define an *extension of μ* to be a character

$$\widetilde{\mu} \in \Pi^*(T_{fl}(\mu))$$

with the property that

$$\widetilde{\mu}|_{T_f} = \mu.$$

There is a natural simply transitive action of $\Pi^*(R(K, \mu)_{dom})$ on extensions of μ, just by tensor product of characters. We call $\widetilde{\mu}$ an *extended dominant weight*.

Lemma 4.8. *The commutator subgroup of $T_{fl}(\mu)$ is contained in the product of the images of the coroots*

$$\beta^\vee : U(1) \to T_f,$$

as β runs over the noncompact imaginary roots in $P_{ess}(\mu)$ (Proposition 4.4). Since μ is trivial on all of these coroots, the quotient $T_{fl}(\mu)/\ker(\mu)$ is abelian.

Example 4.9. Suppose $\mathbf{G} = \mathbf{GL}_{2n}$ with the standard real form (complex conjugation of matrices) $G = GL_{2n}(\mathbb{R})$. We choose the Cartan involution

$\theta(g) = {}^t g^{-1}$, so that $K = O_{2n}$ is the compact orthogonal group. We may choose as maximal torus

$$T_{f,0} = [SO_2]^n,$$

embedded as block-diagonal matrices. If we identify $\mathbb{R}^2$ with $\mathbb{C}$ as usual, then $SO_2 \subset GL_2(\mathbb{R})$ is identified with U_1, the group of multiplications by complex numbers of absolute value 1. It follows that

$$Z_{GL_2(\mathbb{R})}(SO_2) \simeq \mathbb{C}^\times,$$

the multiplicative group of nonzero complex numbers; and that

$$H_f = Z_{GL_{2n}(\mathbb{R})}([SO_2]^n) \simeq [\mathbb{C}^\times]^n,$$

a fundamental Cartan subgroup of G. The maximal compact subgroup of $[\mathbb{C}^\times]^n$ is $[U_1]^n$, so we find

$$T_f = T_{f,0} = [SO_2]^n.$$

Characters of SO_2 may be naturally identified with $\mathbb{Z}$, so $\Pi^*(T_f) = \mathbb{Z}^n$.

If we write $e_1, \dots, e_n$ for the standard basis of $\mathbb{Z}^n$, the compact root system is

$$\Delta(K, T_f) = \{\pm e_i \pm e_j, \mid 1 \le i \ne j \le n\}.$$

As positive roots we may choose

$$\Delta_c^+ = \{e_i \pm e_j, \mid 1 \le i < j \le n\}.$$

Adding these characters gives

$$2\rho_c = (2n - 2, 2n - 4, \dots, 2, 0) \in \mathbb{Z}^n.$$

The roots of K are precisely the restrictions to T_f of the complex roots of H_f. Evidently none of these complex roots is orthogonal to $2\rho_c$. The imaginary roots are $\{\pm 2e_j\}$, all noncompact. Only the last of these is orthogonal to $2\rho_c$, so the set P of Proposition 4.4 is $\{2e_n\}$.

The compact Weyl group $W(K, T_f)$ acts on $[SO_2]^n$ by permuting the coordinates and inverting some of them. The Weyl group $W(K_0, T_{f,0})$ is the subgroup inverting always an even number of coordinates. Consequently the large compact Cartan is

$$T_{fl} = [SO_2]^{n-1} \times O_2,$$

and the R group for K is

$$R(K)_{dom} = \{1, s_{2e_n}\}.$$

(Other choices of Δ_c^+ would replace n by some other coordinate m in this equality.) A weight $\mu = (\mu_1, \ldots, \mu_n)$ is dominant if and only if

$$\mu_1 \geq \mu_2 \geq \cdots \geq \mu_{n-1} \geq |\mu_n|.$$

It is P-positive (remark after Proposition 4.4) if and only if $\mu_n \geq 0$.

Here is a classification of irreducible representations of K.

Theorem 4.10. *Suppose G is the group of real points of a connected reductive algebraic group, K is a maximal compact subgroup (Definition 2.4), and T_f is a small Cartan subgroup of K (Definition 3.1). Fix a positive root system for T_f in K (Definition 3.3, and let $R(K)_{dom}$ be the R-group for K (acting on the character group $\Pi^*(T_f)$). Choose roots $\{\beta_i\}$ as in Proposition 4.4. Then the irreducible representations of K are in natural one-to-one correspondence with $R(K)_{dom}$ orbits of extended dominant weights $\widetilde{\mu}$ (Definition 4.7).*

Write $\tau(\widetilde{\mu})$ for the irreducible representation of K corresponding to the extended weight $\widetilde{\mu}$. Here are some properties of $\tau(\widetilde{\mu})$.

(1) *The restriction of $\tau(\widetilde{\mu})$ to $K^\sharp$ (Definition 3.1) is the direct sum of the irreducible representations of highest weights $r \cdot \mu$ (for $r \in R(K)_{dom}$), each appearing with multiplicity one. These summands remain irreducible on restriction to K_0. If we write μ_0 for the restriction of μ to $T_{f,0}$, and $\tau(\mu_0)$ for the irreducible representation of K_0 of highest weight μ_0, then*

$$\dim \tau(\widetilde{\mu}) = (\dim \tau(\mu_0))(|R(K)_{dom}/R(K, \mu)_{dom}|)$$

The first factor is given by the Weyl dimension formula, and the second is a power of two.

(2) *The highest weight of $\tau(\widetilde{\mu})$ is*

$$(\widetilde{\mu})_l = \operatorname{Ind}_{T_{fl}(\mu)}^{T_{fl}} \widetilde{\mu}.$$

This theorem is still not quite amenable to computers: although the set of P-positive dominant characters μ of T_f can be traversed using the algebraic character lattice $\Pi^*(\mathbf{H}_f)$ and its automorphism θ (Proposition 6.3), the extensions of μ are a bit subtle. We can calculate the group $R(K, \mu)_{dom}$ (Proposition 4.4), whose character group acts in a simply transitive way on the extensions; but there is no natural base point in the set of extensions, so it is not clear how to keep track of which is which. We will resolve this problem in Theorem 11.9 below. For the moment, we will simply reformulate Theorem 4.10 in a way that does not involve a choice of positive roots for K.

Definition 4.11. Suppose $\mu \in \Pi^*(T_f)$ is any character. The Weyl group $W(K, T_f)$ of Definition 4.1 acts on $\Pi^*(T_f)$, so we can define

$$W(K, T_f)^\mu = \{w \in W(K, T_f) \mid w \cdot \mu = \mu\},$$

the stabilizer of μ. Using the isomorphism $R(K) \simeq W(K, T_f)/W(K_0, T_{f,0})$ of Proposition 4.2, this defines a subgroup

$$R(K, \mu) = W(K, T_f)^\mu / W(K_0, T_{f,0})^\mu \subset R(K).$$

It is not difficult to show that this definition agrees with the one in Proposition 4.4 when μ is dominant.

We need a notion of extended weight here, and the absence of a chosen positive root system for K complicates matters slightly. Let us write

$$N_f = \text{normalizer of } T_f \text{ in } K^\sharp,$$

$$N_{fl} = \text{normalizer of } T_f \text{ in } K.$$

Then $N_f/T_f \simeq W(K^\sharp, T_f) = W(K_0, T_{f,0})$, and $N_{fl}/T_f \simeq W(K, T_f)$. The subgroups corresponding to the stabilizer of μ are

$$N_f(\mu) = \{n \in N_f \mid n \cdot \mu = \mu\},$$

$$N_{fl}(\mu) = \{n \in N_{fl} \mid n \cdot \mu = \mu\}.$$

Clearly $N_{fl}(\mu)/T_f \simeq W(K, T_f)^\mu$, and $N_f(\mu)/T_f \simeq W(K^\sharp, T_f)^\mu$.

For each root α of T_f in K there is a three-dimensional root subgroup of K locally isomorphic to $SU(2)$. The simple reflection $s_\alpha \in W(K^\sharp, T_f)$ has a representative σ_α in this root subgroup; the representative σ_α is unique up to multiplication by the coroot circle subgroup $\alpha^\vee(U_1) \subset T_f$.

A *small extension of* μ is a character $\widetilde{\mu}$ of $N_{fl}(\mu)$ subject to the following two conditions.

(1) The restriction of $\widetilde{\mu}$ to T_f is equal to μ.

(2) For each compact root α with $s_\alpha \in W(K^\sharp, T_f)^\mu$, we have $\widetilde{\mu}(\sigma_\alpha) = 1$.

The hypothesis $s_\alpha \in (K^\sharp, T_f)^\mu$ is equivalent to

$$\langle \mu, \alpha^\vee \rangle = 0;$$

that is, to μ being trivial on coroot circle subgroup $\alpha^\vee(U_1)$. Since σ_α is well-defined up to this subgroup, the requirement in (2) is independent of the choice of σ_α.

We will show in a moment (Lemma 4.12) that the quotient of $N_{fl}(\mu)$ by the kernel of μ and the various σ_α is actually abelian; so looking for one-dimensional extensions $\widetilde{\mu}$ is natural. It will turn out that small extensions of

μ must exist. Once that is known, it follows immediately that $\Pi^*(R(K,\mu))$ acts in a simply transitive way on the set of small extensions of μ, just by tensor product of characters.

The Weyl group $W(K,T_f)$ acts naturally on extended weights; the stabilizer of $\widetilde{\mu}$ is equal to $W(K,T_f)^\mu$. Write

$$P_K^e(T_f) = \text{small extensions of characters of } T_f.$$

(This is the first instance of our general scheme of writing P for a set of parameters for representations.) Elements of $P_K^e(T_f)$ are also called *extended weights*.

Lemma 4.12. *In the setting of Definition 4.11, choose a positive root system $(\Delta^+)'(K,T_f)$ making μ dominant. Choose a corresponding set P' of noncompact imaginary roots as in Proposition 4.4. Define $P'(\mu) \subset P'$ as in Proposition 4.4.*

The commutator subgroup of $N_f(\mu)$ is contained in the product of the images of the coroots

$$\gamma^\vee : U(1) \to T_f,$$

as γ runs over the noncompact imaginary roots in $P(\mu)$ and the compact roots orthogonal to μ. Since μ is trivial on all of these coroots, the quotient

$$N_{fl}(\mu)/\langle \ker(\mu), \{\sigma_\alpha\}\rangle$$

is abelian.

Corollary 4.13. *Suppose G is the group of real points of a connected reductive algebraic group, K is a maximal compact subgroup (Definition 2.4), and T_f is a small Cartan subgroup of K (Definition 3.1). Suppose $\widetilde{\mu} \in P_K^e(T_f)$ is an extended weight for K (Definition 4.11). Then there is a unique irreducible representation $\tau(\widetilde{\mu}) \in \Pi^*(K)$, with the following properties.*

(1) *If μ is dominant (Definition 3.3), then $\tau(\widetilde{\mu})$ is equal to the representation $\tau(\widetilde{\mu}|T_{fl}(\mu))$ defined in Theorem 4.10.*

(2) *If $w \in W(K,T_f)$, then $\tau(w \cdot \widetilde{\mu}) = \tau(\widetilde{\mu})$.*

We say that $\tau(\widetilde{\mu})$ has extremal weight μ or $\widetilde{\mu}$.

Two extended weights $\widetilde{\mu}$ and $\widetilde{\gamma}$ define the same irreducible representation of K if and only if $\widetilde{\gamma} \in W(K,T_f) \cdot \widetilde{\mu}$. In this way we get a bijection

$$\Pi^*(K) \leftrightarrow P_K^e(T_f)/W(K,T_f).$$

According to Corollary 4.6, the two conditions in this Corollary specify $\tau(\widetilde{\mu})$ completely; one just has to check that the resulting representation of K is well-defined.

5. Fundamental Series, Limits, and Continuations

The Harish-Chandra parameter for a discrete series representation is perhaps best thought of as a character of a two-fold cover (the "square root of 2ρ cover") of a compact Cartan subgroup. I am going to avoid discussion of coverings by shifting the parameter by ρ, the half sum of a set of imaginary roots.

Definition 5.1. In the setting of Definition 3.1, a *shifted Harish-Chandra parameter for a fundamental series representation* is a pair

$$\Phi = (\phi, \Delta_{im}^+)$$

(with $\phi \in \Pi^*(H_f)$ a character and $\Delta_{im}^+ \subset \Delta_{im}(G, H_f)$ a system of positive roots) subject to the following requirement:

(1-std) For every positive imaginary root $\alpha \in \Delta_{im}^+$, we have

$$\langle d\phi, \alpha^\vee \rangle > 1.$$

This requirement implies that the positive root system is entirely determined by the character ϕ. We keep it in the notation for formal consistency with notation for coherently continued representations, to be defined in a moment.

We define

$$P_G^s(H_f) = \{\text{shifted parameters for fundamental series}\}$$
$$= \{\text{pairs } \Phi = (\phi, \Delta_{im}^+) \text{ satisfying condition (1-std) above.}\}$$

(The superscript s stands for "shifted"; it is a reminder that this parameter differs by a ρ shift from the Harish-Chandra parameter.)

The positivity requirement (1-std) on Φ looks a little peculiar, because ϕ differs by a ρ-shift from the most natural parameter. In terms of the infinitesimal character parameter $\zeta(\Phi)$ defined below, the positivity requirement is just

$$\langle \zeta(\Phi), \alpha^\vee \rangle > 0 \qquad (\alpha \in \Delta_{im}^+).$$

The reason we do not use $\zeta(\Phi)$ instead of ϕ is that $\zeta(\Phi)$ is only in $\mathfrak{h}^*$: it does not remember the values of ϕ off the identity component of H_f.

The imaginary roots divide into compact and noncompact as usual, according to the eigenvalue of θ on the corresponding root space:

$$\Delta_{im,c}^+ = \text{positive imaginary roots in } \mathfrak{k},$$
$$\Delta_{im,n}^+ = \text{remaining positive imaginary roots.}$$

Define

$$2\rho_{im} = \sum_{\alpha \in \Delta_{im}^+} \alpha, \qquad 2\rho_{im,c} = \sum_{\beta \in \Delta_{im,c}^+} \beta \qquad 2\rho_{im,n} = \sum_{\gamma \in \Delta_{im,n}^+} \gamma.$$

Each of these characters may be regarded as belonging either to the group of real characters $\Pi^*(H_f)$ or to the algebraic character lattice $\Pi^*(\mathbf{H}_f)$. The lattice of characters may be embedded in $\mathfrak{h}_f^*$ by taking differentials, so we may also regard these characters as elements of $\mathfrak{h}_f(\mathbb{C})^*$. There they may be divided by two, defining ρ_{im}, $\rho_{im,c}$, and $\rho_{im,n}$ in $\mathfrak{h}_f(\mathbb{C})^*$. We will be particularly interested in the weight

$$\zeta(\Phi) = d\phi - \rho_{im} \in \mathfrak{h}_f^*,$$

called the *infinitesimal character parameter for* Φ. Notice that it is dominant and regular for Δ_{im}^+.

  More generally, a *shifted Harish-Chandra parameter for a limit of fundamental series representations* is a pair

$$\Phi = (\phi, \Delta_{im}^+)$$

(with $\phi \in \Pi^*(H_f)$ a character and $\Delta_{im}^+ \subset \Delta_{im}(G, H_f)$ a system of positive roots) subject to the following requirement:

(1-lim) For every positive imaginary root $\alpha \in \Delta_{im}^+$, we have

$$\langle d\phi, \alpha^\vee \rangle \geq 1.$$

Again this requirement implies that the positive root system is entirely determined by the character ϕ. Again we are interested in the infinitesimal character parameter

$$\zeta(\Phi) = d\phi - \rho_{im} \in \mathfrak{h}_f^*;$$

now it is dominant but possibly singular for Δ_{im}^+.

  Define

$$P_G^{s,lim}(H_f) = \{\text{shifted parameters for limits of fundamental series}\}$$
$$= \{\text{pairs } \Phi = (\phi, \Delta_{im}^+) \text{ satisfying condition (1-lim) above.}\}$$

  We can write the positivity requirement (1-lim) in the equivalent form

$$\langle \zeta(\Phi), \alpha^\vee \rangle \geq 0 \qquad (\alpha \in \Delta_{im}^+).$$

  Finally, a *shifted Harish-Chandra parameter for a continued fundamental series representation* is a pair

$$\Phi = (\phi, \Delta_{im}^+)$$

with $\phi \in \Pi^*(H_f)$, Δ^+_{im} a system of positive imaginary roots, and no positivity hypothesis on ϕ. (This is the setting in which the positive root system needs to be made explicit.) We still use the infinitesimal character parameter

$$\zeta(\Phi) = d\phi - \rho_{im} \in \mathfrak{h}^*_f;$$

which now need not have any positivity property.

We define

$$P^{s,cont}_G(H_f) = \{\text{shifted parameters for continued fundamental series}\}$$
$$= \{\text{pairs } \Phi = (\phi, \Delta^+_{im}).\}$$

For each kind of parameter there is a discrete part, which remembers only the restriction ϕ_d of the character ϕ to the compact torus T_f. We write for example

$$P^{s,lim}_{G,d}(H_f) = \{\text{shifted discrete parameters for limits of fundamental series}\}$$
$$= \{\text{pairs } \Phi_d = (\phi_d, \Delta^+_{im})\}$$

with $\phi_d \in \Pi^*(T_f)$ and Δ^+_{im} a system of positive imaginary roots satisfying condition (1-lim) above.

A given discrete parameter has a distinguished extension to a full parameter, namely the one which is trivial on $A_{f,0}$ (the vector subgroup of H_f). By using this extension, we may regard $P^{s,lim}_{G,d}(H_f)$ as a subset of $P^{s,lim}_G(H_f)$.

Proposition 5.2. *In the setting of Definition 5.1, suppose $\Phi = (\phi, \Delta^+_{im}) \in P^s_G(H_f)$ is a shifted Harish-Chandra parameter for a fundamental series representation. Then there is a fundamental series representation $I(\Phi)$ for G attached to Φ. It is always non-zero, and its restriction to K depends only on the restriction of Φ to T_f. The infinitesimal character of $I(\Phi)$ corresponds (in the Harish-Chandra isomorphism) to the weight $\zeta(\Phi) \in \mathfrak{h}^*_f$. This fundamental series representation has a (necessarily irreducible) Langlands quotient $J(\Phi)$. For Φ and Ψ in $P^s_G(H_f)$, we have*

$$I(\Phi) \simeq I(\Psi) \Longleftrightarrow J(\Phi) \simeq J(\Psi) \Longleftrightarrow \Psi \in W(G, H_f) \cdot \Phi.$$

Define

$$\mu(\Phi) = [\phi - 2\rho_{im,c}]|_{T_f} \in \Pi^*(T_f).$$

Then $\mu(\Phi)$ is dominant with respect to the compact imaginary roots $\Delta^+_{im,c}$. The R-group $R(K, \mu(\Phi))$ (Definition 4.11) is necessarily trivial, so we may identify $\mu(\Phi)$ with an extended weight for K (Definition 4.11). The corresponding representation $\tau(\mu(\Phi))$ of K (Corollary 4.13) is the unique lowest K-type of $I(\Phi)$ and of $J(\Phi)$.

We will discuss the construction of $I(\Phi)$ in Section 11. For now we observe only that $I(\Phi)$ is a discrete series representation (in the strong sense that the matrix coefficients are square integrable) if and only if $H_f = T_f$ is compact. In that case $I(\Phi) = J(\Phi)$ is irreducible, and its Harish-Chandra parameter (from his classification of discrete series) is equal to $\zeta(\Phi)$.

For limits of fundamental series the situation is quite similar; the main problem is that the corresponding representation of G may vanish. Here is a statement.

Proposition 5.3. *In the setting of Definition 5.1, suppose $\Phi = (\phi, \Delta_{im}^{+}) \in P_G^{s,lim}(H_f)$ is a shifted Harish-Chandra parameter for a limit of fundamental series. Then there is a limit of fundamental series representation $I(\Phi)$ for G attached to Φ. Its restriction to K depends only on the restriction of Φ to T_f. The infinitesimal character of $I(\Phi)$ corresponds (in the Harish-Chandra isomorphism) to the weight $\zeta(\Phi) \in \mathfrak{h}_f^*$. This fundamental series representation has a Langlands quotient $J(\Phi)$. We have*

$$\Psi \in W(G, H_f) \cdot \Phi \Longrightarrow I(\Phi) \simeq I(\Psi) \Longleftrightarrow J(\Phi) \simeq J(\Psi).$$

Define

$$\mu(\Phi) = [\Phi - 2\rho_{im,c}]|_{T_f} \in \Pi^*(T_f).$$

Then the following three conditions are equivalent:

(1) *the weight $\mu(\Phi)$ is dominant with respect to the compact imaginary roots $\Delta_{im,c}^{+}$;*

(2) *there is no simple root for Δ_{im}^{+} which is both compact and orthogonal to $\phi - \rho_{im}$; and*

(3) *the standard limit representation $I(\Phi)$ is non-zero.*

Assume now that these equivalent conditions are satisfied. The R-group $R(K, \mu(\Phi))$ (Definition 4.11) is necessarily trivial, so we may identify $\mu(\Phi)$ with an extended weight for K (Definition 4.11). The corresponding representation $\tau(\Phi)$ of K (Corollary 4.13) is the unique lowest K-type of $I(\Phi)$ and of $J(\Phi)$.

Here is the result for coherent continuation.

Proposition 5.4. *In the setting of Definition 5.1, suppose $\Phi = (\phi, \Delta_{im}^{+}) \in P_G^{s,cont}(H_f)$ is a shifted Harish-Chandra parameter for continued fundamental series. Then there is a virtual representation $I(\Phi)$ for G, with the following properties.*

(1) *The restriction of $I(\Phi)$ to K depends only on the restriction of Φ to T_f.*

(2) *We have*

$$I(\Phi) \simeq I(w \cdot \Phi) \qquad (w \in W(G, H_f)).$$

(3) *The virtual representation $I(\Phi)$ has infinitesimal character corresponding to $\zeta(\Phi) \in \mathfrak{h}_f^*$.*

(4) *If $\zeta(\Phi)$ is weakly dominant for Δ_{im}^+, then $I(\Phi)$ is equivalent (as a virtual representation) to the limit of fundamental series attached to Φ in Proposition 5.3.*

(5) *Suppose V is a finite-dimensional representation of G. Recall that $\Delta(V, H_f)$ denotes the multiset of weights of H_f on V (Definition 3.2). Then*

$$I(\Phi) \otimes V \simeq \sum_{\delta \in \Delta(V, H_f)} I(\Phi + \delta).$$

Here $\Phi + \delta$ denotes the continued fundamental series parameter

$$\Phi + \delta = (\phi + \delta, \Delta_{im}^+) \in P_G^{s,cont}(H_f).$$

The result in (4) justifies the ambiguous notation $I(\Phi)$; without it, we would need to specify whether we were regarding Φ as an element of $P^{s,cont}$ or of P^s.

One of the central goals of this paper is writing down the bijection between irreducible representations of K and certain "final standard limit" representations of G. We are well on the way to this already; here is how.

Definition 5.5. A limit parameter $\Phi \in P_G^{s,lim}(H_f)$ (Definition 5.1) is called *final* if the corresponding representation $I(\Phi)$ is non-zero. According to Proposition 5.3, this is equivalent to requiring that $\mu(\Phi)$ be dominant with respect to $\Delta_{im,c}^+$. We write $P_G^{s,finlim}(H_f)$ for the set of final limit parameters for H_f.

Proposition 5.6. *The map $\Phi \mapsto \tau(\Phi)$ (Proposition 5.3) is an injection from $W(G, H_f)$ orbits of discrete final limit parameters for H_f into $\Pi^*(K)$. The image consists precisely of those irreducible representations of K for which any highest weight μ has the following two properties.*

(1) *The weight $\mu + 2\rho_c \in \Pi^*(T_f)$ is regular with respect to the system of restricted roots of T_f in G. Because of this property, we can define $\Delta^+(G, T_f)(\mu)$ to be the unique positive system making $\mu + 2\rho_c$ dominant.*

(2) *The weight $\mu + 2\rho_c - \rho(\mu) \in \mathfrak{t}_f^*$ is weakly dominant for $\Delta^+(G, T_f)(\mu)$.*

There is a similar statement just for fundamental series; the only change is that one imposes the stronger condition that $\mu + 2\rho_c - \rho$ is strictly dominant for $\Delta^+(G, T_f)$.

6. Characters of Compact Tori

Suppose $\mathbf{H}$ is a θ-stable maximal torus in $\mathbf{G}$ that is defined over $\mathbb{R}$. The *compact factor of H* is

$$\mathbf{T} = \mathbf{H}^\theta, \tag{6.1a}$$

the (algebraic) group of fixed points of θ on $\mathbf{H}$. Clearly this is an abelian algebraic group defined over $\mathbb{R}$. Its group of real points is

$$T = H^\theta = H \cap K, \tag{6.1b}$$

the maximal compact subgroup of the real Cartan subgroup H. Write

$$\mathbf{T}_e = (\mathbf{H}^\theta)_e, \quad T_e = \mathbf{T}_e(\mathbb{R}); \tag{6.1c}$$

this last group is a connected compact torus, equal to the identity component T_0.

So far all of this fits well with standard notation for real groups. Now we begin to deviate a little. The *split component of H* is

$$\mathbf{A} = \mathbf{H}^{-\theta}, \tag{6.1d}$$

the (algebraic) group of fixed points of $-\theta$ on $\mathbf{H}$. Again this is an abelian algebraic group defined over $\mathbb{R}$. Its group of real points is

$$A = H^{-\theta}. \tag{6.1e}$$

We have containments

$$A_0 \subset A_e \subset A, \tag{6.1f}$$

usually both proper. The complex torus $\mathbf{A}_e$ is the maximal $\mathbb{R}$-split torus in $\mathbf{H}$, so A_e is a product of copies of $\mathbb{R}^\times$. Consequently A_0 is a vector group, isomorphic to its Lie algebra by the exponential map. There is a direct product of Lie groups (the *Cartan decomposition*)

$$H = T \times A_0, \tag{6.1g}$$

but we will avoid this (first of all on the aesthetic grounds that it is not algebraic). Again in real groups one typically writes A for our A_0; this we will avoid even more assiduously.

The next goal is to describe the set of characters of T:

$$\Pi^*_{adm}(T) = \text{continuous characters } \xi \colon T \to U_1 \tag{6.2}$$

Proposition 6.3.   (1) *In the setting of* (6.1), *restriction to T defines a natural identification*

$$\Pi^*_{adm}(T) \simeq \Pi^*_{alg}(\mathbf{H})/(1-\theta)\Pi^*_{alg}(\mathbf{H})$$

given by restriction of characters of $\mathbf{H}$ to T.

(2) *The identity component $\mathbf{T}_e$ is equal to the image of the homomorphism*

$$\delta(\theta) \colon \mathbf{H} \to \mathbf{H}, \qquad \delta(\theta)(h) = h \cdot \theta(h).$$

(3) *If $\xi \in \Pi^*_{alg}(\mathbf{H})$, then*

$$\xi \circ \delta(\theta) = \xi + \theta(\xi).$$

Consequently ξ is trivial on the image of $\delta(\theta)$ if and only if $\xi + \theta\xi = 0$; that is, if and only if ξ belongs to the -1 eigenspace $\Pi^*_{alg}(\mathbf{H})^{-\theta}$ of θ on the character lattice.

(4) *Restriction to T_0 defines a natural identification*

$$\Pi^*_{adm}(T_0) \simeq \Pi^*_{alg}(\mathbf{H})/\Pi^*_{alg}(\mathbf{H})^{-\theta}.$$

The δ in (3) is meant to stand for "double."

The main point of Proposition 6.3 is the identification of characters of T with cosets in the lattice of characters of $\mathbf{H}$, a lattice to which the `atlas` software has access.

7. Split Tori and Representations of K

In Section 5 we wrote down the lowest K-type correspondence between fundamental series and certain representations of K. In this section we look at the opposite extreme case: lowest K-types of principal series representations. The general case is going to be built from these two extremes in a fairly simple way.

In this section we therefore assume that G is quasisplit, and that H_s is a θ-stable maximally split Cartan subgroup of G. What these assumptions mean is that there are no imaginary roots of H_s in G. Define

$$T_s = H_s^\theta = H_s \cap K, \qquad A_s = H_s^{-\theta} \tag{7.1a}$$

as in (6.1), the compact and split parts of H_s. Again, the quasisplit hypothesis means that every root has a non-trivial restriction to A_s, and even to the real identity component $A_{s,0}$. These restrictions form a (possibly non-reduced) root system

$$\Delta(G, A_s) = \{\alpha|_{\mathfrak{a}_s} \mid \alpha \in \Delta(G, H_s)\}. \tag{7.1b}$$

The Weyl group of this root system is isomorphic to $W(G, H_s)$ and to $W(\mathbf{G}, \mathbf{H}_s)^\theta$:

$$W(\mathbf{G}, \mathbf{H}_s)^\theta \simeq W(G, H_s) \simeq W(G, A_s); \qquad (7.1c)$$

the isomorphisms are given by restriction of the action from $\mathbf{H}$ to H_s to $A_{s,0}$.

There is some possible subtlety or confusion here arising from the disconnectedness of A_s. We have written $\Delta(G, A_s)$ instead of $\Delta(G, A_{s,0})$ simply because the former is shorter; but it is not entirely clear what should be meant by a root system inside the character group of a disconnected reductive abelian group like $\mathbf{A}_s$. One reassuring fact is that if α_1 and α_2 are roots in $\Delta(G, H_s)$ having the same restriction to $\mathfrak{a}_s$, then they also have the same restriction to $\mathbf{A}_s$. (The reason is that the hypothesis implies that either $\alpha_1 = \alpha_2$, or $\alpha_1 = -\theta\alpha_2$.) This means that there can be no confusion about the set $\Delta(G, A_s)$: it is the set of orbits of $\{1, -\theta\}$ on $\Delta(G, H_s)$.

Definition 7.2. In the setting of (7.1), a *(shifted) parameter for a principal series representation* is a pair

$$\Phi = (\phi, \emptyset)$$

with $\phi \in \Pi^*(H_s)$ a character and $\emptyset$ a set of positive roots for the (empty set of) imaginary roots of H_s in G. We retain the word "shifted" and the ordered pair structure for formal consistency with Definition 5.1 even though the shift (by imaginary roots) is now zero. In the same way, we may also call these same pairs *shifted parameters for limits of principal series* or *shifted parameters for continued principal series* even though the distinction among these concepts (which depends on imaginary roots) is empty. We define

$$P_G^s(H_s) = P_G^{s,lim}(H_s) = P_G^{s,cont}(H_s) = \Pi^*(H_s) \times \{\emptyset\},$$

the set of shifted parameters for principal series representations. We will need the weight

$$\zeta(\Phi) = d\phi - \rho_{im} = d\phi \in \mathfrak{h}_s^*,$$

called the *infinitesimal character parameter for* Φ. We may occasionally write $\zeta_\mathfrak{g}(\Phi)$ for clarity.

The *(shifted) discrete parameters for principal series representations* are the restrictions to T_s of parameters for principal series:

$$P_{G,d}^s(H_s) = \Pi^*(T_s) \times \{\emptyset\}.$$

As in the case of fundamental series, each discrete parameter has a distinguished extension to a full parameter (the one that is trivial on $A_{s,0}$), allowing us to regard

$$P^s_{G,d}(H_s) \subset P^s_G(H_s).$$

Proposition 7.3. *In the setting of Definition 7.2, suppose $\Phi = (\phi, \emptyset) \in P^s_G(H_s)$ is a (shifted) Harish-Chandra parameter for a principal series representation. Choose a system of positive roots $\Delta^+(G, A_s)$ making the real part of $d\phi|_{\mathfrak{a}_s}$ weakly dominant, and let $B_s \supset H_s$ be the corresponding Borel subgroup of G. Put*

$$I(\Phi) = \mathrm{Ind}^G_{B_s} \Phi$$

(normalized induction). Then $I(\Phi)$ is a principal series representation of G attached to Φ; its equivalence class is independent of the choice of $\Delta^+(G, A_s)$. It is always non-zero, and its restriction to K is

$$I(\Phi)|_K = \mathrm{Ind}^K_{T_s} \phi|_{T_s}$$

(which depends only on the restriction of Φ to T_s). The infinitesimal character of $I(\Phi)$ corresponds to the weight $\zeta(\Phi) \in \mathfrak{h}^*_s$. This principal series representation has a (possibly reducible) Langlands quotient $J(\Phi)$. We have

$$I(\Phi) \simeq I(\Psi) \Longleftrightarrow J(\Phi) \simeq J(\Psi) \Longleftrightarrow \Psi \in W(G, H_s) \cdot \Phi.$$

Now regard Φ as a shifted parameter for continued principal series, and suppose V is a finite-dimensional representation of G. Recall the multiset $\Delta(V, H_s)$ of weights of H_s on V (Definition 3.2). Then

$$I(\Phi) \otimes V \simeq \sum_{\delta \in \Delta(V, H_s)} I(\Phi + \delta)$$

as virtual representations.

In the coherent continuation formula mentioned last, both sides are actual representations of G. Nevertheless, the equality may be true only as virtual representations: composition factors may be arranged differently on the two sides. The simplest example has $I(\Phi)$ the nonspherical unitary principal series for $SL_2(\mathbb{R})$ (with $J(\Phi) = I(\Phi)$ the sum of two limits of discrete series), and V the two-dimensional representation. The trivial representation appears twice as a quotient on the right side of the formula, but not at all as a quotient on the left. Part of the difficulty is that the Borel subgroup B_s is dominant only for one of the two parameters $\Phi + \delta$ (with δ a weight of V).

In this proposition, in contrast to Definition 3.3, the positive roots are really those in B_s (and not their negatives).

To understand the K-types of the principal series $I(\Phi)$, we need to understand which representations of K can contain the character $\phi|_{T_s}$ of T_s.

Since we understand representations of K in terms of their highest weights, this amounts to understanding the relationship between T_s and the Cartan subgroup T_f of K. In order to discuss this, we need the Knapp-Stein R-group for principal series.

Definition 7.4. In the setting of Definition 7.2, fix a parameter $\Phi = (\phi, \emptyset) \in P_G^s(H_s)$. Define

$$W(G, H_s)^\Phi = \{w \in W(G, H_s) \mid w \cdot \Phi = \Phi\},$$

the stabilizer of Φ (that is, of the character ϕ) in the real Weyl group. The set of *good roots* for A_s in G is

$$\Delta_\Phi(G, A_s) = \{\alpha \in \Delta(G, A_s) \mid \phi \text{ is trivial on } \alpha^\vee\}.$$

Since α is just the restriction of a root, the meaning of the condition "ϕ is trivial on $\alpha^\vee$" requires some explanation. First, we can construct (not uniquely) a homomorphism from $SL_2(\mathbb{R})$ into G using H_s and the root α. (In the special case of a real root, this homomorphism is described a little more precisely in (7.6a) below.) Restricting this homomorphism to the diagonal torus $\mathbb{R}^\times$ of $SL_2(\mathbb{R})$ gives a homomorphism

$$\alpha^\vee : \mathbb{R}^\times \to A_s$$

which *is* uniquely determined; and the first requirement we want to impose is that $\phi \circ \alpha^\vee$ is trivial. When α is not the restriction of a real root, then the simple real rank one subgroup of G corresponding to α is covered by $SU(2,1)$ or $SL_2(\mathbb{C})$. In each of these real groups the maximally split Cartan is $\mathbb{C}^\times$, so the real coroot $\alpha^\vee$ extends to

$$\alpha_\mathbb{C}^\vee : \mathbb{C}^\times \to H_s.$$

(The extension is well-defined up to composition with complex conjugation.) In these cases (to call α good) we impose the requirement that $\phi \circ \alpha_\mathbb{C}^\vee$ is trivial.

The good roots form a subroot system of the restricted roots, and the corresponding Weyl group

$$W_0(G, H_s)^\Phi = W(\Delta_\Phi(G, A_s))$$

is a normal subgroup of $W(G, H_s)^\Phi$. The R-*group of* Φ is by definition the quotient

$$R(\Phi) = W(G, H_s)^\Phi / W_0(G, H_s)^\Phi.$$

Proposition 7.5. *Suppose we are in the setting of Definition 7.4.*

(1) *The irreducible constituents of the Langlands quotient representation $J(\Phi)$ all occur with multiplicity one. There is a natural simply transitive action of the character group $\Pi^*(R(\Phi))$ on these constituents.*

(2) *Write Φ_d for the discrete part of Φ (the restriction of ϕ to T_s, extended to be trivial on $A_{s,0}$). Then there is a natural inclusion $W(G, H_s)^\Phi \hookrightarrow W(G, H_s)^{\Phi_d}$, which induces an inclusion $R(\Phi) \hookrightarrow R(\Phi_d)$.*

(3) *Each irreducible summand of $I(\Phi_d) = J(\Phi_d)$ contains a unique lowest K-type of $I(\Phi_d)$. There is a natural simply transitive action of $\Pi^*(R(\Phi_d))$ on this set $A(\Phi_d)$ of K-types.*

(4) *The sets $A(\Phi_d)$ partition a certain subset of $\Pi^*(K)$. We have*

$$A(\Phi_d) = A(\Psi_d) \iff \Psi_d \in W(G, H_s) \cdot \Phi_d.$$

In order to understand the lowest K-types of these principal series representations, we must therefore describe $A(\Phi_d)$ as a set of (extended) highest weights. The first step is to relate T_f (where extended highest weights are supposed to live) to T_s (where Φ_d lives). So choose a set

$$\beta_1, \ldots, \beta_r \tag{7.6a}$$

of strongly orthogonal real roots of H_s in G, of maximal cardinality. Attached to each real root α we can find an algebraic group homomorphism defined over $\mathbb{R}$

$$\psi_\alpha \colon \mathbf{SL}_2 \to \mathbf{G}, \tag{7.6b}$$

in such a way that

(1) ψ_α carries the diagonal Cartan subgroup of SL_2 into H_s, and the upper triangular subgroup into the α root subgroup; and

(2) $\psi_\alpha({}^t g^{-1}) = \theta(\psi_\alpha(g))$.

Such a homomorphism ψ_α is unique up to conjugation in $\mathbf{SL}_2$ by

$$\begin{pmatrix} i & 0 \\ 0 & -i \end{pmatrix} \tag{7.6c}$$

In particular, the element

$$m_\alpha = \psi_\alpha \begin{pmatrix} -1 & 0 \\ 0 & -1 \end{pmatrix} = \alpha^\vee(-1) \in T_s \tag{7.6d}$$

is well-defined.

Since the roots $\{\beta_i\}$ are strongly orthogonal and real, the groups $\psi_{\beta_i}(SL_2)$ commute with each other and with $T_{s,0}$.

Definition 7.7. A shifted (limit) parameter $\Phi = (\phi, \emptyset) \in P_G^s(H_s)$ (Definition 5.1) is called *final* if for every real root α of H_s in G,

$$\text{either} \quad \langle d\phi, \alpha^\vee \rangle \neq 0 \quad \text{or} \quad \phi(m_\alpha) = 1.$$

(If Φ is discrete, the requirement is that $\phi(m_\alpha) = 1$ for every real root α.) We write $P_G^{s,finlim}(H_f)$ for the set of final limit parameters for H_f.

For the remainder of this section we consider the important special case

$$G \text{ split}, \quad \Phi \in P_{G,d}^{s,finlim} \text{ discrete final limit parameter.} \tag{7.8a}$$

This is the same thing as a character

$$\phi \in \Pi^*(T_s), \qquad \phi(m_\alpha) = 1 \qquad (\text{all } \alpha \in \Delta(G, H_s)). \tag{7.8b}$$

Here m_α is the element of order 2 defined in (7.6d).

Proposition 7.9. *In the setting* (7.8), *the principal series representation* $I(\Phi)$ *has a unique lowest K-type* $\tau(\Phi)$. *This representation of K is trivial on the identity component of $G^{der} \cap K$, with G^{der} the derived group of G. The map $\Phi \mapsto \tau(\Phi)$ is an injection from discrete final limit parameters for H_s into $\Pi^*(K)$. (The group $W(G, H_s)$ acts trivially on discrete final limit parameters for H_s, so we could also say "$W(G, H_s)$ orbits of discrete final limit parameters...") The image consists precisely of those irreducible representations of K which are trivial on $(G^{der} \cap K)_0$. These are the representations of K for which any highest weight μ has the property that $\langle \mu, \beta^\vee \rangle = 0$ for any root $\beta \in \Delta(G, T_f)$.*

8. Parametrizing Extended Weights

We have a computer-friendly parametrization of characters of T_f available from Proposition 6.3:

$$\Pi^*(T_f) = \Pi^*_{alg}(\mathbf{H}_f)/(1 - \theta_f)\Pi^*_{alg}(\mathbf{H}_f). \tag{8.1}$$

Here I have written θ_f for the action of θ on the fundamental Cartan $\mathbf{H}_f$; this is in some sense a compromise with the software point of view, in which the Cartan is always fixed and only the Cartan involution is changing. In order to parametrize representations of K, we need not just a character μ of T_f, but an extension $\tilde{\mu}$ of that character to a slightly larger group. We therefore need a computer-friendly parametrization of such extensions, and this section seeks to provide such a parametrization.

We begin with the special case of Proposition 7.9: assume first of all that

$$G \text{ is split, with } \theta\text{-stable split Cartan } H_s. \tag{8.2a}$$

This means that

$$\theta_s(\alpha) = -\alpha \qquad (\alpha \in \Delta(G, H_s)). \tag{8.2b}$$

On the fundamental Cartan, it is equivalent to assume that there is an orthogonal set of imaginary roots

$$\{\gamma_1, \ldots, \gamma_m\} \subset \Delta_{im,n}(G, H_f) \tag{8.2c}$$

spanning the root system $\Delta(G, T_f)$, and for which the successive Cayley transforms are defined: γ_i is noncompact if and only if it is preceded by an even number of γ_j to which it is not strongly orthogonal. The successive Cayley transforms identify $\Pi^*_{alg}(\mathbf{H}_s)$ with $\Pi^*_{alg}(\mathbf{H}_f)$ in such a way that

$$\theta_f = \prod_{j=1}^{m} s_{\gamma_j} \circ \theta_s. \tag{8.2d}$$

In particular, θ_f acts on the roots by minus the product of the reflections in the γ_j. We also want to impose a condition on the weights of T_f that we consider: we look only at

$$\{\mu \in \Pi^*(T_f) \mid \langle \mu, \alpha^\vee \rangle = 0 \quad (\alpha \in \Delta(G, T_f)\}. \tag{8.2e}$$

This restriction is equivalent (since the $\gamma_j^\vee$ span $\mathfrak{t}_f$ modulo the center of $\mathfrak{g}$) to

$$\{\mu \in \Pi^*(T_f) \mid \langle \mu, \gamma_j^\vee \rangle = 0 \quad (j = 1, \ldots, m)\}. \tag{8.2f}$$

We call these weights "G-spherical," and write the set of them as

$$P_{K,G-sph}(T_f). \tag{8.2g}$$

The corresponding set of extended weights (Definition 4.11) is written

$$P_{K,G-sph}^e(T_f). \tag{8.2h}$$

Here is an explanation of the terminology.

Lemma 8.3. *Suppose G is split, and (τ, E) is an irreducible representation of K having extremal weight $\mu \in \Pi^*(T_f)$ (Corollary 4.13). Recall from (2.5) the decomposition $\mathfrak{g} = \mathfrak{k} + \mathfrak{s}$ of the Lie algebra into the $+1$ and -1 eigenspaces of θ. Then E can be extended to a $(\mathfrak{g}, K)$-module on which $\mathfrak{s}$ acts by zero if and only if μ is G-spherical. That is, μ is G-spherical if and only if any irreducible representation of K of extremal weight μ is trivial on the ideal $[\mathfrak{s}, \mathfrak{s}] \subset \mathfrak{k}$.*

Recall from Proposition 7.9 the discrete final limit parameters for H_s:

$$P_{G,d}^{s,finlim}(H_s) = \{\Phi \in \Pi^*(T_s) \mid \phi(m_\alpha) = 1 \quad (\alpha \in \Delta(G, H_s))\}. \tag{8.4}$$

The element $m_\alpha \in T_s$ was defined in (7.6d).

Proposition 8.5. *Suppose G is split. The following sets of representations of K (called G-spherical) are all the same.*

(1) *Representations having a G-spherical extremal weight (cf. (8.2)).*

(2) *Representations trivial on $[\mathfrak{s}, \mathfrak{s}] \subset \mathfrak{k}$ (cf. Lemma 8.3).*

(3) *Lowest K-types of principal series with discrete final limit parameters (Proposition 7.9).*

This establishes a bijection $\Phi \mapsto \tau(\Phi)$ from discrete final limit parameters for H_s to G-spherical representations of K. Taking into account the parametrization of $\Pi^*(K)$ by extended highest weights gives bijections

$$P_{K,G-sph}^e(T_f) \quad \longleftrightarrow \quad \Pi_{G-sph}^*(K) \quad \longleftrightarrow \quad P_{G,d}^{s,finlim}(H_s),$$
$$\widetilde{\mu}(\Phi) \quad \longleftrightarrow \quad \tau(\Phi) \quad \longleftrightarrow \quad \Phi$$

Proposition 6.3 provides a computer-friendly parametrization of the discrete final limit parameters of H_s:

$$P_{G,d}^{s,finlim}(H_s) = \{\lambda \in \Pi_{alg}^*(H_s) \mid \langle \lambda, \alpha^\vee \rangle \in 2\mathbb{Z} \ (\alpha \in \Delta(G, H_s))\}$$
$$/(1 - \theta_s)\Pi_{alg}^*(H_s). \tag{8.6}$$

The condition on λ in the numerator arises from the fact that $\lambda(m_\alpha) = (-1)^{\langle \lambda, \alpha^\vee \rangle}$. The description of characters of T_f in (8.1) immediately specializes to a description of the G-spherical characters:

$$P_{K,G-sph}(T_f) = \{\lambda_0 \in \Pi_{alg}^*(H_f) \mid \langle \lambda_0, \gamma_j^\vee \rangle = 0 \ (j = 1, \ldots, m)\}$$
$$/(1 - \theta_f)\Pi_{alg}^*(H_f). \tag{8.7}$$

An equivalent description of the numerator is

$$P_{K,G-sph}(T_f) = \{\lambda_0 \in \Pi_{alg}^*(H_f) \mid \langle \lambda_0, \alpha^\vee + \theta\alpha^\vee \rangle = 0 \ (\alpha \in \Delta(G, H_f))\}$$
$$/(1 - \theta_f)\Pi_{alg}^*(H_f). \tag{8.8}$$

What we are going to do is describe a natural surjective map

$$P_{G,d}^{s,finlim}(H_s) \twoheadrightarrow P_{K,G-sph}(T_f) \tag{8.9a}$$

Since the first set is identified by Proposition 8.5 with G-spherical extended weights, this amounts to

$$P_{K,G-sph}^e(H_s) \twoheadrightarrow P_{K,G-sph}(T_f) \tag{8.9b}$$

This will be precisely the map $\widetilde{\mu} \mapsto \mu$ restricting an extended weight to T_f.

In order to describe the map, it is helpful to introduce some auxiliary lattices attached to a maximal torus $\mathbf{H} \subset \mathbf{G}$. To every $\lambda \in \Pi_{alg}^*(\mathbf{H})$ one can attach a $\mathbb{Z}/2\mathbb{Z}$-grading of the coroot system, by

$$\epsilon(\lambda) \colon \Delta^\vee \to \{0, 1\}, \qquad \epsilon(\lambda)(\alpha^\vee) = \langle \lambda, \alpha^\vee \rangle \quad (\text{mod } 2). \tag{8.10}$$

The map of (8.9) is closely related to these gradings. In terms of the gradings, we define

$$\Pi^*_{ev}(\mathbf{H}) = \{\lambda \in \Pi^*_{alg}(\mathbf{H}) \mid \epsilon(\lambda) = 0\}$$
$$R(\mathbf{H}) = \mathbb{Z}\Delta(\mathbf{G}, \mathbf{H}) = \text{root lattice}$$
$$R_{ev}(\mathbf{H}) = \{\phi \in R(\mathbf{H}) \mid \epsilon(\phi) = 0\} \qquad (8.11)$$
$$\Pi^*_R(\mathbf{H}) = \{\lambda \in \Pi^*_{alg}(\mathbf{H}) \mid \exists \phi \in R \text{ with } \epsilon(\lambda) = \epsilon(\phi)\}$$
$$= \epsilon^{-1}(\epsilon(R)).$$

We can immediately use these definitions to rewrite (8.6) (which we know is parametrizing G-spherical representations of K) as

$$P^{s,finlim}_{G,d}(H_s) = \Pi^*_{ev}(\mathbf{H}_s)/(1 - \theta_s)\Pi^*_{alg}(\mathbf{H}_s). \qquad (8.12)$$

To analyze (8.7) in a parallel way, we begin with a lemma.

Lemma 8.13. *Suppose $\{\gamma_1, \ldots, \gamma_m\}$ is a maximal orthogonal set of roots in $\Delta(\mathbf{G}, \mathbf{H})$. Suppose ϵ is a grading of the coroots with values in $\mathbb{Z}/2\mathbb{Z}$, and that*

$$\epsilon(\gamma_j^\vee) = 0, \qquad (j = 1, \ldots, m).$$

Then there is a subset $A \subset \{\gamma_1, \ldots, \gamma_m\}$ so that ϵ is equal to the grading ϵ_A defined by the sum of the roots in this subset:

$$\epsilon(\alpha^\vee) \equiv \sum_{\gamma \in A} \langle \gamma, \alpha^\vee \rangle \pmod 2$$

I am grateful to Jeff Adams and Becky Herb for providing a (fairly constructive) proof of this lemma, which appears in the next section.

Lemma 8.14. *In the setting of (8.2), consider $P_{K,G-sph}(T_f)$ as described by (8.7). According to Lemma 8.13, any element $\lambda_0 \in \Pi^*_{alg}(\mathbf{H}_f)$ representing an element of $P^*_{K,G-sph}$ (that is, orthogonal to all γ_j) must belong to $\Pi^*_R(\mathbf{H}_f)$. This defines a map*

$$P_{K,G-sph}(T_f) \to \Pi^*_R(\mathbf{H}_f)/[(1 - \theta_f)\Pi^*(\mathbf{H}_f) + R].$$

This map is an isomorphism.

Proof. We first prove surjectivity. Suppose $\lambda \in \Pi^*_R(\mathbf{H}_f)$; we need to show that the class of λ is in the image of our map. Since the denominator includes the root lattice R, we may replace λ by some $\lambda_1 = \lambda + \psi$ with $\psi \in R$ and $\lambda_1 \in \Pi^*_{ev}(\mathbf{H}_f)$. Since λ_1 takes even values on all coroots, we can define

$$\lambda_0 = \lambda_1 - \sum_{j=1}^m (\langle \lambda_1, \gamma_j^\vee \rangle/2)\gamma_j$$

(again modifying λ_1 by an element of R). Clearly λ_0 is orthogonal to all the γ_j, so it represents a class in $P_{K,G-sph}(T_f)$ mapping to the class of λ.

Next we prove injectivity. Suppose λ_0 represents a class in $P_{K,G-sph}(T_f)$ mapping to zero; that is, that

$$\lambda_0 = \xi - \theta_f(\xi) + \psi, \qquad (8.15a)$$

with $\psi \in R$.

The term $\xi - \theta_f(\xi)$ is orthogonal to all imaginary roots, including the γ_j, so $\langle \psi, \gamma_j^\vee \rangle = 0$ for all j. By (8.2d), this is equivalent to $\theta_f(\psi) = -\psi$. Choose a θ-stable system of positive roots. List the simple imaginary roots as $\alpha_1, \ldots, \alpha_r$, and the simple complex roots as $\beta_1, \theta_f\beta_1, \ldots, \beta_s, \theta_f\beta_s$. Because these $r + 2s$ roots are a basis for the root lattice R, it is clear that the -1 eigenspace of θ_f on R has as basis the s elements $\beta_i - \theta_f\beta_i$. In particular, ψ is a sum of these elements, so

$$\psi \in (1 - \theta_f)R \subset (1 - \theta_f)\Pi^*(\mathbf{H}_f). \qquad (8.15b)$$

Now (8.15) shows that λ_0 represents 0 in $P_{K,G-sph}(T_f)$, proving injectivity.

$\square$

Proposition 8.16. *Suppose that we use a Cayley transform to identify* $\Pi_{alg}^*(\mathbf{H}_f)$ *with* $\Pi_{alg}^*(\mathbf{H}_s)$. *This identification is unique up to the action of* $W(G, H_s)$. *This identification sends* $P_{K,G-sph}(T_f)$ *(as described in Lemma 8.14) to*

$$\Pi_R^*(\mathbf{H}_s)/[(1 - \theta_s)\Pi^*(\mathbf{H}_s) + R].$$

The Weyl group $W(G, H_s)$ *acts trivially on this quotient, so the identification with* $P_{K,G-sph}(T_f)$ *is well-defined. This quotient in turn may be identified (by the natural inclusion) with*

$$\Pi_{ev}^*(\mathbf{H}_s)/[(1 - \theta_s)\Pi^*(\mathbf{H}_s) + R_{ev}].$$

In this way $P_{K,G-sph}(T_f)$ *is naturally identified as a quotient of*

$$\Pi_{ev}^*(\mathbf{H}_s)/(1 - \theta_s)\Pi^*(\mathbf{H}_s) \simeq P_{G,d}^{s,finlim}(H_s).$$

Taking into account the identification of Proposition 8.5, we have given a surjective map

$$P_{K,G-sph}^e(T_f) \to P_{K,G-sph}(T_f)$$

from extended (extremal) weights of G-spherical representations of K, to (extremal) weights. This map is just restriction to T_f.

·

Proof. The action of Weyl group elements on algebraic characters is by addition of elements of the root lattice R. Since we are dividing by R in the first formula, this shows that the Weyl group action is trivial on the quotient, as claimed. Since θ_s and θ_f differ by the product of reflections in the γ_j, it also follows that

$$(1 - \theta_s)\Pi^*(\mathbf{H}_s) + R = (1 - \theta_f)\Pi^*(\mathbf{H}_f) + R$$

(using the Cayley transform identifications). This provides the first formula of the proposition. The map from the second formula to the first comes from the inclusion of Π^*_{ev} in Π^*_R. Surjectivity is immediate from the definitions, and injectivity is straightforward.

For the last claim, recall that $T_f \subset G^\sharp = G_0 \cdot Z(G)$; and furthermore $G_0 \subset G_{der,0} \cdot Z(G)$. It follows that any G-spherical character of T_f is determined by its restriction to $Z(G) \cap K$. So we only need to verify that the mappings described in the proposition do not change central character. Since at every stage we are simply adding and subtracting roots, or changing Cartans by an inner automorphism of $\mathbf{G}$, this is clear. $\qquad\square$

It is now more or less a routine matter to parametrize extensions of a general weight $\mu \in \Pi^*(T_f)$. To do this, we first choose positive roots

$$\Delta_c^+ \subset \Delta(K, T_f) \tag{8.17a}$$

in such a way that μ is dominant. This defines $2\rho_c$ as in (4.3), the sum of the positive compact roots, and the large Cartan $T_{fl} \supset T_f$ of Definition 3.4. Define $T_{fl}(\mu)$ as in Definition 4.7, the stabilizer of μ in T_{fl}. We now choose a Levi subgroup $L \supset H_f$ large enough so that

$$T_{fl}(\mu) \subset L, \tag{8.17b}$$

(so that we can compute extensions of μ inside L), but small enough that μ is L-spherical:

$$\langle \mu, \gamma^\vee \rangle = 0, \qquad (\gamma \in \Delta(L, T_f)). \tag{8.17c}$$

Finally, we require that

$$L \text{ is split, with split Cartan subgroup } H \tag{8.17d}$$

(so that we can apply Proposition 8.16).

Here is one way to find such an L. Define

$$R(K, \mu) = T_{fl}(\mu)/T_f \subset W(G, H_f) \tag{8.17e}$$

(Proposition 4.4). Recall from Proposition 4.4 the set of strongly orthogonal noncompact imaginary roots

$$P(\mu) = \{\beta_1, \ldots, \beta_s\} \tag{8.17f}$$

with the property that

$$P(\mu) \cup -P(\mu) = \{\beta \in \Delta(G, T_f) \mid \langle \mu, \beta^\vee \rangle = \langle 2\rho_c, \beta^\vee \rangle = 0\}. \tag{8.17g}$$

(We have changed notation a little from Proposition 4.4, by renumbering the roots β_i.) These are the roots of a split Levi subgroup $L_{small}(\mu) \supset H_f$, locally isomorphic to a product of s copies of $SL_2(\mathbb{R})$ and an abelian group. Obviously μ is L_{small}-spherical, and Proposition 4.4 shows that $T_{fl}(\mu) \subset L_{small}$. (We could even have accomplished this using the slightly smaller set $P_{ess}(\mu)$ described in (4.5).) I will describe another (larger) natural choice for L in Section 13.

9. Proof of Lemma 8.13

I repeat that the following argument is due to Adams and Herb. The notation will be a little less burdensome if we interchange roots and coroots, proving instead

Lemma 9.1. *Suppose* $S = \{\gamma_1, \ldots, \gamma_m\}$ *is a maximal orthogonal set of roots in* $\Delta(\mathbf{G}, \mathbf{H})$. *Suppose* ϵ *is a grading of the roots with values in* $\mathbb{Z}/2\mathbb{Z}$, *and that*

$$\epsilon(\gamma_j) = 0, \qquad (j = 1, \ldots, m).$$

Then there is a subset $A \subset S$ *so that* ϵ *is equal to the grading* ϵ_A *defined by the sum of the coroots in this subset:*

$$\epsilon(\alpha) \equiv \sum_{\gamma \in A} \langle \alpha, \gamma^\vee \rangle \pmod 2$$

Proof. We proceed by induction on m. If $m = 0$ then Δ is empty and the lemma is trivial. So suppose $m > 0$. Define

$$\Delta_{m-1} = \{\beta \in \Delta \mid \langle \beta, \gamma_m^\vee \rangle = 0\}, \qquad S_{m-1} = \{\gamma_1, \ldots, \gamma_{m-1}\}. \tag{9.2a}$$

the roots orthogonal to γ_m. Clearly S_{m-1} is a maximal orthogonal set of roots in Δ_{m-1}, and by hypothesis ϵ is trivial on S_{m-1}. By inductive hypothesis there is a subset $A_{m-1} \subset S_{m-1}$ so that

$$\epsilon(\alpha) \equiv \sum_{\gamma \in A_{m-1}} \langle \alpha, \gamma^\vee \rangle \pmod 2 \qquad (\alpha \in \Delta_{m-1}). \tag{9.2b}$$

Define

$$A' = A_{m-1} \subset S, \qquad A'' = A_{m-1} \cup \{\gamma_m\}. \tag{9.2c}$$

It follow immediately from (9.2a) that

the gradings $\epsilon_{A'}$ and $\epsilon_{A''}$ agree with ϵ on Δ_{m-1} and $\pm\gamma_m$. (9.2d)

We are going to show that taking A equal to one of the two subsets A' and A'' satisfies the requirement of Lemma 9.1 that $\epsilon_A = \epsilon$. To do that we need a lemma.

Lemma 9.3. *In the setting of* (9.2), *there are two mutually exclusive possibilities.*

(1) *The coroot* $\gamma_m^\vee$ *takes even values on every root:*

$$\langle \alpha, \gamma_m^\vee \rangle \in 2\mathbb{Z}, \qquad (\alpha \in \Delta).$$

In this case the integer span of γ_m *and* Δ_{m-1} *contains* Δ.

(2) *There is a root* $\alpha_1 \in \Delta$ *so that* $\langle \alpha_1, \gamma_m^\vee \rangle$ *is odd. In this case the integer span of* γ_m, Δ_{m-1}, *and* α_1 *contains* Δ.

In case (1), both of the gradings $\epsilon_{A'}$ *and* $\epsilon_{A''}$ *are equal to* ϵ. *In case (2), exactly one of them is equal to* ϵ.

The last statement of this lemma completes the proof of Lemma 9.1 □

Proof of Lemma 9.3. Since Δ is reduced, the possible values for $\langle \alpha, \gamma_m^\vee \rangle$ are 0, ± 1, ± 2, and ± 3. Obviously the pairings of roots with $\gamma_m^\vee$ are either all even or not all even. Consider the first possibility. If α is any root, then $\langle \alpha, \gamma_m^\vee \rangle$ is equal to 0, 2, or -2. In the first case, α belongs to Δ_{m-1}. In the second, either $\alpha = \gamma_m$ or $\alpha - \gamma_m$ is a root in Δ_{m-1}; so in either case α is in the integer span of Δ_{m-1} and γ_m. In the third case, either $\alpha = -\gamma_m$ or $\alpha + \gamma_m$ is a root in Δ_{m-1}, and we get the same conclusion.

Consider now the second possibility, and fix α_1 having an odd pairing with the coroot $\gamma_m^\vee$. Possibly after replacing α_1 by $\pm\alpha_1 \pm \gamma_m$, we may assume that

$$\langle \alpha_1, \gamma_m^\vee \rangle = 1. \tag{9.4a}$$

Suppose $\beta \in \Delta$; we want to write

$$\beta = p\gamma_m + q\alpha_1 + \delta \qquad (p, q \in \mathbb{Z}, \delta \in \Delta_{m-1}). \tag{9.4b}$$

If $\langle \beta, \gamma_m^\vee \rangle$ is even, then we can achieve (9.4b) as in the first possibility (not using α_1). So we may assume that $\langle \beta, \gamma_m^\vee \rangle$ is odd. Perhaps after replacing β by $-\beta$, we may assume that $\langle \beta, \gamma_m^\vee \rangle$ is 1 or 3. Perhaps after replacing β by $\beta - \gamma_m$, we may assume that

$$\langle \beta, \gamma_m^\vee \rangle = 1. \tag{9.4c}$$

If $\langle \beta, \alpha_1^\vee \rangle > 0$, then $\delta = \beta - \alpha_1$ is a root in Δ_{m-1}, and (9.4b) follows. So we may assume

$$\langle \beta, \alpha_1^\vee \rangle \leq 0. \tag{9.4d}$$

Now it follows from (9.4c) that $\beta - \gamma_m$ is a root taking the value -1 on the coroot $\gamma_m^\vee$. Furthermore

$$\langle \beta - \gamma_m, \alpha_1^\vee \rangle = \langle \beta, \alpha_1^\vee \rangle - \langle \gamma_m, \alpha_1^\vee \rangle.$$

The first term is non-positive by (9.4d) and the second strictly negative by (9.4a). It follows that $\delta = \beta - \gamma_m + \alpha_1$ is a root or zero. In the first case $\delta \in \Delta_{m-1}$ (by (9.4a) and (9.4c)), so in any case $\beta = \gamma_m - \alpha_1 + \delta$ has the form required in (9.4b).

The last assertion of the lemma follows from (9.2d) in the first case. In the second case, the hypothesis on α_1 shows that $\epsilon_{A'}(\alpha_1) = 1 - \epsilon_{A''}(\alpha_1)$. So exactly one of these two gradings agrees with ϵ on α_1, and the assertion now follows from (9.2d). $\qquad\square$

10. Highest Weights for K and θ-Stable Parabolic Subalgebras

The description of $\Pi^*(K)$ in terms of $\Pi^*(T_f)$ in Theorem 4.10 was mediated by the (θ-stable) Borel subalgebra $\mathfrak{b}_c$ of $\mathfrak{k}$. In this section we consider more general correspondences of the same sort, between representations of K and those of certain Levi subgroups $L \cap K$. These correspondences will be used in Section 11 to describe the lowest K-types of standard limit representations.

We therefore fix a θ-stable parabolic subalgebra

$$\mathfrak{q} \supset \mathfrak{u} \tag{10.1a}$$

of $\mathfrak{g}$; here $\mathfrak{u}$ denotes the nil radical of $\mathfrak{q}$. Recall that σ is the antiholomorphic involution of $\mathfrak{g}$ defining the real form. We assume also that $\mathfrak{q}$ is opposite to $\sigma\mathfrak{q}$; equivalently,

$$\mathfrak{q} \cap \sigma\mathfrak{q} = \mathfrak{l} \tag{10.1b}$$

is a Levi subalgebra of $\mathfrak{q}$. Clearly $\sigma\mathfrak{l} = \mathfrak{l}$, so $\mathfrak{l}$ is defined over $\mathbb{R}$. It is also preserved by the Cartan involution θ. We write L for the corresponding real Levi subgroup of G. It is not difficult to show that

$$L = \{g \in G \mid \mathrm{Ad}(g)(\mathfrak{q}) = \mathfrak{q}\}. \tag{10.1c}$$

Like G, L is the group of real points of a connected reductive algebraic group, and $\theta|_L$ is a Cartan involution for L; so $L \cap K$ is a maximal compact subgroup of L. Choose a θ-stable fundamental Cartan subgroup H_f for L

as in Definition 3.1. The use of the same notation as for G is justified by the following result.

Proposition 10.2. *In the setting of (10.1), the fundamental Cartan H_f for L is also a fundamental Cartan subgroup for G.*

Now let us fix a choice of positive roots

$$\Delta_c^+(L) \subset \Delta(L \cap K, T_f). \tag{10.3a}$$

Define a maximal nilpotent subalgebra of $\mathfrak{l} \cap \mathfrak{k}$ by the requirement

$$\Delta(\mathfrak{n}_c(\mathfrak{l}), T_f) = -\Delta_c^+(L) \tag{10.3b}$$

as in Definition 3.3, so that

$$\mathfrak{b}_c(\mathfrak{l}) = \mathfrak{t}_f + \mathfrak{n}_c(\mathfrak{l}) \tag{10.3c}$$

is a Borel subalgebra of $\mathfrak{l} \cap \mathfrak{k}$. We now get a Borel subalgebra of $\mathfrak{k}$ by defining

$$\mathfrak{b}_c = \mathfrak{b}_c(\mathfrak{l}) + \mathfrak{u} \cap \mathfrak{k}, \qquad \mathfrak{n}_c = \mathfrak{n}_c(\mathfrak{l}) + \mathfrak{u} \cap \mathfrak{k} \tag{10.3d}$$

The corresponding positive root system is

$$\Delta_c^+ = \Delta_c^+(L) \cup -\Delta(\mathfrak{u} \cap \mathfrak{k}, T_f). \tag{10.3e}$$

In the setting of (10.1), we will refer to such choices of fundamental Cartan and positive root systems as *compatible*.

The identifications

$$R(K) \simeq W(K, T_f)/W(K_0, T_{f,0}),$$
$$R(L \cap K) \simeq W(L \cap K, T_f)/W((L \cap K)_0, T_{f,0})$$

and the obvious inclusion of Weyl groups define a natural map

$$R(L \cap K) \to R(K) \tag{10.3f}$$

It is not difficult to show that this map is an inclusion.

We will be using Proposition 4.4 for both L and G, so we need to choose the sets $P(K)$ and $P(L \cap K)$ defined there in a compatible way. It turns out that we need not have $P(L \cap K) \subset P(K)$. For example, if L is locally isomorphic to $SL_2(\mathbb{R})$, then $P(L \cap K)$ is always a single root; but this root will be in P only under special circumstances. Because of this fact, it is not clear what "compatible" ought to mean for P.

Proposition 10.4. *In the setting of (10.3), suppose $A \subset P(L \cap K)$ is such that $w_A \in R(L \cap K)$ (cf. Proposition 4.4). Then $A \subset P(K) \cup -P(K)$, and $w_A \in R(K)$.*

It follows from this proposition that

$$P_{ess}(L \cap K) \subset P_{ess}(K) \cup -P_{ess}(K).$$

(notation as in (4.5)). It therefore makes sense to require

$$P_{ess}(L \cap K) \subset P_{ess}(K) \tag{10.5}$$

as the compatibility requirement between $P(K)$ and $P(L \cap K)$.

Theorem 10.6. *Suppose we are in the setting of (10.1); choose compatible positive root systems $\Delta^+(L \cap K, T_f) \subset \Delta^+(K, T_f)$ as in (10.3), and compatible sets $P(K)$ and $P(L \cap K)$ as in (10.5). Suppose that (τ_q, E_q) is an irreducible representation of $L \cap K$, of extended highest weight $\widetilde{\mu_q}$ (Definition 4.11). Define a generalized Verma module*

$$M_K(\tau_q) = U(\mathfrak{k}) \otimes_{\mathfrak{q} \cap \mathfrak{k}} E_q,$$

which is a $(\mathfrak{k}, L\cap K)$-module. Write $\mathbb{L}_k$ for the kth Bernstein derived functor carrying $(\mathfrak{k}, L \cap K)$-modules to K-modules, and $S = \dim \mathfrak{u} \cap \mathfrak{k}$. Define

$$\zeta_{\mathfrak{k}} = d\mu_q + \rho(\mathfrak{u} \cap \mathfrak{k}) + \rho(\mathfrak{l} \cap \mathfrak{k}) \in \mathfrak{t}_f^*,$$

a weight parametrizing the infinitesimal character of the generalized Verma module $M_K(\tau_q)$.

(1) *If $\zeta_{\mathfrak{k}}$ vanishes on any coroot of K, then $\mathbb{L}_k(M_K(\tau_q)) = 0$ for every k. Suppose henceforth that $\zeta_{\mathfrak{k}}$ is regular for K.*

(2) *If $\zeta_{\mathfrak{k}}$ is dominant for $\Delta^+(K, T_f)$, then the generalized Verma module $M_K(\tau_q)$ is irreducible. In this case $\mathbb{L}_k(M_K(\tau_q)) = 0$ for $k \neq S$. The K representation $\mathbb{L}_S(M_K(\tau_q))$ is generated by the highest weight*

$$\mu = \mu_q \otimes 2\rho(\mathfrak{u} \cap \mathfrak{k}).$$

The corresponding extended group $T_{fl}(\mu)$ (Proposition 4.4) contains $T_{fl}(\mu_q)$ (via the inclusion (10.3f)), and the representation of $T_{fl}(\mu)$ on the μ weight space is

$$\mathrm{Ind}_{T_{fl}(\mu_q)}^{T_{fl}(\mu)} \widetilde{\mu};$$

here $\widetilde{\mu}$ denotes the representation $\widetilde{\mu_q} \otimes 2\rho(\mathfrak{u} \cap \mathfrak{k})$ of $T_{fl}(\mu_q)$. In particular, if the R-groups $R(L \cap K, \mu_q)$ and $R(K, \mu)$ are equal (under the inclusion (10.3f)), then $\mathbb{L}_S(M_K(\tau_q))$ is the irreducible representation of K of extended highest weight $\widetilde{\mu}$ (Definition 4.11).

(3) *In general, let $w \in W(K_0, T_{f,0})$ be the unique element so that $\zeta_{\mathfrak{k}}$ is dominant for $w(\Delta^+(K, T_f))$. Define $k_0 = k(\mu_q) = S - l(w)$ (with*

l(w) the length of the Weyl group element w). The number k_0 is equal to the cardinality of the set of roots

$$B = \{\alpha \in \Delta(\mathfrak{u}, T_f) \cap -w\Delta^+(K, T_f)\}.$$

and

$$\mu = \mu_{\mathfrak{q}} + \sum_{\alpha \in B} \alpha;$$

this weight has differential $\zeta_{\mathfrak{k}} - w\rho_c$. In this case $\mathbb{L}_k(M_K(\tau_{\mathfrak{q}})) = 0$ for $k \neq k_0$. The K representation $\mathbb{L}_{k_0}(M_K(\tau_{\mathfrak{q}}))$ includes the extremal weight μ.

11. Standard Representations, Limits, and Continuations

Throughout this section we will fix a θ-stable real Cartan subgroup H as in (6.1), with compact part $T = H^\theta$ and split part $A = H^{-\theta}$. (Recall that in our notation A is usually not connected: it is the identity component A_0 that is the vector group of traditional notation, figuring in the direct product Cartan decomposition $H = TA_0$.) We will be constructing representations of G attached to characters of H. The constructions will rely in particular on two Levi subgroups between H and G. In Harish-Chandra's work the most important is

$$M = Z_G(A_0). \tag{11.1a}$$

This is a reductive algebraic subgroup of G, equal to the group of real points of $\mathbf{M} = Z_{\mathbf{G}}(\mathbf{A}_e)$. The roots of H in M are precisely the imaginary roots of H in G:

$$\Delta(M, H) = \Delta_{im}(G, H). \tag{11.1b}$$

Once again our notation is a little different from the classical notation, in which M usually denotes the interesting factor in the direct product decomposition (often called "Langlands decomposition")

$$Z_G(\text{classical } A) = (\text{classical } M) \times (\text{classical } A).$$

This notation is inconvenient for us because the classical M need not be the real points of a connected reductive algebraic group.

Clearly the Cartan subgroup H is fundamental in M, so we can use the results of Section 5 to make representations of M from characters of H. The great technical benefit of using M is that M contains a maximally split Cartan subgroup H_s of G. The two Cartan decompositions

$$G = K(A_{s,0})K, \qquad M = (M \cap K)(A_{s,0})(M \cap K)$$

allow one to compare "behavior at infinity" on M and on G directly. (Such analysis is at the heart of Harish-Chandra's work, and of Langlands' proof of his classification of $\Pi^*(G)$. We will not make explicit use of it here.)

At the same time we will use

$$L = Z_G(T_0). \tag{11.1c}$$

This is a reductive algebraic subgroup of G, equal to the group of real points of $\mathbf{L} = Z_{\mathbf{G}}(\mathbf{T}_e)$. The roots of H in L are precisely the real roots of H in G:

$$\Delta(L, H) = \Delta_{re}(G, H). \tag{11.1d}$$

The Cartan subgroup H is split in L, so the results of Section 7 tell us about representations of L attached to characters of H. The great technical benefit of using L is that L contains a fundamental Cartan subgroup H_f of G. For that reason representations of $L \cap K$ are parametrized (approximately) by characters of T_f, which in turn (approximately) parametrize representations of K.

Definition 11.2. In the setting of (11.1), a *shifted Harish-Chandra parameter for a standard representation* is a pair

$$\Phi = (\phi, \Delta_{im}^+)$$

(with $\phi \in \Pi^*(H)$ a character and $\Delta_{im}^+ \subset \Delta_{im}(G, H_f)$ a system of positive roots) subject to the following requirement:

(1-std) For every positive imaginary root $\alpha \in \Delta_{im}^+$, we have

$$\langle d\phi, \alpha^\vee \rangle > 1.$$

We define

$$P_G^s(H) = \text{shifted parameters for standard representations}$$

$$= \text{pairs } (\phi, \Delta_{im}^+) \text{ satisfying condition (1-std) above.}$$

(The superscript s stands for "shifted"; it is a reminder that this parameter differs by a ρ shift from the Harish-Chandra parameter.)

The imaginary roots divide into compact and noncompact as usual, according to the eigenvalue of θ on the corresponding root space:

$$\Delta_{im,c}^+ = \text{positive imaginary roots in } \mathfrak{k},$$

$$\Delta_{im,n}^+ = \text{remaining imaginary roots.}$$

Define

$$2\rho_{im} = \sum_{\alpha \in \Delta_{im}^+} \alpha, \qquad 2\rho_{im,c} = \sum_{\beta \in \Delta_{im,c}^+} \beta \qquad 2\rho_{im,n} = \sum_{\gamma \in \Delta_{im,n}^+} \gamma.$$

Each of these characters may be regarded as belonging either to the group of real characters $\Pi^*_{adm}(H)$ or to the algebraic character lattice $\Pi^*_{alg}(\mathbf{H})$. The lattice of characters may be embedded in $\mathfrak{h}^*$ by taking differentials, so we may also regard these characters as elements of $\mathfrak{h}^*$. There they may be divided by two, defining ρ_{im}, $\rho_{im,c}$, and $\rho_{im,n}$ in $\mathfrak{h}^*$.

The *infinitesimal character parameter for* Φ is

$$\zeta(\Phi) = d\phi - \rho_{im} \in \mathfrak{h}^*;$$

we write

$$\zeta(\Phi) = \lambda(\Phi) + \nu(\Phi) \qquad (\lambda(\Phi) \in \mathfrak{t}^*, \nu(\Phi) \in \mathfrak{a}^*)$$

More generally, a *shifted Harish-Chandra parameter for a standard limit representation* is a pair

$$\Phi = (\phi, \Delta^+_{im})$$

(with $\phi \in \Pi^*(H)$ a character and $\Delta^+_{im} \subset \Delta_{im}(G, H_f)$ a system of positive roots) subject to the following requirement:

(1-lim) For every positive imaginary root $\alpha \in \Delta^+_{im}$, we have

$$\langle d\phi, \alpha^\vee \rangle \geq 1.$$

We define

$$P^{s,lim}_G(H) = \text{shifted parameters for limits of standard representations}$$

$$= \text{pairs } \Phi = (\phi, \Delta^+_{im}) \text{ satisfying condition (1-lim) above.}$$

Finally, a *shifted Harish-Chandra parameter for a continued standard representation* is a pair

$$\Phi = (\phi, \Delta^+_{im})$$

with $\phi \in \Pi^*(H)$, Δ^+_{im} a system of positive imaginary roots, and no positivity hypothesis. We define

$$P^{s,cont}_G(H) = \text{shifted parameters for continued standard representations}$$

$$= \text{pairs } (\phi, \Delta^+_{im}).$$

For limit and continued parameters we can define the infinitesimal character parameter $\zeta(\Phi)$ in the obvious way.

For each kind of parameter there is a discrete part, which remembers only the restriction ϕ_d of the character ϕ to the compact torus T. We write for example

$$P^{s,lim}_{G,d}(H) = \text{shifted disc. params. for limits of standard reps.}$$

$$= \text{pairs } \Phi_d = (\phi_d, \Delta^+_{im}) \text{ satisfying condition (1-lim) above.}$$

A given discrete parameter has a distinguished extension to a full parame-ter, namely the one which is trivial on A_0 (the vector subgroup of H). By using this extension, we may regard $P_{G,d}^{s,lim}(H)$ as a subset of $P_G^{s,lim}(H)$.

Proposition 11.3. *In the setting of Definition 11.2, suppose* $\Phi \in P_G^s(H)$ *is a shifted Harish-Chandra parameter for a standard representation. Then there is a standard representation* $I(\Phi)$ *for* G *attached to* Φ, *which may be constructed as follows. Choose a real parabolic subgroup*

$$P = MN \subset G$$

in such a way that the real part of the differential $d\phi|_\mathfrak{a}$ *is weakly dominant for all the roots of* H *in* N. *Let* $I_M(\Phi)$ *be the fundamental series represen-tation of* M *specified by Proposition 5.2; this is in fact a (relative) discrete series representation of* M. *Define*

$$I(\Phi) = \operatorname{Ind}_{MN}^G I_M(\Phi) \otimes 1$$

(normalized induction), a standard representation of G. *It is always non-zero, and its restriction to* K, *which is*

$$I(\Phi)|_K = \operatorname{Ind}_{M \cap K}^K I_M(\Phi)|_{M \cap K},$$

depends only on the restriction of Φ *to* T. *The infinitesimal character of* $I(\Phi)$ *corresponds to the weight* $\zeta(\Phi) \in \mathfrak{h}^*$ *of Definition 11.2. This standard representation has a Langlands quotient* $J(\Phi)$, *which is always non-zero but may be reducible. We have*

$$I(\Phi) \simeq I(\Psi) \iff J(\Phi) \simeq J(\Psi) \iff \Psi \in W(G,H) \cdot \Phi.$$

For limits of standard representations the situation is quite similar; the main problem is that the corresponding representation of G may vanish. Here is a statement.

Proposition 11.4. *In the setting of Definition 11.2, suppose* $\Phi \in P_G^{s,lim}(H)$ *is a shifted Harish-Chandra parameter for a standard limit rep-resentation. Then there is a standard limit representation* $I(\Phi)$ *for* G *at-tached to* Φ, *which may be constructed by parabolic induction exactly as in Proposition 11.3. The infinitesimal character of* $I(\Phi)$ *corresponds to the weight* $\zeta(\Phi) \in \mathfrak{h}^*$ *of Definition 11.2. Its restriction to* K *depends only on the restriction of* Φ *to* T. *This standard limit representation representation has a Langlands quotient* $J(\Phi)$. *We have*

$$\Psi \in W(G,H) \cdot \Phi \implies I(\Phi) \simeq I(\Psi) \iff J(\Phi) \simeq J(\Psi).$$

Define

$$\mu(\Phi) = [\phi - 2\rho_{im,c}]|_T \in \Pi^*(T).$$

Then the following three conditions are equivalent:

(1) *the weight $\mu(\Phi)$ is dominant with respect to the compact imaginary roots $\Delta^+_{im,c}$;*

(2) *there is no simple root for Δ^+_{im} which is both compact and orthogonal to $\phi - \rho_{im}$; and*

(3) *the standard limit representation $I(\Phi)$ is non-zero.*

Here is the result for coherent continuation.

Proposition 11.5. *In the setting of Definition 11.2, suppose $\Phi \in P^{s,cont}_G(H)$ is a shifted Harish-Chandra parameter for continued standard representations. Then there is a virtual representation $I(\Phi) = I_G(\Phi)$ for G with the following properties.*

(1) *If $P = MN$ is any parabolic subgroup of G with Levi factor M, then*

$$I(\Phi) = \mathrm{Ind}^G_{MN} I_M(\Phi),$$

where the inducing (virtual) representation on the right is the one described in Proposition 5.4.

(2) *The restriction of $I(\Phi)$ to K depends only on Δ^+_{im} and the restriction of ϕ to T. Explicitly,*

$$I(\Phi)|_K = \mathrm{Ind}^K_{M\cap K} I_M(\Phi)|_{M\cap K}.$$

(3) *We have*

$$I(\Phi) \simeq I(w \cdot \Phi) \qquad (w \in W(G, H)).$$

(4) *The virtual representation $I(\Phi)$ has infinitesimal character corresponding to $\zeta(\Phi, \Delta^+_{im}) \in \mathfrak{h}^*$.*

(5) *If $\zeta(\Phi)$ is weakly dominant for Δ^+_{im}, then $I(\Phi)$ is equivalent to the standard limit representation attached to Φ in Proposition 11.4.*

(6) *Suppose V is a finite-dimensional representation of G. Recall that $\Delta(V, H)$ denotes the multiset of weights of H on V (Definition 3.2). Then*

$$I(\Phi) \otimes V \simeq \sum_{\delta \in \Delta(V,H)} I(\Phi + \delta).$$

In order to describe lowest K-types of standard representations, it is convenient to give a completely different construction of them, by cohomological induction. We begin with coherent continuations of standard representations, later specializing to the standard representations themselves.

Fix a parameter

$$\Phi = (\phi, \Delta^+_{im}) \in P^{s,cont}_G(H) \tag{11.6a}$$

(Definition 11.2). Recall the Levi subgroup $L \supset H$ defined in (11.1c), corresponding to the real roots. A θ-stable parabolic subalgebra $\mathfrak{q} = \mathfrak{l} + \mathfrak{u}$ is called *weakly Φ-compatible* if

$$\Delta_{im}^+ \subset -\Delta(\mathfrak{u}, H). \qquad (11.6b)$$

We may also say that Φ is *weakly $\mathfrak{q}$-compatible*, or that Φ and $\mathfrak{q}$ are *weakly compatible*. Here is a way to construct (any) such $\mathfrak{q}$. Fix a generic element

$$\tau \in i\mathfrak{t}_0^* \qquad (11.6c)$$

which is dominant for Δ_{im}^+. The "genericity" we require of τ is that the only roots to which τ is orthogonal are the real roots of H in G. Attached to any such generic τ there is a weakly compatible $\mathfrak{q} = \mathfrak{q}(\tau) = \mathfrak{l} + \mathfrak{u}(\tau)$ characterized by

$$\Delta(\mathfrak{u}(\tau), H) = \{\alpha \in \Delta(G, H) \mid \langle \tau, \alpha^\vee \rangle < 0\}. \qquad (11.6d)$$

When Φ is actually a standard limit parameter, we will sometimes wish to require more of $\mathfrak{q}$. Recall the weight $\lambda = \lambda(\Phi) \in \mathfrak{t}^*$ attached to Φ (Definition 11.2). We say that $\mathfrak{q}$ is *strongly Φ-compatible* if it is weakly Φ-compatible, and in addition

$$\langle \lambda, \alpha \rangle \leq 0, \qquad (\alpha \in \Delta(\mathfrak{u}, H)). \qquad (11.6e)$$

Again we may say that Φ is *strongly $\mathfrak{q}$-compatible*, or that the pair is *strongly compatible*. Here is a way to construct (any) strongly Φ-compatible parabolic. Since λ is weakly dominant for Δ_{im}^+, the weight

$$\tau_1 = \lambda + \epsilon\tau \qquad (11.6f)$$

is dominant for Δ_{im}^+ and generic as long as ϵ is a small enough positive real; and in that case $\mathfrak{q}(\tau_1)$ is strongly Φ-compatible. A little more explicitly,

$$\Delta(\mathfrak{u}(\tau_1), H) = \{\alpha \in \Delta \mid \langle \lambda, \alpha^\vee \rangle < 0, \text{ or } \langle \lambda, \alpha^\vee \rangle = 0 \text{ and } \langle \tau, \alpha^\vee \rangle < 0\}. \qquad (11.6g)$$

Notice that, just as in the definition of Borel subalgebra for $\mathfrak{k}$ in Definition 3.3, we have put the *negative* roots in the nil radical. One effect (roughly speaking) is that Verma modules constructed using $\mathfrak{q}$ and (standard limit) Φ tend (in the strongly compatible case) to be irreducible. (This statement is not precisely true, for example because the definition of strongly Φ-compatible ignores $\nu(\Phi)$ (Definition 11.2).)

The involution θ preserves $\Delta(\mathfrak{u}, H)$. The fixed points are precisely the imaginary roots Δ_{im}^+. The remaining roots therefore occur in pairs

$$\{\alpha, \theta\alpha\} \qquad (\alpha \text{ complex in } \Delta(\mathfrak{u}, H)). \qquad (11.6h)$$

Evidently the two roots α and $\theta\alpha$ have the same restriction to $T = H^\theta$. In this way we get a well-defined character

$$\rho_{cplx} = \sum_{\substack{\text{pairs } \{\alpha, \theta\alpha\} \\ \text{cplx in } \Delta(\mathfrak{u}, H)}} -\alpha|_T \quad \in \Pi^*_{adm}(T). \qquad (11.6\text{i})$$

Extending this character to be trivial on A_0, we get $\rho_{cplx} \in \Pi^*_{adm}(H)$. As the notation suggests, the differential of this character is equal to half the sum of the complex roots in $\mathfrak{u}^{op}$, which we are thinking of as positive.

Now define

$$\Phi_{\mathfrak{q}} = (\phi \otimes \rho_{cplx}, \emptyset) \in P^s_L(H), \qquad (11.6\text{j})$$

a (shifted) Harish-Chandra parameter for a principal series representation of L. The corresponding infinitesimal character parameter for L is

$$\zeta_{\mathfrak{l}}(\Phi_{\mathfrak{q}}) = d\phi_{\mathfrak{q}} = d\phi + \rho_{cplx} = \zeta(\Phi) + \rho_{im} + \rho_{cplx} = \zeta_{\mathfrak{g}}(\Phi) - \rho(\mathfrak{u}); \quad (11.6\text{k})$$

the last equality is equivalent to $\mathfrak{q}$ being weakly Φ-compatible. This is the infinitesimal character of the standard principal series representation $I_L(\Phi_{\mathfrak{q}})$.

We can now form the generalized Verma module

$$M(\Phi_{\mathfrak{q}}) = U(\mathfrak{g}) \otimes_{\mathfrak{q}} I_L(\Phi_{\mathfrak{q}}), \qquad (11.6\text{l})$$

which is a $(\mathfrak{g}, L \cap K)$-module. Its infinitesimal character is given by adding the half sum of the roots of $\mathfrak{u}$ to the infinitesimal character of $I_L(\Phi_{\mathfrak{q}})$; it is therefore equal (still in the weakly Φ-compatible case) to $\zeta(\Phi)$.

Theorem 11.7. *Suppose* $\Phi \in P^{s,cont}_G(H)$ *is a parameter for a continued standard representation (Definition 11.2) and* $\mathfrak{q} = \mathfrak{l} + \mathfrak{u}$ *is a weakly Φ-compatible parabolic (see (11.6)). Write* $\mathbb{L}_k$ *for the* k*th Bernstein derived functor carrying* $(\mathfrak{g}, L \cap K)$*-modules to* $(\mathfrak{g}, K)$*-modules, and* $S = \dim \mathfrak{u} \cap \mathfrak{k}$*. Then the virtual representation* $I(\Phi, \Delta^+_{im})$ *may be constructed from the generalized Verma module* $M(\Phi_{\mathfrak{q}})$ *of (11.6l):*

$$I(\Phi) = \sum_{k=0}^{S} (-1)^k \mathbb{L}_{S-k}(M(\Phi_{\mathfrak{q}}))$$

Suppose from now on that Φ *is a standard limit parameter (Definition 11.2), and that* $\mathfrak{q}$ *is strongly Φ-compatible (cf. (11.6)). Then*

$$\mathbb{L}_k(M(\Phi_{\mathfrak{q}})) = \begin{cases} I(\Phi) & k = S \\ 0 & k \neq S. \end{cases}$$

Write $A(\Phi_{\mathfrak{q}})$ for the set of lowest $L \cap K$-types of the principal series representation $I(\Phi_{\mathfrak{q}})$ (Proposition 7.5), and $A(\Phi)$ for the set of lowest K-types of the standard limit representation $I(\Phi)$. If $\widetilde{\mu_{\mathfrak{q}}}$ is an extended highest weight of $\tau_{\mathfrak{q}} \in A(\Phi_{\mathfrak{q}})$ (Definition 4.11), define $\widetilde{\mu} = \widetilde{\mu_{\mathfrak{q}}} \otimes 2\rho(\mathfrak{u} \cap \mathfrak{k})$ as in Theorem 10.6. There are two possibilities.

(1) *If μ is dominant for Δ_c^+, then $\widetilde{\mu}$ is an extended highest weight of the irreducible K-representation*

$$\tau = \mathbb{L}_S(M_K(\tau_{\mathfrak{q}})),$$

In this case the bottom layer map of [5] exhibits τ as a lowest K-type of $I(\Phi)$.

(2) *If μ is not dominant for Δ_c^+, then*

$$\mathbb{L}_S(M_K(\tau_{\mathfrak{q}})) = 0.$$

In both cases $\mathbb{L}_k(M_K(\tau_{\mathfrak{q}})) = 0$ for $k \neq S$. This construction defines an inclusion

$$A(\Phi) \hookrightarrow A(\Phi_{\mathfrak{q}}), \qquad \tau \mapsto \tau_{\mathfrak{q}}.$$

The first assertions of the theorem (those not referring to lowest K-types) are special cases of Theorem 12.8, which will be formulated and proved in section 12.

This theorem (in conjunction with Theorem 10.6) effectively computes the highest weight parameters of the lowest K-types of $I(\Phi)$ by reduction to the special case of principal series for split groups, which was treated in Section 7.

Definition 11.8. Suppose $\Phi \in P^{s,lim}(G)$ is a standard limit parameter (Definition 11.2). We say that Φ is *final* if

(1) the standard limit representation $I(\Phi)$ (Proposition 11.3) is not zero; and

(2) if we choose a strongly Φ-compatible θ-stable parabolic subalgebra $\mathfrak{q}$ as in (11.6), then $\Phi_{\mathfrak{q}}$ is a final limit parameter for L (Definition 7.7)

Conditions equivalent to (1) are given in Proposition 11.4. The second condition is equivalent to

(2') for every real root α of H in G,

$$\text{either} \quad \langle d\phi_{\mathfrak{q}}, \alpha^{\vee} \rangle \neq 0 \quad \text{or} \quad \phi_{\mathfrak{q}}(m_{\alpha}) = 1.$$

If Φ is discrete, the requirement is that $\phi_{\mathfrak{q}}(m_{\alpha}) = 1$ for every real root α.)
We write $P_G^{s,finlim}(H)$ for the set of final limit parameters for H.

Theorem 11.9. *Suppose* $\Phi \in P_{G,d}^{s,lim}(H)$ *is a discrete standard limit parameter (Definition 11.2) and* $\mathfrak{q} = \mathfrak{l} + \mathfrak{u}$ *is a strongly* Φ-*compatible parabolic (see (11.6)). Assume that* Φ *satisfies condition (2) in the definition of final (cf. Definition 11.8): that is, that* $\phi_{\mathfrak{q}}(m_\alpha) = 1$ *for every real root* α *of* H *in* G. *Then* $A(\Phi_{\mathfrak{q}})$ *(Proposition 7.5) consists of a single irreducible representation* $\tau_{\mathfrak{q}}$ *of* $L \cap K$, *which is trivial on* $L^{der} \cap K$ *and therefore one-dimensional. Write* $\widetilde{\mu_{\mathfrak{q}}}$ *for the (unique) extended highest weight of* $\tau_{\mathfrak{q}}$ *(Definition 4.11), and define* $\widetilde{\mu} = \widetilde{\mu_{\mathfrak{q}}} \otimes 2\rho(\mathfrak{u} \cap \mathfrak{k})$ *as in Theorem 10.6. Then* Φ *is final—that is,* $I(\Phi) \neq 0$—*if and only if* μ *is dominant with respect to* Δ_c^+.

Assume now that Φ *is final (so that* μ *is dominant for* K). *Then the irreducible* K *representation* $\tau(\Phi)$ *of extended highest weight* $\widetilde{\mu}$ *is the unique lowest* K-*type of* $I(\Phi)$.

The correspondence $\Phi \mapsto \tau(\Phi)$ *is a bijection from* K-*conjugacy classes of discrete final limit parameters for* G *onto* $\Pi^*(K)$.

12. More Constructions of Standard Representations

Proposition 11.3 and Theorem 11.7 describe two constructions of standard representations of G (and, implicitly, of their restrictions to K). In order to get an algorithm for branching from G to K, we need to generalize these constructions. In fact we will make use only of the generalized construction by cohomological induction; but the generalization for real parabolic induction is easier to understand and serves as motivation, so we begin with that.

As in (6.1) and Section 11, we fix a θ-stable real Cartan subgroup

$$H \subset G, \qquad T = H^\theta, \qquad A = H^{-\theta}. \tag{12.1a}$$

Fix also a shifted Harish-Chandra parameter for a continued standard representation

$$\Phi = (\phi, \Delta_{im}^+) \in P_G^{s,cont}(H) \tag{12.1b}$$

(Definition 11.2). Recall that ϕ is a character of H, and Δ_{im}^+ a system of positive roots for the imaginary roots of H in G. We will make extensive use of the infinitesimal character parameter

$$\zeta(\Phi) = d\phi - \rho_{im} \in \mathfrak{h}^* \tag{12.1c}$$

(Definition 11.2).

We now fix the parabolics we will use to construct continued standard representations. If P is any real parabolic subgroup of G, then $M = P \cap \theta(P)$ is a Levi subgroup. If N is the unipotent radical of P, it follows that

$P = MN$ is a Levi decomposition. We want to fix such a subgroup with the property that H is contained in M:

$$P = MN \text{ real parabolic}, \qquad H \subset M. \qquad (12.1\mathrm{d})$$

The smallest possible choice for M is the subgroup

$$M_1 = Z_G(A_0) \subset M \qquad (12.1\mathrm{e})$$

introduced in (11.1). One way to see this containment is to notice that complex conjugation must permute the roots of H in $\mathfrak{n}$, but complex conjugation sends the roots of H in $\mathfrak{m}_1$ to their negatives. Because all the imaginary roots of H occur in M, we may regard our fixed parameter as a parameter for M:

$$\Phi = (\phi, \Delta_{im}^+) \in P_M^{s,cont}(H). \qquad (12.1\mathrm{f})$$

In the same way, we fix a θ-stable parabolic subalgebra as in (10.1) subject to two requirements: first, that L contain H

$$\mathfrak{q} = \mathfrak{l} + \mathfrak{u} \qquad \theta\text{-stable parabolic}, \qquad H \subset L; \qquad (12.1\mathrm{g})$$

and second, that

$$\Delta_{im}(\mathfrak{u}, \mathfrak{h}) \subset -\Delta_{im}^+. \qquad (12.1\mathrm{h})$$

The smallest possible choice for L is the subgroup

$$L_1 = Z_G(T_0) \subset L \qquad (12.1\mathrm{i})$$

introduced in (11.1). To see this containment, notice that the action of θ must permute the roots of H in $\mathfrak{u}$, but sends the roots of H in $\mathfrak{l}_1$ to their negatives.

A θ-stable parabolic subalgebra satisfying conditions (12.1g) and (12.1h) is called *weakly Φ-compatible*. In case $L = L_1$, this is precisely the notion introduced in (11.6); all that has changed here is that we are allowing L to be larger than L_1. Just as in the earlier setting, we may say instead that Φ is *weakly $\mathfrak{q}$-compatible*, or that Φ and $\mathfrak{q}$ are *weakly compatible*.

Just as in (11.6i), the non-imaginary roots of H in $\mathfrak{u}$ occur in complex pairs $\{\alpha, \theta\alpha\}$, and we define

$$\rho_{cplx} = \sum_{\substack{\text{pairs } \{\alpha, \theta\alpha\} \\ \text{cplx in } \Delta(\mathfrak{u}, H)}} -\alpha|_T \quad \in \Pi_{adm}^*(T). \qquad (12.2\mathrm{a})$$

Extending this character to be trivial on A_0, we get $\rho_{cplx} \in \Pi_{adm}^*(H)$. Put

$$\phi_{\mathfrak{q}} = \phi \otimes \rho_{cplx} \in \Pi^*(H), \qquad \Delta_{im}^+(L) = \Delta_{im}^+ \cap \Delta(\mathfrak{l}, H). \qquad (12.2\mathrm{b})$$

We have

$$\Phi_{\mathfrak{q}} = (\phi_{\mathfrak{q}}, \Delta_{im}^{+}(L)) \in P_{L}^{s,cont}(H), \qquad \zeta_{\mathfrak{l}}(\Phi_{\mathfrak{q}}) = \zeta_{\mathfrak{g}}(\Phi) - \rho(\mathfrak{u}). \qquad (12.2c)$$

Here $\rho(\mathfrak{u}) = -d\rho_{cplx} - \rho_{im} + \rho_{im}(L)$ is half the sum of the roots of $\mathfrak{h}$ in $\mathfrak{u}$.

We will have occasion later to invert the correspondence $\Phi \to \Phi_{\mathfrak{q}}$. Suppose therefore that $\mathfrak{q} = \mathfrak{l} + \mathfrak{u}$ is a θ-stable parabolic subalgebra as in (10.1), that $H \subset L$ is a θ-stable Cartan subgroup, and that

$$\Psi = (\psi, \Delta_{im}^{+}(L)) \in P_{L}^{s,cont}(H) \qquad (12.3a)$$

is a Harish-Chandra parameter for a continued standard representation of L. Define

$$\Delta_{im}^{+} = \Delta_{im}^{+}(L) \cup -\Delta_{im}(\mathfrak{u}, H); \qquad (12.3b)$$

this is a set of positive imaginary roots for H in G. Write

$$\psi_{\mathfrak{g}}^{\mathfrak{q}} = \psi \otimes \rho_{cplx}^{-1}. \qquad (12.3c)$$

Then

$$\Psi_{\mathfrak{g}}^{\mathfrak{q}} = (\psi_{\mathfrak{g}}^{\mathfrak{q}}, \Delta_{im}^{+}) \in P_{G}^{s,cont}(H). \qquad (12.3d)$$

The infinitesimal character parameters are related by

$$\zeta_{\mathfrak{g}}(\Psi_{\mathfrak{g}}^{\mathfrak{q}}) = \zeta_{\mathfrak{l}}(\Psi) + \rho(\mathfrak{u}). \qquad (12.3e)$$

In order to discuss standard limit representations, we will sometimes want to impose stronger hypotheses on the relationship between Φ and $\mathfrak{q}$. Recall from Definition 11.2 the weight $\lambda(\Phi) \in \mathfrak{t}^{*}$. We say that $\mathfrak{q}$ is *strongly Φ-compatible* if it is weakly Φ-compatible (cf. (12.1g), (12.1h)), and in addition

$$\langle \lambda(\Phi), \alpha^{\vee} \rangle \le 0, \qquad \alpha \in \Delta(\mathfrak{u}, H). \qquad (12.4a)$$

We will also say that Φ is *strongly $\mathfrak{q}$-compatible*, or that Φ and $\mathfrak{q}$ are *strongly compatible*.

Proposition 12.5. *In the setting* (12.1), *suppose that* $\Phi = (\phi, \Delta_{im}^{+})$ *is a shifted Harish-Chandra parameter for a continued standard representation. Then the corresponding virtual representation* $I_{G}(\Phi)$ *(Proposition 11.5) may be realized as*

$$I_{G}(\Phi) = \text{Ind}_{MN}^{G} I_{M}(\Phi).$$

In particular, its restriction to K *is*

$$I_{G}(\Phi)|_{K} = \text{Ind}_{M \cap K}^{K} I_{M}(\Phi, \Delta_{im}^{+})|_{M \cap K}.$$

Proof. Choose a parabolic subgroup $P_1 = M_1 N_1 \subset M$ with Levi factor $M_1 = Z_G(A_0)$ as in Proposition 11.3. Then $Q_1 = M_1(N_1 N)$ is a parabolic subgroup of G. According to Proposition 11.5 (applied first to G and then to M) we have

$$I_G(\Phi) = \operatorname{Ind}_{M_1(N_1 N)}^G I_{M_1}(\Phi),$$

$$I_M(\Phi) = \operatorname{Ind}_{M_1 N_1}^M I_{M_1}(\Phi).$$

Applying induction by stages to these two formulas gives the main claim of the proposition. The second follows because

$$G = KP, \qquad P \cap K = M \cap K.$$

$\square$

Some care is required on one point. If Φ is a standard limit parameter, then the standard limit representations $I_G(\Phi)$ and $I_M(\Phi)$ are both actual representations (not merely virtual representations). The main identity of Proposition 12.5 therefore makes sense as an identity of representations. *It need not be true.* The theorem asserts only that it is true on the level of virtual representations; that is, that both sides have the same irreducible composition factors, appearing with the same multiplicities.

To see an example of this, suppose $G = GL_4(\mathbb{R})$, and H is the split Cartan subgroup consisting of diagonal matrices. There are no imaginary roots, so we omit the empty set of positive imaginary roots from the notation. We have

$$H = \{h = (h_1, h_2, h_3, h_4) \mid h_i \in \mathbb{R}^\times\}; \qquad (12.6a)$$

the h_i are the diagonal entries of h. We consider the character ϕ of H defined by

$$\phi(h) = |h_1|^{3/2} |h_2|^{-3/2} |h_3|^{1/2} |h_4|^{-1/2}. \qquad (12.6b)$$

Regarded as an element of $P_G^s(H)$, this is the Harish-Chandra parameter of the trivial representation of G. The corresponding standard representation $I_G(\Phi)$ is the space of densities on the full flag variety of G (the space of complete flags in $\mathbb{R}^4$). In order to realize this, we need to use a non-standard Borel subgroup B containing H (consisting neither of upper triangular nor of lower triangular matrices).

Now define $M = GL_2 \times GL_2$. The representation $I_M(\Phi)$ has its unique irreducible quotient the finite-dimensional representation $\mathbb{C}^3 \otimes \mathbb{C}^1$ of $GL_2 \times GL_2$. The first of these is the quotient of the adjoint representation of GL_2 by the center. This quotient carries an invariant Hermitian form (related to the Killing form of the first GL_2 factor. It follows that the induced representation

$$\text{Ind}_{MN}^{G} I_M(\Phi) \qquad (12.6c)$$

has as a quotient the unitarily induced representation

$$\text{Ind}_{MN}^{G} \mathbb{C}^3 \otimes \mathbb{C}^1. \qquad (12.6d)$$

This quotient carries a non-degenerate invariant Hermitian form (induced from the one for M).

It is an easy consequence of the Langlands classification that only the Langlands quotient of a standard representation (and not some larger quotient) can carry a non-degenerate invariant Hermitian form. For this reason, the induced representation $\text{Ind}_{MN}^{G} I_M(\Phi)$ cannot be isomorphic to the standard representation $I_G(\Phi)$.

Our next task is to realize continued standard representations by cohomological induction from the θ-stable parabolic subalgebra $\mathfrak{q}$ of (12.1) (assumed always to be weakly Φ-compatible). Just as in (11.6), we begin with the virtual representation

$$I_L(\Phi_\mathfrak{q}) \qquad (12.7a)$$

of L. (Whatever construction we use for this virtual representation will construct it as a difference of two $(\mathfrak{l}, L \cap K)$-modules of finite length; for example the even and odd degrees respectively of some cohomologically induced representations. We will have no need to be very explicit about this.) We can therefore construct the (virtual) generalized Verma module

$$M(\Phi_\mathfrak{q}) = U(\mathfrak{g}) \otimes_\mathfrak{q} I_L(\Phi_\mathfrak{q}), \qquad (12.7b)$$

which is a (virtual) $(\mathfrak{g}, L \cap K)$-module. Its infinitesimal character is given by adding the half sum of the roots of $\mathfrak{u}$ to the infinitesimal character of $I_L(\Phi_\mathfrak{q})$; it is therefore (in light of (12.2c)) equal to $\zeta(\Phi)$.

Theorem 12.8. *Suppose* $\Phi = (\phi, \Delta_{im}^+) \in P_G^{s,cont}(H)$ *is a parameter for a continued standard representation (Definition 11.2) and* $\mathfrak{q} = \mathfrak{l}+\mathfrak{u}$ *is a weakly* Φ-*compatible parabolic (see (12.1)). Write* $\mathbb{L}_k$ *for the kth Bernstein derived functor carrying* $(\mathfrak{g}, L \cap K)$-*modules to* $(\mathfrak{g}, K)$-*modules, and* $S = \dim \mathfrak{u} \cap \mathfrak{k}$. *Then the virtual representation* $I_G(\Phi)$ *may be constructed from the virtual generalized Verma module* $M(\Phi_\mathfrak{q})$ *of (12.7b):*

$$I_G(\Phi) = \sum_{k=0}^{S} (-1)^k \mathbb{L}_{S-k}(M(\Phi_\mathfrak{q}))$$

Suppose in addition that Φ is a standard limit parameter, and that $\mathfrak{q}$ is strongly Φ-compatible (cf. (12.4)). Then

$$\mathbb{L}_k(M(\Phi_{\mathfrak{q}})) = \begin{cases} I_G(\Phi) & k = S \\ 0 & k \neq S. \end{cases}$$

Proof. We begin with the last assertion of the theorem, concerning the case when $\mathfrak{q}$ is strongly Φ-compatible. In this case the claim amounts to Theorem 11.225 of [5].

For the first assertions, we look at the family of parameters

$$\Phi + \gamma = (\phi \otimes \gamma, \Delta_{im}^+), \qquad (\gamma \in \Pi^*(\mathbf{H}))$$

The family of virtual representations $I_G(\Phi + \gamma)$ is a coherent family ([8], Definition 7.2.5); this is the how continued standard representations are defined. The family of virtual representations appearing on the right side of the first formula in the proposition is also coherent; this is proved in the same way as [8], Corollary 7.2.10. Two coherent families coincide if and only if they agree for at least one choice of γ with the property that the infinitesimal character $\zeta(\Phi + \gamma)$ is regular. If γ_0 is sufficiently negative for the roots of H in $\mathfrak{u}$—for example, if γ_0 is a large enough multiple of $-2\rho(\mathfrak{u})$—then $\mathfrak{q}$ is strongly $(\Phi+\gamma)$-compatible whenever γ is close to γ_0. In this case the equality of the two sides is a consequence of the last assertion of the theorem. Most such choices of γ will make $\zeta(\Phi + \gamma)$ regular, and so force the two coherent families to coincide. $\square$

Theorem 12.8 is phrased to construct a given (continued) standard representation of G by cohomological induction. We will also want to use it to identify the result of applying cohomological induction to a given (continued) standard representation of L, and for this a slight rephrasing is convenient.

Theorem 12.9. *Suppose $\mathfrak{q} = \mathfrak{l} + \mathfrak{u}$ is a θ-stable parabolic subalgebra of $\mathfrak{g}$, $H \subset L$ a θ-stable Cartan subgroup, and $\Psi = (\psi, \Delta_{im}^+(L)) \in P_L^{s,cont}(H)$ is the Harish-Chandra parameter for a continued standard representation of L. Define*

$$\Psi_{\mathfrak{g}}^{\mathfrak{q}} = (\psi_{\mathfrak{g}}^{\mathfrak{q}}, \Delta_{im}^+) \in P_G^{s,cont}(H)$$

as in (12.3).

(1) *The parabolic $\mathfrak{q}$ is weakly $\Psi_{\mathfrak{g}}^{\mathfrak{q}}$-compatible (see (12.1)). We may therefore transfer $\Psi_{\mathfrak{g}}^{\mathfrak{q}}$ to a parameter for L as in (12.2), and we have*

$$(\Psi_{\mathfrak{g}}^{\mathfrak{q}})_{\mathfrak{q}} = \Psi.$$

(2) *Write $I_L(\Psi)$ for the virtual $(\mathfrak{l}, L \cap K)$-module attached to the parameter Ψ, and*

$$M(\Psi) = U(\mathfrak{g}) \otimes_{\mathfrak{q}} I_L(\Psi)$$

for the corresponding virtual $(\mathfrak{g}, L \cap K)$-module. Write $\mathbb{L}_k$ for the kth Bernstein derived functor carrying $(\mathfrak{g}, L \cap K)$-modules to $(\mathfrak{g}, K)$-modules, and $S = \dim \mathfrak{u} \cap \mathfrak{k}$. Then

$$\sum_{k=0}^{S} (-1)^k \mathbb{L}_{S-k}(M(\Psi)) = I_G(\Psi_{\mathfrak{g}}^{\mathfrak{q}}).$$

(3) *Suppose now that Ψ is a standard limit parameter for L, so that $I_L(\Psi)$ is an actual representation, and $M(\Psi)$ is an actual $(\mathfrak{g}, L \cap K)$-module. Write $\lambda_L(\Psi)$ for the weight introduced in Definition 11.2, and*

$$\lambda_G(\Psi_{\mathfrak{g}}^{\mathfrak{q}}) = \lambda_L(\Psi) + \rho(\mathfrak{u}).$$

Then $\mathfrak{q}$ is strongly $\Psi_{\mathfrak{g}}^{\mathfrak{q}}$-compatible if and only if $\lambda_G(\Psi_{\mathfrak{g}}^{\mathfrak{q}})$ is weakly antidominant for the roots of H in $\mathfrak{u}$. If this is the case, then

$$\mathbb{L}_k(M(\Psi)) = \begin{cases} I_G(\Psi_{\mathfrak{g}}^{\mathfrak{q}}) & k = S \\ 0 & k \neq S. \end{cases}$$

Proof. The assertions in (1) are formal and very easy. With these in hand, the rest of the theorem is just Theorem 12.8 applied to the parameter $\Psi_{\mathfrak{g}}^{\mathfrak{q}} \in P_G^{s,cont}(H)$. $\qquad\square$

We conclude this section with some results on the restrictions to K of cohomologically induced representations. Generalizing (12.7), we begin with a virtual representation Z of L of finite length. The restriction of Z to $L \cap K$ decomposes as a (virtual) direct sum

$$Z = \sum_{\tau \in \Pi^*(L \cap K)} Z(\tau); \tag{12.10a}$$

here the finite-dimensional virtual representation $Z(\tau)$ is a (positive or negative) multiple of the irreducible representation τ. From Z we construct a virtual generalized Verma module

$$M(Z) = U(\mathfrak{g}) \otimes_{\mathfrak{q}} Z. \tag{12.10b}$$

This is a virtual $(\mathfrak{g}, L \cap K)$-module. We will need to know the restriction of $M(Z)$ to $(\mathfrak{k}, L \cap K)$. This will be described in terms of the generalized

Verma modules of Theorem 10.6: if E is a (virtual) $(\mathfrak{q} \cap \mathfrak{k}, L \cap K)$ module, we write

$$M_K(E) = U(\mathfrak{k}) \otimes_{\mathfrak{q} \cap \mathfrak{k}} E, \qquad (12.10c)$$

which is a (virtual) $(\mathfrak{k}, L \cap K)$-module. The restriction formula we need is

$$M(Z)|_{(\mathfrak{k}, L \cap K)} = \sum_{m \geq 0} M_K(Z \otimes S^m(\mathfrak{g}/(\mathfrak{k} + \mathfrak{q}))). \qquad (12.10d)$$

Even if Z is a an actual representation, the restriction formula (12.10d) is true only on the level of virtual representations. The Verma module $M(Z)$ has a $(\mathfrak{k}, L \cap K)$-stable filtration for which the mth level of the associated graded module is given by the mth term in (12.10d).

It is worth noticing that the restriction formula (12.10d) depends only on $Z|_{L \cap K}$. We will deduce from this a formula for $\sum_k (-1)^k \mathbb{L}_{S-k}(M(Z))|_K$, again depending only on $Z|_{L \cap K}$. If Z is an actual representation, one might guess that each individual representation $\mathbb{L}_{S-k}(M(Z))|_K$ depends only on $Z|_{L \cap K}$. This guess is *incorrect*.

One of the main steps in our program to get branching laws for standard representations is to write irreducible representations of K explicitly as integer combinations of cohomologically induced representations. For that we will need to get rid of the infinite sum in (12.10d), and this we do using a Koszul complex. Recall from (2.5) the decomposition $\mathfrak{g} = \mathfrak{k} + \mathfrak{s}$. This is compatible with the triangular decomposition

$$\mathfrak{g} = \mathfrak{u}^{op} + \mathfrak{l} + \mathfrak{u}, \qquad (12.10e)$$

so we get

$$\mathfrak{g}/(\mathfrak{k} + \mathfrak{q}) \simeq \mathfrak{u}^{op} \cap \mathfrak{s} \qquad (12.10f)$$

as representations of $L \cap K$. Now consider the $(\mathfrak{q} \cap \mathfrak{k}, L \cap K)$-modules

$$E^+(\mathfrak{q}) = \sum_{p \text{ even}} \bigwedge^p \mathfrak{g}/(\mathfrak{k} + \mathfrak{q}), \qquad E^-(\mathfrak{q}) = \sum_{p \text{ odd}} \bigwedge^p \mathfrak{g}/(\mathfrak{k} + \mathfrak{q}) \qquad (12.10g)$$

and the virtual $(\mathfrak{q} \cap \mathfrak{k}, L \cap K)$-module

$$E^{\pm}(\mathfrak{q}) = E^+ - E^- = \sum_p (-1)^p \bigwedge^p \mathfrak{g}/(\mathfrak{k} + \mathfrak{q}). \qquad (12.10h)$$

Proposition 12.11. *Suppose we are in the setting of* (12.10), *so that* $\mathfrak{q}$ *is a θ-stable parabolic subalgebra of* $\mathfrak{g}$, *and* Z *is a virtual representation of* L

of finite length. Then

$$\sum_k (-1)^k \mathbb{L}_{S-k}(M(Z))|_K$$

$$= \sum_{m \geq 0} \sum_k (-1)^k \mathbb{L}_{S-k} \left(M_K(Z \otimes S^m(\mathfrak{g}/(\mathfrak{k}+\mathfrak{q}))) \right)$$

Suppose now that we can find a virtual $(\mathfrak{l}, L \cap K)$-*module of finite length* $Z^{\pm}$ *with the property that*

$$Z^{\pm}|_{L \cap K} = (Z|_{L \cap K}) \otimes E^{\pm}$$

(see (12.10h)). Then

$$\sum_k (-1)^k \mathbb{L}_{S-k}(M(Z^{\pm}))|_K = \sum_k (-1)^k \mathbb{L}_{S-k} \left(M_K(Z) \right).$$

Proof. The first assertion (which is due to Zuckerman) is a version of [8], Theorem 6.3.12. For the second, the Koszul complex for the vector space $\mathfrak{g}/(\mathfrak{k}+\mathfrak{q})$ provides an equality of virtual representations of $L \cap K$

$$\sum_p (-1)^p \bigwedge^p \left(\mathfrak{g}/(\mathfrak{k}+\mathfrak{q})\right) \otimes S^{m-p}\left(\mathfrak{g}/(\mathfrak{k}+\mathfrak{q})\right) = \begin{cases} 0, & m > 0 \\ \mathbb{C}, & m = 0. \end{cases} \quad (12.12)$$

More precisely, the Koszul complex has the pth summand on the left in degree p, and a differential of degree -1; its cohomology is zero for $m > 0$, and equal to $\mathbb{C}$ in degree zero for $m = 0$. Tensoring with Z and applying the first assertion, we get the second assertion. $\qquad \square$

Information about how to compute $\mathbb{L}_k(M_K(E))$ (for E an irreducible representation of $L \cap K$) may be found in Theorem 10.6. We will need only the very special case of "bottom layer K-types" for standard representations, described in Theorem 11.7.

It is by no means obvious *a priori* that the virtual L representation $Z^{\pm}$ should exist. The $L \cap K$ representations E^+ and E^- need not extend to L, so we cannot proceed just by tensoring with finite-dimensional representations of L. We will be applying the proposition to continued standard representations Z, and in that case we are saved by the following result (applied to L instead of G).

Lemma 12.13. *Suppose* $\Phi = (\phi, \Delta_{im}^+) \in P_G^{s,cont}(H)$ *is a shifted Harish-Chandra parameter for a continued standard representation, and that* (τ, E) *is a finite-dimensional representation of* K. *Write* $T = H \cap K$, *and*

$$\tau|_T = \sum_{\gamma_T \in \Pi^*(T)} m_E(\gamma_T) \gamma_T;$$

here the integer multiplicities $m_E(\gamma_T)$ are finite and non-negative. For each γ_T, choose an extension $\gamma \in \Pi^*(H)$ of the character γ_T to H; for example, one can choose the unique extension trivial on A_0. Write

$$\Phi + \gamma = (\phi \otimes \gamma, \Delta_{im}^+) \in P_G^{s,cont}(H).$$

Then

$$\big(I_G(\Phi) \otimes E\big)|_K \simeq \sum_{\gamma_T \in \Pi^*(T)} m_E(\gamma_T) I_G(\Phi + \gamma)|_K.$$

What the lemma says is that we can find an expression writing

(continued standard)⊗(representation of K)

as the restriction to K of a finite sum of continued standard representations of G. To make it explicit, we need to be able to compute the restriction of the representation of K to T. To compute this restriction for a general representation of K is beyond what standard results about compact Lie groups can provide; think of the example of $K = O(n)$ and $T = O(1)^n$ (arising from the split Cartan in $GL_n(\mathbb{R})$). In the present setting we are interested only in restricting the representations $E^\pm(\mathfrak{q})$ defined in (12.10) to $T \subset H \subset L$. This can be done explicitly in terms of the root decomposition of H acting $\mathfrak{g}$.

Proof. If all of the roots of H in G are imaginary, then $I_G(\Phi, \Delta_{im}^+)$ is cohomologically induced from a character of H (Theorem 11.7), and the lemma follows from the first formula in Proposition 12.11. The general case may be reduced to this case by part (2) of Proposition 11.5. (Alternatively, one can begin with the case when all roots of H in G are real, and reduce to that case using Theorem 11.7 and Proposition 12.11.) $\qquad\square$

13. From Highest Weights to Discrete Final Limit Parameters

Theorem 11.9 describes in a more or less algorithmic way how to pass from a discrete final limit parameter Φ to the highest weight of an irreducible representation $\tau(\Phi)$ of K. We need to be able to reverse this process: to begin with the highest weight of an irreducible representation, and extract a discrete final limit parameter. The algorithm of Theorem 11.9 uses a θ-stable parabolic subalgebra $\mathfrak{q}$ constructed (not quite uniquely) from Φ. Once we have $\mathfrak{q}$, the algorithm is easily reversible. Our task therefore is to construct $\mathfrak{q}$ directly from τ. This is very close to the basic algorithm underlying the classification of $\Pi^*(G)$ as described in [8]. In this section we will describe the small modification of that algorithm that we need.

We begin with a Δ_c^+-dominant weight

$$\mu \in \Pi_{dom}^*(T_f) \tag{13.1a}$$

(Definition 3.3), and with $2\rho_c \in \Pi_{dom}^*(T_f)$ the modular character of (4.3). We will need the set of roots

$$Q_0^\pm(\mu) = \{\beta \in \Delta(G, T_f) \mid \langle \mu + 2\rho_c, \beta^\vee \rangle = 0\}. \tag{13.1b}$$

These roots are (the restrictions to T_f of) noncompact imaginary roots of H_f in G; they are strongly orthogonal, and are the roots of a split Levi subgroup $L_0(\mu)$, locally isomorphic to a product of copies of $SL_2(\mathbb{R})$ and an abelian group. The roots in $Q_0(\mu)$ are going to be in the Levi subgroup $L(\mu)$ that we ultimately construct. It is convenient to choose one root from each pair $\pm\beta$ in $Q_0^\pm(\mu)$, and to call the resulting set $Q_0(\mu)$. Consider the Weyl group

$$W_0(\mu) = \{w_A = \prod_{\beta \in A} s_\beta \mid A \subset Q_0(\mu)\}, \tag{13.1c}$$

a product of copies of $\mathbb{Z}/2\mathbb{Z}$.

We now define

$$\Delta^+(G, T_f)(\mu) = \{\alpha \in \Delta(G, T_f) \mid \langle \mu + 2\rho_c, \alpha^\vee \rangle > 0\} \cup Q_0(\mu). \tag{13.1d}$$

This is the restriction to T_f of a θ-stable positive root system $\Delta^+(G, H_f)(\mu)$. Of course it depends on the choice of $Q_0(\mu)$; changing the choice replaces $\Delta^+(\mu)$ by $w_A\Delta^+(\mu)$ for some $A \subset Q_0(\mu)$. This positive root system defines a modular character

$$2\rho(\mu) = \sum_{\alpha \in \Delta^+(\mu)} \alpha \in \Pi_{alg}^*(\mathbf{H}_f); \tag{13.1e}$$

this character is fixed by θ, and therefore trivial on $A_{f,0}$. We write $2\rho(\mu)$ also for the restriction to T_f. Differentiating and dividing by two gives

$$\rho(\mu) \in \mathfrak{t}_f^* \subset \mathfrak{h}_f^*.$$

Construction of $\mathfrak{q}$: project the weight $\mu + 2\rho_c - \rho(\mu)$ on the $\Delta^+(G, T_f)(\mu)$ positive Weyl chamber, getting a weight $\lambda(\mu)$. Write

$$\mu + 2\rho_c - \rho(\mu) = \lambda(\mu) - \phi,$$

with ϕ a non-negative rational combination of $\Delta^+(\mu)$-simple roots. Let $\mathfrak{q}$ be the parabolic corresponding to the set of simple roots whose coefficient in ϕ is strictly positive. (Necessarily this includes the roots in $Q_0(\mu)$.) This is the one.

14. Algorithm for Projecting a Weight on the Dominant Weyl Chamber

The algorithm of Section 13 depends on projecting a certain rational weight on a positive Weyl chamber. In general (for example in the papers of Carmona and Aubert-Howe) this projection is defined in terms of orthogonal geometry. Since the language of root data avoids the orthogonal form (instead keeping roots and coroots in dual spaces) it is useful to rephrase their results in this language. At the same time I will sketch an algorithm to carry out the projection. So fix a root system in a rational or real vector space. It is convenient to label this space as V^*, and to have the coroots in V:

$$\Delta \subset V^*, \qquad \Delta^\vee \subset V. \tag{14.1a}$$

We fix also sets of positive and simple roots

$$\Delta^+ \supset \Pi. \tag{14.1b}$$

These give rise to fundamental weights and coweights

$$\begin{aligned}
\chi_\alpha &\in \mathbb{Q}\Delta \subset V^*, &\langle \chi_\alpha, \beta^\vee \rangle &= \delta_{\alpha,\beta}, &(\alpha, \beta \in \Pi) \\
\chi_{\alpha^\vee}^\vee &\in \mathbb{Q}\Delta^\vee \subset V, &\langle \beta, \chi_{\alpha^\vee}^\vee \rangle &= \delta_{\alpha,\beta} &(\alpha, \beta \in \Pi).
\end{aligned} \tag{14.1c}$$

We also need a small generalization: if $B \subset \Pi$, then

$$\Delta(B) = \mathbb{Z}B \cap \Delta \tag{14.1d}$$

is a root system in V with simple roots B. (If Δ is the root system for a reductive algebraic group $\mathbf{G}$, then $\Delta(B)$ corresponds to a Levi subgroup $\mathbf{L}(B)$.) Write $B^\vee$ for the set of simple coroots corresponding to B. We therefore get

$$\begin{aligned}
\chi_\alpha(B) &\in \mathbb{Q}\Delta(B) \subset V^*, &\langle \chi_\alpha(B), \beta^\vee \rangle &= \delta_{\alpha,\beta} &(\alpha, \beta \in B) \\
\chi_{\alpha^\vee}^\vee(B) &\in \mathbb{Q}\Delta^\vee(B) \subset V, &\langle \beta, \chi_{\alpha^\vee}^\vee(B) \rangle &= \delta_{\alpha,\beta} &(\alpha, \beta \in B).
\end{aligned} \tag{14.1e}$$

The question of constructing these elements explicitly will arise in the course of the algorithm; of course the defining equations to be solved are linear with integer coefficients.

We are interested in the geometry of cones

$$C = \{\lambda \in V^* \mid \langle \lambda, \alpha^\vee \rangle \geq 0, \alpha \in \Pi\}$$
$$C^\vee = \{v \in V \mid \langle \alpha, v \rangle \geq 0, \alpha \in \Pi\}$$
$$P = \{ \sum_{a_\alpha \geq 0} a_\alpha \alpha \} \subset V^* \qquad (14.1\text{f})$$
$$P^\vee = \{ \sum_{b_{\alpha^\vee} \geq 0} b_{\alpha^\vee} \alpha^\vee \} \subset V$$

the *positive Weyl chambers* and the *positive root (coroot) cone* respectively. Sometimes it will be convenient as well to consider the two subspaces

$$Z^* = \{\lambda \in V^* \mid \langle \lambda, \alpha^\vee \rangle = 0, \alpha \in \Pi\}$$
$$Z = \{v \in V \mid \langle \alpha, v \rangle = 0, \alpha \in \Pi\}. \qquad (14.1\text{g})$$

These are dual vector spaces under the restriction of the pairing between V and V^*.

Here is the general theoretical statement, due to Langlands. For a variety of slightly different and illuminating perspectives, see [6], Lemma 4.4; [2], Lemma IV.6.11; [3], Proposition 1.2; and [1], Proposition 1.16.

Theorem 14.2. *In the setting* (14.1), *there is for each* $\nu \in V^*$ *a unique subset* $A \subset \Pi$ *and a unique expression*

$$\nu = \lambda - \phi \qquad (\lambda \in C, \phi \in P)$$

subject to the following requirements.

(1) *The element* ϕ *is a combination of roots* $\alpha \in A$ *with strictly positive coefficients:*

$$\phi = \sum_{\alpha \in A} a_\alpha \alpha, \qquad a_\alpha > 0.$$

(2) *The element* λ *vanishes on* $\alpha^\vee$ *for all* $\alpha \in A$:

$$\lambda = \sum_{\alpha \notin A} c_\alpha \chi_\alpha + z \qquad (c_\alpha \geq 0, z \in Z).$$

Of course there is a parallel statement for V. This formulation looks a little labored: if one is willing to fix a W-equivariant identification of V with V^* (and so to think of the pairing between them as a symmetric form on V) then the two requirements imposed on λ and ϕ may be written simply as $\langle \lambda, \phi \rangle = 0$, and there is no need to mention A. But for our purposes the most interesting output of the algorithm *is* the set A, which defines the θ-stable parabolic $\mathfrak{q} = \mathfrak{q}(\mu)$. So mentioning this set in the statement of the theorem, and constructing it directly, seem like reasonable ideas.

Proof. The uniqueness of the decomposition is a routine exercise; what is difficult is existence, and for that I will give a constructive proof. All that we will construct is the set A. This we will do inductively. At each stage of the induction, we will begin with a subset $B_m \subset A$. We will then test the simple roots outside B_m. Any root that fails the test must belong to A, and will be added to the next set B_{m+1}. If all the roots outside of B_m pass the test, then $B_m = A$, and we stop. The algorithm begins with the empty set $B_0 = \emptyset$, which is certainly a subset of A. Here is the inductive test. For each simple root $\beta \notin B_m$, construct the fundamental coweight

$$\chi^\vee = \chi_{\beta^\vee}^\vee(B_m \cup \{\beta\})$$

with respect to the simple roots consisting of B_m and β. This coweight vanishes on the roots in B_m, but takes the value 1 on the root β. (I will say a word about constructing this coweight in a moment.) The test we apply is to ask whether $\langle \nu, \chi^\vee \rangle$ is non-negative. If it is, then β passes. If β fails, then I claim that it must belong to A, and we add it to B_{m+1}. Here is why. As a fundamental coweight, $\chi^\vee$ is a nonnegative rational combination of $\beta^\vee$ and the various simple coroots in $B_m^\vee \subset A^\vee$. It therefore has a non-negative pairing with $\lambda \in C$; so if it is strictly negative on ν, then

$$0 < \langle \phi, \chi^\vee \rangle \tag{14.3}$$

$$= \sum_{\alpha \in B_m} a_\alpha \langle \alpha, \chi^\vee \rangle + \sum_{\gamma \in A - B_m} a_\gamma \langle \gamma, \chi^\vee \rangle \tag{14.4}$$

The first sum is zero by construction of $\chi^\vee$. Because $\chi^\vee$ is a non-negative combination of $\beta^\vee$ and coroots from B_m, every term in the second sum is non-positive except the term with $\gamma = \beta$. Of course that term is present only if $\beta \in A$, as we wished to show.

It remains to show that if every simple root outside B_m passes this test, then $B_m = A$. Suppose $B_m \neq A$. We may therefore find a simple root β so that

$$\langle \sum_{\gamma \in A - B_m} a_\gamma \gamma, \beta^\vee \rangle > 0.$$

Because a simple root and a distinct simple coroot have non-positive pairing, this equation shows first of all that $\beta \in A - B_m$. We show that β fails the test. Define

$$\chi_0^\vee = \beta^\vee + \sum_{\alpha \in B_m} (-\langle \alpha, \beta^\vee \rangle) \chi_{\alpha^\vee}^\vee(B_m) \tag{14.5}$$

$$= \beta^\vee + \delta^\vee \tag{14.6}$$

This element is evidently in the $\mathbb{Q}$-span of $B_m^\vee \cup \{\beta^\vee\}$, and its pairing with the roots in B_m is zero. For its pairing with β, the coroot $\beta^\vee$ contributes 2, and the terms from the sum are all non-negative (since the elements $\chi_{\alpha^\vee}^\vee(B_m)$ are non-negative combinations of simple roots which are in B_m, and therefore distinct from β. So the pairing with β is at least two. Consequently

$$\chi_0^\vee = c\chi^\vee, \qquad \text{some } c \geq 2.$$

We are trying to prove that $\chi^\vee$ has a strictly negative pairing with ν, so it suffices to prove that $\langle \nu, \chi_0^\vee \rangle < 0$. We have already seen that $\chi_0^\vee$ is a combination of roots in A, so its pairing with λ must be zero by condition (2) of Theorem 14.2. Since $\nu = \lambda - \phi$, what we want to prove is

$$0 < \langle \phi, \chi_0^\vee \rangle. \tag{14.7}$$

Since $\chi_0^\vee$ is orthogonal to the roots in B_m, the left side is

$$= \langle \sum_{\gamma \in A - B_m} a_\gamma \gamma, \chi_0^\vee \rangle$$

$$= \langle \sum_{\gamma \in A - B_m} a_\gamma \gamma, \beta^\vee \rangle + \sum_{\gamma \in A - B_m} a_\gamma \langle \gamma, \delta^\vee \rangle.$$

It follows from (14.5) that $\delta^\vee$ is a non-negative combination of roots in B_m. Consequently every term in the second sum is non-negative. The first term is strictly positive by the choice of β, proving (14.7). $\qquad\square$

15. Making a List of Representations of K

Here is a way to make a list of the irreducible representations of K that is compatible with the Langlands classification. As usual begin with a fundamental Cartan H_f, with $T_f = H_f \cap K$, and a fixed system

$$\Delta_c^+ \subset \Delta(K, T_f) \tag{15.1a}$$

of positive roots for K. As in Definition 3.3, this choice provides a Borel subalgebra

$$\mathfrak{b}_c = \mathfrak{t}_f + \mathfrak{n}_c \tag{15.1b}$$

corresponding to the negative roots.

The first serious step is to pick a representative of each T_{fl}-conjugacy class of θ-stable parabolic subalgebras $\mathfrak{q} = \mathfrak{l} + \mathfrak{u}$ such that

(1) the Levi subalgebra $\mathfrak{l}$ contains $\mathfrak{h}_f$;
(2) the Borel subalgebra $\mathfrak{b}_c$ is contained in $\mathfrak{q}$; and
(3) the Levi subgroup L has split derived group.

This is the same as picking representatives of K-conjugacy classes of θ-stable $\mathfrak{q}$ opposite to their complex conjugates, subject only to the last requirement on the (unique) θ-stable real Levi subgroup L. These parabolic subalgebras are in one-to-one correspondence with the corresponding Zuckerman unitary representations $A(\mathfrak{q})$. The larger set of T_{fl} conjugacy classes of θ-stable $\mathfrak{q}$ subject only to conditions (1), (2), and

(3′) the group L has no compact simple factor

is already picked out by the software `blocku` command. Recognizing condition (3) is equally easy.

So we fix a collection

$$\mathcal{Q} = \{\mathfrak{q}\} \tag{15.1c}$$

of representatives for the T_{fl}-conjugacy classes of parabolic subalgebras satisfying conditions (1)–(3) above. For each $\mathfrak{q}$, fix a maximally split Cartan subgroup

$$H \subset L, \tag{15.1d}$$

with $T = H \cap K$ as usual.

The goal is to list all the irreducible representations of K having a highest weight μ so that the construction of Section 13 gives the parabolic $\mathfrak{q}$. The disjoint union over $\mathfrak{q} \in \mathcal{Q}$ of these lists will be a parametrization of $\Pi^*(K)$. The list corresponding to $\mathfrak{q}$ is a certain subset of the discrete final limit parameters for H. Precisely,

$$P_{G,d}^{s,\mathfrak{q},finlim}(H) = \{\Phi \in P_{G,d}^{s,finlim}(H) \mid \Phi \text{ is strongly } \mathfrak{q}\text{-compatible}\} \tag{15.1e}$$

The notion of "strongly compatible" was introduced in (11.6). We will see that the set described by (15.1e) is essentially a set of characters of T, subject to certain parity and positivity conditions. We want to make this more explicit. First of all we invoke the notation of (11.6), writing in particular

$$\Delta_{im}^+ = \Delta_{im}(G, H) \cap (-\Delta(\mathfrak{u}, H)). \tag{15.1f}$$

The assumption of weak compatibility from (11.6) means that

$$P_{G,d}^{s,\mathfrak{q},finlim}(H) \subset \{\Phi = (\phi, \Delta_{im}^+) \mid \phi \in \Pi^*(T)\}; \tag{15.1g}$$

that is, that we are looking only at parameters involving this particular choice of positive imaginary roots. Recall also from (11.6) the character ρ_{cplx} of T and the shift $\phi_{\mathfrak{q}} = \phi \otimes \rho_{cplx}$. The corresponding parameters for L are

$$\Phi_{\mathfrak{q}} = (\phi_{\mathfrak{q}}, \emptyset) \in P_{L,d}^s(H). \tag{15.1h}$$

With this notation, the characters ϕ of T that contribute to $P_{G,d}^{s,\mathfrak{q},finlim}(H)$ have the following characteristic properties.

(1) For each (necessarily real) root $\alpha \in \Delta(L, H)$, $\phi_{\mathfrak{q}}(m_\alpha) = 1$. Equivalently, $\phi(m_\alpha) = \rho_{cplx}(m_\alpha)$.

This is the characteristic property of $P_{L,d}^{s,finlim}(H)$. It guarantees that the principal series representation $I_L(\Phi_{\mathfrak{q}})$ has a unique lowest $L \cap K$-type $\tau_{\mathfrak{q}}(\Phi)$, which is one-dimensional and L-spherical (Proposition 8.5). Consequently $\tau_{\mathfrak{q}}(\Phi)$ has a unique weight $\mu_{\mathfrak{q}}(\Phi) \in \Pi^*(T_f)$. We write

$$\widetilde{\mu}_{\mathfrak{q}}(\Phi) \in \Pi^*(T_{fl}(\mu_{\mathfrak{q}}(\Phi)))$$

for the extended highest weight of $\tau_{\mathfrak{q}}(\Phi)$; recall (again from Proposition 8.5) that this extended highest weight may be parametrized just by $\Phi_{\mathfrak{q}}$ (and so by Φ). Define

$$\widetilde{\mu}(\Phi) = \widetilde{\mu}_{\mathfrak{q}}(\Phi) \otimes 2\rho(\mathfrak{u} \cap \mathfrak{k}) \in \Pi^*(T_{fl}(\mu_{\mathfrak{q}}(\Phi))), \qquad \mu(\Phi) = \widetilde{\mu}(\Phi)|_{T_f}$$

as in Theorem 10.6.

(2) The weight

$$\zeta(\Phi) = d\phi_{\mathfrak{q}} + \rho(\mathfrak{u})$$

takes non-positive values on every coroot of $\mathfrak{h}$ in $\mathfrak{u}$. Equivalently, Φ is strongly $\mathfrak{q}$-compatible.

This second requirement ensures (among other things) that Φ is a discrete standard limit parameter. To ensure that Φ is final, we just need to arrange $I(\Phi) \neq 0$. This is equivalent to either of the next two requirements.

(3) The weight $\mu(\Phi) \in \Pi^*(T_f)$ is dominant for $\Delta^+(K)$.
(3') For each simple imaginary root $\alpha \in \Delta_{im}^+(G, H)$ which is compact, we have

$$\langle \zeta(\Phi), \alpha^\vee \rangle > 0.$$

Theorem 15.2. *Suppose K is a maximal compact subgroup of the reductive algebraic group G, T_f is a small Cartan subgroup of K, Δ_c^+ a choice of positive roots for T_f in K, and other notation is as in (15.1). Fix a collection $\mathcal{Q} = \{\mathfrak{q}\}$ of representatives for the θ-stable parabolic subalgebras satisfying conditions (1)–(3) there. For each $\mathfrak{q} \in \mathcal{Q}$, fix a maximally split Cartan subgroup $H \subset L$, and define $P_{G,d}^{s,\mathfrak{q},finlim}(H)$ as in (15.1e). Then there is a bijection*

$$\coprod_{\mathfrak{q} \in \mathcal{Q}} P_{G,d}^{s,\mathfrak{q},finlim}(H) \leftrightarrow \Pi^*(K), \qquad \Phi \leftrightarrow \tau(\Phi).$$

Here the first term is a disjoint union of sets of discrete final limit parameters, and the bijection is that of Theorem 11.9 (described also in (15.1) above).

Proof. That every irreducible representation of K appears in this correspondence is clear from Theorem 11.9: to show that every θ-stable $\mathfrak{q}$ can be conjugated by K to be compatible with a fixed T_f and $\mathfrak{b}_c$ is elementary. To see that the correspondence is one-to-one, suppose that $(\mathfrak{q}_1, \Phi_1)$ and $(\mathfrak{q}_2, \Phi_2)$ are two parameters in this theorem, with $\tau(\Phi_1) = \tau(\Phi_2)$. The algorithm of Sections 13 and 14, for any $\Phi \in P_{G,d}^{s,\mathfrak{q},finlim}(H))$, reconstructs $\mathfrak{q}$ from one of the highest weights of $\tau(\Phi)$. It follows from this algorithm that $\mathfrak{q}_1$ and $\mathfrak{q}_2$ must be conjugate by T_{fl}/T_f (since this finite group acts transitively on the highest weights of an irreducible representation τ of K). By the choice of $\mathcal{Q}$, necessarily $\mathfrak{q}_1 = \mathfrak{q}_2$. $\square$

Example 15.3. Suppose $G = Sp_4(\mathbb{R})$. In this case $K = U_2$. We can choose $T_f = H_f = U_1 \times U_1$, so that $\Pi^*(T_f) \simeq \mathbb{Z}^2$. Write $\{e_1, e_2\}$ for the standard basis elements of $\mathbb{Z}^2$. The root system is then

$$\Delta_{im,c} = \{\pm(e_1 - e_2)\}, \qquad \Delta_{im,n} = \{\pm 2e_1, \pm 2e_2, \pm(e_1 + e_2)\}.$$

We choose

$$\Delta_c^+ = \{e_1 - e_2\}.$$

Since K is connected, irreducible representations are parametrized precisely by their highest weights, which are

$$\{\mu = (\mu_1, \mu_2) \in \mathbb{Z}^2 \mid \mu_1 \geq \mu_2\}.$$

There are exactly eight θ-stable parabolic subalgebras $\mathfrak{q}$ having split Levi factor and compatible with Δ_c^+. I will list them, list the dominant weights μ associated to each, and finally list the corresponding discrete final limit parameters.

The first four are Borel subalgebras containing $\mathfrak{t}_f$. For each I will list only the two simple roots in the Borel subalgebra; these are the *negatives* of the simple roots for the corresponding positive system.

$$\mathfrak{q}_1 \leftrightarrow \{-e_1 + e_2, -2e_2\}$$
$$\{\mu \mid \mu_1 \geq \mu_2 \geq 2\}$$
$$\{\phi \mid \phi_1 - 1 > \phi_2 \geq 1\}$$
$$\mu = \phi - (1, -1).$$

$$\mathfrak{q}_2 \leftrightarrow \{-e_1 - e_2, 2e_2\}$$
$$\{\mu \mid \mu_1 - 1 \geq -\mu_2 \geq 0\}$$
$$\{\phi \mid \phi_1 > -\phi_2 > 0\}$$
$$\mu = \phi - (1, -1).$$

$$\mathfrak{q}_3 \leftrightarrow \{e_1 + e_2, -2e_1\}$$
$$\{\mu \mid 0 \geq -\mu_1 \geq \mu_2 + 1\}$$
$$\{\phi \mid 0 > -\phi_1 > \phi_2\}$$
$$\mu = \phi - (1, -1).$$

$$\mathfrak{q}_4 \leftrightarrow \{-e_1 + e_2, 2e_1\}$$
$$\{\mu \mid -2 \geq \mu_1 \geq \mu_2\}$$
$$\{\phi \mid -1 \geq \phi_1 > \phi_2 + 1\}$$
$$\mu = \phi - (1, -1).$$

These four families of characters ϕ of T_f together exhaust the final standard limit characters (Definition 5.1) which are K-dominant.

The large regions corresponding to $\mathfrak{q}_1$ and $\mathfrak{q}_2$ are separated by the line $\mu_2 = 1$, and this line essentially corresponds to the next parabolic. The Levi factor is $L = U_1 \times Sp_2$, $H_{12} = U_1 \times GL_1$, so a (discrete) character ϕ of H_{12} is a pair (ϕ_1, ϵ), with $\phi_1 \in \mathbb{Z}$ and $\epsilon \in \mathbb{Z}/2\mathbb{Z}$. The corresponding parabolic and dominant weights are

$$\mathfrak{q}_{12} = \mathfrak{q}_1 + \mathfrak{q}_2, \qquad \Delta(\mathfrak{u}_1) = \{-e_1 \pm e_2, -2e_1\}$$
$$\{\mu \mid \mu_1 \geq \mu_2 = 1\}$$
$$\{\phi \mid \phi_1 \geq 1, \epsilon = 0\}$$
$$\mu = (\phi_1, 1).$$

In exactly the same way, the regions corresponding to $\mathfrak{q}_3$ and $\mathfrak{q}_4$ are separated by the line $\mu_1 = -1$, and this line essentially corresponds to the next parabolic. The Levi factor is $L_{34} = Sp_2 \times U_1$, with split Cartan $H_{34} = GL_1 \times U_1$, so a (discrete) character ϕ of H_{34} is a pair (ϵ, ϕ_2), with $\phi_2 \in \mathbb{Z}$ and $\epsilon \in \mathbb{Z}/2\mathbb{Z}$. The corresponding parabolic and dominant weights are

$$\mathfrak{q}_{34} = \mathfrak{q}_3 + \mathfrak{q}_4, \qquad \Delta(\mathfrak{u}_1) = \{\pm e_1 - e_2, -2e_2\}$$
$$\{\mu \mid -1 = \mu_1 \geq \mu_2\}$$
$$\{\phi \mid \phi_2 \leq -1, \epsilon = 0\}$$
$$\mu = (-1, \phi_1).$$

The two Cartans H_{12} and H_{34} are conjugate; the conjugation identifies the character (ϕ_1, ϵ) of H_{12} with (ϵ, ϕ_1) of H_{34}. We have therefore accounted for all of the even characters of H_{12} except the one with $\phi_1 = 0$. Since the imaginary root is $2e_1$, this missing character violates condition (1) of Definition 11.2 to be a standard limit character.

The regions corresponding to $\mathfrak{q}_2$ and $\mathfrak{q}_3$ are separated by the line $\mu_1 + \mu_2 = 0$, and (most of) this line corresponds to a parabolic with Levi factor $L_{23} = U_{1,1}$, with split Cartan $H_{23} = \mathbb{C}^\times$. A (discrete) character Φ of H_{23} corresponds to an integer j, best written in our root system coordinates as $(j/2, -j/2)$. The corresponding parabolic and dominant weights are

$$\mathfrak{q}_{23} = \mathfrak{q}_2 + \mathfrak{q}_3, \qquad \Delta(\mathfrak{u}_{23}) = \{-e_1 + e_2, -2e_1, 2e_2\}$$
$$\{\mu \mid \mu_1 = -\mu_2 > 0\}$$
$$\{\phi = (j/2, j/2) \mid j > 0 \text{ even}\}$$
$$\mu = (j/2, j/2).$$

The condition on ϕ that j should be even is just $\phi_{\mathfrak{q}_{23}}(m_\alpha) = 1$ for the real root α of H_{23}. The condition $j \neq 0$ is condition (1) of Definition 11.2, required to make ϕ a standard limit character. The Weyl group of H_{23} has order 4 (generated by the involutions $z \mapsto \bar{z}$ and $z \mapsto \bar{z}^{-1}$ of $\mathbb{C}^\times$), so it includes elements sending j to $-j$. The characters attached to $\mathfrak{q}_{23}$ therefore exhaust the discrete final limit characters of H_{23} up to the conjugacy.

There remains only the trivial representation of K, corresponding to $\mu = (0,0)$. The corresponding parabolic is $\mathfrak{q}_0 = \mathfrak{g}$, with Levi factor $L_0 = Sp_4$ and split Cartan $H_0 = GL_1 \times GL_1$. The only discrete characters of H_0 are the four characters of order two; and only the trivial one is final (since the elements m_α include all the non-trivial $(\pm 1, \pm 1) \in GL_1 \times GL_1$). The trivial character ϕ corresponds to $\mu = 0$.

16. G-Spherical Representations as Sums of Standard Representations

In order to complete the algorithm for branching laws from G to K, we need explicit formulas expressing each irreducible representation of K as a finite integer combination of discrete final standard limit representations. We will get such formulas by reduction to the G-spherical case (using cohomological induction as in Section 10). In the G-spherical case, we will use a formula of Zuckerman which we now recall.

We begin as in Section 8 by assuming that

$$G \text{ is split, with } \theta\text{-stable split Cartan } H_s. \tag{16.1a}$$

Fix a discrete final limit parameter

$$\Phi_0 \in P_{G,d}^{s,finlim}(H_s), \tag{16.1b}$$

with

$$\tau(\Phi_0) \in \Pi_{G-sph}^*(K) \tag{16.1c}$$

the corresponding G-spherical representation of K (Proposition 8.5). According to Lemma 8.3, there is a unique extension

$$\tau_0 = \tau_G(\Phi_0) \in \Pi_{adm}^*(G) \tag{16.1d}$$

of $\tau(\Phi_0)$ to a $(\mathfrak{g}, K)$-module (acting on the same one-dimensional space) in which $\mathfrak{s}$ acts by zero. Zuckerman's formula actually writes τ_0 as an integer combination of standard representations of G (in the Grothendieck group of virtual representations); ultimately we will be interested only in the restriction of this formula to K.

At this point we can drop the hypothesis that G be split, and work with any one-dimensional character τ_0 of G. (In fact Zuckerman's identity is for virtual representations of G, and it can be formulated with τ_0 replaced by any finite-dimensional irreducible representation of G.) We will make use of the results only for G split, but the statements and proofs are no more difficult in general.

Lemma 16.2. *Suppose τ_0 is a one-dimensional character of the reductive group G. Suppose that H is a θ-stable maximal torus in G, and that $\Delta^+ \subset \Delta(G, H)$ is a system of positive roots. Write $T = H^\theta$ for the compact part of H. Attached to this data there is a unique shifted Harish-Chandra parameter*

$$\Phi(\Delta^+, \tau_0) = (\phi(\Delta^+, \tau_0), \Delta_{im}^+)$$

characterized by the following requirements.

(1) Δ_{im}^+ consists of the imaginary roots in Δ^+.
Define

$$\rho_{im} \in \mathfrak{t}^* \subset \mathfrak{h}^*$$

as in Definition 11.2 to be half the sum of the roots in Δ_{im}^+, and $\rho \in \mathfrak{h}^*$ to be half the sum of Δ^+. Define

$$\rho_{cplx} = \sum_{\substack{pairs \ \{\alpha, \theta\alpha\} \\ cplx \ in \ \Delta^+}} \alpha|_T \quad \in \Pi_{adm}^*(T).$$

(2) The differential of $\phi(\Delta^+, \tau_0)$ is equal to $\rho + \rho_{im} + d\tau_0|_{\mathfrak{h}}$.

(3) The restriction of $\phi(\Delta^+, \tau_0)$ to T is equal to

$$\rho_{cplx} + 2\rho_{im} + \tau_0|_T.$$

The notation ρ_{cplx} is not inconsistent with that introduced in (11.6i), but it is a bit different. In the earlier setting, we were interested in a positive root system related to a θ-stable parabolic subalgebra, and therefore as close to θ-stable as possible. For every complex root α, either α and $\theta\alpha$ were both positive (in which case α contributed to ρ_{cplx}) or $-\alpha$ and $-\theta\alpha$ were both positive (in which case $-\alpha$ contributed to ρ_{cplx}). In the present setting there is a third possibility: that α and $\theta\alpha$ have opposite signs. In this third case there is no contribution to ρ_{cplx}. The character defined here may therefore be smaller than the one defined in (11.6i).

Proof. The first point is to observe that the two roots α and $\theta\alpha$ have the same restriction to $T = H^\theta$, so ρ_{cplx} is a well-defined character of T. It is very easy to check that

$$\rho|_{\mathfrak{t}} = d\rho_{cplx} + \rho_{im} : \qquad (16.3)$$

the point is that complex pairs of positive roots $\{\beta, -\theta\beta\}$ contribute zero to $\rho|_{\mathfrak{t}}$.

Now a character χ of H is determined by its differential $d\chi \in \mathfrak{h}^*$ and by its restriction $\chi_T \in \Pi^*(T)$ to T; the only compatibility required between these two things is $d(\chi_T) = (d\chi)|_{\mathfrak{t}}$. The two requirements in the lemma specify the differential of $\phi(H, \Delta^+, \tau_0)$ and its restriction to T, and (16.3) says that these two requirements are compatible. It follows that $\phi(H, \Delta^+, \tau_0)$ is a well-defined character of H. To see that it is shifted parameter for a standard representation, use Definition 11.2 and the fact that ρ and ρ_{im} are both dominant regular for Δ_{im}^+. (Because τ_0 is a one-dimensional character of G, its differential is orthogonal to all roots.) $\square$

In order to state Zuckerman's formula, we need one more definition: that of "length" for the shifted parameters in Lemma 16.2. All that appears in an essential way in the formulas is differences of two lengths; so there is some choice of normalization in defining them. From the point of view of the Beilinson-Bernstein picture of Harish-Chandra modules, the length is related to the dimension of some orbit of **K** on the flag variety of **G**. The most natural normalizations would define the length to be *equal* to the dimension of the orbit, or equal to its codimension. In fact the original definition is equal to neither of these (but rather to the difference between the dimension of the given **K** orbit and the minimal **K** orbits). Changing the definition now would surely introduce more problems than it would solve.

Definition 16.4 ([8], Definition 8.1.4). Suppose H is a θ-stable maximal torus in G (cf. (6.1)), and that $\Delta^+ \subset \Delta(G, H)$ is a system of positive roots. The *length of* Δ^+, written $\ell(\Delta^+)$, is equal to

$$\dim_{\mathbb{C}} \mathbf{B}_{c,0} - [(1/2)\#\{\alpha \in \Delta^+ \text{ complex } \mid \theta\alpha \in \Delta^+\} + \#\Delta^+_{im,c} + \dim T].$$

If $\mathfrak{b}$ is the Borel subalgebra corresponding to Δ^+, then the term in square brackets is the dimension of $\mathfrak{b} \cap \mathfrak{k}$: that is, of the isotropy group at $\mathfrak{b}$ for the action of **K** on the flag variety of **G**. The first term is the maximum possible value of this dimension, attained precisely for H fundamental and Δ^+ θ-stable.

The definition in [8] is written in a form that is more difficult to explain; rewriting it in this form is a fairly easy exercise in the structure of real reductive groups. Here is Zuckerman's formula.

Theorem 16.5 (Zuckerman; cf. [8], Proposition 9.4.16). *Suppose G is a reductive group and τ_0 is a one-dimensional character of G. Define*

$$r(G) = \text{(number of positive roots for } G) \\ - \text{(number of positive roots for } K).$$

Then there is an identity in the Grothendieck group of virtual admissible representations of finite length

$$\tau_0 = (-1)^{r(G)} \sum_H \sum_{\Delta^+} (-1)^{\ell(\Delta^+)} I(\Phi(\Delta^+, \tau_0)).$$

The outer sum is over θ-stable Cartan subgroups of G, up to K-conjugacy; and the inner sum is over positive root systems for $\Delta(G, H)$, up to the action of the Weyl group $W(G, H)$. The shifted Harish-Chandra parameter $\Phi(\Delta^+, \tau_0)$ is defined in Lemma 16.2, and the standard representation $I(\Phi)$ in Proposition 11.3.

The number $r(G)$ is the codimension of the minimal orbits of **K** on the flag manifold of **G**; so $r(G) - \ell(\Delta^+)$ is the codimension of the **K** orbit of the Borel corresponding to Δ^+. The left side of the formula is (according to the Borel-Weil theorem) the cohomology of the flag variety with coefficients in a (nearly) trivial bundle. The terms on the right side are local cohomology groups along the **K** orbits, with coefficients in this same (nearly) trivial bundle. The equality of the two sides—more precisely, the existence of a complex constructed from the terms on the right with cohomology the representation on the left—can be interpreted in terms of the "Grothendieck-Cousin complex" introduced by Kempf in [4].

Proof. From the definition of the standard representations in Proposition 11.3, it is clear that $I(\Phi + (\tau_0|_H)) \simeq I(\Phi) \otimes \tau_0$. In this way we can reduce to the case when τ_0 is trivial. That is the case treated in [8], Proposition 9.4.16. $\square$

The terms in this sum correspond precisely to the orbits of **K** on the flag variety for **G**. For the split real form of E_8, there are $320, 206$ orbits. (The command `kgb` in the present version of the `atlas` software first counts these orbits, then provides some descriptive information about them.)

Using the branching laws in Section 12, we can immediately get a version of Zuckerman's formula for arbitrary representations of K.

Theorem 16.6. *Suppose G is a reductive group and (τ, E) is an irreducible representation of K. Let $\Phi \in P_{G,d}^{s,finlim}(H)$ be a discrete final limit parameter so that $\tau(\Phi) = \tau$ is the unique lowest K-type of $I_G(\Phi)$ (Theorem 11.9). Choose a strongly Φ-compatible parabolic subalgebra $\mathfrak{q} = \mathfrak{l} + \mathfrak{u}$ as in (11.6), so that $\tau_{\mathfrak{q}}$ is a one-dimensional L-spherical character of $L \cap K$. In particular, $\tau_{\mathfrak{q}}$ is the unique lowest $L \cap K$-type of the principal series representation $I_L(\Phi_{\mathfrak{q}})$.*

$$\tau(\Phi) = (-1)^{r(L)} \sum_{H_1 \subset L} \sum_{\Delta^+} \sum_{A \subset \Delta(\mathfrak{u}^{op} \cap \mathfrak{s})} (-1)^{|A| + \ell(\Delta^+)} I_G(\Phi(\Delta^+, \tau_{\mathfrak{q}})_{\mathfrak{g}}^{\mathfrak{q}} + 2\rho(A))|_K$$

Here the outer sum is over θ-stable Cartan subgroups of L, up to conjugacy by $L \cap K$, and the next sum is over positive root systems for $\Delta(L, H_1)$, up to the action of $W(L, H_1)$. The translation from the parameter $\Phi(\Delta^+, \tau_{\mathfrak{q}})$ for L to a parameter for G is accomplished by (12.3). The character $2\rho(A)$ is any extension to H_1 (for example, the one trivial on $A_{1,0}$) of the character of T_1 on the sum of the weights in A.

Since the left side in this theorem is a representation only of K and not of G, this is not an identity (as in Zuckerman's formula) of virtual representations of G. The parameters for G appearing on the right need not be standard limit parameters (although the "first term," corresponding to

$H_1 = H$, is just $(\Phi_{\mathfrak{q}})^{\mathfrak{q}}_{\mathfrak{g}} = \Phi$). In order to get a formula expressing the irreducible representation $\tau(\Phi)$ of K in terms of standard limit representations of G, we therefore need to rewrite each term on the right as an integer combination of standard limit representations. The character identities of Hecht and Schmid in [7] provide algorithms for doing exactly that, which appear to be reasonable to implement on a computer.

References

[1]   Anne-Marie Aubert and Roger Howe, *Géométrie des cônes aigus et application à la projection euclidienne sur la chambre de Weyl positive*, J. Algebra **149** (1992), 472–493.

[2]   Armand Borel and Nolan Wallach, *Continuous Cohomology, Discrete Subgroups, and Representations of Reductive Groups*, Princeton University Press, Princeton, New Jersey, 1980.

[3]   Jacques Carmona, *Sur la classification des modules admissibles irréductibles*, Noncommutative Harmonic Analysis and Lie Groups, 1983, pp. 11–34.

[4]   George Kempf, *The Grothendieck-Cousin complex of an induced representation*, Adv. in Math. **29** (1978), 310–396.

[5]   Anthony W. Knapp and David A. Vogan Jr., *Cohomological Induction and Unitary Representations*, Princeton University Press, Princeton, New Jersey, 1995.

[6]   Robert P. Langlands, *On the classification of representations of real algebraic groups*, Representation Theory and Harmonic Analysis on Semisimple Lie Groups, 1989, pp. 101–170.

[7]   Wilfried Schmid, *Two character identities for semisimple Lie groups*, Noncommutative Harmonic Analysis and Lie groups, 1977, pp. 196–225.

[8]   David A. Vogan Jr., *Representations of Real Reductive Lie Groups*, Birkhäuser, Boston-Basel-Stuttgart, 1981.

[9]   ———, *Unitary Representations of Reductive Lie Groups*, Annals of Mathematics Studies, Princeton University Press, Princeton, New Jersey, 1987.

[10] ———, *The method of coadjoint orbits for real reductive groups*, Representation Theory of Lie Groups, 1999.

SMALL SEMISIMPLE SUBALGEBRAS OF SEMISIMPLE LIE ALGEBRAS

JEB F. WILLENBRING

University of Wisconsin - Milwaukee
Department of Mathematical Sciences
P.O. Box 0413, Milwaukee WI 53201-0413, USA
E-mail: jw@uwm.edu

GREGG J. ZUCKERMAN

Yale University Mathematics Department
P.O. Box 208283, New Haven, CT 06520-8283, USA
E-mail: gregg@math.yale.edu

To Roger Howe, with friendship and admiration.

Let $\mathfrak{g}$ denote a semisimple Lie algebra with the property that none of its simple factors is of type A_1. Suppose that $\mathfrak{k} \subseteq \mathfrak{g}$ is a Lie subalgebra isomorphic to $\mathfrak{sl}_2$. The goal of this paper is to prove the existence of a positive integer $b(\mathfrak{k}, \mathfrak{g})$ such that if V is any irreducible finite dimensional $\mathfrak{g}$-module then when restricted to $\mathfrak{k}$, the decomposition of V will contain some irreducible $\mathfrak{k}$-module with dimension less than $b(\mathfrak{k}, \mathfrak{g})$. Beyond proving the theorem, we show how it may be generalized by introducing the notion of a *small* subalgebra.

1. Introduction

The main goal of this paper is to provide a proof of the following:

Theorem (see [18]). *Let $\mathfrak{k}$ be an $\mathfrak{sl}_2$-subalgebra of a semisimple Lie algebra $\mathfrak{g}$, none of whose simple factors is of type A_1. Then there exists a positive integer $b(\mathfrak{k}, \mathfrak{g})$, such that for every irreducible finite dimensional $\mathfrak{g}$-module V, there exists an injection of $\mathfrak{k}$-modules $W \to V$, where W is an irreducible $\mathfrak{k}$-module of dimension less than $b(\mathfrak{k}, \mathfrak{g})$.*

The first author was supported in part by NSA Grant # H98230-05-1-0078.

Note that the above statement (announced in [18]) is not asserting that *every* irreducible representation of $\mathfrak{k}$ of dimension less then $b(\mathfrak{k}, \mathfrak{g})$ occurs. Rather, the theorem asserts a *uniform* upper bound on the dimension of the smallest irreducible representation of $\mathfrak{k}$ that occurs. One might say we consider the *lowest* highest weight of a $\mathfrak{g}$-module relative to the restricted action of $\mathfrak{k}$.

More precisely, let W_i denote the irreducible $i+1$ dimensional representation of $\mathfrak{sl}_2$. If V is an irreducible finite dimensional $\mathfrak{g}$-module then, upon restriction to a fixed subalgebra isomorphic to $\mathfrak{sl}_2$ we have a decomposition:

$$V \cong \bigoplus_{i=1}^{\infty} \underbrace{W_i \oplus \cdots \oplus W_i}_{m_i(V)}$$

with $m_i(V) \geq 0$ being the multiplicity of W_i in V. The above theorem asserts that the number:

$$\max \left\{ \min_{m_i(V) \neq 0} i \mid V \text{ an irreducible finite dimensional } \mathfrak{g}\text{-module} \right\}$$

is finite provided that $\mathfrak{g}$ has no simple factor of type A_1.

In fact, the result is more general than $\mathfrak{k} \cong \mathfrak{sl}_2$. In Section 3.5.2 (and then in Theorem 4.0.11) we discuss the situation when $\mathfrak{k}$ is replaced by a general semisimple Lie algebra. The general result involves an analogous hypothesis that the simple factors of $\mathfrak{g}$ not belong to a certain finite set of simple Lie algebras determined by the subalgebra $\mathfrak{k}$. The proof in the general situation uses essentially the same machinery as the $\mathfrak{sl}_2$ case. We emphasize the $\mathfrak{sl}_2$ case in some non-trivial examples.

As the title suggests, we introduce the terminology that a semisimple Lie algebra $\mathfrak{k}$ is said to be a *small* subalgebra of a semisimple Lie algebra $\mathfrak{g}$ if there exists a bound $b(\mathfrak{k}, \mathfrak{g})$ such that for every irreducible finite dimensional $\mathfrak{g}$-module V, there exists an injection of $\mathfrak{k}$-modules $W \to V$, where W is an irreducible $\mathfrak{k}$-module having dimension less than $b(\mathfrak{k}, \mathfrak{g})$. For example, the above theorem asserts that any $\mathfrak{sl}_2$-subalgebra of (say) E_8 is small in E_8. Note that the rank n symplectic Lie algebra is not small in $\mathfrak{sl}_{2n}$, under the standard embedding. This can be seen by observing that the symmetric powers of the standard representation of sl_{2n} remain irreducible when restricted to sp_{2n}.

Although technically different, we remark that it would not significantly harm the content of the result if we regard $b(\mathfrak{k}, \mathfrak{g})$ as a bound on a norm of the highest weight of the representation of $\mathfrak{k}$ rather than its dimension. Clearly, in the $\mathfrak{sl}_2$-case these are essentially the same notion.

We consider the results of this paper interesting in their own right. However, we should note that the original motivation stems from a study of generalized Harish-Chandra modules as described in [18]. We recall the situation here.

Let $\mathfrak{k}$ denote a reductive subalgebra of a semisimple Lie algebra $\mathfrak{g}$. By a $(\mathfrak{g}, \mathfrak{k})$-module M we mean a $\mathfrak{g}$-module such that $\mathfrak{k}$ acts locally finitely. That is, for all $v \in M$, $\dim \mathcal{U}(\mathfrak{k})v < \infty$ (here $\mathcal{U}(\mathfrak{k})$ is the universal enveloping algebra of $\mathfrak{k}$). We say that the $(\mathfrak{g}, \mathfrak{k})$-module is *admissible* if every irreducible finite dimensional representation of $\mathfrak{k}$ occurs with finite multiplicity in M.

The question arises: Given an infinite dimensional, admissible $(\mathfrak{g}, \mathfrak{k})$-module M, does there exist a semisimple subalgebra $\mathfrak{k}' \subset \mathfrak{k}$ such that M is an admissible $(\mathfrak{g}, \mathfrak{k}')$-module? Indeed if $\mathfrak{k}'$ is a small subalgebra of $\mathfrak{k}$ then the answer is *no*. To see this, let $W_1, W_2, \cdots, W_m$ denote all of the irreducible $\mathfrak{k}'$-modules with dimension less than $b(\mathfrak{k}', \mathfrak{k})$ (it is an easy exercise using the Weyl dimension formula to see that there are finitely many). Next, consider any infinite set S of pairwise inequivalent irreducible finite dimensional $\mathfrak{k}$-modules occurring in M. As $\mathfrak{k}'$-modules, each $V \in S$ will have an irreducible constituent equivalent to at least one W_i. By the pigeonhole principle there must exist j (with $1 \leq j \leq m$) such that *infinitely* many elements of S contain W_j as a $\mathfrak{k}'$-submodule. Thus M has an irreducible $\mathfrak{k}'$-module with infinite multiplicity.

Another source of motivation comes from the theory of induced representations. Let G and K denote a linear algebraic group over $\mathbb{C}$ with Lie algebra $\mathfrak{g}$ and $\mathfrak{k}$ respectively. Assume that $K \subseteq G$. If V is a G-representation then we may restrict the G-action to K to obtain a representation of K. Restriction is in fact a functor, denoted Res, from the category of regular G-representations to regular K-representations. The right adjoint of Res is the induction functor, denoted Ind. The relationship between these two functors is the *Frobenius Reciprocity* theorem asserting the linear isomorphism,

$$\mathrm{Hom}_G(V, \mathrm{Ind}_K^G W) \cong \mathrm{Hom}_K(\mathrm{Res}_G^K V, W)$$

where V and W are representations of G and K respectively.

As is widely done, results about the restriction of a G-representation may be reinterpreted in terms of induced K-representations. In our context one can ask: Does there exist a *finite dimensional* K-representation, W, such that any irreducible representation of G may be embedded in $Ind_K^G W$? Using Frobenius reciprocity, one can easily see that the question is answered in the affirmative exactly when $\mathfrak{k}$ is a small subalgebra of $\mathfrak{g}$ by taking W to be the direct sum all irreducible representations of K with dimension less than $b(\mathfrak{k}, \mathfrak{g})$.

Historically, an important variation of this question concerns the theory of *models* (see [1]), which have played a crucial role in representation theory of real groups (see for example [24]). In the setting of compact Lie groups, one seeks a K-representation, W such that every G-representation occurs in $\operatorname{Ind}_K^G W$ with multiplicity one. Note that in this setting one assumes that K is the maximal compact subgroup of the split real form corresponding to G. Thus one is motivated to understand induced representations outside of this setting.

These theorems might be classified in the area of branching rules. However, it is fair to say that they are of a more qualitative nature than other results in the subject. In fact, it is important to note that the standard algorithms relying on the Weyl character formula or Kostant's partition function do not seem to shed light on this type of question. On the other hand, the results presented here may be interpreted combinatorially as statements about characters of representations. For example, let s_λ denotes the Schur polynomial in the variables $x_0, x_1, x_2, \cdots, x_n$ indexed by a partition λ as in [17]. We may make the substitution $x_i = t^{n-2i}$ and then expand

$$s_\lambda = \sum_{k \geq 0} m_k \chi^k(t),$$

where $\chi^k(t) = \frac{t^{k+1} - t^{-k-1}}{t - t^{-1}}$. The non-negative integers m_k are combinatorially complicated, although there is a bound $b(n)$ independent of λ such that $m_k > 0$ for some $k < b(n)$. This fact is an application of the above theorem with $\mathfrak{g} = \mathfrak{sl}_n$ and $\mathfrak{k}$ a principally embedded $\mathfrak{sl}_2$-subalgebra of $\mathfrak{g}$.

In this paper we do not obtain a bound on the numbers $b(\mathfrak{k}, \mathfrak{g})$ in the most general situation. We do however, compute the smallest possible value for some specific cases such as when $\mathfrak{g} = G_2$ and $\mathfrak{k}$ is a principal $\mathfrak{sl}_2$-subalgebra. Certainly, any example can be dealt with computationally using our methods, and a future direction of research would be to obtain a bound for $b(\mathfrak{k}, \mathfrak{g})$ for large families of pairs $(\mathfrak{g}, \mathfrak{k})$.

The proofs in this paper require the theory of algebraic groups in a fairly strong way. It is important to note that our results ultimately rely on Theorem 2.0.1, which is proved in Section 2. This theorem involves the question of when the coordinate ring of a linear algebraic group has unique factorization.

The geometric nature of the problem leads one to try other approaches involving geometric ideas that have, in other settings, shed light on problems of representation theory. One approach that was suggested to us by Alan Huckleberry was to cast the problem in the context of symplectic geometry. In this setting the theory of the moment map may provides enough information about the support of a $\mathfrak{g}$-module when restricted to a non-abelian subalgebra $\mathfrak{k}$. This approach may be very fruitful in that it may also provide information regarding the multiplicities arising from the branching problem. This will be a future direction of research.

The paper is organized as follows: Section 2 opens with a statement of Theorem 2.0.1, which turns out to be an important step in the proof of the main result. Section 3 introduces the necessary facts about Lie algebras and representation theory, with the goal being the proof of Proposition 3.5.7 (as an application of Theorem 2.0.1), and Proposition 3.3.1. In Section 4 we prove the main theorem, using Propositions 3.3.1 and 3.5.7. In Section 5, we apply the theorem to the special case where $\mathfrak{g}$ is the exceptional Lie algebra G_2, and $\mathfrak{k}$ is a principal $\mathfrak{sl}_2$-subalgebra of $\mathfrak{g}$. We obtain a sharp estimate of $b(\mathfrak{k}, \mathfrak{g})$ (in this case).

2. Invariant Theory

The context for this section lies in the theory of algebraic group actions on varieties. A good general reference for our terminology and notation is [21] which contains translations of works, [20] and [22]. For general notation and terminology from commutative algebra and algebraic geometry see [8] and [11]. For the general theory of linear algebraic groups, see [2].

All varieties are defined over $\mathbb{C}$, although we employ some results that are valid in greater generality. Unless otherwise stated, all groups are assumed to have the structure of a connected linear algebraic group. Of particular interest is the situation where a group, G, acts on an affine or quasiaffine variety. Given a (quasi-) affine variety, X, we denote the ring of regular functions on X, by $\mathbb{C}[X]$, and $\mathbb{C}[X]^G$ denotes the ring of G-invariant functions.

We now turn to the following:

Problem. *Let X be an irreducible (quasi-) affine variety. Let G be a algebraic group acting regularly on X. When is $\mathbb{C}[X]^G \not\cong \mathbb{C}$? That is to say, when do we have a non-trivial invariant?*

The general problem may be too hard in this generality. We begin by investigating a more restrictive situation which we describe next.

For our purposes, a generic orbit, $\mathcal{O} \subset X$, is defined to be an orbit of a point $x \in X$ with minimal isotropy group (ie: $G_{x_0} = \{g \in G | g \cdot x_0 = x_0\}$ for $x_0 \in X$).

Theorem 2.0.1. *Assume that X is an irreducible quasiaffine variety with a regular action by a linear algebraic group H such that:*

    (1) *A generic H-orbit, $\mathcal{O}$, in X has $\dim \mathcal{O} < \dim X$,*

    (2) $\mathbb{C}[X]$ *is factorial, and*

    (3) H *has no rational characters[1].*

Then, $\mathbb{C}[X]^H \ncong \mathbb{C}$. Furthermore, $\operatorname{trdeg} \mathbb{C}[X]^H = \operatorname{codim} \mathcal{O}$.

This theorem will be used to prove Proposition 3.5.1. In order to provide a proof of the above theorem we require some preparation. Let $\mathbb{C}(X)$ denote the field of complex valued rational functions on X. Our plan will be to first look at the ring of rational invariants, $\mathbb{C}(X)^H$. Given an integral domain, R, let QR denote the quotient field. Clearly $Q\left(\mathbb{C}[X]^H\right) \subseteq \mathbb{C}(X)^H$. Under the assumptions of Theorem 2.0.1 we have equality. As is seen by:

Theorem 2.0.2 ([21] p. 165). *Suppose $k(X) = Qk[X]$. If either,*

    (a) *the group G^0 is solvable[2], or*

    (b) *the algebra $k[X]$ is factorial,*

then any rational invariant of the action $G : X$ can be represented as a quotient of two integral semi-invariants (of the same weight). If, in addition, G^0 has no nontrivial characters (which in case (a) means it is unipotent), then $k(X)^G = Q\left(k[X]^G\right)$.

As one might expect from the hypothesis (2) of Theorem 2.0.1, our applications of Proposition 2.0.2 will involve condition (b). We will then use:

Proposition 2.0.3 ([21] p. 166). *Suppose the variety X is irreducible. The algebra $k[X]^G$ separates orbits in general position if and only if $k(X)^G = Qk[X]^G$, and in this case there exists a finite set of integral invariants that separates orbits in general position and the transcendence degree of $k[X]^G$ is equal to the codimension of an orbit in general position.*

Proof of Theorem 2.0.1. In our situation X is a quasiaffine variety, thus $k(X) = Qk[X]$. By assumptions (2) and (3) of Theorem 2.0.1 and Theorem 2.0.2 (using part (b)) we have that $k(X)^G = Q\left(k[X]^G\right)$. $\dim \mathbb{C}[X]^H =$

---

[1]A rational character of H is defined to be a regular function $\chi : H \to \mathbb{C}^\times$ such that $\chi(xy) = \chi(x)\chi(y)$ for all $x, y \in H$.

[2]Notation: As usual, G^0 denotes the connected component of the identity in G.

codim $\mathcal{O}$ follows from the irreducibility of X and Proposition 2.0.3. By assumption (1), codim $\mathcal{O} > 0$. Thus, $\mathbb{C}[X]^H \not\cong \mathbb{C}$. $\qquad\Box$

Question. For our purposes, we work within the context of the assumptions of Theorem 2.0.1, but to what extent may we relax the assumptions to keep the conclusion of the theorem?

Consider a triple (G, S, H) such that the following conditions $(*)$ hold[3]

Conditions $(*)$**.**

(1) G is a connected, simply-connected, semisimple linear algebraic group over $\mathbb{C}$.

(2) S and H are connected algebraic subgroups of G such that:

(a) $S \subseteq G$, is a connected algebraic subgroup with no non-trivial rational characters.

(b) $H \subseteq G$, is semisimple (and hence has no non-trivial rational characters).

For our situation we will require the following result (used in Proposition 3.5.1):

Theorem 2.0.4 (Voskresenskiĭ (see [19], [25], [26]))**.** *If G is a connected, simply connected, semisimple linear algebraic group then the ring of regular functions, $\mathbb{C}[G]$, is factorial.*

Remark 2.0.5. From the theory of the big cell, we can easily deduce that a connected algebraic group is always *locally* factorial. One should note however that there are obstructions to the factorial property of $\mathbb{C}[G]$. For example, if G is a connected linear algebraic group then $\mathbb{C}[G]$ is factorial iff the Picard group of G (denoted $Pic(G)$) is trivial (see [11] Proposition 6.2 and Corollary 6.16). Using this fact we can deduce Theorem 2.0.4 from:

Proposition 2.0.6 (see [13] Proposition 4.6)**.** *Let G be a connected linear algebraic group. Then, there exists a finite covering $G' \to G$ of algebraic groups such that $Pic(G') = 0$.*

Remark 2.0.7. If we assume that G is semisimple (and connected), then we can deduce that the dual of the fundamental group of G is isomorphic to $Pic(G)$ from Proposition 3.2 of [14].

We now return to our situation. Later, in the proof of Proposition 3.5.1 we will combine Theorem 2.0.4 with the following result:

---

[3]Throughout this article, we denote the Lie algebras of G and H by $\mathfrak{g}$ and $\mathfrak{k}$, but we do not need this notation at present.

Theorem 2.0.8 (see [21], page 176). *If $\mathbb{C}[X]$ is factorial and the group S is connected and has no nontrivial characters, then $\mathbb{C}[X]^S$ is factorial.*

We will apply the above result to the variety $X := G/S$. Regarding the geometric structure of the quotient we refer the reader to the excellent survey in [20] Section 4.7. We briefly summarize the main points: if G is an algebraic group with an algebraic subgroup $L \subseteq G$ then the quotient G/L has the structure of a quasiprojective variety. If G is reductive, then G/L is affine iff L is reductive; G/L is projective iff L is a parabolic subgroup (ie: L contains a Borel subgroup of G).

The condition of G/L being quasiaffine is more delicate, but includes the case where $L = S$ from the conditions $(*)$. More specifically, G/L is quasiaffine when L is quasiparabolic[4], which will be the situation in Section 3.4 of the present paper.

Remark 2.0.9. In general, a subgroup S of G is called *observable* if G/S is quasiaffine. A theorem of Sukhanov (see [20] page 194) asserts that a connected algebraic subgroup $H \subseteq G$ is observable if and only if it is tamely embedded [5] in some quasiparabolic subgroup. Also note that by a result of Weisfeiler, any algebraic subgroup of G may be tamely embedded in some parabolic subgroup.

A consequence of Theorems 2.0.4 and 2.0.8 is that the algebra $\mathbb{C}[X]$ is factorial, where $X = G/S$. The group H then acts regularly on G/S by left translation (ie: $x \cdot gS = xgS$, for $g \in G$, $x \in H$). And therefore, we are in a position to apply Theorem 2.0.1, as in the next section.

3. Representation Theory

This section begins by recalling some basic notation, terminology and results of Lie theory. We refer the reader to [3], [4], [9], [10], and [12] for this material.

3.1. Notions from Lie theory. Let G denote a semisimple, connected, complex algebraic group. We will assume that G is simply connected. T will denote a maximal algebraic torus in G. Let $r = \dim T$. Let B be a Borel subgroup containing T. The unipotent radical, U, of B is a maximal unipotent subgroup of G such that $B = T \cdot U$. Let $\mathfrak{g}$, $\mathfrak{h}$, $\mathfrak{b}$, $\mathfrak{n}^+$ denote the

---

[4]A *quasiparabolic* subgroup of G is one of the form $\{g \in P | \lambda(g) = 1\}$ where P is parabolic subgroup of G and λ is a dominant character of P.

[5]An algebraic group H is tamely embedded in an algebraic group F if the unipotent radical of H is contained in the unipotent radical of F.

Lie algebras of G, T, B, and U respectively. Let $W := N_G(T)/T$ denote the Weyl group corresponding to G and T.

The Borel subalgebra, $\mathfrak{b}$, contains the Cartan subalgebra, $\mathfrak{h}$ and is a semidirect sum[6] $\mathfrak{b} = \mathfrak{h} \mathbin{⋫} \mathfrak{n}^+$.

The weights of $\mathfrak{g}$ are the linear functionals $\xi \in \mathfrak{h}^*$. For $\alpha \in \mathfrak{h}^*$, set:

$$\mathfrak{g}_\alpha = \{X \in \mathfrak{g} \mid [H, X] = \alpha(H)X \ \forall H \in \mathfrak{h}\}.$$

For $0 \neq \alpha \in \mathfrak{h}^*$, we say that α is a root if $\mathfrak{g}_\alpha \neq (0)$. For such α, we have $\dim \mathfrak{g}_\alpha = 1$. Let Φ denote the set of roots. We then have the decomposition:

$$\mathfrak{g} = \mathfrak{h} \oplus \sum_{\alpha \in \Phi} \mathfrak{g}_\alpha.$$

The choice of B defines a decomposition $\Phi = \Phi^+ \cup -\Phi^+$ so that $\mathfrak{n}^+ = \sum_{\alpha \in \Phi^+} \mathfrak{g}_\alpha$. We refer to Φ^+ (resp. $\Phi^- := -\Phi^+$) as the positive (resp. negative) roots. Set: $\mathfrak{n}^- = \sum_{\alpha \in \Phi^+} \mathfrak{g}_{-\alpha}$. Let $\overline{B}$ denote the (opposite) Borel subgroup of G with Lie algebra $\mathfrak{h} \oplus \mathfrak{n}^-$. There is a unique choice of simple roots $\Pi = \{\alpha_1, \cdots, \alpha_r\}$ contained in Φ^+, such that each $\alpha \in \Phi^+$ can be expressed as a non-negative integer sum of simple roots. Π is a vector space basis for $\mathfrak{h}^*$. Given $\xi, \eta \in \mathfrak{h}^*$ we write $\xi \preceq \eta$ if $\eta - \xi$ is a non-negative integer combination of simple (equiv. positive) roots. $\preceq$ is the dominance order on $\mathfrak{h}^*$.

For each positive root α, we may choose a triple: $X_\alpha \in \mathfrak{g}_\alpha$, $X_{-\alpha} \in \mathfrak{g}_{-\alpha}$ and $H_\alpha \in \mathfrak{h}$, such that $H_\alpha = [X_\alpha, X_{-\alpha}]$ and $\alpha(H_\alpha) = 2$. Span $\{X_\alpha, X_{-\alpha}, H_\alpha\}$ is then a three dimensional simple (TDS) subalgebra of $\mathfrak{g}$, and is isomorphic to $\mathfrak{sl}_2$.

The adjoint representation, $ad : \mathfrak{g} \to \mathrm{End}(\mathfrak{g})$ allows us to define the Killing form, $(X, Y) = \mathrm{Trace}(ad\, X \ ad\, Y)$ $(X, Y \in \mathfrak{g})$. The semisimplicity of $\mathfrak{g}$ is equivalent to the non-degeneracy of the Killing form. By restriction, the form defines a non-degenerate form on $\mathfrak{h}$, also denoted $(,)$. Using this form we may define $\iota : \mathfrak{h} \to \mathfrak{h}^*$ by $\iota(X)(-) = (X, -)$ $(X \in \mathfrak{h})$, which allows us to identify $\mathfrak{h}$ with $\mathfrak{h}^*$. Under this identification, we have $\iota(H_\alpha) = \frac{2\alpha}{(\alpha, \alpha)} =: \alpha^\vee$.

By definition, the Weyl group, W acts on T. By differentiating this action we obtain an action on $\mathfrak{h}$, which is invariant under $(,)$. Via ι, we obtain an action of W on $\mathfrak{h}^*$. In light of this, we view W as a subgroup of the orthogonal group on $\mathfrak{h}^*$. W preserves Φ. For each $\alpha \in \Phi$, set $s_\alpha(\xi) = \xi - (\xi, \alpha^\vee)\alpha$ (for $\xi \in \mathfrak{h}^*$) to be the reflection through the hyperplane defined by $\alpha^\vee$. We have $s_\alpha \in W$. For $\alpha_i \in \Pi$, let $s_i := s_{\alpha_i}$, be the simple

---

[6]The symbol $\mathbin{⋫}$ stands for the semidirect sum of Lie algebras. If $\mathfrak{g} = \mathfrak{g}_1 \mathbin{⋫} \mathfrak{g}_2$ then $\mathfrak{g}_2$ is an ideal in $\mathfrak{g}$ and $\mathfrak{g}/\mathfrak{g}_2 \cong \mathfrak{g}_1$.

reflection defined by α_i. W is generated by the simple reflections. For $w \in W$, let $w = s_{i_1} s_{i_2} \cdots s_{i_\ell}$ be a reduced expression (ie: an expression for w with shortest length). The number ℓ is independent of the choice of reduced expression. We call $\ell =: \ell(w)$ the length of w. Note that $\ell(w) = |w(\Phi^+) \cap \Phi^-|$. There is a unique longest element of W, denoted w_0 of length $|\Phi^+|$.

The fundamental weights, $\{\omega_1, \cdots, \omega_r\}$ are defined by the conditions $(\omega_i, \alpha_j^\vee) = \delta_{i,j}$. We fix the ordering of the fundamental weights to correspond with the usual numbering of the nodes in the Dynkin diagram as in [4]. Set $\rho = \frac{1}{2} \sum_{\alpha \in \Phi^+} \alpha = \sum_{i=1}^r \omega_i$. A weight $\xi \in \mathfrak{h}^*$ is said to be dominant if $(\lambda, \alpha) \geq 0$ for all $\alpha \in \Pi$. The weight lattice $P(\mathfrak{g}) = \{\xi \in \mathfrak{h}^* | (\xi, \alpha^\vee) \in \mathbb{Z}\} = \sum_{i=1}^r \mathbb{Z}\omega_i$. We define the dominant integral weights to be those $\xi \in P(\mathfrak{g})$ such that $(\xi, \alpha) \geq 0$ for all $\alpha \in \Pi$. The set of dominant integral weights, $P_+(\mathfrak{g})$, parameterizes the irreducible finite dimensional representations of $\mathfrak{g}$ (or equivalently, of G). We have [7] $P_+(\mathfrak{g}) = \sum_{i=1}^r \mathbb{N}\omega_i$.

3.2. Notions from representation theory. $\mathcal{U}(\mathfrak{g})$ denotes the universal enveloping algebra of $\mathfrak{g}$. The category of Lie algebra representations of $\mathfrak{g}$ is equivalent to the category of $\mathcal{U}(\mathfrak{g})$–modules. A $\mathfrak{g}$-representation (equiv. $\mathcal{U}(\mathfrak{g})$–module), M, is said to be a weight module if $M = \bigoplus M(\xi)$, where:

$$M(\xi) = \{v \in M | Hv = \xi(H)v \, \forall H \in \mathfrak{h}\}.$$

Among weight modules are the modules admitting a highest weight vector. That is to say, a unique (up to scalar multiple) vector, $v_0 \in M$ such that:

(1) $\mathbb{C}v_0 = M^{\mathfrak{n}^+} := \{v \in M | \mathfrak{n}^+ v = 0\}$,
(2) $M(\lambda) = \mathbb{C}v_0$ for some $\lambda \in \mathfrak{h}^*$, and
(3) $\mathcal{U}(\mathfrak{n}^-)v_0 = M$.

Such a module is said to be a highest weight module (equiv highest weight representation). λ is the highest weight of M. Given $\xi \in \mathfrak{h}^*$ with $M(\xi) \neq (0)$ we have $\xi \preceq \lambda$.

For $\lambda \in \mathfrak{h}^*$, we let $\mathbb{C}_\lambda$ be the 1-dimensional representation of $\mathfrak{h}$ defined by λ, then extended trivially to define a representation of $\mathfrak{b}$ by requiring $\mathfrak{n}^+ \cdot \mathbb{C}_\lambda = (0)$. Let $N(\lambda) := \mathcal{U}(\mathfrak{g}) \otimes_{\mathcal{U}(\mathfrak{b})} \mathbb{C}_\lambda$ denote the Verma module defined by λ. Let $L(\lambda)$ denote the irreducible quotient of $N(\lambda)$. For $\lambda, \mu \in \mathfrak{h}^*$, $L(\lambda) \cong L(\mu)$ iff $\lambda = \mu$. $L(\lambda)$ (and $N(\lambda)$) are highest weight representations. Any irreducible highest weight representation is equivalent to $L(\lambda)$ for a

---

[7]As usual, $\mathbb{N} = \{0, 1, 2, \cdots\}$ (the non-negative integers).

unique $\lambda \in \mathfrak{h}^*$. The theorem of the highest weight asserts that $\dim L(\lambda) < \infty$ iff $\lambda \in P_+(\mathfrak{g})$.

Each $\mu \in P(\mathfrak{g})$, corresponds to a linear character of T, denoted e^μ. For $\lambda \in P_+(\mathfrak{g})$, the character of $L(\lambda)$ defines a complex valued regular function on T. This character may be expressed as in the following:

Theorem 3.2.1 (Weyl).

$$ch\, L(\lambda) = \frac{\sum_{w \in W} (-1)^{\ell(w)} e^{w(\lambda+\rho)-\rho}}{\prod_{\alpha \in \Phi^+} (1 - e^{-\alpha})}.$$

It becomes necessary for us to refer to representations of both the group G and the Lie algebra of G (always denoted $\mathfrak{g}$). By our assumptions on G (in conditions $(*)$), every finite dimensional complex representation of $\mathfrak{g}$ integrates to a regular representation of G. The differential of this group representation recovers the original representation of the Lie algebra. We will implicitly use this correspondence.

For our purposes, of particular importance is the decomposition of the regular [8] representation of G. That is,

Theorem 3.2.2. *For $f \in \mathbb{C}[G]$, $(g,h) \in G \times G$ define: $(g,h) \cdot f(x) = f(g^{-1}xh)$ $(x \in G)$. Under this action we have the classical Peter-Weyl decomposition:*

$$\mathbb{C}[G] \cong \bigoplus_{\lambda \in P_+(\mathfrak{g})} L(\lambda)^* \otimes L(\lambda), \qquad (3.1)$$

as a representation of $G \times G$. Here the superscript $*$ denotes the dual representation. Note that $L(\lambda)^*$ is an irreducible highest weight representation of highest weight $-w_0(\lambda)$.

We introduce the following notation:

Definition 3.2.3. Let $\sigma : G \to GL(V)$ (resp. $\tau : H \to GL(W)$) be a representation of a group G (resp. H). If $H \subseteq G$ we may regard (σ, V) as a representation of H by restriction. We set:

$$[V, W] := \dim \mathrm{Hom}_H(V, W).$$

If either V or W is infinite dimensional $\mathrm{Hom}_H(V,W)$ may be infinite dimensional. In this case $[V,W]$ should be regarded as an infinite cardinal. (We will not encounter this situation in what is to follow.) If V is completely reducible as an H-representation, and W is irreducible, $[V,W]$ is the multiplicity of W in V (By Schur's lemma).

---

[8]We use the word "regular" in two senses, the other being in the context of algebraic geometry.

Note: We will use the same (analogous) notation in the category of Lie algebra representations.

3.3. On the $\chi(T)$-gradation of $\mathbb{C}[U\backslash G]$. We apply a philosophy taught to us by Roger Howe. As before, U is a maximal unipotent subgroup of G, T a maximal torus (normalizing U).

As in Theorem 3.2.2, $U \times G \subseteq G \times G$ acts on $\mathbb{C}[G]$. We have $\mathbb{C}[G]^U \cong \mathbb{C}[U\backslash G]$. As T normalizes U we have an action of T on $\mathbb{C}[U\backslash G]$ via $t \cdot f(x) = f(t^{-1}x)$ for $t \in T$ and $x \in U\backslash G$. We call this action the *left* action, since the multiplication is on the left. By Theorem 3.2.2, we have:

$$\mathbb{C}[U\backslash G] = \bigoplus_{\lambda \in P_+(\mathfrak{g})} (L(\lambda)^*)^U \otimes L(\lambda). \tag{3.2}$$

A consequence of the theorem of the highest weight is that $\dim (L(\lambda)^*)^U = 1$. We let, $\chi(T) \cong \mathbb{Z}^r$, denote the character group of T. Each $\lambda \in P_+(\mathfrak{g})$ defines a character, e^λ, of T. Set: $\mathbb{C}[U\backslash G]_\lambda := \{f \in \mathbb{C}[U\backslash G] \mid f(t^{-1}x) = e^\lambda(t)f(x) \ \forall x \in U\backslash G, t \in T\}$. G then acts (by right multiplication) on $\mathbb{C}[U\backslash G]$, and under this action we have: $\mathbb{C}[U\backslash G]_\lambda \cong L(\lambda)$. We then obtain a $\chi(T)$-gradation of the algebra. That is to say, $\mathbb{C}[U\backslash G]_\xi \cdot \mathbb{C}[U\backslash G]_\eta \subseteq \mathbb{C}[U\backslash G]_{\xi+\eta}$. We exploit this phenomenon to obtain:

Proposition 3.3.1. *Let W be an irreducible finite dimensional representation of a reductive subgroup, H, of G. Let $\lambda, \mu \in P_+(\mathfrak{g})$. If $L(\mu)^H \neq (0)$ and $[W, L(\lambda)] \neq 0$ then $[W, L(\lambda + \mu)] \neq 0$.*

Remark 3.3.2. In Proposition 3.3.1, the G-representations $L(\lambda)$ and $L(\lambda + \mu)$ are regarded as H-representations by restriction.

Proof of Proposition 3.3.1. Let $f \in \mathbb{C}[U\backslash G]_\mu^H \cong L(\mu)^H$ and $\tilde{W} \subseteq \mathbb{C}[U\backslash G]_\lambda \cong L(\lambda)$ such that $W \cong \tilde{W}$ as a representation of H. Then $f \cdot W \subseteq \mathbb{C}[U\backslash G]_{\lambda+\mu}$. Under the (right) action of H we have, $f \cdot \tilde{W} \cong W$. Therefore, $[W, \mathbb{C}[U\backslash G]_{\lambda+\mu}] \neq 0$. $\qquad\square$

3.4. The maximal parabolic subgroups of G. A connected algebraic subgroup, P, of G containing a Borel subgroup is said to be parabolic. There exists an inclusion preserving one-to-one correspondence between parabolic subgroups and subsets of Π. We will recall the basic set-up.

Let $\mathfrak{p} = Lie(P)$ denote the Lie algebra of a parabolic subgroup P. Then $\mathfrak{p} = \mathfrak{h} \oplus \sum_{\alpha \in \Gamma} \mathfrak{g}_\alpha$, where $\Gamma := \Phi^+ \cup \{\alpha \in \Phi \mid \alpha \in \mathrm{Span}(\Pi')\}$ for a unique $\Pi' \subseteq \Pi$. Set:

$$\mathfrak{l} = \mathfrak{h} \oplus \sum_{\alpha \in \Gamma \cap -\Gamma} \mathfrak{g}_\alpha, \quad \mathfrak{u}^+ = \sum_{\alpha \in \Gamma, \, \alpha \notin -\Gamma} \mathfrak{g}_\alpha, \text{ and } \mathfrak{u}^- = \sum_{\alpha \in \Gamma, \, \alpha \notin -\Gamma} \mathfrak{g}_{-\alpha}.$$

Then we have, $\mathfrak{p} = \mathfrak{l} \oplus \mathfrak{u}^+$ and $\mathfrak{g} = \mathfrak{u}^- \oplus \mathfrak{p}$. The subalgebra $\mathfrak{l}$ is the Levi factor of $\mathfrak{p}$, while $\mathfrak{u}^+$ is the nilpotent radical of $\mathfrak{p}$. $\mathfrak{l}$ is reductive and hence $\mathfrak{l} = \mathfrak{l}_{ss} \oplus \mathfrak{z}(\mathfrak{l})$, where $\mathfrak{z}(\mathfrak{l})$ and $\mathfrak{l}_{ss}$ denote the center and semisimple part of $\mathfrak{l}$ respectively.

The following result is well known (see for example, [16] page 196). This form is a slight modification of Exercise 12.2.4 in [10] (p. 532).

Proposition 3.4.1. *For $0 \neq \lambda \in P_+(\mathfrak{g})$, let v_λ be a highest weight vector in $L(\lambda)$. Let $X = G \cdot v_\lambda \subseteq L(\lambda)$ denote the orbit of v_λ and let $G_{v_\lambda} = \{g \in G | g \cdot v_\lambda = v_\lambda\}$ denote the corresponding isotropy group. Then, X is a quasiaffine variety stable under the action of $\mathbb{C}^\times$ on $L(\lambda)$ defined by scalar multiplication. This $\mathbb{C}^\times$-action defines a graded algebra structure on $\mathbb{C}[X] = \bigoplus_{d=0}^\infty \mathbb{C}[X]^{(d)}$.*

The action of G on $\mathbb{C}[X]$, defined by $g \cdot f(x) = f(g^{-1} \cdot x)$ (for $g \in G$ and $x \in X$) commutes with the $\mathbb{C}^\times$-action and therefore each graded component of $\mathbb{C}[X]$ is a representation of G. Furthermore, we have:

$$\mathbb{C}[X]^{(n)} \cong L(-nw_0(\lambda))$$

for all $n \in \mathbb{N}$.

Proof. Set $V = L(\lambda)$, and let v_λ be a highest weight vector in V. Let $\mathbb{P}V = \{[v] | v \in V\}$ denote the complex projective space on V. G then acts on $\mathbb{P}V$ by $g \cdot [v] = [g \cdot v]$ for $g \in G$ and $v \in V$. The isotropy group of $[v_\lambda]$ contains the Borel subgroup, B and therefore is a parabolic subgroup, $P \subseteq G$. G/P is projective so, $G \cdot [v_\lambda] \subset \mathbb{P}V$ is closed. This means that the affine cone, $\mathbb{A} := \bigcup_{g \in G}[g \cdot v_\lambda] \subset V$ is closed. Let $a \in \mathbb{A}$. Then, $a = z(g \cdot v_\lambda)$ for some $z \in \mathbb{C}$ and $g \in G$. The action of G on V is linear so, $a = g \cdot zv_\lambda$. By the assumption that $\lambda \neq 0$, T acts on $\mathbb{C}v_\lambda$ by a non-trivial linear character. All non-trivial linear characters of tori are surjective (this is because the image of a (connected) algebraic group homomorphism is closed and connected). Therefore, if $z \neq 0$, then $zv_\lambda = t \cdot v_\lambda$ for some $t \in T$. This fact implies[9] that either $a \in X$ or $a = 0$. And so, $\overline{X} = X \cup \{0\} = \mathbb{A}$. X is therefore quasiaffine since $X = \mathbb{A} - \{0\}$ ($0 \notin X$ since $\lambda \neq 0$). We have also shown that X is stable under scalar multiplication by a non-zero complex number.

Let $v_\lambda^* \in L(\lambda)^*$ be a highest weight vector. Upon restriction v_λ^* defines a regular function on X in $\mathbb{C}[X]^{(1)}$, which is a highest weight vector for the left action of G on $\mathbb{C}[X]$. Then, $(v_\lambda^*)^n \in \mathbb{C}[X]^{(n)}$. $(v_\lambda^*)^n$ is a highest weight vector, hence, $[\mathbb{C}[X]^{(n)}, L(n\lambda)^*] \neq 0$. That is to say, we have an injective

---

[9]If $a \neq 0$, $a = g \cdot (t \cdot v_\lambda) = (gt) \cdot v_\lambda \in X$.

G-equivariant map, $\psi : L(n\lambda)^* \hookrightarrow \mathbb{C}[X]^{(n)}$. It remains to show that ψ is an isomorphism.

Since X is quasiaffine, $\mathbb{C}[X] \cong \mathbb{C}[G]^{G_{v_\lambda}}$. By restriction of the regular representation, $G \times G_{v_\lambda}$ acts on $\mathbb{C}[G]$. And so by, Theorem 3.2.2:

$$\mathbb{C}[X] \cong \bigoplus_{\xi \in P_+(\mathfrak{g})} L(\xi)^* \otimes L(\xi)^{G_{v_\lambda}}$$

Set $\mathcal{L}_\xi := L(\xi)^{G_{v_\lambda}}$. $U \subseteq G_{v_\lambda}$ since v_λ is a highest weight vector. Therefore, $\mathcal{L}_\xi \subseteq L(\xi)^U$ and $\dim \mathcal{L}_\xi \leq 1$. And so, $\mathbb{C}[X]$ is multiplicity free as a representation of G (under the action of left multiplication).[10] We will show that the only possible ξ for which $\dim \mathcal{L}_\xi > 0$ are the non-negative integer multiples of λ. By the fact that $\mathbb{C}[X]$ is multiplicity free we will see that ψ must be an isomorphism.

If $\mathcal{L}_\xi \neq (0)$ then $\mathcal{L}_\xi = L(\xi)^U$ since they are both 1-dimensional. Assume that $\mathcal{L}_\xi \neq (0)$. Choose, $0 \neq v_\xi \in \mathcal{L}_\xi$. Note that v_ξ is a highest weight vector. T acts on $\mathbb{C}v_\xi$ by $t \cdot v_\xi = e^\xi(t)v_\xi$ for all $t \in T$. Set $T_\xi := T \cap G_{v_\lambda}$ and $\mathfrak{h}_\lambda := Lie(T_\lambda) = \{H \in \mathfrak{h} \mid \lambda(H) = 0\}$.

$H \in \mathfrak{h}_\lambda$ implies both $H \cdot v_\xi = 0$ and $H \cdot v_\xi = \xi(H)v_\xi$. Hence we have $\xi(H) = 0$ when $H \in \mathfrak{h}_\lambda$. This statement is equivalent to λ and ξ being linearly dependent. Furthermore, we have $n_1\xi = n_2\lambda$ for $n_1, n_2 \in \mathbb{N}$ since ξ and λ are both dominant integral weights.

If $\dim \mathcal{L}_\xi = 1$ then $[\mathbb{C}[X]^{(n)}, L(\xi)^*] \neq 0$ for some $n \in \mathbb{N}$. This forces $[\mathbb{C}[X]^{(nn_1)}, L(n_1\xi)^*] \neq 0$ and therefore, $[\mathbb{C}[X]^{(nn_1)}, L(n_2\lambda)^*] \neq 0$. As before, $[\mathbb{C}[X]^{(n_2)}, L(n_2\lambda)^*] \neq 0$. Using the fact that $\mathbb{C}[X]$ is multiplicity free, we have $nn_1 = n_2$. And so, $\xi = n\lambda$. $\square$

Remark. The closure of the variety X in Proposition 3.4.1 is called the *highest weight variety* in [23] (see [10]).

If $\Pi' = \Pi - \{\alpha\}$ for some simple root α, then the corresponding parabolic subgroup is maximal (among proper parabolic subgroups). Consequently, the maximal parabolic subgroups of G may be parameterized by the nodes of the Dynkin diagram, equivalently, by fundamental weights of G. Set: $\lambda = -w_0(\omega_k)$. Let $v_k \in L(\lambda)^U$ be a highest weight vector. Define:

$$X_{\mathfrak{g}}^{(k)} := G \cdot v_k \subseteq L(\lambda) \qquad (1 \leq k \leq r),$$

the orbit of v_k under the action of G. (When there is no chance of confusion, we write $X^{(k)}$ for $X_{\mathfrak{g}}^{(k)}$.)

---

[10]Alternatively, $\overline{B}$ has a dense orbit in X (since $\overline{B}U$ is dense in G (ie: the big cell)). A dense orbit under a Borel subgroup is equivalent to having a multiplicity free coordinate ring.

We have seen that this orbit has the structure of a quasiaffine variety. It is easy to see that the isotropy group, $S^{(k)}$ of v_k is of the form $S^{(k)} = [P_k, P_k]$, where P_k denotes a maximal parabolic subgroup of G.[11] Let $\mathfrak{p} := Lie(P_k)$ and $\mathfrak{p} = \mathfrak{l} \oplus \mathfrak{u}^+$ denote the Levi decomposition. We have $\mathfrak{p} = \mathfrak{l} + n^+$ and $n^+ = [\mathfrak{b}, \mathfrak{b}] \subset [\mathfrak{p}, \mathfrak{p}]$ and $\mathfrak{l}_{ss} = [\mathfrak{l}, \mathfrak{l}] \subseteq [\mathfrak{p}, \mathfrak{p}]$. Therefore, $[\mathfrak{p}, \mathfrak{p}] = \mathfrak{l}_{ss} + n^+ = \mathfrak{l}_{ss} \oplus \mathfrak{u}^+$. (Note that all that is lost is $\mathfrak{z}(\mathfrak{l})$.) We say that P_k is the parabolic subgroup corresponding to the fundamental weight ω_k.

We have $\dim X^{(k)} = \dim \mathfrak{g}/[\mathfrak{p}, \mathfrak{p}]$ and $2 \dim \mathfrak{g}/[\mathfrak{p}, \mathfrak{p}] = \dim(\mathfrak{g}/\mathfrak{l}_{ss}) + 1$ (Note that $\dim \mathfrak{z}(\mathfrak{l}) = 1$ since P_k is maximal.) In light of these facts, we see that the dimension of $X^{(k)}$ may be read off of the Dynkin diagram. The dimension is important for the proof of Corollary 3.5.2. In Section 6, we explicitly compute $\dim \mathfrak{g}/\mathfrak{l}_{ss}$ for the exceptional Lie algebras and low rank classical Lie algebras.

Corollary 3.4.2. *For $X = X^{(k)}$ where $1 \le k \le r$ we have:*

$$\mathbb{C}[X] \cong \bigoplus_{n=0}^{\infty} L(n\omega_k) \tag{3.3}$$

as a representation of G.

Proof. The result is immediate from Proposition 3.4.1. □

3.5. A consequence of Theorem 2.0.1.

The goal of Section 2 was to prove Theorem 2.0.1. We now apply this theorem to obtain the following:

Proposition 3.5.1. *Assume G and H satisfy conditions (∗), and we take $S = S^{(k)}$, for some $1 \le k \le r$. Set: $X := X^{(k)}$.*

$$\dim H < \dim X \implies \mathbb{C}[X]^H \ncong \mathbb{C}.$$

Proof. If $\mathcal{O}$ is a generic H-orbit in X then $\dim \mathcal{O} \le \dim H < \dim X$. By Theorems 2.0.4 and 2.0.8, the algebra $\mathbb{C}[X]$ is factorial, because $\mathbb{C}[X] = \mathbb{C}[G]^S$. The result follows from Theorem 2.0.1. □

We now provide a representation theoretic interpretation of this proposition as it relates to $\mathfrak{sl}_2$.

---

[11]This fact in not true for the orbit of an arbitrary dominant integral weight. In general, $[\mathfrak{p}, \mathfrak{p}] \ne Lie(G_{v_\lambda})$, for any parabolic subalgebra $\mathfrak{p}$.

3.5.1. The $\mathfrak{sl}_2$-case. Consider a triple (G, S, H) such that the following conditions $(**)$ hold:

Conditions $(**)$.

    (1) (G, S, H) satisfy conditions $(*)$, and:
    (2) $\mathfrak{g}$ has no simple factor of Lie type A_1.
    (3) $S = S^{(k)}$, where $1 \leq k \leq r$. Set: $X := X^{(k)}$ $(:= G/S)$.
    (4) $\mathfrak{sl}_2 \cong Lie(H) =: \mathfrak{k} \subseteq \mathfrak{g}$

Corollary 3.5.2. *Assume conditions $(**)$, and that $\mathfrak{g}$ has no simple factor of Lie type $A2$. Then, $\mathbb{C}[X]^H \ncong \mathbb{C}$.*

Proof. The statement can be reduced to the case where $\mathfrak{g}$ is simple. For G simple and not of type A_1 or A_2, we appeal to the classification of maximal parabolic subgroups (see the tables in Section 6) to deduce that $\dim X > 3$. as $H \cong SL_2(\mathbb{C})$ (locally). Hence, $\dim H < \dim X$. We are then within the hypothesis of Proposition 3.5.1. $\hfill\square$

We next address the case when $\mathfrak{g}$ does have a simple factor of Lie type $A2$. For this material we need to analyze the set of $\mathfrak{sl}_2$-subalgebras of $\mathfrak{g} = \mathfrak{sl}_3$, up to a Lie algebra automorphism. For general results on the subalgebras of $\mathfrak{g}$ we refer the reader to [6] and [7].

In the case of $\mathfrak{g} = \mathfrak{sl}_3$ there are two such $\mathfrak{sl}_2$ subalgebras. One being the root sl_2-subalgebra corresponding to any one of the three positive roots of $\mathfrak{g}$. The other is the famous principal sl_2-subalgebra.

A principal $\mathfrak{sl}_2$-subalgebra ([7], [15]) of $\mathfrak{g}$ is a subalgebra $\mathfrak{k} \subseteq \mathfrak{g}$ such that $\mathfrak{k} \cong sl_2$ and contains a regular nilpotent element. These subalgebras are conjugate, so we sometimes speak of "the" principal $\mathfrak{sl}_2$-subalgebra. There is a beautiful connection between the principal $\mathfrak{sl}_2$-subalgebra and the cohomology of G, (see [15]). There is a nice discussion of this theory in [5].

Lemma 3.5.3. *Assume conditions $(**)$. Let G and H be such that $\mathfrak{g} \cong \mathfrak{sl}_3$, and $Lie(H)$ is a principal $\mathfrak{sl}_2$-subalgebra of $\mathfrak{g}$. Then, $\dim \mathbb{C}[X]^H > 0$, for $X = X^{(1)}$ or $X = X^{(2)}$.*

Proof. A principal $\mathfrak{sl}_2$-subgroup in $\mathfrak{sl}_3$ is embedded as a symmetric subalgebra. More precisely, let $H := SO_3(\mathbb{C}) \subseteq G$, then $Lie(H)$ is a principal $\mathfrak{sl}_2$-subalgebra of $\mathfrak{sl}_3$.

In general, if $G = SL_n(\mathbb{C})$ and $K = SO_n(\mathbb{C})$ with H embedded in G in the standard way, then the pair (G, H) is symmetric (ie: H is the fixed point set of a regular involution on G). As before, let r denote the rank of

G. In order the prove the lemma (for $r = 2$), it suffices to observe that by the Cartan-Helgason theorem (see [10], Chapter 11) we have:

$$\dim L(\lambda)^K = \begin{cases} 1, & \lambda \in \sum_{i=1}^{r} 2\mathbb{N}\omega_i; \\ 0, & \text{Otherwise.} \end{cases}$$

$\square$

Lemma 3.5.4. *Assume conditions* (∗∗). *Let G and H be such that $\mathfrak{g} \cong \mathfrak{sl}_3$, and $Lie(H)$ is a root $\mathfrak{sl}_2$-subalgebra of $\mathfrak{g}$. Then, $\dim \mathbb{C}[X]^H > 0$, for $X = X^{(1)}$ or $X = X^{(2)}$.*

Proof. It is the case that $L(\omega_1)$ and $L(\omega_2)$ both have H-invariants, as they are equivalent to the standard representation of G and its dual respectively.

$\square$

Summarizing we obtain:

Proposition 3.5.5. *Under assumptions* (∗∗), *$\dim \mathbb{C}[X]^H > 0$.*

Proof. The statement reduces to the case where $\mathfrak{g}$ is simple, because a maximal parabolic subalgebra of $\mathfrak{g}$ must contain all but one simple factor of $\mathfrak{g}$. Therefore, assume that $\mathfrak{g}$ is simple without loss of generality. For $\mathfrak{g} \not\cong \mathfrak{sl}_3$, apply Corollary 3.5.2. If $\mathfrak{g} \cong \mathfrak{sl}_3$, then $Lie(H)$ is embedded as either a root $\mathfrak{sl}_2$-subalgebra, or as a principal $\mathfrak{sl}_2$-subalgebra. Apply Lemmas 3.5.4 and 3.5.3 to the respective cases.

$\square$

The following will be of fundamental importance in Section 4.

Definition 3.5.6. For G and H as in conditions (∗), we consider the following set of positive integers [12]:

$$M(G, H, j) := \left\{ n \in \mathbb{Z}^+ \mid \dim [L(n\omega_j)]^H \neq (0) \right\}$$

where j is a positive integer with $1 \leq j \leq r$. Set:

$$m(G, H, j) := \begin{cases} \min M(G, H, j), & \text{if } M(G, H, j) \neq \emptyset; \\ 0, & \text{if } M(G, H, j) = \emptyset. \end{cases}$$

We will also write $m(\mathfrak{g}, \mathfrak{k}, j)$ (resp. $M(\mathfrak{g}, \mathfrak{k}, j)$) where (as before) $\mathfrak{k} = Lie(H)$ and $\mathfrak{g} = Lie(G)$.

---

[12]As always, $\mathbb{Z}^+ = \{1, 2, 3, \cdots\}$ (the positive integers).

Proposition 3.5.7. *For G and H as in conditions (∗∗),*

$$m(G, H, k) > 0 \text{ for all, } 1 \le k \le r.$$

Proof. Apply Corollary 3.4.2 and Proposition 3.5.5. □

Proposition 3.5.1 applies to a much more general situation than $\mathfrak{k} \cong \mathfrak{sl}_2$.

3.5.2. The semisimple case. As it turns out, what we have done for the $\mathfrak{sl}_2$-subalgebras can be done for any semisimple subalgebra.

Proposition 3.5.8. *If G and H are as in condition (∗), then for each $1 \le k \le r$,*

$$\dim H < \dim X^{(k)} \implies m(G, H, k) > 0.$$

Proof. Apply Proposition 3.5.1 and Corollary 3.4.2. □

In order to effectively apply Proposition 3.5.8 we will want to guarantee that $\dim H < \dim X^{(k)}$ for all $1 \le k \le r$. This will happen if all simple factors of $\mathfrak{g}$ have sufficiently high rank. This idea motivates the following definition. Consider a group G (as in condition (∗)) and define:

$$e(\mathfrak{g}) := \min_{1 \le k \le r} \dim X_{\mathfrak{g}}^{(k)}$$

For a semisimple complex Lie algebra $\mathfrak{k}$, define[13]:

$$E(\mathfrak{k}) = \left\{ \mathfrak{s} \; \middle| \; \begin{array}{l} (1) \; \mathfrak{s} \text{ is a simple complex Lie algebra} \\ (2) \; \dim \mathfrak{k} \ge e(\mathfrak{s}) \end{array} \right\}.$$

Corollary 3.5.9. *Assume G and H as in conditions (∗). If $\mathfrak{g}$ has no simple factor that is in the set $E(\mathfrak{k})$ then $m(G, H, k) > 0$ for all k with $1 \le k \le r$.*

Proof. Immediate from Proposition 3.5.8 and the definition of $E(\mathfrak{k})$. □

Example.

$$E(\mathfrak{sl}_3) = \{A_1, A_2, A_3, A_4, A_5, A_6, A_7, B_2, B_3, B_4, C_3, C_4, D_4, G_2\}$$

By the tables in Section 6, we can determine:

$\mathfrak{s}$	A_1	A_2	A_3	A_4	A_5	A_6	A_7	B_2	B_3	B_4	C_3	C_4	D_3	D_4	G_2
$e(\mathfrak{s})$	2	3	4	5	6	7	8	4	6	8	6	8	4	7	6

---

[13]Explanation for notation: E and e are chosen with the word "exclusion" in mind.

4. A Proof of a Theorem in Penkov-Zuckerman

This section is devoted to the proof of the following theorem:

Theorem 4.0.10. *Let $\mathfrak{k}$ be an $\mathfrak{sl}_2$-subalgebra of a semisimple Lie algebra $\mathfrak{g}$, none of whose simple factors is of type A_1. Then there exists a positive integer $b(\mathfrak{k}, \mathfrak{g})$, such that for every irreducible finite dimensional $\mathfrak{g}$-module V, there exists an injection of $\mathfrak{k}$-modules $W \to V$, where W is an irreducible $\mathfrak{k}$-module of dimension less than $b(\mathfrak{k}, \mathfrak{g})$.*

Proof. We assume all of the structure of Sections 3.1 and 3.2 (ie: $\mathfrak{h}$, Φ, W, etc.). Consider a fixed $\mathfrak{k}$ and $\mathfrak{g}$. For $k \in \mathbb{N}$, let $V(k)$ denote the irreducible, finite dimensional representation of $\mathfrak{k}$ of dimension $k + 1$. Each irreducible representation of $\mathfrak{g}$ may be regarded as a $\mathfrak{k}$ representation by restriction. As before, for $\lambda \in P_+(\mathfrak{g})$, we let $[L(\lambda), V(k)]$ denote the multiplicity of $V(k)$ in $L(\lambda)$. Set:

$$g_0(\lambda) := \min\{\dim V(k) | k \in \mathbb{N} \text{ and } [L(\lambda), V(k)] \neq 0\}.$$

For each fundamental weight ω_i $(1 \leq i \leq r)$, let $m_i := m(\mathfrak{g}, \mathfrak{k}, i)$ (as in Definition 3.5.6). By Proposition 3.5.7, $m_i \neq 0$ for all i. Set $\delta_i := m_i \omega_i$, and define $\mathcal{C}_0 := \{\sum_{i=1}^r a_i \omega_i | 0 \leq a_i < m_i\}$. We set:

$$b(\mathfrak{k}, \mathfrak{g}) := \max\{g_0(\lambda) | \lambda \in \mathcal{C}_0\} + 1.$$

For $\mathbf{q} \in \mathbb{N}^r$, let $\mathcal{C}_\mathbf{q} := (\sum_{i=1}^r q_i \delta_i) + \mathcal{C}_0$. By the division algorithm, the collection of sets, $\{\mathcal{C}_\mathbf{q} \mid \mathbf{q} \in \mathbb{N}^r\}$ partitions $P_+(\mathfrak{g})$. We claim that for every $\mathbf{q} \in \mathbb{N}^r$, $\max\{g_0(\lambda) | \lambda \in \mathcal{C}_q\} < b(\mathfrak{k}, \mathfrak{g})$.

The result follows from this claim. Indeed, let $V = L(\lambda)$, for $\lambda \in P_+(\mathfrak{g})$. There exists (a unique) $\mathbf{q} \in \mathbb{N}^r$ such that $\lambda \in \mathcal{C}_\mathbf{q}$. Let W be the irreducible $\mathfrak{k}$-representation of dimension $g_0(\lambda)$. By definition of g_0, there exists an injection of W into $L(\lambda)$. By the claim, $\dim W = g_0(\lambda) < b(\mathfrak{k}, \mathfrak{g})$. The result follows.

We now will establish the claim by applying Proposition 3.3.1. Let $\lambda' \in \mathcal{C}_\mathbf{q}$ with $\mathbf{q} \in \mathbb{N}^r$. Set: $\lambda := \lambda' - \mu$ for $\mu = \sum_{i=1}^r q_i \delta_i$. By definition of $\mathcal{C}_\mathbf{q}$, we have $\lambda \in \mathcal{C}_0$, and so $g_0(\lambda) < b(\mathfrak{k}, \mathfrak{g})$. This means that there exists an irreducible $\mathfrak{k}$-representation, W, such that $\dim W < b(\mathfrak{k}, \mathfrak{g})$ and $[W, L(\lambda)] \neq 0$. By definition, $\lambda' = \lambda + \mu$. We see that since $L(\delta_i)^\mathfrak{k} \neq (0)$ for all i, we have $L(\mu)^\mathfrak{k} \neq (0)$. Applying Proposition 3.3.1 we see that $[W, L(\lambda')] \neq 0$. This means that $g_0(\lambda') \leq b(\mathfrak{k}, \mathfrak{g})$. The claim follows. $\square$

Theorem 4.0.11. *Let $\mathfrak{k}$ be a semisimple subalgebra of a semisimple Lie algebra $\mathfrak{g}$, none of whose simple factors is in the set $E(\mathfrak{k})$. Then there exists a positive integer $b(\mathfrak{k}, \mathfrak{g})$, such that for every irreducible finite dimensional $\mathfrak{g}$-module V, there exists an injection of $\mathfrak{k}$-modules $W \to V$, where W is an irreducible $\mathfrak{k}$-module of dimension less than $b(\mathfrak{k}, \mathfrak{g})$.*

Proof. The proof is essentially the same as the proof of Theorem 4.0.10. The only changes are a substitution of Proposition 3.5.8 for Proposition 3.5.7, and we index irreducible representation of $\mathfrak{k}$ by $P_+(\mathfrak{k})$ rather than $\mathbb{N}$. The result follows from the fact that for a given positive integer d there are only finitely many $\mathfrak{k}$-modules with dimension equal to d (here we use that $\mathfrak{k}$ is semisimple!).

We leave it to the reader to fill in the details. $\square$

The above theorems begs us to compute the smallest value of $b(\mathfrak{k}, \mathfrak{g})$. This is the subject of the Section 5 for the case when $\mathfrak{g}$ is the exceptional Lie algebra G_2 and $\mathfrak{k}$ is a principal $\mathfrak{sl}_2$-subalgebra of $\mathfrak{g}$. Other examples will follow in future work.

We remark that the number $b(\mathfrak{k}, \mathfrak{g})$ clearly depends on $\mathfrak{k}$ (as the notation suggests). Of course, there are only finitely many $\mathfrak{sl}_2$-subalgebras in $\mathfrak{g}$, up to automorphism of $\mathfrak{g}$. We can therefore, consider the maximum value of $b(\mathfrak{k}, \mathfrak{g})$ as $\mathfrak{k}$ ranges over this finite set. We will call this number $b(\mathfrak{g})$. With this in mind, one might attempt to estimate $b(\mathfrak{g})$ for a given semisimple Lie algebra $\mathfrak{g}$.

On the other hand, there is a sense that one could fix $\mathfrak{k}$ to be (say) a principal $\mathfrak{sl}_2$-subalgebra in some $\mathfrak{g}$. We could then consider the question of whether $b(\mathfrak{k}, \mathfrak{g})$ is bounded as $\mathfrak{g}$ varies (among semisimple Lie algebras with no simple A_1 factor).

Even more impressive would be allowing *both* $\mathfrak{g}$ and $\mathfrak{k}$ to vary. It is certainly not clear that a bound would even exists for $b(\mathfrak{k}, \mathfrak{g})$. If it did, we would be interested in an estimate.

5. Example: G_2

In this section we consider an example which illustrates the result of Section 4.

Let G be a connected, simply connected, complex algebraic group with $\mathfrak{g} \cong G_2$. Let K be a connected principal SL_2-subgroup of G. As before, we set $\mathfrak{k} = Lie(K)$.

We order the fundamental weights of $\mathfrak{g}$ so that $L(\omega_1) = 7$ and $\dim L(\omega_2) = 14$. For the rest of this section, we will refer to the representation $L(a\omega_1 + b\omega_2)$ (for $a, b \in \mathbb{N}$) as $[a, b]$. In Table 1, the entry in row i column j is $\dim[i, j]^K$:

$i\backslash j$	0	1	2	3	4	5	6	7	8	9	10	11	12	13	14	15	16	17	18	19
0	1	0	1	0	1	0	2	0	2	0	3	0	4	0	4	1	5	1	6	1
1	0	0	0	1	0	1	1	1	2	2	2	3	3	4	4	5	5	6	7	7
2	0	0	1	0	2	0	3	1	4	2	5	3	7	4	9	5	11	7	13	9
3	0	0	0	1	1	2	2	3	3	5	5	6	7	8	9	11	11	13	14	16
4	1	0	2	1	3	2	5	3	7	5	9	7	12	9	15	12	18	15	22	18
5	0	1	1	2	2	4	4	6	6	8	9	11	12	14	15	18	19	22	23	26
6	1	0	2	2	4	3	7	5	10	8	13	11	17	15	21	19	26	23	32	28
7	0	1	1	3	3	5	6	8	9	12	12	16	17	20	22	25	27	31	33	37
8	1	1	3	2	6	5	9	8	13	12	18	16	23	21	29	27	35	33	42	40
9	0	1	2	4	4	7	8	11	12	16	17	21	23	27	29	34	36	41	44	49
10	2	1	4	4	7	7	12	11	17	16	23	22	30	28	37	36	45	44	54	52
11	0	2	2	5	6	9	10	14	16	20	22	27	29	35	37	43	46	52	56	62
12	2	2	5	5	9	9	15	14	21	21	28	28	37	36	46	45	56	55	67	66
13	0	2	3	6	7	11	13	17	19	25	27	33	36	42	46	53	56	64	68	76
14	2	2	6	6	11	11	17	18	25	25	34	34	44	44	55	55	67	67	80	80
15	1	3	4	8	9	14	16	21	24	30	33	40	44	51	55	64	68	77	82	91
16	3	3	7	8	13	14	21	21	30	31	40	41	52	53	65	66	79	80	95	95
17	0	4	5	9	11	16	19	25	28	35	39	47	51	60	65	74	80	90	96	107
18	3	3	8	9	15	16	24	25	34	36	46	48	60	61	75	77	91	93	109	111
19	1	4	6	11	13	19	22	29	33	41	45	54	60	69	75	86	92	104	111	123

Table 1. $\dim[i, j]^K$ where K is a principle SL_2-subgroup in G_2.

Table 1 was generated by an implementation of the Weyl character formula (see Theorem 3.2.1) for the group G_2 using the computer algebra system MAPLE®. The characters were restricted to a maximal torus in K, thus allowing us to find the character of $[i, j]$ as a representation of $SL_2(\mathbb{C})$. This character was used to compute the dimension of the invariants for K.

Using the same implementation we can compute the values of $g_0(\lambda)$ for the pair $(\mathfrak{k}, \mathfrak{g})$. We display these data in the table below for $0 \le i, j \le 19$.

Of particular interest is the 7 in row 1 column 0. The irreducible G_2-representation $[1, 0]$ is irreducible when restricted to a principal $\mathfrak{sl}_2$ subalgebra. And therefore, $g_0(\omega_1) = \dim[1, 0] = 7$. Note that most entries in Table 2 are 1. Following the proof in Section 4 we see that $b(\mathfrak{k}, \mathfrak{g}) = 7+1 = 8$. That is, every finite dimensional representation of G_2 contains an irreducible $\mathfrak{k}$-representation of dimension less than 8.

i\j	0	1	2	3	4	5	6	7	8	9	10	11	12	13	14	15	16	17	18	19
0	1	3	1	3	1	3	1	3	1	3	1	3	1	3	1	1	1	1	1	1
1	7	5	3	1	3	1	1	1	1	1	1	1	1	1	1	1	1	1	1	1
2	5	3	1	3	1	3	1	1	1	1	1	1	1	1	1	1	1	1	1	1
3	3	3	3	1	1	1	1	1	1	1	1	1	1	1	1	1	1	1	1	1
4	1	3	1	1	1	1	1	1	1	1	1	1	1	1	1	1	1	1	1	1
5	3	1	1	1	1	1	1	1	1	1	1	1	1	1	1	1	1	1	1	1
6	1	3	1	1	1	1	1	1	1	1	1	1	1	1	1	1	1	1	1	1
7	3	1	1	1	1	1	1	1	1	1	1	1	1	1	1	1	1	1	1	1
8	1	1	1	1	1	1	1	1	1	1	1	1	1	1	1	1	1	1	1	1
9	3	1	1	1	1	1	1	1	1	1	1	1	1	1	1	1	1	1	1	1
10	1	1	1	1	1	1	1	1	1	1	1	1	1	1	1	1	1	1	1	1
11	3	1	1	1	1	1	1	1	1	1	1	1	1	1	1	1	1	1	1	1
12	1	1	1	1	1	1	1	1	1	1	1	1	1	1	1	1	1	1	1	1
13	3	1	1	1	1	1	1	1	1	1	1	1	1	1	1	1	1	1	1	1
14	1	1	1	1	1	1	1	1	1	1	1	1	1	1	1	1	1	1	1	1
15	1	1	1	1	1	1	1	1	1	1	1	1	1	1	1	1	1	1	1	1
16	1	1	1	1	1	1	1	1	1	1	1	1	1	1	1	1	1	1	1	1
17	3	1	1	1	1	1	1	1	1	1	1	1	1	1	1	1	1	1	1	1
18	1	1	1	1	1	1	1	1	1	1	1	1	1	1	1	1	1	1	1	1
19	1	1	1	1	1	1	1	1	1	1	1	1	1	1	1	1	1	1	1	1

Table 2. Value of $g_0(i\omega_1 + j\omega_2)$ for a principal SL_2-subgroup in G_2.

Even more interesting is the fact that Table 2 suggests that there are only 26 ordered pairs (a, b) such that $g_0(a\omega_1 + b\omega_2) > 1$. This is indeed the case:

Theorem 5.0.12. *For all* $a, b \in \mathbb{N}$, $[a, b]^K \neq (0)$ *except for the following list of 26 exceptions:*

$$[0, 1], [0, 3], [0, 5], [0, 7], [0, 9], [0, 11], [0, 13], [1, 0], [1, 1],$$
$$[1, 2], [1, 4], [2, 0], [2, 1], [2, 3], [2, 5], [3, 0], [3, 1], [3, 2],$$
$$[4, 1], [5, 0], [6, 1], [7, 0], [9, 0], [11, 0], [13, 0], [17, 0].$$

Proof. By inspection of Table 1 (or Table 2), it suffices to show that for all $a, b \in \mathbb{N}$ such that $a > 24$ or $b > 24$ we have $\dim[a, b]^K > 0$. Let $\mu_1, \cdots, \mu_6$ denote the highest weights of the representations $[0, 2], [0, 17], [4, 0], [15, 0], [5, 1]$, and $[1, 3]$. From the table we see that each of these representations has a K-invariant. Let $Z = \sum_{i=1}^{6} \mathbb{N}\mu_i \subset P_+(\mathfrak{g})$. By Proposition 3.3.1, (for W trivial), each element of Z has a K-invariant. It is easy to see that $E = P_+(\mathfrak{g}) - Z$ is a finite set. In fact, computer calculations show that E consists of the following 194 elements:

[0, 1], [0, 3], [0, 5], [0, 7], [0, 9], [0, 11], [0, 13], [0, 15], [1, 0], [1, 1], [1, 2], [1, 4], [1, 6], [1, 8], [1, 10], [1, 12], [1, 14], [1, 16], [1, 18], [2, 0], [2, 1], [2, 2], [2, 3], [2, 4], [[0, 1], [0, 3], [0, 5], [0, 7], [0, 9], [0, 11], [0, 13], [0, 15], [1, 0], [1, 1], [1, 2], [1, 4], [2, 0], [2, 1], [2, 2], [2, 3], [2, 4], [2, 5], [2, 7], [3, 0], [3, 1], [3, 2], [3, 3], [3, 4], [3, 5], [3, 6], [3, 7], [3, 8], [3, 10], [4, 1], [4, 3], [4, 5], [4, 7], [4, 9], [4, 11], [4, 13], [5, 0], [5, 2], [5, 4], [6, 1], [6, 3], [6, 5], [7, 0], [7, 2], [7, 4], [8, 1], [8, 3], [8, 5], [9, 0], [9, 2], [9, 4], [10, 1], [10, 3], [10, 5], [11, 0], [11, 2], [11, 4], [12, 1], [12, 3], [12, 5], [13, 0], [13, 2], [13, 4], [14, 1], [14, 3], [14, 5], [16, 1], [17, 0], [17, 2], [17, 4], [18, 1], [18, 3], [18, 5], [2, 5], [2, 7], [2, 9], [2, 11], [2, 13], [2, 15], [2, 17], [2, 19], [2, 21], [3, 0], [3, 1], [3, 2], [3, 3], [3, 4], [3, 5], [3, 6], [3, 7], [3, 8], [3, 10], [3, 12], [3, 14], [3, 16], [3, 18], [3, 20], [3, 22], [3, 24], [4, 1], [4, 3], [4, 5], [4, 7], [4, 9], [4, 11], [4, 13], [4, 15], [5, 0], [5, 2], [5, 4], [5, 6], [5, 8], [5, 10], [5, 12], [5, 14], [5, 16], [6, 0], [6, 1], [6, 2], [6, 3], [6, 5], [6, 7], [6, 9], [6, 11], [6, 13], [6, 15], [6, 17], [6, 19], [7, 0], [7, 1], [7, 2], [7, 3], [7, 4], [7, 5], [7, 6], [7, 8], [7, 10], [7, 12], [7, 14], [7, 16], [7, 18], [7, 20], [7, 22], [8, 1], [8, 3], [8, 5], [8, 7], [8, 9], [8, 11], [8, 13], [8, 15], [9, 0], [9, 2], [9, 4], [9, 6], [9, 8], [9, 10], [9, 12], [9, 14], [9, 16], [10, 0], [10, 1], [10, 3], [10, 5], [10, 7], [10, 9], [10, 11], [10, 13], [10, 15], [10, 17], [11, 0], [11, 1], [11, 2], [11, 3], [11, 4], [11, 6], [11, 8], [11, 10], [11, 12], [11, 14], [11, 16], [11, 18], [11, 20], [12, 1], [12, 3], [12, 5], [12, 7], [12, 9], [12, 11], [12, 13], [12, 15], [13, 0], [13, 2], [13, 4], [13, 6], [13, 8], [13, 10], [13, 12], [13, 14], [13, 16], [14, 0], [14, 1], [14, 3], [14, 5], [14, 7], [14, 9], [14, 11], [14, 13], [14, 15], [14, 17], [15, 1], [16, 1], [17, 0], [17, 2], [17, 4], [18, 0], [18, 1], [18, 3], [18, 5], [18, 7], [19, 1], [21, 0], [21, 2], [22, 0], [22, 1], [22, 3], [22, 5], [23, 1], [25, 0], [26, 0], [26, 1], [26, 3], [27, 1], [29, 0], [30, 1], [31, 1], [33, 0], [34, 1], [37, 0], [38, 1], [41, 0], [42, 1], [46, 1].

If one checks, we see that each of these has an invariant except for the 26 exceptional values in the statement of the theorem.

Unfortunately, our tables (in this paper) are not big enough to see all of these weights. For this reason, we can alternately consider the set $Z' = \sum_{i=1}^{9} \mathbb{N}\mu_i'$, where $\mu_1', \cdots, \mu_9'$ are the highest weights of the representations $[0, 2]$, $[0, 17]$, $[4, 0]$, $[6, 0]$, $[15, 0]$, $[5, 1]$, $[1, 3]$, $[7, 1]$, and $[1, 6]$. Again, each of these has a K-invariant (from the tables) and therefore by Proposition 3.3.1, each element of Z' has a K-invariant. Set $E' = P_+(\mathfrak{g}) - Z'$. E' has 73 elements. They are: [0, 1], [0, 3], [0, 5], [0, 7], [0, 9], [0, 11], [0, 13], [0, 15], [1, 0], [1, 1], [1, 2], [1, 4], [2, 0], [2, 1], [2, 2], [2, 3], [2, 4], [2, 5], [2, 7], [3, 0], [3, 1], [3, 2], [3, 3], [3, 4], [3, 5], [3, 6], [3, 7], [3, 8], [3, 10], [4, 1], [4, 3], [4, 5], [4, 7], [4, 9], [4, 11], [4, 13], [5, 0], [5, 2], [5, 4], [6, 1], [6, 3], [6, 5], [7, 0], [7, 2], [7, 4], [8, 1], [8, 3], [8, 5], [9, 0], [9, 2], [9, 4], [10, 1], [10, 3], [10, 5], [11, 0], [11, 2], [11, 4], [12, 1], [12, 3], [12, 5], [13, 0], [13, 2], [13, 4], [14, 1], [14, 3], [14, 5], [16, 1], [17, 0], [17, 2], [17, 4], [18, 1], [18, 3], [18, 5].

This time, each of these representations is on our tables. And so, the reader may (and should) check that, apart from the 26 exceptional values in the statement of the theorem, all of these have a positive dimension of K-invariants. □

6. Tables for Maximal Parabolic Subalgebras

In the proof of Corollary 3.5.2 we needed to see that if $\mathfrak{g}$ was a simple Lie algebra not of type A2, then $\dim(\mathfrak{g}/[\mathfrak{p},\mathfrak{p}]) > 3$ for a maximal parabolic subalgebra $\mathfrak{p}$. If $\mathfrak{p} = \mathfrak{l} \oplus \mathfrak{u}^+$, $[\mathfrak{l},\mathfrak{l}] = \mathfrak{l}_{ss}$ then we can deduce that this inequality is equivalent to $\dim(\mathfrak{g}/\mathfrak{l}_{ss}) > 5$ (using the formula $\dim(\mathfrak{g}/[\mathfrak{p},\mathfrak{p}]) = \frac{1}{2}(\dim \mathfrak{g}/\mathfrak{l}_{ss} + 1)$). It is possible to exhaust over all possible cases to establish this fact. The exceptional cases are in the following table:

Dimension of $\mathfrak{g}/\mathfrak{l}_{ss}$ for the exceptional groups.

$G \setminus \mathfrak{l}_{ss}$	1	2	3	4	5	6	7	8
G2	A1	A1						
dim $\mathfrak{g}/\mathfrak{l}_{ss}$	11	11						
F4	C3	A2A1	A2A1	B3				
	31	41	41	31				
E6	D5	A5	A4A1	A2A2A1	A4A1	D5		
	33	43	51	59	51	33		
E7	D6	A6	A1A5	A1A2A3	A2A4	D5A1	E6	
	67	85	95	107	101	85	55	
E8	D7	A7	A1A6	A1A2A4	A4A3	D5A2	E6A1	E7
	157	185	197	213	209	195	167	115

For the classical cases, the situation reduces to the examination of several families of parabolic subalgebras. In each family, we can determine the dimension of $\mathfrak{g}/\mathfrak{l}_{ss}$ for any of parabolic $\mathfrak{p} = \mathfrak{l} \oplus \mathfrak{u}^+$, from the formulas:

$$\dim A_n = (n+1)^2 - 1$$
$$\dim B_n = n(2n+1)$$
$$\dim C_n = n(2n+1)$$
$$\dim D_n = n(2n-1)$$

For example, if $\mathfrak{g} = A_n$ ($n \geq 2$) and $\mathfrak{l}_{ss} = A_p \oplus A_q$ for $p+q = n-1$ (set $\dim A_0 := 0$) we have $\dim(\mathfrak{g}/\mathfrak{l}_{ss}) = \dim A_n - \dim A_p - \dim A_q = n^2 - p^2 - q^2 + 2n + 2p + 2q - 2$. Upon examination of the possible values of this polynomial over the parameter space we see that the smallest value is 5. This case occurs only for $n = 2$. All other values of p and q give rise to larger values. In fact, for $\mathfrak{g}$ classical and not of type A the dimension is also greater than 5. The low rank cases are summarized in the following tables:

Dimension of $\mathfrak{g}/\mathfrak{l}_{ss}$ for type A.

$G \setminus \mathfrak{l}_{ss}$	1	2	3	4	5
A2	A1	A1			
	5	5			
A3	A2	A1A1	A2		
	7	9	7		
A4	A3	A1A2	A1A2	A3	
	9	13	13	9	
A5	A4	A1A3	A2A2	A1A3	A4
	11	17	19	17	11

Dimension of $\mathfrak{g}/\mathfrak{l}_{ss}$ for type D.

G \ $\mathfrak{l}_{ss}$	1	2	3	4	5	6
D3	A1A1	A2	A2			
	9	7	7			
D4	A3	A1A1A1	A3	A3		
	13	19	13	13		
D5	D4	A1D3	A2A1A1	A4	A4	
	17	27	31	21	21	
D6	D5	A1D4	A2D3	A1A1A3	A5	A5
	21	35	43	45	31	31

Dimension of $\mathfrak{g}/\mathfrak{l}_{ss}$ for type B.

Group \ $\mathfrak{l}_{ss}$	1	2	3	4	5
B2	A1	A1			
	7	7			
B3	B2	A1A1	A2		
	11	15	13		
B4	B3	A1B2	A1A2	A3	
	15	23	25	21	
B5	B4	A1B3	A2B2	A1A3	A4
	19	31	37	37	31

Dimension of $\mathfrak{g}/\mathfrak{l}_{ss}$ for type C.

Group \ $\mathfrak{l}_{ss}$	1	2	3	4	5
C2	A1	A1			
	7	7			
C3	C2	A1A1	A2		
	11	15	13		
C4	C3	A1C2	A1A2	A3	
	15	23	25	21	
C5	C4	A1C3	A2C2	A1A3	A4
	19	31	37	37	31

References

[1]   I. N. Bernšteĭn, I. M. Gel′fand, and S. I. Gel′fand, *Models of representations of compact Lie groups*, Funkcional. Anal. i Priložen. **9** (1975), no. 4, 61–62 (Russian). MR 0414792 (54 #2884) ↑406

[2]   Armand Borel, *Linear algebraic groups*, Second, Graduate Texts in Mathematics, vol. 126, Springer-Verlag, New York, 1991. MR **92d**:20001 ↑407

[3]   Nicolas Bourbaki, *Lie groups and Lie algebras. Chapters 1–3*, Elements of Mathematics (Berlin), Springer-Verlag, Berlin, 1989. Translated from the French, Reprint of the 1975 edition. MR **89k**:17001 ↑410

[4]   _____ , *Lie groups and Lie algebras. Chapters 4–6*, Elements of Mathematics (Berlin), Springer-Verlag, Berlin, 2002. Translated from the 1968 French original by Andrew Pressley. MR **2003a**:17001 ↑410, 412

[5]   David H. Collingwood and William M. McGovern, *Nilpotent orbits in semisimple Lie algebras*, Van Nostrand Reinhold Mathematics Series, Van Nostrand Reinhold Co., New York, 1993. MR **94j**:17001 ↑418

[6]   E. B. Dynkin, *Maximal subgroups of the classical groups*, Trudy Moskov. Mat. Obšč. **1** (1952), 39–166. MR 14,244d ↑418

[7]   _____ , *Semisimple subalgebras of semisimple Lie algebras*, Mat. Sbornik N.S. **30(72)** (1952), 349–462 (3 plates). MR 13,904c ↑418

[8]   David Eisenbud, *Commutative algebra, with a view toward algebraic geometry*, Graduate Texts in Mathematics, vol. 150, Springer-Verlag, New York, 1995. MR **97a**:13001 ↑407

[9] William Fulton and Joe Harris, *Representation theory*, Springer-Verlag, New York, 1991. A first course, Readings in Mathematics. MR **93a:**20069 ↑410

[10] R. Goodman and N.R. Wallach, *Representations and invariants of the classical groups*, Cambridge University Press, Cambridge, 1998. MR **99b:**20073 ↑410, 415, 416, 419

[11] Robin Hartshorne, *Algebraic geometry*, Springer-Verlag, New York, 1977. Graduate Texts in Mathematics, No. 52. MR 57 #3116 ↑407, 409

[12] Anthony W. Knapp, *Lie groups beyond an introduction*, Second, Progress in Mathematics, vol. 140, Birkhäuser Boston Inc., Boston, MA, 2002. MR **2003c:**22001 ↑410

[13] Friedrich Knop, Hanspeter Kraft, Domingo Luna, and Thierry Vust, *Local properties of algebraic group actions*, Algebraische Transformationsgruppen und Invariantentheorie, 1989, pp. 63–75.MR1044585 ↑409

[14] Friedrich Knop, Hanspeter Kraft, and Thierry Vust, *The Picard group of a G-variety*, Algebraische Transformationsgruppen und Invariantentheorie, 1989, pp. 77–87.MR1044586 ↑409

[15] Bertram Kostant, *The principal three-dimensional subgroup and the Betti numbers of a complex simple Lie group*, Amer. J. Math. **81** (1959), 973–1032. MR 22 #5693 ↑418

[16] Hanspeter Kraft, *Geometrische Methoden in der Invariantentheorie*, Aspects of Mathematics, D1, Friedr. Vieweg & Sohn, Braunschweig, 1984 (German). MR **768181 (86j:**14006) ↑415

[17] I. G. Macdonald, *Symmetric functions and Hall polynomials*, 2nd ed., Oxford Mathematical Monographs, The Clarendon Press Oxford University Press, New York, 1995. With contributions by A. Zelevinsky; Oxford Science Publications.MR1354144 (96h:05207) ↑406

[18] Ivan Penkov and Gregg Zuckerman, *Generalized Harish-Chandra Modules: A New Direction*, Acta Applicandae Mathematicae **81** (2004), no. 1, 311–326. ↑403, 404, 405

[19] V. L. Popov, *Picard groups of homogeneous spaces of linear algebraic groups and one-dimensional homogeneous vector fiberings*, Izv. Akad. Nauk SSSR Ser. Mat. **38** (1974), 294–322. MR 50 #9867 ↑409

[20] V. L. Popov and È. B. Vinberg, *Invariant theory*, Algebraic geometry, 4 (russian), 1989, pp. 137–314, 315. MR **92d:**14010 ↑407, 410

[21] I. R. Shafarevich (ed.), *Algebraic geometry. IV*, Encyclopaedia of Mathematical Sciences, vol. 55, Springer-Verlag, Berlin, 1994. Linear algebraic groups. Invariant theory, A translation of *Algebraic geometry. 4* (Russian), Akad. Nauk SSSR Vsesoyuz. Inst. Nauchn. i Tekhn. Inform., Moscow, 1989 [MR 91k:14001], Translation edited by A. N. Parshin and I. R. Shafarevich. MR **95g:**14002 ↑407, 408, 410

[22] T. A. Springer, *Linear algebraic groups*, Algebraic geometry, 4 (russian), 1989, pp. 5–136, 310–314, 315. Translated from the English. MR **92g:**20061 ↑407

[23] È. B. Vinberg and V. L. Popov, *A certain class of quasihomogeneous affine varieties*, Izv. Akad. Nauk SSSR Ser. Mat. **36** (1972), 749–764. MR 47 #1815 ↑416

[24] David A. Vogan Jr., *Representations of real reductive Lie groups*, Progress in Mathematics, vol. 15, Birkhäuser Boston, Mass., 1981. MR **632407** (**83c:**22022) ↑406

[25] V. E. Voskresenskiĭ, *Picard groups of linear algebraic groups*, Studies in number theory, no. 3 (russian), 1969, pp. 7–16. MR 42 #5999 ↑409

[26] _____ , *Algebraic groups and their birational invariants*, Translations of Mathematical Monographs, vol. 179, American Mathematical Society, Providence, RI, 1998. Translated from the Russian manuscript by Boris Kunyavski [Boris È. Kunyavskiĭ]. MR **99g:**20090 ↑409